COMPUTER SCIENCE, TECHNOLOGY AND APPLICATIONS

ROBOTICS

STATE OF THE ART AND FUTURE TRENDS

Computer Science, Technology and Applications

Additional books in this series can be found on Nova's website under the Series tab.

Additional E-books in this series can be found on Nova's website under the E-book tab.

COMPUTER SCIENCE, TECHNOLOGY AND APPLICATIONS

ROBOTICS

STATE OF THE ART AND FUTURE TRENDS

GIOVANNI LEGNANI
AND
IRENE FASSI
EDITORS

Nova Science Publishers, Inc.
New York

Library of Congress Cataloging-in-Publication Data

Robotics : state of the art and future trends / editors, Giovanni Legnani and Irene Fassi.
p. cm.
Includes index.
ISBN 978-1-62100-403-5 (hardcover)
1. Robots, Industrial. 2. Robotics. 3. Manipulators (Mechanism) I. Legnani, Giovanni. II. Fassi, Irene.
TS191.8.R625 2011
629.8'92--dc23
2011034222

Published by Nova Science Publishers, Inc. † New York

CONTENTS

PREFACE

Robotics, although being a relatively new discipline, has however reached a good maturity in many domains, both from the research and market point of view. Over the years, many books, textbooks, handbooks have been published, covering all different aspects of this challenging field. So, why another book on robotics?

Of course because robotics is in great evolution and evolution must be studied, and understood to be governed.

This book holds a very exciting title. We imagined to start with a critical analysis of the state of the market, to understand which type of robots are present, which operations they perform, which are the unresolved requests of the market. Another important aspect is the analysis of the more promising research fields and the results which are "ready" in the labs and are waiting for a successful industrialization and commercialization process.

Many people would think that such a challenging analysis could be performed only by a set of high qualified experts with a strong experience. And the result would be one among the many other books in the field of robotics. We decided to afford the challenge using a different approach based on bright promising investigators maybe with less experience but a lot of enthusiasm.

Each Chapter is structured with a first section addressing the fundamentals (a brief review of the state of the art and current practices on the assigned topic) and a second part discussing an interesting application.

The book is opened by a Chapter which reviews the current status of the industrial automation. The global market and the different application fields are analyzed both from a dimension point of view as well as available technologies. This Chapter reviews the general robot structures and highlights the manipulator typologies dominating the scene (anthropomorphic, SCARA, delta, cartesian). However many different practical versions dedicated to specific fields (e.g. foundry, assembly, food, medical) have been developed to fulfill special requirements. The different versions differ for size, mechanical performances (payload, velocity, accuracy) as well as for external sensorization and programmability features. Intelligence and sensorization are important ingredients of many modern installations. The myth of a general purpose manipulator able to perform any task with success is a bit obfuscated by the necessity of high specialization in many fields.

While most of the installed industrial robots are serial, the only diffused exception being Delta type manipulators, the research on innovative kinematic structures is still active. Research from a mechanical point of view includes the modeling and the kinematics analysis of the manipulators. Chapter 2 and Chapter 3 address the kineto-static analysis and workspace optimization of various types of parallel manipulators with 3 and 4 degrees of freedom (dof).

The necessity of improving the dynamical performances of the manipulators calls for an extension of the classical analysis: kinematics cannot be restricted to position, velocity, and acceleration but it should be extended also to jerk. To better study this topic, an approach based on screw theory may be very successful and efficient, especially when dealing with parallel kinematic machines. Chapter 4 addresses the mobility analysis of 3 dof parallel manipulators using the screw theory, while Chapter 5 discusses the application of screw algebra to jerk analysis of 6 dof manipulators.

Chapter 6 aims at providing the fundamentals for dynamic modeling, analysis and system identification of parallel robots and proposes a model based identification method to identify the dynamic parameters value. This methodology can be applied also to serial manipulators. The necessity to improve the control performances of the manipulators in term of stability and path accuracy requires new tools. In some applications, standard PID controllers seem not adequate and several non conventional techniques have been proposed in the last decades. In Chapter 7, a controller based on fractional derivative is presented. Chapter 8 is devoted to "visual feedback" techniques, which allow controlling a robot by a vision system. The camera located on the manipulator analyzes a scene and commands the robot to reach a predefined pose with respect to a fixed or moving body. This is one of the most common modern techniques to give flexibility to a working cell.

An emerging research field deals with robots interacting with loosely structured environments. In the industrial foreground, manipulators and humans may share the same working space to cooperate in the execution of a task. This is the destruction of the traditional idea that human and robots should be kept separated for safety issues. Chapter 9 presents a new robotic system based on multiple sensors and human cooperation to gain more flexibility in various industrial tasks.

A strict interaction between human and robot is required also is the medical environment, where the use of robotic means is becoming every day more and more important. Chapter 10 and Chapter 11 provide the rationale for the use of robots in the rehabilitation field. The main issues in designing and implementing such systems are reviewed and two different applications based either on the customization of an industrial robot or on the designing of an exoskeleton for the upper limb rehabilitation, are presented.

Another challenging task is robotics at the micro-world. Chapter 12 reviews the main challenges in automating picking and precisely place of parts with dimensions of few micrometers and presents some interesting case studies.

Of course a single book cannot analyze all the modern robotic issues. A large encyclopedia would be necessary. But the present book can be considered like a balcony from which we can observe a wide panorama of some of the challenging robotic issues of today – an instrument to understand the place from where we came, where we are, where we are going.

Finally, we want to acknowledge the efforts of all the contributors, who answered promptly and patiently to all our requests and we wish that the readers will find interesting and fruitful hints for their future work. Enjoy the reading!

Irene Fassi, Giovanni Legnani

In: Robotics: State of the Art and Future Trends
Editors: G. Legnani and I. Fassi
ISBN: 978-1-62100-403-5

Chapter 1

ROBOT IN INDUSTRIAL APPLICATIONS: STATE OF THE ART AND CURRENT TRENDS

***L. Tirloni*[1,*], *I. Fassi*[2,4,†] *and G. Legnani*[3,4,‡]**
[1]REA Robotics s.r.l, Via Volta 35, 35030 Veggiano Padova, Italy
[2]Istituto Tecnologie Industriali ed Automazione, Consiglio Nazionale delle Ricerche
Via Bassini 15, 20131 Milano, Italy
[3]Università di Brescia, Dip. Ingegneria Meccanica e Automazione,
Via Branze 38, 25123 Brescia, Italy
[4]SIRI – Italian Robotic and Automation Association, Italy

Abstract

This Chapter aims at analyzing the state of the art in industrial robotics, in terms of successful industrial applications, from the perspective of robotic manufacturers and system integrators. It is worth to note that some of the solutions presented hereafter as innovative may be well known among the scientific community. However, due to many reasons, they only recently reached the market.

The idea is to highlight the gap between recent research results and industrial applications to stimulate the technology transfer from academia to common practice.

Firstly, the current market of industrial robotics will be analyzed and future trends discussed, in term of number and type of installed robots as well as in term of available technology. At the beginning, the paper presents official statistics updated to 2010. These statistics are derived from those collected by IFR (International Federation of Robotics) in cooperation with OECD and UNECE. The selected data describe the size of the market, the more consolidated applications and the new trends.

The second part of the chapter describes the robotic technology available on the market, in term of kinematical structure of the manipulators, programmability of robotized cells, and sensors to improve flexibility. The different robot typologies are reviewed with respect to the robot applications and commercially available models.

* E-mail address: lorenzo.tirloni@reagroup.it
† E-mail address: irene.fassi@itia.cnr.it
‡ E-mail address: giovanni.legnani@ing.unibs.it

The topics are discussed with several references to standard and innovative applications highlighting where technology transfer and/or some degree of innovation is still necessary.

1. Introduction

1.1 The Market

Nowadays robots are widely used in industrial applications, spanning from food to pharmaceutical to naval industry to perform either simple tasks, as loading/unloading of work-pieces on machine tools or more complex processes, such as assembly, soldering, painting, deburring, cutting. Figure 1 shows the distribution of industrial robots in the different sectors and their target application.

Major robot manufacturers are in Japan and Europe. Among them, the most important in term of volume of produced robots and product range are: Fanuc (J), Nachi (J), Kawasaki (J), Motoman (J), Toshiba (J), Panasonic (J), Denso (J), ABB (S), Kuka (D), Reiss (D), Staubli (CH), Hyundai (ROK), and Comau (I).

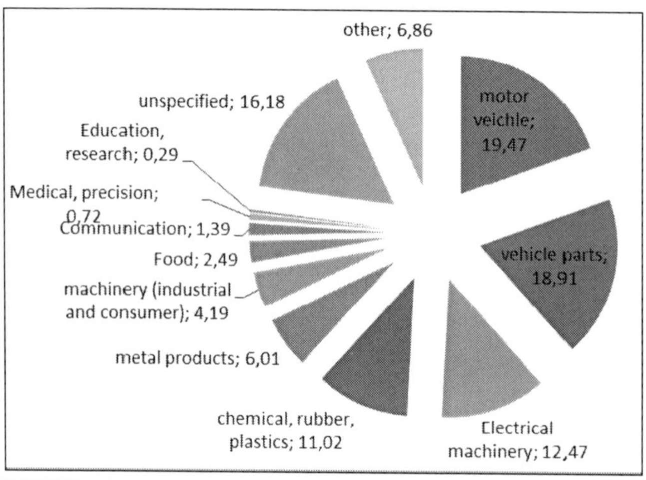

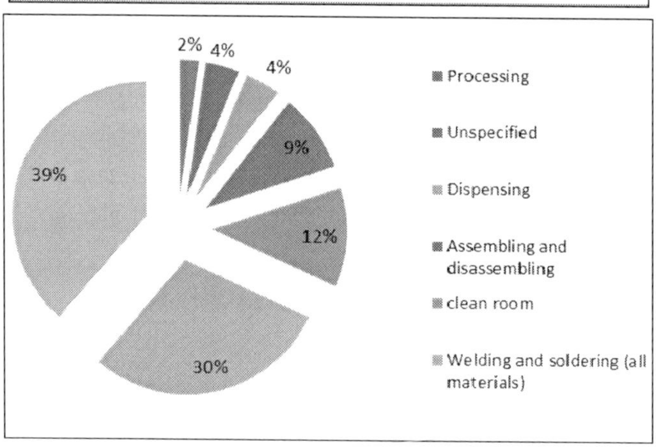

Figure 1. Relative size of the more diffused robotics applications and sectors.

The automotive industry has been the main driver to the use of robots in its production plants introducing their use also along the whole supply chain.

Motivations on the use of robots include mainly the requirement to perform tasks in hazardous environments (e.g. powders, foams, explosive or inflammable environments) or hazardous work-pieces (e.g. high temperatures, toxic materials, heavy loads), the requirement to improve quality and reduce defects and also to increase the production.

Statistics show a positive growing trend for the robotics market till 2008; in 2009, the global economic breakdown led to a significant crisis also in this sector. However, as shown in Figure 2, during 2010 various regions experienced good recovery rates in robot sales. Asia was on top with an increase of 127%, the second highest level ever recorded. About 17,000 units were shipped to Americas, 87% more than in 2009, reaching almost the level of 2008. In Europe, about 30,000 units were sold, 45% more than in 2009. This is however still about 15% lower than the peak levels of 2007 and 2008.

The most dynamic markets, as highlighted by the IFR Statistical Department [1], were China, the Republic of Korea and the ASEAN countries. Sales to these markets almost tripled. In 2010, the Republic of Korea was on top with almost 23,000 robots sold. Japan recovered with a lower growth rate of 66% to about 21,000 units. This is followed by North America which recovered by 90% to about 16,000 units and China with almost 15,000 units sold (+170%). Germany ranked 5th with about 13,400 units sold (+57%). The first quarters of 2011 registered another substantial rise of 18% in robot sales. The electronics industry, the automotive industry and the metal industry were the main drivers of the high increase of robot sales in 2010, as well as in 2011.

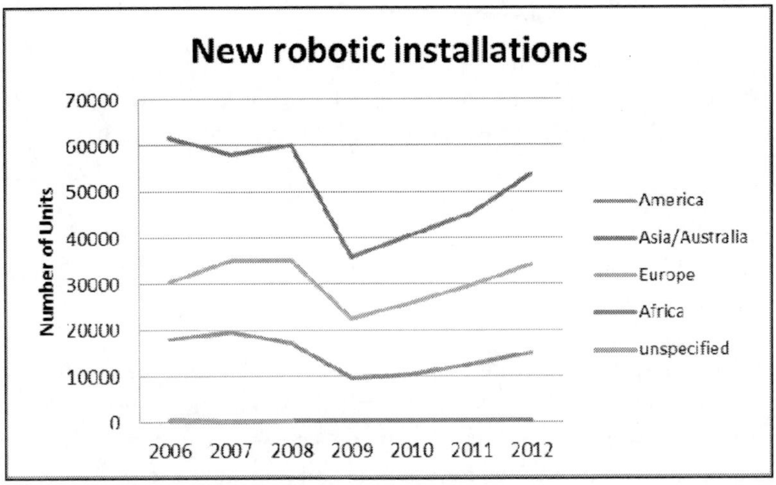

Figure 2. New robotics installations: current trend and forecast.

IFR and EUROP ETP observatory [1, 2] forecast that robot installations will continuously increase also in Europe and in the Americas. The automotive industry is continuing to implement new technologies and materials thus requiring new manufacturing lines. The application of robots in other industries i.e. the food and beverage industry, the pharmaceutical and solar cells industries will further increase. It is estimated that the stock of robots operating in the factories world-wide will increase to about 1.3 million at the end of 2014. Robots will penetrate areas with a still low rate of automation, such as small companies and service companies, due to the improvements in safety, flexibility, accuracy, ease of use of robots and thanks also to the decrease of robot installation costs over labor costs (Figure 3).

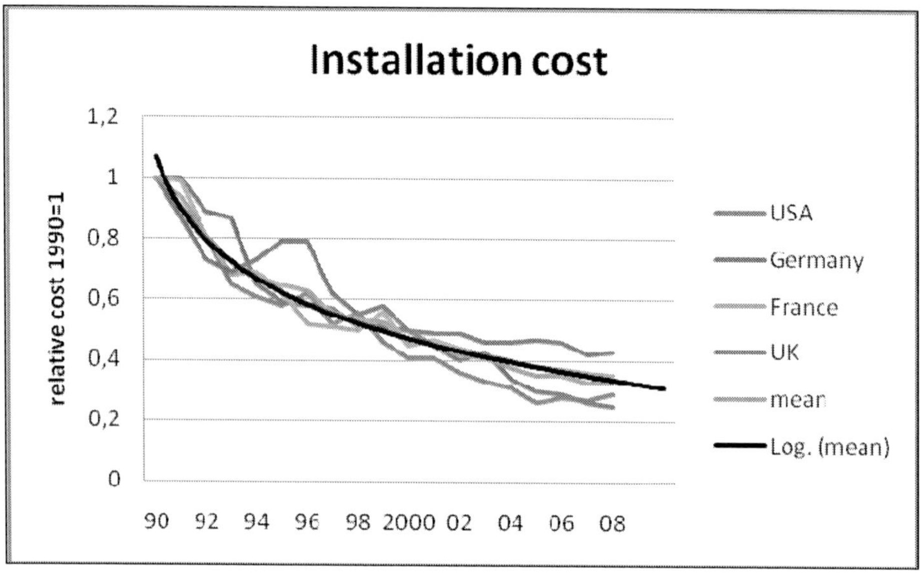

Figure 3. Average robot price with respect to labor compensation.

1.2. Robot Typologies

To better understand how a robot is chosen, the different typologies of commercial robots currently available on the market have to be described. Figure 4 shows the most popular robot geometries. Industrially used robots are classified into four main families:

- a) Cartesian robots (also known as Gantry Robot);
- b) Cylindrical/polar, and SCARA robots;
- c) parallel kinematics manipulators (e.g. "delta" robot);
- d) Anthropomorphic robots.

Each kinematic structure determines a corresponding shape of the working space.

Manipulators of type a), b), and d) are called "serial manipulators" because they are realized connecting in series several links each one actuated by a revolute or a prismatic joint. Each joint is actuated by a motor; the term axis usually denotes a joint and its actuator (linear or revolute motor). The first three joints (main axes) determine the motion of the arm and so the XYZ position of the gripper whose orientation can be modified by 1, 2 or 3 "secondary axes" located in the wrist. Usually industrial manipulators have between 4 and 6 degrees of freedom (DOF). Parallel manipulators are characterized by several links connected in parallel to move the mobile base. Some examples will be described further in this chapter and through the whole book.

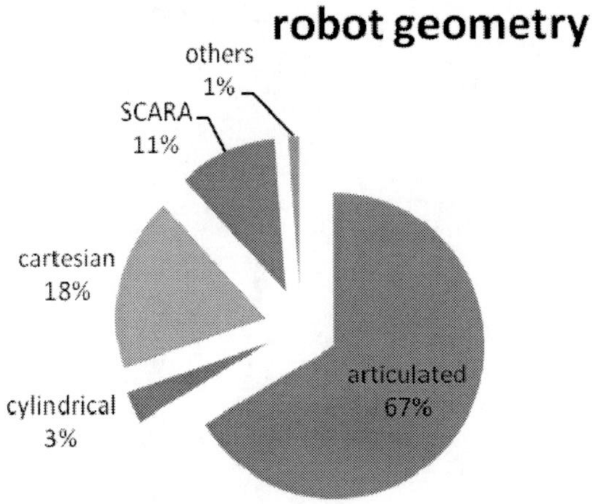

Figure 4. Robot geometries currently on the market.

The name of the manipulator structure depends on the configuration of the three main axes.

Cartesian robots employ linear axes, along X, Y, Z direction. Usually they have additional axes for rotations (wrist with A, B, C rotations). Cartesian manipulators may have the structure of Figure 5; small manipulators are generally of type (a), while the biggest ones have usually a gantry structure type (b). Cartesian manipulators are not commonly used in industrial applications except for machining. They have high machining precision and accuracy due to their intrinsic rigidity. Big Cartesian robots are sometime used to carry and move anthropomorphic manipulators in order to enlarge their working space. The result is a redundant robotic system having the dexterity of an anthropomorphic manipulator with a bigger working space. In this case the anthropomorphic manipulator is generally of a small size with a payload not greater than 30kg. These combinations of manipulators are practically applied to work on large size products like containers or boats. Generally both robots are synchronously controlled by a unique controller. Actually, the controllers of all the major producers of anthropomorphic manipulators are able to control also additional "external axes". Some controllers are able to control up to 36 external axes; however, a number of 9 external axes is common. External axes are also often used to control two or more synchronized working cells.

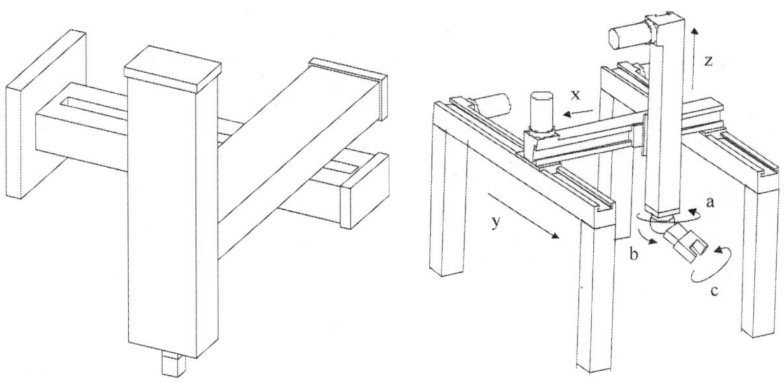

(a) (b)

Figure 5. Cartesian robot architectures.

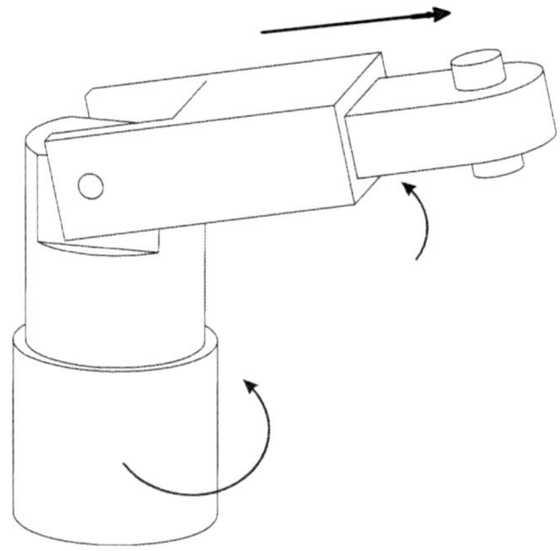

Figure 6. Polar robot.

The term ***polar robot*** properly refers to a robot composed by a series of 2 revolute joints, followed by a prismatic one (Figure 6). Sometimes, in the common industrial practice, also cylindrical robots (Figure 7) are improperly called polar.

The acronym **SCARA** stands for Selective Compliance Assembly Robot Arm. Its basic structure (Figure 8) is constituted by two revolute axes which position the gripper in the XY plane. A further linear axis may be used to impose the Z coordinate. In few cases the Z axis is the first one. An additional axis is sometime employed to rotate the gripper around the Z axis. The two revolute axes incorporate a controlled compliance while the Z axis is rigid. This design allows compensating for inaccuracy in the XY position during *peg in hole* assembly operation, in which small pieces are inserted in bigger ones from the top.

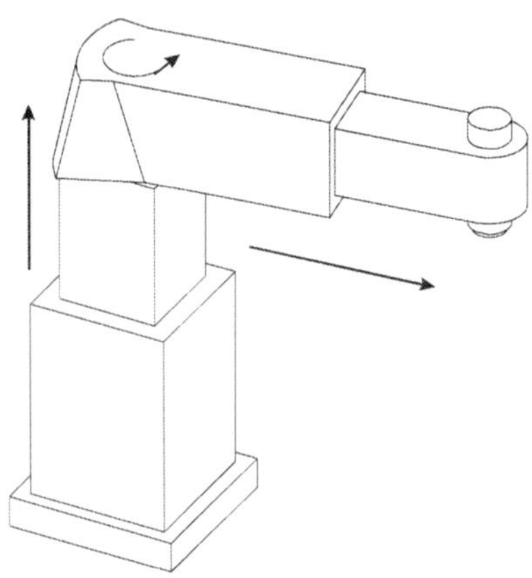

Figure 7. Cylindrical robot.

This kind of robot had a big success for its performance (accuracy, speed) and its limited cost. In the recent years it partially lost popularity due to the success of small size anthropomorphic and delta manipulators. However, due to new design trends related to direct drive technology and price reduction, the number of new installed SCARA robots is still significant. However it is useful only for planar applications with limited payloads (20kg maximum); the working space is generally smaller than 2 meters. Classical applications (90% of the cases) are fast Pick and Place.

A particular pick and place working cycle (called SCARA cycle) has been standardized to compare the performance of manipulators of different models and different firms. This cycle includes a return motion composed by three segments: vertical *pick* motion of 25mm, a linear displacement of 300mm and a *place* motion of 25mm. Generally a SCARA robot may perform the cycle in a period between 0.5 and 0.9 seconds depending on the payload. Delta manipulators may perform the same cycle in a shorter time (0.3-0.4 seconds) but with a smaller payload (usually less than 1 kg).

Typical SCARA applications include: fast assembly of electronic products (it is worth to note that many producers of SCARA manipulators are also important electronics producers like Toshiba, Panasonic, Bosch, Mitsubishi), packaging for the Pharmaceutical field or the food industry.

A modification of the SCARA architecture is the cylindrical one. A pure cylindrical structure (1st rotative axis followed by two orthogonal prismatic joints) is seldom utilized in industrial field. On the contrary a combination of rotative and prismatic joints (cylindrical pair) around a vertical direction followed by 2 rotative axes (total 4 axes) is often used for fast and delicate manipulations.

This robot typology is primary used to manipulate silicon wafers in semiconductor industry; some robot producers have created a specific division dedicated to this application.

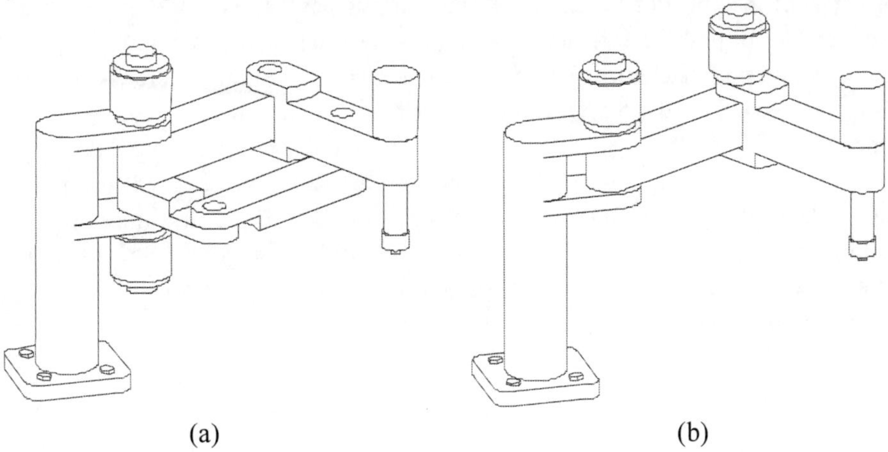

(a) (b)

Figure 8. Two version of SCARA Robot; a): the actuators of axis 1 and 2 are located on the fixed base, b): the actuator of the second joint is located on the axis motion.

While generally industrial robots have a serial structure, special manipulators may have a parallel kinematic configuration. The most common manipulator having this type of configuration is the DELTA robot (Figure 9). Its structure consists in 3 cranks (parallelograms) which can rotate with respect to the fixed base. The mobile platform has 3 translational degrees of freedom. A fourth leg may be used to transmit a rotary motion about the vertical axis from the base to the end-effector mounted on the mobile platform. The three rotational axes are coplanar and oriented 120° degrees one from the other. At the end of each parallelogram there are two spherical joints. Six rods connect the mobile platform to the cranks by spherical joints. Actuation may be obtained with rotational (DC or AC servo) motors (the most common) or with linear actuators.

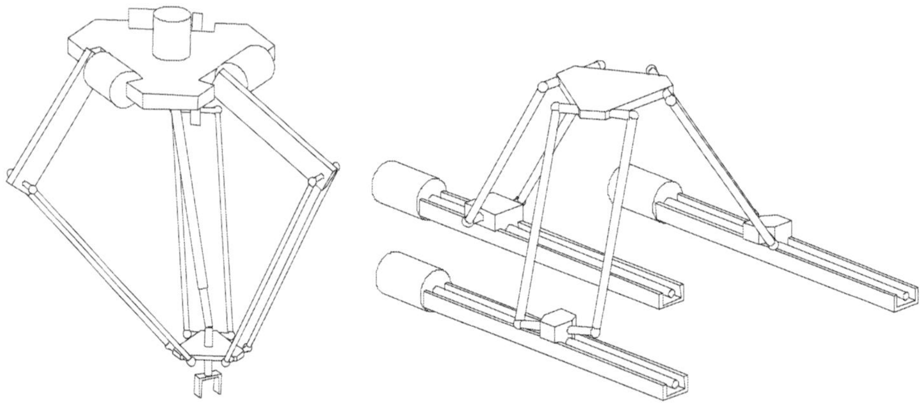

Figure 9. Delta architecture: classical and linear structure (Triaglide).

This manipulator was invented by R. Clavel at EPFL and then patented in the late ‘80. The license was then bought by ABB Flexible Automation (1999) which sold several thousand manipulators of this type. In 2009 the patent expired and many other companies have developed similar robots with excellent performances. A model by Adept, named "Quattro", obtains the 4th degree of freedom (rotation around z axis) using a 4th crank in parallel to the others.

This type of manipulators is extremely fast, but its payload is quite low. For this reason, they are suitable for high performance pick and place tasks in specific application fields like food industry and pharmaceutics. The payload is between few grams to few kilograms with a cycle time between 0.3 and 0.5 seconds (considering a standard SCARA cycle).

Other types of parallel manipulators (or hybrid parallel-serial manipulators) have been produced (e.g. the Tricept, Figure 10), but no one gained significant diffusion since now. Triaglide (Figure 9 b) is a 3 DOF parallel manipulator able to perform translations in XYZ; the working space may be very large in the direction of the slide joints. Some 6 DOF parallel structures based on the Stewart-Gough architecture (Figure 11) have been proposed for 5 axes machining but their industrial diffusion is very low.

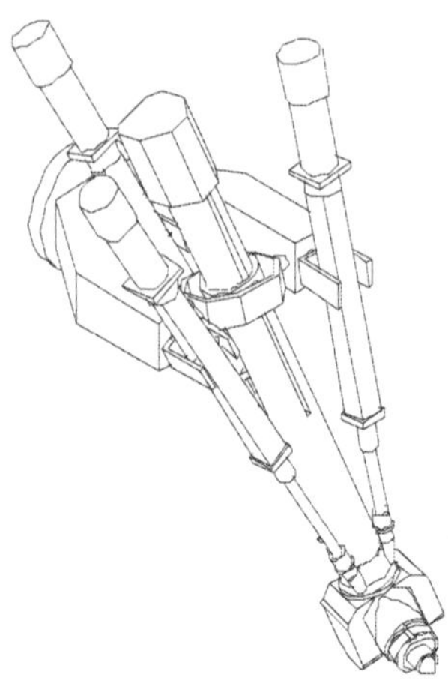

Figure 10. The Tricept architecture.

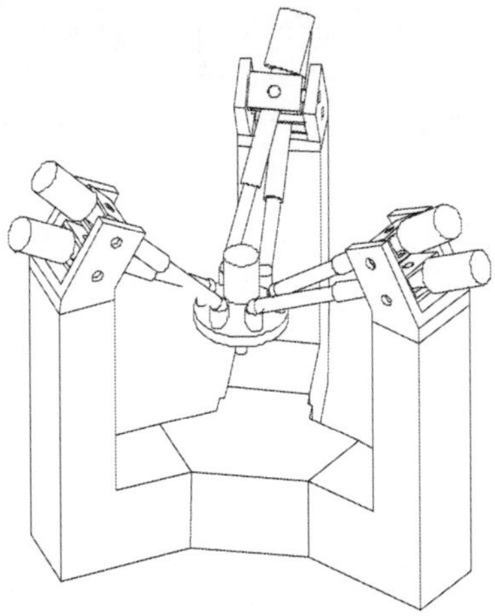

Figure 11. The Stewart-Gough platform.

The Tricept exhibits stiffness and accuracy suitable for machining applications. The 3 dof parallel structure is completed by a 2 dof serial wrist thus allowing 5 axis machining. Its behavior is not isotropic, exhibiting a very high stiffness in one direction. Even if there are some samples implemented on industrial field, the Tricept is still a niche product.

The most diffuse manipulators have anthropomorphic geometry (Figure 11). They generally have 5 or 6 degrees of freedom, thus reaching any point of the working space with the desired orientation of the end-effector. A large variety of dimension, payload and specialization for particular applications is available. The most common varieties of anthropomorphic manipulators are:

- general manipulation robot
- foundry robot
- welding robot (wire welding and spot welding)
- painting robot
- palletizing robot
- clean room robot

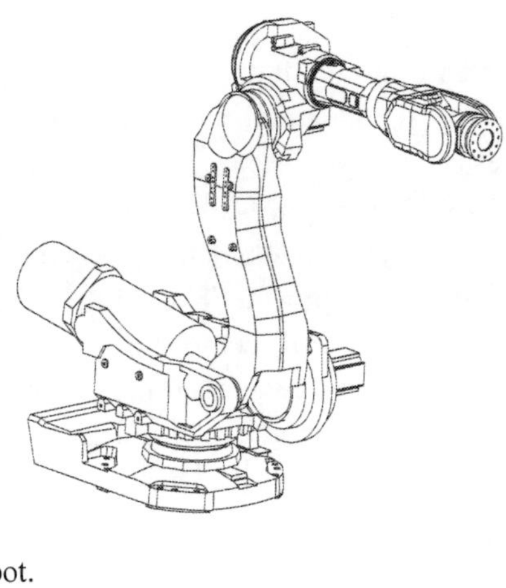

Figure 12. Antropomorphic robot.

Each variety of anthropomorphic manipulators has specific construction details in terms of mechanical structure, actuators, servo-drives and/or controller. Criteria for their selection will be addressed in the next paragraphs.

2. Performances and Criteria for Selection

The choice of the most suitable manipulator for a specific application depends on several requirements:

- application field
- payload
- shape and dimension of the working space
- speed
- repeatability and accuracy

Each of the abovementioned requirements is discussed in the following.

2.1. Application Field

Within a single industrial company many different robotized working cells may be present, each one requiring different robot characteristics. The same type of specific application may be present in different industries. The most common specific processes to be automatized include:

- production of raw or intermediate materials
- treatment and processing of intermediate materials
- welding (spot welding or wire welding)
- assembly
- painting
- packaging and palletizing

Each of the listed application may include quality control. An analysis of the working condition for each application is fundamental to determine the best manipulator geometry useful to automatize the process.

Let us consider a metal die-cast or gravity cast process where a robot feeds the machines and unloads the realized pieces. The robot manipulates cups containing high temperature molten aluminium or steel in a hot, harsh environment. Thus it must have specific characteristics to protect its own mechanical and electrical components. Protections may include special corrosion resistant seals, grease, and even pressurized motors, protection carters, to prevent the penetration of dust, liquids and gasses; sometime total IP67 protection is required.

Sometimes it is sufficient to protect the wrist, but in some circumstances it is necessary to protect the whole manipulator and control unit.

For this reason the main manufacturers produce special manipulator versions denominated "foundry". The casted pieces are generally manipulated again during other processes employing lathes, mills, working centers or machining centers.

Generally in this case it is not necessary the adoption of special robot versions and general “handling” robots are suitable. The manufacturing process may include welding applications. In this case, two robot alternatives are available: robot for spot welding and robot for wire welding. The robots for arc welding (wire welding) belong to a specific category of light weight manipulators with hollow wrist to simplify the passage of the electrical wires as well as the welding wire. More over their controller implements specific trajectories as well as adaptive control algorithms like touch sensing and trajectory adaption by current measure or by laser tracking (see specific sections in the following). On the contrary, spot welding tasks may require manipulators able to move heavy welding tools and simple point-to-point trajectories may be sufficient. The process is slow (generally less than 50 cm/m), the execution is difficult due to complex welding paths, it is difficult to reach the welding zones, human operators must use protections to avoid the risk of touching melt metal and to protect their eyes from excess of light produced by the welding process. Welding applications are commonly found in the automotive industry, who has boosted the development of specialized robots. Welding robots nowadays cover almost the 30% of the robotic market (Figure 1).

To realize a final product it may be necessary to assemble several pieces by friction fitting or by screws or riveting. In this case the robot does not need special performances except the possibility to hold the weight of pieces or the tools and to have a sufficient working space. General handling robot can be used. For a correct manipulation, the gripper geometry, shape and dimensions play a major role.

Very often at the end of the production cycle painting is necessary. These phase has been optimized to fulfill the needs of some industries like automotive ones because it is dangerous for human operators (varnish contains toxic components) and because an automatic painting procedure is more repetitive and permits the reduction of wasted paint. Painting robots have special protection to prevent that inflammable paint elements enter inside the manipulator elements and become in contact with electrical components. For this purpose manipulators may be pressurized. Moreover the manipulator may have special mechanical characteristics like hollow wrists to simplify the use of the piping transporting pigments, paint and solvents from the tank to the painting gun. They are controlled by dedicated software able to memorize the trajectory and the process parameters of the painting gun. The robot end effector is firstly manually moved by an operator that teaches the painting trajectory to the robot which is then able to reproduce the learned trajectory.

The last operation is packaging and palletizing of the pieces for transportation and delivery to the final users. Palletizing is often realized by special robot having generally just 4 degrees of freedom (only few applications require 5 or 6 dof robots). Some of these manipulators have a large payload (up to 500 kg) and may perform high speed.

Special mention is required for manipulators working in specially protected environments like “clean room“. They are generally requested in semiconductor industries to handle silicon wafers and sometimes in the food industry. These robot versions require components realized in stainless steel with special features to prevent the leakage of grease or lubricating. These additional costs are not justified in any other industrial sector.

2.2. Payload

The weight, the shape and the inertia tensor of the pieces to be manipulated influence the choice of the robot. The mentioned data must incorporate the characteristics of the piece and those of the gripper and tools.

Anthropomorphic manipulators are produced in a large variety of sizes able to manipulate objects from 2kg up to 1200kg.

Standard manipulators must be installed on the floor, but some of them (generally the small size ones) accept wall or ceiling mounting to better reach the working space from the top.

2.3. Working Space and Dimensions

The geometrical dimensions and shape of the manipulators are important characteristics in their choice. Although geometrical size and payload grows almost proportionally, this is not always true and models with the same size but different payloads are sometime present.

The different versions are based on different sizes of the links, of the speed reducers, transmission gears and of the actuators.

Sometimes, adopting a larger size manipulator reduces the number of installed manipulators, because a single unit may serve two or more working cells.

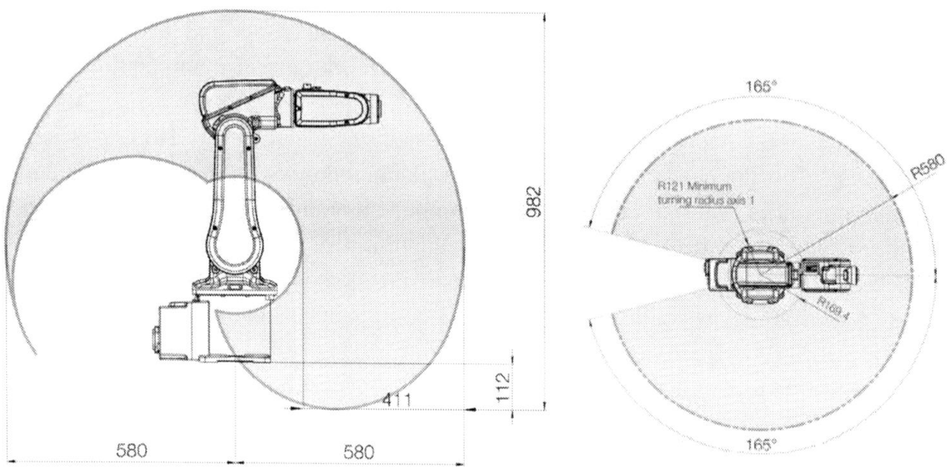

Figure 13. Example of working space of anthropomorphic manipulators.

2.4. Velocity

For several applications, like pick and place, the manipulator velocity is a critical issue, especially when considering small size robots (generally up to 20kg payload).

Sometimes the wrist velocity is not expressed in mm/s but as degrees/s for each axis; this is commonly found for robots exhibiting angular motions (SCARA, polar, anthropomorphic), where the real wrist velocity depends on the position of the end-effector inside the working space. The actual velocity of the end-effector will be the vector summation of the components of all the axes.

To determine if a manipulator is suitable to the given application, it is often mandatory to simulate its behavior using specialized SW products like those used to graphically off-line programming.

Applications using small size anthropomorphic manipulators interfaced to vision systems may work on pick and place cycles at a frequency up to 45 cycles/minute (about 1.3 s/cycle).

2.5. Repeatability and Accuracy

Repeatability and accuracy are two parameters describing the motion precision of machines and manipulator. Accuracy measures the difference between the requested manipulator position and the actually reached point. Repeatability measures the differences in the actual position reached within different repetitions of the same movement. Robots generally exhibit very good repeatability, while their accuracy is not always satisfactory. Typically, the specification data sheet provided by robot manufacturers does not specify the accuracy. When a good accuracy is requested, special calibration techniques must be implemented. However for many standard applications a good repeatability is sufficient to successfully program the working cell.

The accuracy of each manipulator (expressed in millimeters) may vary from few hundredths of micrometers (for small manipulators with payloads of 3-5 kg) and may reach 0.5mm for big robots of 400 kg payload. This value may increase depending on the manipulator use and age. This is mainly due to wear of mechanical parts (joints and gears).

3. Manipulator Controllers

3.1. The architecture

All robot manufacturers use brushless motors installed near the link joints and make use of speed reducers (Harmonic Drive or gear reducers) for the axes 1, 2 and 3. The wrist axes (4, 5 and 6) may use tooth belt. Motors are equipped with incremental encoders made absolute by a battery that supply the step counter even if the manipulator is disconnected by the electrical supply line for months or even few years. The motor size is carefully chosen to reduce weight and clearance.

The electrical panel is generally inserted in a closed case containing the electrical converters (one for each motor), numerical control or cards for trajectory generation, interfaces to external devices, CPU and other digital devices. Each producer has developed its own controller implementing their "know how". Although general control concepts may be similar, many details are different among different products.

General digital components are:

- one CPU (Central Process Unit), that manages all the peripherals (Ethernet ports, digital I/O, fieldbus, USB ports, monitors and teach pendant), the editing, the storage and execution of motion programs including the generation of trajectories. The CPU may be based on several microprocessors.

- each axis card, that may control up to 3 axes implementing the position feedback loop with sample time faster than 1 millisecond. Each manufacturer has dedicated algorithms to control velocity and jerk profiles.
- interfacing cards for Ethernet ports, digital I/O, field buses, USB ports, monitors and teach pendant.
- wireless interfacing cards: in some manipulators the connection between the controller and the teach pendant is wireless. This solution was firstly commercialized by Comau Robotics (2007) and it is now provided by the majority of robot manufacturers. Academic research in this field has been active for many years but the technology transfer towards the industrial playground has only recently succeeded.

3.2. Trajectories

Robot trajectories are usually obtained by a sequence of basic motion elements:

- point to point
- linear segments
- circle arcs

A point-to-point movement is used when the motion must be performed at high speed and the execution of a predefined trajectory is not requested. In this case only the final position to be reached is assigned together with a gross specification of the velocity. The different motors start and finish their motion at the same time.

Linear motion is a modality during which a point of the TCP (tool center point) moves exactly on a straight line. This is obtained by a sophisticated mathematical model of the robot and special algorithms which coordinate the individual motion of each motor. Different approaches may determine different performances in term of trajectory precision. Generally the manufacturer does not explicitly declare the trajectory error which in any case groves with the programmed speed and the payload. Generally the deviation from the theoretical trajectory is below few tenths of millimeters.

Circular movements are specified by 3 points (actual, intermediate, and final) which are connected by the unique circle arc defined by them. For the trajectory precision, considerations similar to those expressed for linear motion are also valid.

3.3. Trajectories Interpolated with External Axes

Many applications require a coordinated motion between the robot and some external axis. In some cases, the external axis is a mobile base on which the manipulator is mounted to extend the working space. Another application deals with the robotic welding of large tanks using a small size robot. An external axis continuously rotates the tank presenting the zone to be welded with the correct orientation. The robot welds the tank whose position and attitude are continuously adjusted by the external axis. A correct execution of the task requires a correct relative motion between the robot and the work-piece. The relative motion depends on the combination of the motion of the robot and of the external axes. It is thus essential that they are correctly interpolated. Some process parameters are also adjusted in real time in dependence of the relative velocity between the tank and the manipulator.

Specific calibration procedures are required to determine the exact rotation axis of each revolute joints and the exact location and direction of each linear axis. This procedure is experimentally performed on the real working cell.

In welding and in gluing applications the motion speed must be proportional to the dispensing velocity of the filler/glue to have a constant quality process.

The availability of specific software options able to deal with these tasks and the easiness to activate them may be determinant in the choice of the manipulator brand.

3.4. Traditionlal Programming Versus sensor Guided Programming

Traditional robot control strategies are based on "trajectory programming". The operator decides the trajectories and the robot execute them. The main drawback of this methodology is that it can be applied only in structured situations where any detail is known from the beginning. Moreover, there are situations where a small error in the trajectory may influence the task very much. For example, when the robot is requested to perform a surface working on a piece, a small error in the tool location may determine the lost of contact or an excessive contact force. Nowadays, manipulator tasks may be programmable in a more flexible way due to their external sensors and improved adaptive control and planning algorithms. Robots are able to learn. The more common adaptive strategies are based on vision systems and force sensors.

3.5. Force Sensors

Force sensors permit to improve productivity in some application fields, mainly related to metal working. Other applications are currently under study in research labs worldwide, but they have not yet reached an industrial maturity due to the complexity of the programming algorithms and to the still high costs of the hardware.

Force sensors based techniques are implemented to obtain fast velocity and stability contact between tool and work-piece. The contact force is measured by a force sensor and special feedback algorithms allow to keep constant the value and intensity of the interacting force while the robot moves at the requested tangential velocity.

Further applications concern grinding, deburring, polishing of pieces produced in foundry. The operator may guide the robot by hand to roughly specify the trajectory to be followed; the robot will adapt the behavior using the sensors in order to improve the trajectory and it will record the exact path to be followed.

This application is suitable for all those processes which require a high quality finishing: in fact the robot is able to sense the exact object shape and follow it applying a constant pression even if the piece geometry is not exactly known. Since the contact force is uniform, the irregularity in the surfaces and edges are removed uniformly.

The robot operates at the maximum possible velocity, slowing down when the working forces become too high to minimizing deflections of the robot arm, breaking of the tool, alteration of the piece produced by excessive stress or heating.

3.6. Vision Systems: 2D and 3D

Vision applied to robotics is one of the first automatic adaptive systems. The vision system supplies the information necessary to adapt the robot behavior to reach the correct position and perform the correct action.

In the last 10 years, the use of 2D vision systems has supported a large development of automated applications, allowing more flexibility and easiness of use. The developers of hardware and software vision tools (electronics, software libraries) have made a huge research effort in the past years, thus reducing the total cost of the final systems while improving their performance. 2D vision systems are able to identify the position and orientation of an object in a planar surface. The systems are also able to identify different objects selecting them from a list of possible candidates.

Fixed cameras are now able to elaborate images in few milliseconds, to compare them with data stored in archives and calculate all the data necessary to position the robot and to grasp the target objects. All these operations may be performed in "mask time" while the robot is still performing the previous working cycle.

These characteristics – together with the growing power of computers in term of speed, mass storage dimension, connectivity and the availability on robot controller units of more open communication ports – have lead to many new, vision based, industrial applications, that have become cheaper and feasible. Fast connections between computers, smart cameras and robot controllers may be performed by high speed dedicated fieldbus (ie: via Ethernet).

Vision systems are commonly used in robots loading/unloading machine tools and machining centers, where working cycles of few seconds are required.

The new 3D vision applications are now able to recognize complex pieces randomly placed in bins. The most used techniques are: stereovision and laser-scanning. Stereovision systems employ two or more cameras watching the bin from different point of view. Laser scanners use a camera watching the light reflected by the object while a laser beam is used to scan the scene to be analyzed. In both cases an object can be recognized even if it is partially covered by others. With respect to 2D systems, 3D systems require more elaborating time and the costs are higher. Many successful demos have been developed, but real industrial applications are still rare.

3.7. Touch Sensing

Touch sensing is typically used in welding applications to overcome several problems related to the geometrical inaccuracy of the process. Touch sensing means that the robot senses some characteristics of the external world by touching it.

Welding applications are among the driving applications for the introduction of automation, as already said elsewhere in this chapter, mainly due to: slowness of the process and potential dangers to the human worker during the manual execution of the task. Welding is a low added value application with limited precision of the metal sheets to be welded; thus the operator, human or robot, needs to compensate for this difficulty using additional 'carefulness'.

Historically, the first method used to compensate for limited precision and thermal deformation was the use of mechanical fixtures to lock the pieces to be welded in predefined positions. This methodology is still largely used, but it is too expensive for small production lots due to the high cost of the locking mean. However, when the location of the features to be welded is not precisely known from the beginning, it is necessary to implement some strategy to identify it. An applicative solution was developed creating a sort of touching sense for the robot employing the welding torch itself.

An electrical circuit is realized connecting the pieces to be welded to the earth circuit of the welding system and supplying a low voltage to the welding torch. When the torch (or the welding wire) touches the object, the electrical circuit is closed and the robot localizes the object. Knowing the geometry of the piece, few touches are sufficient to exactly determine some key points like the beginning and the end of the welding zone. The nominal welding trajectory is then adapted to the experimental measures. Systems working at high voltage (up to some hundred Volts) are also commercially available for applications in which the iron sheet may be dirty or greasy.

3.8. On Line Correction of the Welding Path by Current Measurement

Welding produces a lot of heat that may deform the objects. Sometimes the identification of the beginning and the end of the welding track path is not sufficient, especially for large objects (a frame may measure up to several meters). It is necessary to continuously measure the position of the iron sheet and adapt the robot trajectory in real-time. This is possible by special algorithms based on the measure of the welding currents. In fact the current decreases when the distance between torch and sheet increase. This methodology is known as RTPM (real-time performance management).

3.9. On Line Correction of the Welding Path by Laser Sensor

An innovative methodology to update in real time the welding path is the seam laser tracking (Fig 15, 16). It is the only 3D system employing cameras, which is able to real-time correct the execution path. A camera monitors the area where the welding should be executed; a “plane” of light, produced by a laser source and a special lens, is projected in that area. The shape and the location of the object to be welded is detected by the analysis of the image and by the shape assumed by the light trace reflected by the object. The analysis is continuously updated in real time. As the welding procedure evolves, the coordinate of the new tracking points are evaluated and sent to the robot controller which produces the correct robot motion.

The camera and the laser system are located on the robot end effector nearby the welding torch. They are oriented in such a way to monitor the zone of the welding to be reached next. The laser trace projected on the pieces is larger less then 90mm and correlated to the maximum tolerated geometrical error. The analysis on the image permits the identification of the edges of the two pieces to be welded and of their distance. It is also possible to compensate several welding parameters like speed and electrical current to obtain high quality connections (Fig 16).

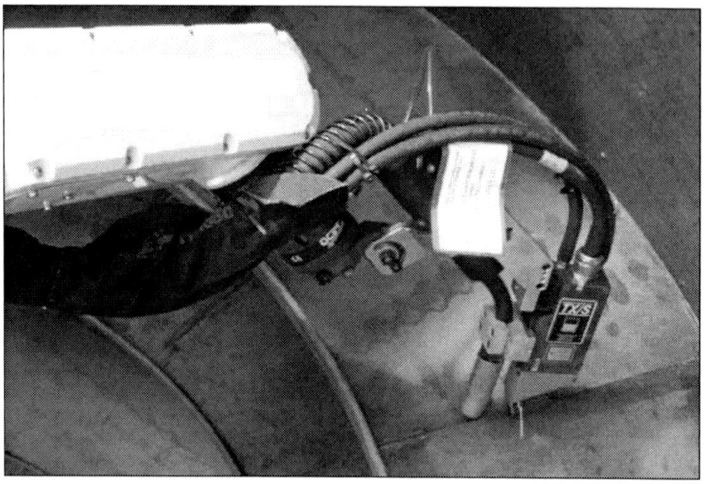

Figure 15. Correction of welding trajectory by laser tracker [4], [6], [7].

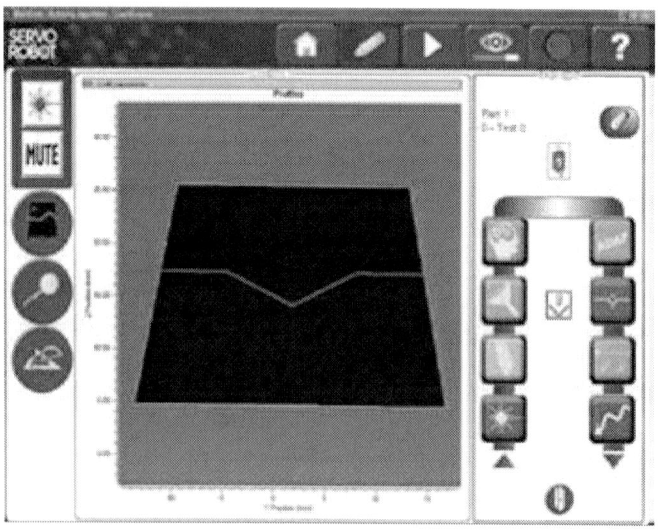

Figure 16. Example of image of the virtual panel for the management of a laser tracker sensor [4].

4. Programming

4.1 Teaching the task

Programming a robot, means teaching it the task expected by it.

While from a mechanical and electronic point of view the different manufacturers have developed similar products, wider differences are present in the programming systems.

In any case three different approaches are generally present:

1) programming by language (on or off line);
2) programming by teaching pendant (on line);

3) programming by graphical system (off line).

All these methodologies always assume that the robot is programmed to follow a predefined trajectory. In limited cases the trajectory may be modified by data collected by some external sensor (force sensor, laser tracker) but a nominal trajectory is always present. In some cases the controller accepts an external offset to be superimposed, in real time, to the programmed trajectory. This input may be used to modify the robot behavior in function of external sensors.

The term 'on-line programming' means that it is performed using the robot controller. In this case, the manipulator cannot operate during the programming. The term 'off-line' means that the programming is developed on a different system and later downloaded on the manipulator controller. In this case, the robot may work during programming.

Programming by language means that the operator must write a list of operations that the robot will perform. The instructions include the path to be followed as well as additional data like execution speed. Special instructions permit the interaction with external devices (sensors, actuators, other machines) in order to coordinate the robot motion with the rest of the working cell and to detect any anomaly.

Basically, any robot is programmed by its own language, but different programming procedures are available. In some cases the programmer must explicitly write the instructions and the value of the coordinates of the points to be reached (programming by language), in other cases the list of the instructions is generated using some graphical interactive program, or the coordinates of the points are generated moving the actual robot to the correct position (teaching mode).

4.1. Programming Languages

Each manufacturer has developed its own programming language. Some of them have taken inspiration from the well-known Basic language; others have developed more sophisticated proprietary languages that show similarities to other programming languages like Pascal or C/C++.

Each language differs from the others for some philosophical details and for the syntax used to express the instructions. However some basic similarities are present in all the languages, that are discussed in the following:

- Motion instructions
- functions
- data type

Some basic motion instructions are always present. In the following we adopt the name assumed in a particular programming system (ABB Rapid [5]). Their name and some details may be different on other systems, but similar concepts are implemented.

Basic motion instructions are:

- MoveAbsJ Move the robot to an absolute position specified by joint coordinates.
- MoveC Move the robot along a circular path
- MoveCDO Move the robot along a circular path and set a digital output

- MoveCSync Move the robot along a circular path and execute a procedure
- MoveExtJ Move an external axis in joint coordinate
- MoveJ Move the j-th joint
- MoveJDO Move the J-th joint and set a digital output
- MoveJSync Move the J-th joint and execute a procedure
- MoveL Move the robot along a linear trajectory
- MoveLSync Move the robot along a linear trajectory and execute a procedure

Other instructions permit the conditional execution of groups or portion of the program (if/then/else), the generation of instruction loops, and the data exchange with other robots or external devices.

Common mathematical functions that can be used to generate the path are for instance:

Abs	The absolute value of a number
ACos	Arc cosine
AOutput	Reads the value of an analogical data
ArgName	Gets the name of an argument
ASin	Arc sine
ATan	Arc tangent
ATan2	Arc tangent of the ratio between two numbers

Different data types may be elaborated, for instance:

bool	logical values true/false
btnres	data resulted by a key press (yes/no)
busstate	State of an I/O bus
byte	Integer values in the range 0-255
integer	Integer values in the range -32768-32767
real	Data in floating point
tool	pose of the robot TCP (xyz plus attitude)
clock	Time interval
confdata	Configuration data of the robot
corrdescr	Descriptor of the corrector generator
dionum	Digital input/output (0 – 1)

As already mentioned, the program instructions can be written directly by the programmer in different ways using the portable programming unit of the robot controller (also called teach pendant) or using an external PC. Instructions can be typed using a keyboard or chosen from a menu showing the available instructions. The data representing robot positions or orientation can be assigned by a numeric keyboard or collected after moving the manipulator to the corresponding pose and memorized by pressing a key (teaching procedure).

The *teach pendant* (Figure 17) is a specialized programming unit with keyboard and screen that the programmer may hold in his/her hands that allows to move directly the robot (jog mode) just pressing few keys or actuating a joystick. The teach pendant is directly connected to the robot controller by a wire or a secure wireless connection.

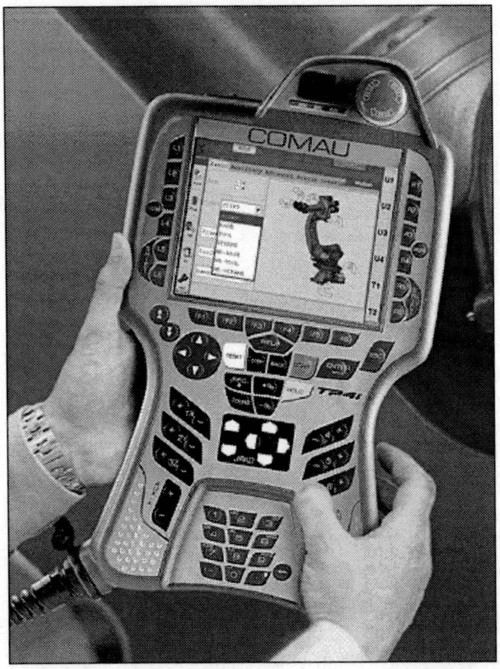

Figure 17. A modern teach pendant; versions with touch screen functionalities are also offered by some manufacturers [3].

4.2. Programming by Teaching

In this programming modality the operator moves the robot to the desired position using the teach pendant. The operator verifies the attained position by visual inspection. When the manipulator has reached the desiderated pose, the pose is memorized by pressing a confirmation key. The manipulator is then moved by to the next pose and the procedure is repeated for each point of the movement cycle. When switched to "automatic mode" the robot will be able to visit in sequence all the memorized poses.

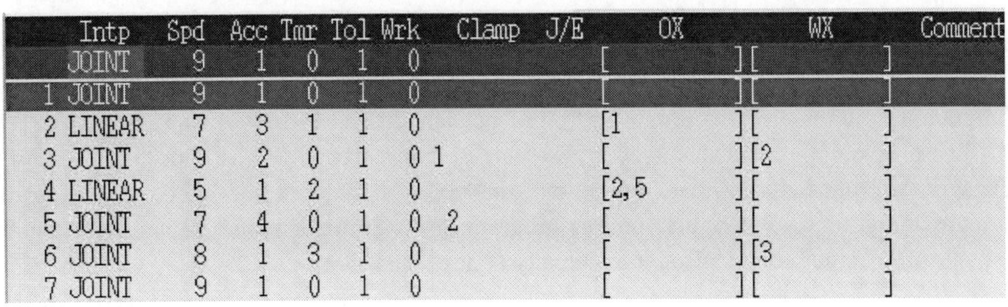

Intp	Spd	Acc	Tmr	Tol	Wrk	Clamp	J/E	OX	WX	Comment
JOINT	9	1	0	1	0			[ ]	[ ]	
1 JOINT	9	1	0	1	0			[ ]	[ ]	
2 LINEAR	7	3	1	1	0			[1]	[ ]	
3 JOINT	9	2	0	1	0	1		[ ]	[2]	
4 LINEAR	5	1	2	1	0			[2,5]	[ ]	
5 JOINT	7	4	0	1	0	2		[ ]	[ ]	
6 JOINT	8	1	3	1	0			[ ]	[3]	
7 JOINT	9	1	0	1	0			[ ]	[ ]	

Figure 18. Screen shot of the status menu in the 'programming by teaching' mode. In this case, a motion sequence of 7 points has been programmed [7].

In some cases a special sensorized device is fixed on the end-effector, and the programmer moves the robot by pushing or pulling this sensor. Thus, the programming is more intuitive.

When a point is memorized it is also necessary to specify the modality to reach it (circular, linear, or joint trajectory), the motion velocity and other parameters. A suitable screen menu summarizes the status during the programming (an example in Figure 18).

The first line is the "title line" which specifies the meaning of the data reported in the next lines. Each of the other lines describes the modality to reach a point.

The meaning of the parameters is summarizes in the following:

Interp: Interpolation, defines the modality to reach the point (linear motion, circular motion, joint motion).

Spd: Speed, it defines the speed during the trajectory. Sometime the values can be expressed in explicit values (e.g. mm/s) otherwise it is expressed just by "qualitative" data (e.g. 1=slow, 9=fast).

Acc: Accuracy, specifies the degree of accuracy required to reach a point (accuracy and speed are in competition).

Tmr: Timer, permits to specify the waiting time in the reached point before proceeding to the next instruction.

Tol: Tool, specifies which tool is mounted on the end-effector and thus the offsets the controller needs to set (each tool has a different shape and size).

Wrk: Work, specifies the coordinates of the reference system related to the working piece

Clamp: permits to open and close the grasping tool.

J/E: permits to specify the jump to a predefined program if a suitable external signal is received.

O/X: permits to send a signal to an external device.

W/X: permits to read a signal arriving from an external device.

Comment: permits to specify a comment to document the program.

datapos: Block of data exchanged with external devices (e.g. cameras)

4.3. Graphical Simulation and Programming

All the robot manufacturers propose a graphical simulation environment to simulate working cells and program their robots (Figure 19). Geometrical models of their products, along with a controller emulator are included in the software environment. Simulation and programming tools are also proposed by independent firms. These software products allow off-line programming, because they are executed on an external PC without the necessity to engage the robot during the simulation; the robot may continue working during the programming time. When the motion programs are terminated and debugged, they are downloaded to the robot controller to actually control the robot.

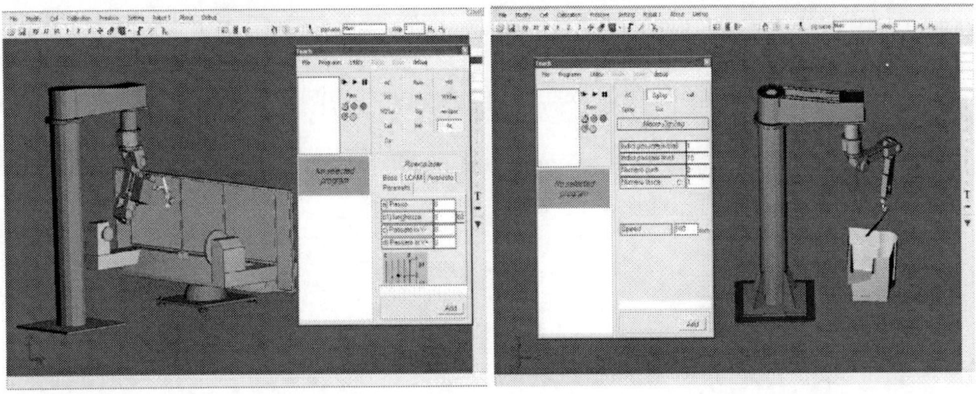

Figure 19. Example of SW for the 3D graphical programming of manipulators [6], [7].

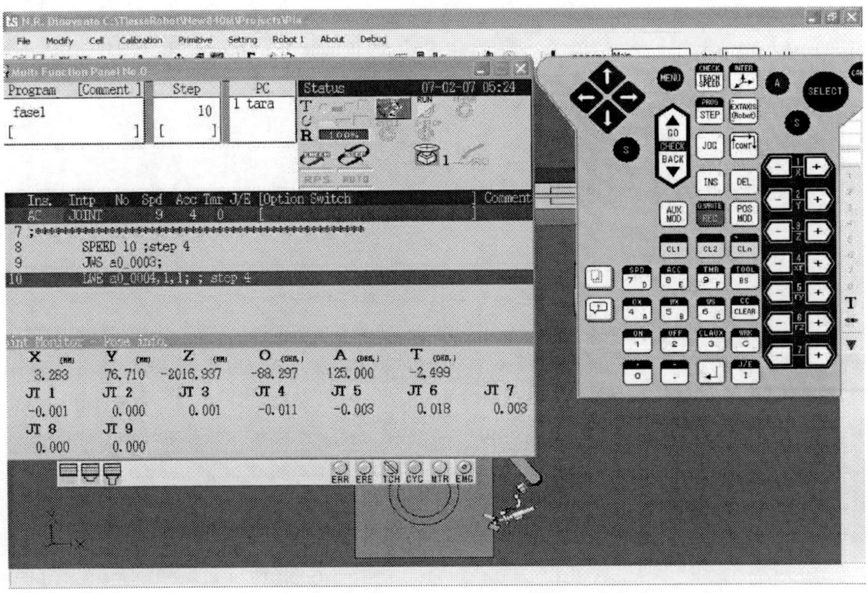

Figure 20. The 3D graphical programming tools emulate also the functionalities of the teach pendant [6], [7].

The graphical simulation software manages 3D components of the robots and the whole working cell:

- all the mechanical components of the working cell
- all the tools present in the working cell
- the workpiece

All the models of the components of the working cell may be developed using the programming SW or imported from standard 3D modeling SW.

This kind of software is useful for the programming of the working cell, but also during the design of the cell itself because of the following advantages:

- it is possible to simulate the layout of all the components of the working cell and check of dimensions and collisions
- it is possible to predict the execution time (the simulator uses the same trajectory generator of the real robot)
- the program is realized in teaching mode inside the simulator and then automatically converted in programming language
- the motion program can be verified before building the working cell to check it feasibility
- the programming is very easy
- the motion programming can be directly downloaded to the robot controller
- the motion programs can be developed and debugged without risk of mechanical failures on the work-cell.

Once the working cell is created, the robot motion can be programmed with graphical tools. The programming procedures may be personalized by the users. It is also possible to simulate the use of the teach pendant (Figure 20).

5. Other Advanced Applications

5.1 Vision based tasks

The already described off-line graphical programming and the adaptive welding procedures are examples of advanced robotic applications. They are both examples of different methodology to simplify and automatize the generation of the robot path. Other application dependent methodologies are based on 3D laser scanner and CAD/CAM postprocessors. In both cases the shape of an object is automatically analyzed to generate the tool path to perform the desired process. The path can be automatically downloaded in robot controller for the actual execution of the task.

The "time-of-flight" laser scanner is a device that directing a laser beam on an object and scanning its surface is able to measure its three-dimensional shape. This is used in many application fields (e.g. marble working). The device is able to evaluate the distance between the laser source and any point of the object by measuring the time employed by the lights to travel from the source to the object and coming back to a suitable sensor after being reflected. Since the speed of the light is known, the distance is proportional to half the traveling time.

Other alternative methodologies to optically measure the shape of objects are based on the projection of structured light (strips of light) whose shape is deformed by the object shape. A camera takes images of the deformed lights, elaborates them and reconstructs the 3D shape of the object. There are several possibilities, but the most common one utilizes linear strips of light whose direction and shape may be modified by active projectors. Alternative choices utilize white and dark chessboards or other patterns.

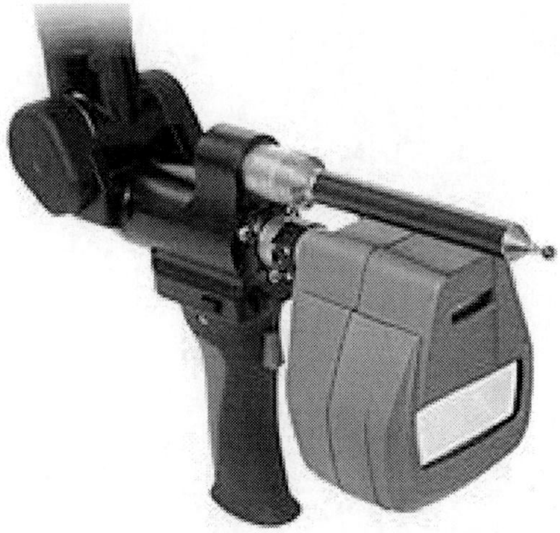

Figure 21. Laser scanner for 3D measurement with a mechanical probe [8].

Figure 22. Projection of structured strips of light to measure the object shape.

Using the described technologies several working cells have been realized for machining stones and marble pieces (such as, basins, ponds, mantelpieces, gravestones, columns, statues, sculptures, bas-relieves) thus reducing the working time and simplifying the installation and the programming of the system (Figure 23). With respect to traditional process, costs are also reduced. The programming of the surface machining may be derived by copying dimes or samples manually realized with other technology or softer material (e.g. wood, plastic). The acquisition of the shape may be utilized also to perform surface tooling like contouring, deburring, brushing, cleaning or drilling. The robotized cells may be used to work also stone, concrete, corian, glass, carbon fiber, bricks, wood and resins.

Figure 23. Robotic milling of a sculpture [9].

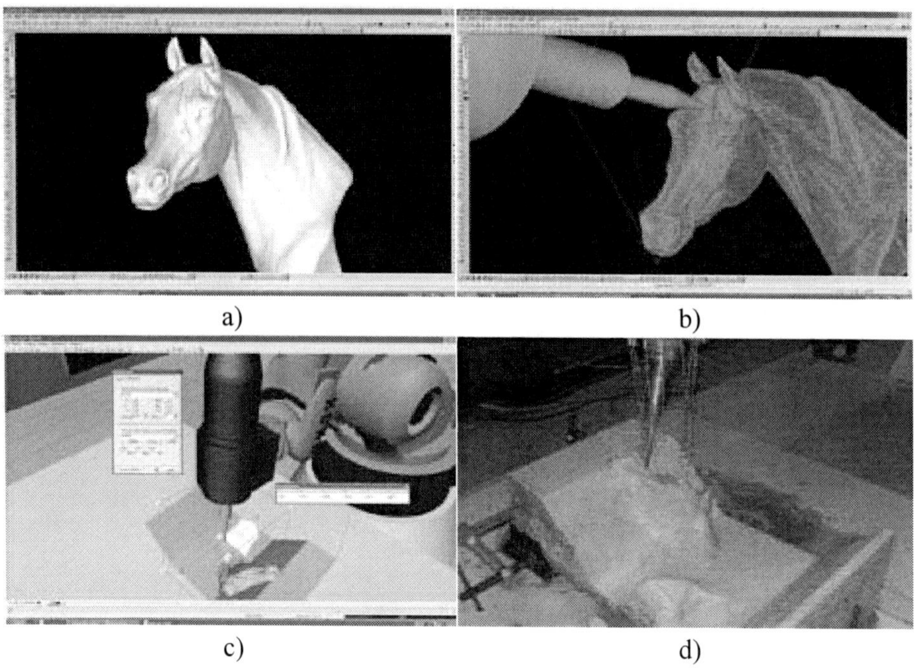

Figure 24. Simulation of the milling process; d) manufacturing [9].

The software permits a complete control over all the phases of design and working, preventing errors by operator and supplying forecast of the working cycle duration. Any operator with minimal informatics knowledge and design skill can professionally utilize the system. The generation of the complex tool trajectories may be performed using specialized 3D CAD-CAM programs which simulate all the phases of the producing cycle. The shape of the objects can be defined importing 3D scanned shapes or generating them by the modeling software (Figure 24). The time necessary to model the working cycle depends on the shape complexity.

Once the virtual model has been realized, the tool path is elaborated by the operator using general or dedicated procedure libraries. The software system is able to show a simulation of all the producing process to check that everything is programmed properly. If necessary, the operator may modify the procedure to optimize the cycle (Figure 24 c). Once the simulation cycle has been confirmed by the operator, the cycle is performed by the real machine on the real piece utilizing the simulated trajectory and resulting working parameters (Figure 24 d).

5.2. Remote Laser Welding

A relatively new industrially available technology is the laser welding and cutting performed by remote laser tools managed by anthropomorphic manipulators. An industrial robot carries a laser source on its end effector. The laser beam is focalized at a certain distance (e.g. 1 meter) where the cut or the weld must be performed. The manipulator directs the laser beam toward the correct working location. With small changes in the position and attitude of the wrist, the focalized laser beam may be moved very fast where necessary. The result is a fast laser tool machine combined with the robot flexibility.

5.3. Other Emerging Applications

Many new robotics applications are proposed every day. However, only some of them successfully penetrate the industrial field, due to costs, reliability, implementation of safety issues. One of the emerging fields is the food. Manipulators may be utilized combined with vision systems to sort and palletizing many food products. Classical working cells works on packed parts like snacks, chocolate, but new applications deal also with chops of meat with irregular and highly deformability shapes. Vision systems are used to identify their mass, special software sort them tacking into account shape and size, manipulators equipped with special grippers grasp the chops and arrange them in pack of suitable size.

New advanced applications include a smart use of sensors. For instance a Spanish company uses a robotized line to detect the sex of fish by color analysis of the internal part of the abdomen. Female fish must be treated to extract eggs which will be sold separately. Fish arrives on a belt and its location and orientation is detected by a vision system. The manipulator punctures it in the abdomen with a needle integrated with an optical sensor. Special software detects the sex by color analysis of the interior and the manipulator sends the fish to the correct production line. The whole cycle requires fractions of seconds.

6. Conclusions

Robotics is an available technology already affirmed in many application fields but always in evolution. The increasing performance of mechanics, electronics, control and software create new opportunities every day. While robotics is an affirmed opportunity in all the structured realities where standard processes are performed, the new trend is to spread into loosely structured environments where the use and implementation of sensors and "artificial intelligence" make manipulators more flexible and able to safely interact with humans without a fence. This Chapter has discussed only affirmed and emerging fields of industrial settings, but it is worth to highlight that robotics has already started spreading other important fields like entertainment, personal care, medical (surgery, rehabilitation).

References

[1] IFR World Robotics 2009 – Industrial Robots
[2] Robotics Vision to 2020 and beyond – The Strategic Research Agenda for Robotics in Europe, 7/9/2009
[3] http://www.comau.com
[4] http://www.servorobot.com/
[5] ABB Flexible Automation, RAPID Reference manual, RAPID Overview On line, http://rab.ict.pwr.wroc.pl/irb1400/overviewrev1.pdf
[6] http://www.tiesserobot.it
[7] http:// www.kawasakirobotics.com
[8] http://www.sixdindia.com/
[9] http://www.tdrobotics.com

In: Robotics: State of the Art and Future Trends
Editors: G. Legnani and I. Fassi

ISBN: 978-1-62100-403-5

Chapter 2

A PARTIALLY-DECOUPLED 4-DOF GENERAL 3SPS/PS PARALLEL MANIPULATOR

***Chin-Hsing Kuo*[1,*] *and Jian S. Dai*[2,†]**
[1]Department of Mechanical Engineering,
National Taiwan University of Science and Technology, Taiwan
[2]Division of Engineering,
King's College London, University of London, United Kingdom

Abstract

In terms of motion strategies, the parallel manipulator that generates three rotational and one translational (3R1T) degree-of-freedom (dof) has found task-oriented applications in some paradigms, e.g., the robotically-assisted laparoscopic surgery [1]. However, there are very few 3R1T in-parallel actuated mechanisms that have been identified and extensively studied. In this chapter, a general 4-dof, 3R1T parallel manipulator is extensively studied. The manipulator is structured with three SPS (spherical-prismatic-spherical joints) legs and one generally placed PS strut where a prismatic joint is attached to the fixed base. Accordingly, the translational part of the output motion is completely independent of the three rotational ones so that the rotational and translational motions are decoupled. We will present the general solutions of inverse and forward kinematics for the manipulator. We will further present the singularity analysis and workspace analysis of the manipulator and conclude this chapter with discussion on its potential applications for minimally invasive surgery.

Introduction

Since the six degree-of-freedom (dof), Stewart-Gough platform was introduced in 1960s, parallel manipulators had caught many attentions in the past decades. For many industrial applications, however, the actual, effective dofs required by the working tasks are not as many as those provided by the six-dof hexapod. For example, an ankle rehabilitation device only requires a 2-dof orientation; a laparoscopy surgical operation maneuvers a 4-dof motion;

* E-mail address: chkuo717@mail.ntust.edu.tw
† E-mail address: jian.dai@kcl.ac.uk

a five-axis machining center specifies a 5-dof spindle motion, etc. Therefore, the lower-dof parallel manipulators, which possess it mobility fewer than six, have recently started to surface and, meanwhile, have drawn extensive research interests to the community.

Among the various lower-dof parallel manipulators, the 4-dof ones belong to one specific class. Up to date, some 4-dof parallel manipulators have been synthesized by systematic approaches [2-8]. In terms of motion strategies, there are generally two types of 4-dof parallel manipulators. One is the 3T1R (the moving platform is with three translational and one rotational degrees of freedom, also known as the Schöenflies motion) type and the other one is the 3R1T (the moving platform is with three rotational and one translational degrees of freedom) type. In the past decade, the 3T1R type raised much interest from researchers. Roland [9] modified the 3-dof Delta manipulator into two 4-dof parallel manipulators, namely Manta and the Kanuk, both of which have three translational and one rotational degrees of freedom. Carricato [10] presented a family of fully isotropic 4-dof parallel manipulators for Schöenflies motion. Company *et al.* [11] designed a 4-dof parallel manipulator that also produces Schöenflies motion for high-speed handling and machining. Richard *et al.* [12] studied and prototyped a 3T1R parallel manipulator. Salgado *et al.* [13] synthesized a new fully-parallel manipulator with Schöenflies motion. Unfortunately, only a few studies were for the 3R1T parallel mechanisms. In 2001, Zlatanov and Gosselin [14] proposed a class of 3R1T 4-dof parallel manipulators with four identical five-revolute legs. In their mechanism, each leg comprises of three intersecting and two parallel revolute joints such that the parallel manipulator forms an overconstrained mechanism. On the other hand, Lu and Hu [15] presented a 3R1T 4-dof parallel manipulator with three SPS legs and one SP leg. The fixed base of their manipulator is attached with four spherical joints of the four legs so that the position and orientation of its moving platform are mutually dependent. In effect, the above two 3R1T manipulators have the coupled output degrees of freedom where the translation and the rotations are dependent on each others.

In 2000 and 2003, a decoupled 3R1T 4-dof parallel manipulator was exploited by Lenarčič *et al.* [16-18] for simulating the movement of human shoulder. The used parallel manipulator consists of four driven legs, which are made up of three outer SPS legs and one central PS leg and are disposed at a specific symmetric geometry. The proposed 3R1T 4-dof parallel manipulator has been studied through several kinematic issues together with their design application for simulating human shoulder. However, some other interesting kinematic properties such as forward kinematics, singularity, workspace, etc. for this mechanism have not been explored due to the different focus. Further, while the illustrated analysis work had been done based on a specific symmetric mechanism configuration, a generalized analysis to the mechanism with its general configuration is more than expected.

Further, though the central strutted Stewart platform structure was presented in 1994 by Dai *et al.* [19] and has been integrating at this decade in the structure of the lower mobility parallel mechanisms, the generally placed strut in the mechanisms has not yet been investigated.

Therefore, the work presented in this chapter is organized to exhaustively study the kinematic properties of the 4-dof general 3SPS/PS parallel manipulator with a generally placed strut. Unlike the previous studies [16-18] which position this 3SPS/PS mechanism at a symmetrical pose, here we study this manipulator with a general configuration. We will investigate the mobility and the structural and motion characteristics of the 3SPS/PS parallel manipulator with a generally placed strut, present the general solutions to the inverse and forward kinematics of the manipulator, and explore the singularity including inverse singularity, direct singularity, and combined singularity analysis. Further, we will examine its orientation workspace, reachable workspaces, and the dexterous workspace.

The Manipulator

The studied 4-dof 3SPS/PS parallel manipulator is made up of a fixed base, a moving platform, three legs and a generally placed strut which connects the moving platform to the fixed base as in Fig. 1. The fixed base $A_1A_2A_3$ is formed as a plane, while the moving platform $B_0B_1B_2B_3$ is a tetrahedron. The three legs are connected to the fixed base and the moving platform with spherical joints at A_1B_1, A_2B_2, and A_3B_3, respectively. The generally placed strut is mounted perpendicularly on the fixed base at O and connected to the moving platform at B_0 by a spherical joint. Since this manipulator is studied with its general form, the strut can be mounted at anywhere on plane $A_1A_2A_3$, including, of course, the centroid O' of the triangle $\Delta A_1A_2A_3$. Also, each leg including the generally placed strut comprises of an upper member and a lower member connected by a prismatic joint which is served as the actuation.

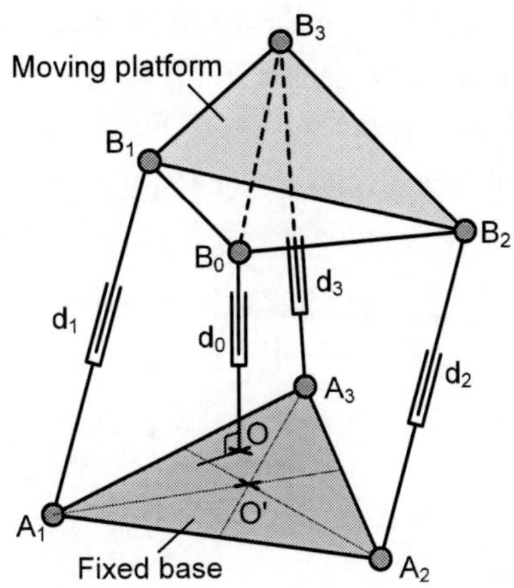

Figure 1. The partially decoupled 4-dof general 3SPS/PS parallel manipulator.

It can be simply observed that there is no redundant constraint [20] in this mechanism so the general Kutzbach-Grübler mobility criterion is exploitable for this mechanism. Therefore, the mobility m of the manipulator can be calculated as:

$$m = \lambda(n - j - 1) + \sum_i f_i - f_p = 6(9 - 11 - 1) + (3 \times 7 + 1 \times 4) - 3 = 4 \quad (1)$$

where λ is the degrees of freedom in space, n the number of links in the manipulator, j the number of joints in the manipulator, f_i the degrees of relative motion permitted by joint i, and f_p the number of passive degrees of freedom in the manipulator. From the structure of the manipulator, we know that there are nine links, seven spherical joints, and four prismatic joints in the mechanism. Furthermore, there is a passive degree of freedom existed in each of the three SPS legs. Therefore, the mobility of this manipulator is obtained as Eq. (1) by four, which is consistent with the number of total legs/actuations in the mechanism.

From its mechanism configuration, some motion characteristics of the parallel manipulator can be easily observed as well. First, the translation of the moving platform is permanently directed perpendicular to the fixed base. Second, the translation of the moving platform is completely controlled by the prismatic joint of the generally placed strut. Besides, the three rotations are pivoted on the motion path of the translation at the center of joint B_0. In other words, the position of the moving platform is only determined by the length of the generally placed strut, whereas the orientation is decided together by the three SPS legs. This fact implies that whatever the lengths of the three SPS legs stretch, the position of the moving platform will be held as long as the length of the generally placed strut keeps constant. This arrangement structurally decouples the rotation and the translation of the moving platform at the pivoted point B_0. It thus dramatically simplifies the kinematic analysis of this parallel manipulator as we will discuss later.

Geometry Analysis

For analyzing the geometry of the manipulator, two Cartesian coordinate systems $A(x, y, z)$ and $B(u, v, w)$ are attached to the fixed base and moving platform, respectively, as shown in Fig. 2. Without losing the generality, it is assumed that the origin of frame A is coincident to the mount point O on plane $A_1A_2A_3$ of the fixed base, the x-axis points along the direction of $\overline{OA_1}$, the z-axis directs along the generally placed strut, and the y-axis is defined by the right-hand rule. On the other hand, the origin of frame B is assumed at the center of the spherical joint B_0, the w-axis points along the direction $\overline{B_0B_1}$, the v-axis lies on plane $B_0B_1B_2$, and the u-axis is defined by the right-hand rule.

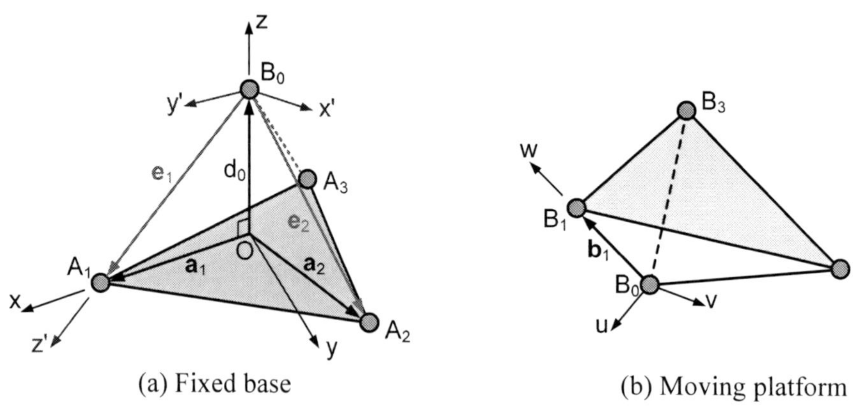

Figure 2. Coordinate systems of the manipulator.

The transformation from the moving frame B to the fixed frame A can be described by a position vector $\mathbf{p}$ defined by the distance between frames A and B and a rotation matrix ${}^{A}R_{B}$ defined by the Euler angle representation. Assume that the initial location of the moving frame B coincides with the fixed frame A and that the final location is obtained by a rotation φ about the w-axis, followed by a second rotation θ about the displaced u-axis, followed by a third rotation ψ about the displaced w-axis, and followed by a linear displacement along the strut from O to B_0. The consecutive rotations, known as the w-u-w Euler angles rotation, is here denoted as $R(w, \varphi)R(u, \theta)R(w, \psi)$. Then the position vector $\mathbf{p}$ of the linear displacement $\overline{OB_0}$ with respect to frame A and the rotation matrix ${}^AR_B$ can be obtained as:

$$\mathbf{p} = \mathbf{d}_0 = [0,\ 0,\ d_0]^{\mathrm{T}} \tag{2}$$

$${}^AR_B = \begin{bmatrix} c\phi c\psi - s\phi c\theta s\psi & -c\phi s\psi - s\phi c\theta c\psi & s\phi s\theta \\ s\phi c\psi + c\phi c\theta s\psi & -s\phi s\psi + c\phi c\theta c\psi & -c\phi s\theta \\ s\theta s\psi & s\theta c\psi & c\theta \end{bmatrix} \tag{3}$$

where $\mathbf{d}_0 = \overline{OB_0}$, $c\varphi$ is a shorthand notation for $\cos\varphi$ and $s\theta$ for $\sin\theta$, and so on. Let the position vector ${}^B\mathbf{b}_i$ of point B_i with respect to the moving frame B be given by

$${}^B\mathbf{b}_i = [b_{iu},\ b_{iv},\ b_{iw}]^{\mathrm{T}} \tag{4}$$

Hence the position vector $\mathbf{b}_i$ of $\overline{B_0B_i}$ expressed in the fixed frame A can be written as

$$\mathbf{b}_i = [b_{ix}, b_{iy}, b_{iz}]^{\mathrm{T}} = {}^AR_B\,{}^B\mathbf{b}_i = \begin{bmatrix} b_{iu}(c\varphi c\psi - s\varphi c\theta s\psi) - b_{iv}(c\varphi s\psi + s\varphi c\theta c\psi) + b_{iw}\, s\varphi s\theta \\ b_{iu}(s\varphi c\psi + c\varphi c\theta s\psi) - b_{iv}(s\varphi s\psi - c\varphi c\theta c\psi) - b_{iw}\, c\varphi s\theta \\ b_{iu}\, s\theta s\psi + b_{iv}\, s\theta c\psi + b_{iw}\, c\theta \end{bmatrix} \tag{5}$$

Then, as shown in Fig. 3, the position vector $\mathbf{q}_i$ of point B_i expressed in the fixed frame A is

$$\mathbf{q}_i = \mathbf{p} + {}^AR_B\,{}^B\mathbf{b}_i \tag{6}$$

Also, let the position vector of point A_i with respect to the fixed frame A be given by

$$\mathbf{a}_i = [a_{ix},\ a_{iy},\ a_{iz}]^{\mathrm{T}} \tag{7}$$

Then a loop-closure equation can be written for leg i as

$$\mathbf{d}_i - \mathbf{d}_0 = \mathbf{b}_i - \mathbf{a}_i \tag{8}$$

where $\mathbf{d}_i = \overline{A_iB_i}$.

Dot-multiplying Eq. (8) with itself produces an equation of constraint imposed by leg i as follows:

$$d_i^2 + d_0^2 - 2\mathbf{d}_i^{\mathrm{T}}\mathbf{d}_0 = \mathbf{a}_i^2 + \mathbf{b}_i^2 - 2\mathbf{a}_i^{\mathrm{T}}\mathbf{b}_i \tag{9}$$

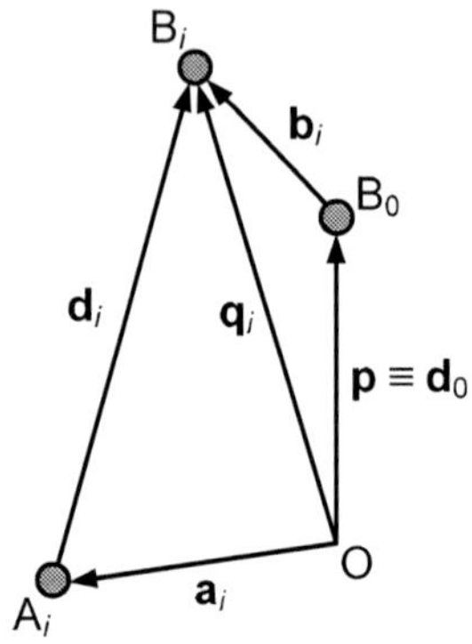

Figure 3. Vector loop closure of leg *i*.

From the chosen coordinate systems as shown in Fig. 2, we have $a_{iz} = 0$. So, from Eqs. (2), (5), and (8), we get

$$\mathbf{d}_i^T \mathbf{d}_0 = \begin{bmatrix} b_{ix} - a_{ix} \\ b_{iy} - a_{iy} \\ b_{iz} + d_0 \end{bmatrix}^{\mathrm{T}} \begin{bmatrix} 0 \\ 0 \\ d_0 \end{bmatrix} = d_0^2 + d_0 (b_{iu}\, \mathrm{s}\theta\, \mathrm{s}\,\psi + b_{iv}\, \mathrm{s}\theta\, \mathrm{c}\,\psi + b_{iw}\, \mathrm{c}\theta) \tag{10}$$

Substituting Eq. (10) into (9) yields

$$d_i^2 = \mathbf{a}_i^2 + \mathbf{b}_i^2 - 2\mathbf{a}_i^{\mathrm{T}} \mathbf{b}_i + d_0^2 + 2d_0 (b_{iu}\, \mathrm{s}\theta\, \mathrm{s}\,\psi + b_{iv}\, \mathrm{s}\theta\, \mathrm{c}\,\psi + b_{iw}\, \mathrm{c}\theta) \tag{11}$$

Kinematic Analysis

Inverse Kinematics

For the inverse kinematics problem, the position vector **p** and the rotation matrix ${}^A R_B$ (or the rotation angles φ, θ, ψ of the moving platform) are given and the leg lengths d_i, $i = 0$, 1, 2, and 3 are to be found. The solution of d_0 is straightforward that can be derived from Eq. (2). Once d_0 is obtained, the other leg lengths d_i, $i = 1$, 2, and 3 can be computed from Eq. (11) as:

$$d_i = \pm\sqrt{\mathbf{a}_i^2 + \mathbf{b}_i^2 - 2\mathbf{a}_i^{\mathrm{T}} \mathbf{b}_i + d_0^2 + 2d_0 (b_{iu}\, \mathrm{s}\theta\, \mathrm{s}\,\psi + b_{iv}\, \mathrm{s}\theta\, \mathrm{c}\,\psi + b_{iw}\, \mathrm{c}\theta)}\,,\ i = 1, 2, 3 \tag{12}$$

As a result, there are one solution of d_0 and generally two solutions of d_i for each given ${}^A R_B$ and **p**. The solution of d_0 with positive/negative sign means that the moving platform is located above/below the fixed base. And, for the two solutions of d_i, only the positive leg length is physically reasonable.

Forward Kinematics

For the forward kinematics problem, the leg lengths d_i, i = 0, 1, 2, 3 are given and the location of the moving platform, expressed by the position vector **p** and the rotation matrix ${}^{A}R_{B}$, is to be found. Before the theoretical derivation, it has been observed that the movements of the position and orientation of the moving platform are decoupled at the spherical joint B_0. The position of the moving platform can be completely decided by the length of the strut. This fact leads us to reach a way that simplifies the forward kinematics problem as two separated, position and orientation problems. In what follows, we first solve the position of the moving platform. Subsequently, the orientation problem is dealt by treating the manipulator as a pure orientation platform.

Position of the moving platform

The solution of the position vector **p** of the moving platform is straightforward. Since the strut length d_0 is given, the vector **p** can be easily identified by Eq. (2). Hence, for each given set of leg lengths d_i, i = 0, 1, 2, and 3, there is only one corresponding position for the moving platform.

Orientation of the moving platform

While solving the orientation problem of the 3SPS/PS manipulator, it can be shifted as the solving of the forward kinematics of a pure orientation platform. For forward kinematics analysis, when the length of the strut d_0 is given, the fixed base $A_1A_2A_3$ together with the strut can be virtually treated as a tetrahedron $B_0A_1A_2A_3$ (the six side lengths are fixed). Then the overall structure of this manipulator can be shifted as a 3SPS pure orientation platform. This makes a simple way to solve the orientation of the manipulator from which the rotation matrix referencing the moving platform to the pseudo tetrahedron $B_0A_1A_2A_3$ can be solved first and the coordinate transformation between the pseudo tetrahedron $B_0A_1A_2A_3$ and the fixed frame A can be followed.

To facilitate this analysis, an intermediate coordinate system $C(x', y', z')$ is introduced as shown in Fig. 2(a). The origin of frame C is coincident to the origin of frame B at B_0, the z'-axis directs along $\overline{B_0A_1}$, the x'-axis lies on plane $B_0A_1A_2$, and the y'-axis is defined by the right-hand rule. Accordingly, we assume that the coordinate transformation from the moving platform to the fixed base is achieved by a position vector **p** and two successive rotation matrices ${}^{A}R_{C}$ and ${}^{C}R_{B}$. The moving frame B originally coincides with the fixed frame A. It is then manipulated by a translation of vector **p** and a rotation of matrix ${}^{A}R_{C}$ to reach frame C, followed by a rotation of matrix ${}^{C}R_{B}$ to transform frame C to B. For simplifying the mathematic derivation, the rotation matrix ${}^{A}R_{C}$ is described by direction cosine method, while the matrix ${}^{C}R_{B}$ is denoted by the Euler angle rotation in terms of $R(x',\varphi')R(y',\theta')R(z',\psi')$ about which the intermediate frame C is rotated. Therefore, let $\mathbf{e}_i$ be the position vector of $\overline{B_0A_i}$ with respect to the fixed frame A. The position vector $\mathbf{q}_i$ of point B_i with respect to the fixed frame A can be obtained by rewriting Eq. (6) as

$$\mathbf{q}_i = \mathbf{p} + {}^{A}R_{C}\,{}^{C}R_{B}\,{}^{B}\mathbf{b}_i \tag{13}$$

where

$$\mathbf{p} = [0,\ 0,\ d_0]^{\mathrm{T}} \tag{14}$$

$${}^{A}R_{C} = \begin{bmatrix} \mathbf{u}'\cdot\hat{\mathbf{x}} & \mathbf{v}'\cdot\hat{\mathbf{x}} & \mathbf{w}'\cdot\hat{\mathbf{x}} \\ \mathbf{u}'\cdot\hat{\mathbf{y}} & \mathbf{v}'\cdot\hat{\mathbf{y}} & \mathbf{w}'\cdot\hat{\mathbf{y}} \\ \mathbf{u}'\cdot\hat{\mathbf{z}} & \mathbf{v}'\cdot\hat{\mathbf{z}} & \mathbf{w}'\cdot\hat{\mathbf{z}} \end{bmatrix} \tag{15}$$

$${}^{C}R_{B} = \begin{bmatrix} \mathrm{c}\varphi'\mathrm{c}\psi' - \mathrm{s}\varphi'\mathrm{c}\theta'\mathrm{s}\psi' & -\mathrm{c}\varphi'\mathrm{s}\psi' - \mathrm{s}\varphi'\mathrm{c}\theta'\mathrm{c}\psi' & \mathrm{s}\varphi'\mathrm{s}\theta' \\ \mathrm{s}\varphi'\mathrm{c}\psi' + \mathrm{c}\varphi'\mathrm{c}\theta'\mathrm{s}\psi' & -\mathrm{s}\varphi'\mathrm{s}\psi' + \mathrm{c}\varphi'\mathrm{c}\theta'\mathrm{c}\psi' & -\mathrm{c}\varphi'\mathrm{s}\theta' \\ \mathrm{s}\theta'\mathrm{s}\psi' & \mathrm{s}\theta'\mathrm{c}\psi' & \mathrm{c}\theta' \end{bmatrix} \tag{16}$$

$$\mathbf{u}' = \frac{(\mathbf{e}_1 \times \mathbf{e}_2) \times \mathbf{e}_1}{\left\|(\mathbf{e}_1 \times \mathbf{e}_2) \times \mathbf{e}_1\right\|} \tag{17}$$

$$\mathbf{v}' = \frac{\mathbf{e}_1 \times \mathbf{e}_2}{\left\|\mathbf{e}_1 \times \mathbf{e}_2\right\|} \tag{18}$$

$$\mathbf{w}' = \frac{\mathbf{e}_1}{\left\|\mathbf{e}_1\right\|} \tag{19}$$

$$\hat{\mathbf{x}} = [1,\ 0,\ 0]^{\mathrm{T}} \tag{20}$$

$$\hat{\mathbf{y}} = [0,\ 1,\ 0]^{\mathrm{T}} \tag{21}$$

$$\hat{\mathbf{z}} = [0,\ 0,\ 1]^{\mathrm{T}} \tag{22}$$

$$\mathbf{e}_1 = \mathbf{a}_1 - \mathbf{d}_0 \tag{23}$$

$$\mathbf{e}_2 = \mathbf{a}_2 - \mathbf{d}_0 \tag{24}$$

It should be noted that the three Euler rotation angles φ, θ, and ψ which relate the moving frame B to the fixed base A can be computed from the multiplication of the two successive rotational transformations ${}^{A}R_{C}\,{}^{C}R_{B}$. In what follows, the solving procedure is arranged by two stages. In the first stage, the coordinate transformation from the pseudo tetrahedron to the fixed base is performed. In the second stage, the orientation relationship between the moving platform and the pseudo tetrahedron is solved.

Stage 1: Coordinate transformation in the fixed base

The purpose of this stage is to solve the rotation matrix ${}^{A}R_{C}$ for finding the coordinate transformation between the intermediate frame C and the fixed frame A. Because the strut length d_0 is given, the vectors $\mathbf{e}_1$ and $\mathbf{e}_2$ of the pseudo tetrahedron can be obtained by Eqs. (23) and (24), respectively. Once $\mathbf{e}_1$ and $\mathbf{e}_2$ are obtained, the rotation matrix ${}^{A}R_{C}$ can be computed through Eqs. (15) and (17)-(22). Since there is always one single solution for $\mathbf{e}_1$ and $\mathbf{e}_2$, the ${}^{A}R_{C}$ has only one corresponding solution.

Stage 2: Pure orientation between two tetrahedrons

The purpose of this stage is to solve the three Euler rotation angles φ', θ', and ψ' to find the orientation relationship between the moving frame B and the intermediate frame C. By taking account of the rotation matrix ${}^{C}R_{B}$ as addressed in Eq. (16), the position vector ${}^{C}\mathbf{b}_i$ of point B_i with respect to the intermediate frame C can be given by

$$
\begin{aligned}
{}^{C}\mathbf{b}_i &= [b_{ix'},\ b_{iy'},\ b_{iz'}]^{\mathrm{T}} = {}^{C}R_{B}\,{}^{B}\mathbf{b}_i \\
&= \begin{bmatrix} b_{iu}(\mathrm{c}\varphi'\mathrm{c}\psi' - \mathrm{s}\varphi'\mathrm{c}\theta'\mathrm{s}\psi') - b_{iv}(\mathrm{c}\varphi'\mathrm{s}\psi' + \mathrm{s}\varphi'\mathrm{c}\theta'\mathrm{c}\psi') + b_{iw}\,\mathrm{s}\varphi'\mathrm{s}\theta' \\ b_{iu}(\mathrm{s}\varphi'\mathrm{c}\psi' + \mathrm{c}\varphi'\mathrm{c}\theta'\mathrm{s}\psi') - b_{iv}(\mathrm{s}\varphi'\mathrm{s}\psi' - \mathrm{c}\varphi'\mathrm{c}\theta'\mathrm{c}\psi') - b_{iw}\,\mathrm{c}\varphi'\mathrm{s}\theta' \\ b_{iu}\,\mathrm{s}\theta'\mathrm{s}\psi' + b_{iv}\,\mathrm{s}\theta'\mathrm{c}\psi' + b_{iw}\,\mathrm{c}\theta' \end{bmatrix}
\end{aligned}
\tag{25}
$$

Also, let the position vector ${}^{C}\mathbf{e}_i$ of $\overline{B_0A_i}$ with respect to the intermediate frame C be given as:

$$ {}^{C}\mathbf{e}_i = [e_{ix'},\ e_{iy'},\ e_{iz'}]^{\mathrm{T}} \tag{26} $$

Then a loop-closure equation can be written for leg i as

$$ {}^{C}\mathbf{d}_i = {}^{C}\mathbf{b}_i - {}^{C}\mathbf{e}_i \tag{27} $$

where ${}^{C}\mathbf{d}_i = \overline{A_iB_i}$ expressed in frame C.

Dot-multiplying Eq. (27) with itself produces a constraint equation imposed by leg i as follows:

$$ d_i^2 = {}^{C}\mathbf{e}_i^2 + {}^{C}\mathbf{b}_i^2 - 2({}^{C}\mathbf{e}_i)^{\mathrm{T}}({}^{C}\mathbf{b}_i) \tag{28} $$

Then, by substituting Eqs. (25) and (26) into (28), the following equation is obtained:

$$\alpha_i \,\mathrm{c}\varphi' + \beta_i \,\mathrm{s}\varphi' + \gamma_i = 0 \tag{29}$$

where

$$\alpha_i = \kappa_{i1} \,\mathrm{c}\,\psi' + \kappa_{i2} \,\mathrm{s}\,\psi' + \kappa_{i3} \tag{30}$$

$$\beta_i = \mu_{i1} \,\mathrm{c}\,\psi' + \mu_{i2} \,\mathrm{s}\,\psi' + \mu_{i3} \tag{31}$$

$$\gamma_i = \eta_{i1} \,\mathrm{c}\,\psi' + \eta_{i2} \,\mathrm{s}\,\psi' + \eta_{i3} \tag{32}$$

$$\kappa_{i1} = e_{iy'} b_{iv} \,\mathrm{c}\theta' + e_{ix'} b_{iu}$$

$$\kappa_{i2} = e_{iy'} b_{iu} \,\mathrm{c}\theta' - e_{ix'} b_{iv}$$

$$\kappa_{i3} = -e_{iy'} b_{iw} \,\mathrm{s}\theta'$$

$$\mu_{i1} = -e_{ix'} b_{iv} \,\mathrm{c}\theta' + e_{iy'} b_{iu}$$

$$\mu_{i2} = -e_{ix'} b_{iu} \,\mathrm{c}\theta' - e_{iy'} b_{iv}$$

$$\mu_{i3} = e_{ix'} b_{iw} \,\mathrm{s}\theta'$$

$$\eta_{i1} = e_{iz'} b_{iv} \,\mathrm{s}\theta'$$

$$\eta_{i2} = e_{iz'} b_{iu} \,\mathrm{s}\theta'$$

$$\eta_{i3} = e_{iz'} b_{iw} \,\mathrm{c}\theta' - \frac{1}{2}({}^{C}\mathbf{e}_i^2 + {}^{C}\mathbf{b}_i^2 - d_i^2)$$

In the above equations, the geometric parameters $e_{ix'}$, $e_{iy'}$, and $e_{iz'}$ can be derived according to the following equation as long as the rotation matrix ${}^{A}R_{C}$ is obtained in the previous stage:

$${}^{C}\mathbf{e}_i = [e_{ix'},\ e_{iy'},\ e_{iz'}]^{\mathrm{T}} = \left({}^{A}R_{C}\right)^{-1}(\mathbf{e}_i - \mathbf{p}) \tag{33}$$

From the coordinate systems chosen, we have $a_{1x'} = a_{1y'} = a_{2y'} = 0$, $b_{1u} = b_{1v} = b_{2u} = 0$, $a_{1z'} = l_1 = \overline{B_0 A_1}$, and $b_{1w} = l_2 = \overline{B_0 B_1}$. Equation (29) written for $i = 1$ yields

$$\theta' = \cos^{-1}\frac{l_1^2 + l_2^2 - d_1^2}{2l_1l_2} \tag{34}$$

Once θ' is found, Eq. (29) can be written twice for $i = 2$ and 3 and solve them for $c\varphi'$ and $s\varphi'$ as:

$$c\varphi' = \frac{\beta_2\gamma_3 - \beta_3\gamma_2}{\alpha_2\beta_3 - \alpha_3\beta_2} \tag{35}$$

$$s\varphi' = \frac{\alpha_3\gamma_2 - \alpha_2\gamma_3}{\alpha_2\beta_3 - \alpha_3\beta_2} \tag{36}$$

provided that $\alpha_2\beta_3 - \alpha_3\beta_2 \neq 0$. Then, according to the trigonometric identity $c\varphi'^2 + s\varphi'^2 = 1$, we have

$$(\beta_2\gamma_3 - \beta_3\gamma_2)^2 + (\alpha_3\gamma_2 - \alpha_2\gamma_3)^2 - (\alpha_2\beta_3 - \alpha_3\beta_2)^2 = 0 \tag{37}$$

Substituting Eqs. (30)-(32) into (37) and making use of the trigonometric identities $c\psi' = \frac{1-t^2}{1+t^2}$ and $s\psi' = \frac{2t}{1+t^2}$ where $t = \tan\frac{\psi'}{2}$, we obtain an eighth-degree polynomial in t:

$$\tau_8 t^8 + \tau_7 t^7 + \square + \tau_1 t + \tau_0 = 0 \tag{38}$$

where τ_i, i = 0, 1, 2, …, 8, are functions of the geometric parameters $\mathbf{e}_i$ and $\mathbf{b}_i$, leg length d_i, and the angle θ'. It follows that, corresponding to each solution of θ', there are at most eight solution of ψ'. Once θ' and ψ' are known, Eqs. (35) and (36) yields a single value of φ'.

Finally, as ${}^AR_C$ and θ', ψ', φ' are respectively obtained in stages 1 and 2, the rotation matrix ${}^AR_B$ can be computed by the fact ${}^AR_B = {}^AR_C\,{}^CR_B$. Furthermore, the rotation parameters θ, ψ, and φ for the moving platform with respect to the fixed coordinate system can be simply calculated, provided that $\sin\theta \neq 0$, by

$$\theta = \cos^{-1} r_{33} \tag{39}$$

$$\varphi = \text{Atan2}(r_{13}/\sin\theta, -r_{23}/\sin\theta) \tag{40}$$

$$\psi = \text{Atan2}(r_{31}/\sin\theta, r_{32}/\sin\theta) \tag{41}$$

where r_{ij} is the element (i, j) of matrix ${}^AR_B$.

In summary, since there are at most eight solution sets for (θ', ψ', φ') and one solution for ${}^AR_C$, the manipulator has at most eight possible orientations. Recalling that the moving platform has only one position for one given set of leg length d_i, it can be concluded that the forward kinematics of the 3SPS/PS manipulator has at most eight feasible configurations.

Numerical Example

Here we give an example for demonstrating the forward kinematics analysis of the 3SPS/PS parallel manipulator. Give that the geometry of the manipulator are $\mathbf{a}_1=[35,\ 0,\ 0]^T$, $\mathbf{a}_2=[-30,\ 10,\ 0]^T$, $\mathbf{a}_3=[-25,\ -30,\ 0]^T$, ${}^B\mathbf{b}_1=[0,\ 0,\ 30]^T$, ${}^B\mathbf{b}_2=[0,\ 20,\ 15]^T$, and ${}^B\mathbf{b}_3=[-10,\ -15,\ 10]^T$ and assume that $d_0=10$, $d_1=35$, $d_2=47$, and $d_3=42$. Then we can obtain an eighth-order equation of variable t as

$$\begin{aligned}&0.39855t^8-9.73476t^7+68.3072t^6-93.6466t^5-35.4675t^4\\&+84.5578t^3-22.7068t^2-9.63453t+6.45935=0\end{aligned} \tag{42}$$

Table 1. Numerical solutions of the example for forward kinematics analysis

No.	Solution			p	ϕ	θ	ψ
t_1	−0.860227			[0, 0, 10]	157.5600	123.9142	10.0893
t_2	−0.436601			[0, 0, 10]	23.7913	109.7357	−145.4245
t_3	0.731362			[0, 0, 10]	102.5235	44.2283	−121.2030
t_4	1.723890			[0, 0, 10]	22.4937	130.9880	31.9512
$t_{5,6}$	0.374887	±	0.353727i	-	-	-	-
$t_{7,8}$	11.258600	±	1.441840i	-	-	-	-

Note: The unit of angles is *degree*.

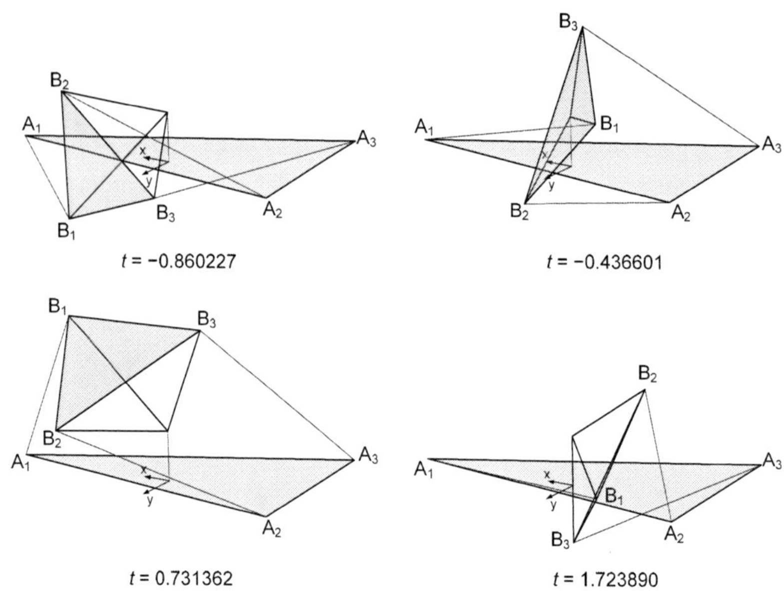

Figure 4. Four possible configurations of the forward kinematic analysis.

The numerical solutions of Eq. (42) are given in Table 1. As one can see, only solutions t_1 to t_4 are feasible to the given problem. Therefore, the corresponding configuration parameters of the mechanism for solutions t_1 to t_4 are obtained in Table 1. It is concluded that there are four possible configurations to this problem. The four possible configurations are depicted in Fig. 4.

Singularity Analysis

For singularity analysis, the velocity equation derived from the vector-loop $\overline{OB_0} + \overline{B_0B_i} = \overline{OA_i} + \overline{A_iB_i}$ can be written as:

$$\dot{d}_0\mathbf{s}_0 + \boldsymbol{\omega}_B \times \mathbf{b}_i = d_i\boldsymbol{\omega}_i \times \mathbf{s}_i + \dot{d}_i\mathbf{s}_i \text{ for } i = 1, 2, 3 \tag{43}$$

where $\mathbf{s}_0$ is the unit vector pointing along $\overline{OB_0}$, $\mathbf{s}_i$ is a unit vector pointing along $\overline{A_iB_i}$, $\boldsymbol{\omega}_B$ is the angular velocity of moving platform with respect to the fixed frame, $\boldsymbol{\omega}_i$ is the angular velocity of leg i with respect to the fixed frame, and $\dot{d}_i$ is the time derivative of the length of leg i. Dot-multiplying both sides of Eq. (43) by $\mathbf{s}_i$ and rearranging the resulting equation, we get

$$(\mathbf{b}_i \times \mathbf{s}_i)\cdot\boldsymbol{\omega}_B = \dot{d}_i - \dot{d}_0(\mathbf{s}_i \cdot \mathbf{s}_0) \text{ for } i = 1, 2, 3 \tag{44}$$

Besides, we can observe that the linear velocity of moving platform is contributed only by the prismatic joint of the strut. Hence an additional velocity equation can be written as

$$\mathbf{v}_p = \dot{d}_0\mathbf{s}_0 \tag{45}$$

where $\mathbf{v}_p$ is the linear velocity of the moving platform. Since the strut is only extensible along the z-axis of the fixed frame, Eq. (45) can be further simplified as

$$v_{p,z} = \dot{d}_0 \tag{46}$$

where $v_{p,z}$ represents the z-axis component of $\mathbf{v}_p$. By combination of Eqs. (44) and (46), the Jacobian equation of the 3SPS/PS manipulator can be written as follows:

$$J_x\dot{\mathbf{x}} = J_q\dot{\mathbf{q}} \tag{47}$$

where

$$J_x = \begin{bmatrix} (\mathbf{b}_1 \times \mathbf{s}_1)^{\mathrm{T}} & 0 \\ (\mathbf{b}_2 \times \mathbf{s}_2)^{\mathrm{T}} & 0 \\ (\mathbf{b}_3 \times \mathbf{s}_3)^{\mathrm{T}} & 0 \\ \mathbf{0} & 1 \end{bmatrix}_{4\times4} \tag{48}$$

$$J_q = \begin{bmatrix} -(\mathbf{s}_1 \cdot \mathbf{s}_0) & 1 & 0 & 0 \\ -(\mathbf{s}_2 \cdot \mathbf{s}_0) & 0 & 1 & 0 \\ -(\mathbf{s}_3 \cdot \mathbf{s}_0) & 0 & 0 & 1 \\ 1 & 0 & 0 & 0 \end{bmatrix}_{4\times4} \tag{49}$$

$$\dot{\mathbf{x}} = \begin{bmatrix} \boldsymbol{\omega}_B \\ v_{p,z} \end{bmatrix}_{4\times1} \tag{50}$$

$$\mathbf{q} = \begin{bmatrix} \dot{d}_0 \\ \dot{d}_1 \\ \dot{d}_2 \\ \dot{d}_3 \end{bmatrix}_{4\times1} \tag{51}$$

From Eqs. (47) to (51), $J = J_q^{-1} J_x$ is the Jacobian matrix, $\mathbf{x}$ denotes the angular velocity and the linear velocity of the moving platform, and $\mathbf{q}$ denotes the elongation rate of the leg lengths. Note that each row of J_x represents a vector that is normal to a plane defined by the triangle $\Delta A_i B_0 B_i$. We define a vector $\mathbf{n}_i = \mathbf{b}_i \times \mathbf{s}_i$ to help the following explanation.

Inverse Kinematic Singularities

It is well-known that an inverse kinematic singularity of the manipulator occurs when the determinant of J_q goes to zeros [21], denoted by $\det(J_q) = 0$. As the manipulator belongs to its inverse singularity configuration, it loses extra dofs. From Eq. (49), we observe that J_q will not be degenerate whatever the values of $\mathbf{s}_0$ and $\mathbf{s}_i$ are (i.e., the four row vectors are always linearly independent). So there exists no inverse kinematic singularity within the workspace of the manipulator. However, inverse kinematic singularities can occur at the workspace boundary where one or more legs are in fully stretched or retracted positions.

Direct Kinematic Singularities

An inverse kinematic singularity of the manipulator occurs when $\det(J_x)=0$ [21]. As the manipulator belongs to its direct singularity configuration, it gains extra dofs. From Eq. (48), we observe that $\det(J_x)$ will become zeros when $\mathbf{n}_1$, $\mathbf{n}_2$, or $\mathbf{n}_3$ vanishes or when they are linearly dependent. Physically, the direct singularity is dominated by the three vectors $\mathbf{n}_i$ which represent the transitory motion contributed by the three SPS leg *i*. It can be identified geometrically from the planes A_iOB_i, i = 1, 2, and 3. Therefore, three different direct kinematic singularities are concluded as follows:

Type 1: $\mathbf{n}_i = \mathbf{0}$. This type stands for the geometry in which points A_i, B_0, and B_i lie on a straight line. Under this condition, the manipulator will gain one dof. That is, when all actuators are locked, the moving platform can make an infinitesimal rotation about a line of intersection of the two planes defined by triangles $\Delta A_jB_0B_j$ and $\Delta A_kB_0B_k$, $i \neq j \neq k$. An example for this type of singularity is shown in Fig. 5(a).

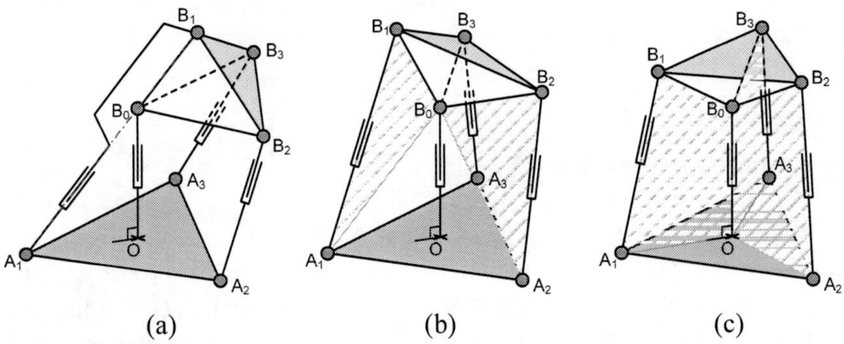

Figure 5. Examples of the three types of direct singular configurations. (a) A_1, B_0, and B_1 lie on a straight line; (b) $\Delta A_1B_0B_1$ and $\Delta A_2B_0B_2$ are co-plane; (c) $\Delta A_1B_0B_1$, $\Delta A_2B_0B_2$, and $\Delta A_3B_0B_3$ intersect in a common line.

Type 2: $\mathbf{n}_i$ and $\mathbf{n}_j$ are linearly dependent ($i \neq j$). This type stands for the geometry in which the two planes defined by triangles $\Delta A_iB_0B_i$ and $\Delta A_jB_0B_j$ are coincident. Under this condition, the manipulator will gain one dof. That is, when all actuators are locked, the moving platform can make an infinitesimal rotation about a line of intersection of the two planes defined by triangles $\Delta A_iB_0B_i$ and $\Delta A_kB_0B_k$, $i \neq j \neq k$. An example for this type of singularity is shown in Fig. 5(b).

Type 3: $\mathbf{n}_i$, $\mathbf{n}_j$, and $\mathbf{n}_k$ are linearly dependent ($i \neq j \neq k$). This type stands for the geometry in which the three planes defined by $\Delta A_1B_0B_1$, $\Delta A_2B_0B_2$, and $\Delta A_3B_0B_3$ intersect in a common line. Under this condition, the manipulator will gain one dof. That is, when all actuators are locked, the moving platform can make an infinitesimal rotation about a line of intersection of the three planes. An example for this type of singularity is shown in Fig. 5(c).

Combined Singularities

Combined singularity occurs when both $\det(J_x)$ and $\det(J_q)$ are zeros [21]. For the manipulator analyzed, it cannot be happened within the workspace of the manipulator. But it can occur at the workspace boundary when one of the three legs or the strut is at its extreme reach associated with one of the following three conditions being satisfied: (1) points A_i, B_0, and B_i, for $i = 1,\ 2,\ 3$, lie on a straight line; (2) the two planes of triangles $\Delta A_i B_0 B_i$ and $\Delta A_j B_0 B_j$, for $i, j = 1,\ 2,\ 3$ ($i \neq j$), are coincident; or (3) the three planes of triangle $\Delta A_i B_0 B_i$, for $i = 1,\ 2,\ 3$, intersect at a common line.

Workspace Analysis

In addition to the kinematic and singularity analyses, determining the workspace of the manipulator is also an important design issue since it defines the volume that the moving platform can reach. In general, the workspace of a manipulator is classified by three categories: *reachable workspace*, *dexterous workspace*, and *orientation workspace*. In what follows, these three kinds of workspace for the 3SPS/PS parallel manipulator are discussed.

Kinematic Limitations to Workspace

To determine the workspace of a manipulator, the major problem is to consider the kinematic constraints within the manipulator's range of motion. Specifically, there are three major kinematic constraints to the workspace of parallel manipulators: the limited length of legs, the limited motion range of the passive joints, and the link interference [22]. Each constraint is described and formulated as follows.

Limited length of legs

Each leg including the strut has its limited range of extraction and contraction. The length of each leg can be expressed as follows:

$$d_{i,\min} \le d_i \le d_{i,\max},\ i = 0,\ 1,\ 2,\ 3 \tag{52}$$

where $d_{i,\min}$ and $d_{i,\max}$ are the minimum and maximum allowable lengths of leg *i* ($\overline{OB_0}$ for *i* = 0), respectively.

Limited motion range of the passive joints

Each passive joint has a limited range of motion. For spherical joints as we chosen here, its limited range of motion is constrained together by the leg strut and the spherical cover of the joint, Fig. 6. This constraint can be written as

$$\cos^{-1}\left(\frac{{}^{v}\mathbf{a}_i \cdot \mathbf{d}_i}{\left\|{}^{v}\mathbf{a}_i\right\|\left\|\mathbf{d}_i\right\|}\right) \le \xi_{i,\max},\ i=1,\ 2,\ 3 \tag{53a}$$

$$\cos^{-1}\left(\frac{{}^{v}\mathbf{b}_i \cdot (-\mathbf{d}_i)}{\left\|{}^{v}\mathbf{b}_i\right\|\left\|-\mathbf{d}_i\right\|}\right) \le \delta_{i,\max},\ i=0,\ 1,\ 2,\ 3 \tag{53b}$$

where ${}^{v}\mathbf{a}_i$ is the vector directed from the sphere center to the axis of symmetry of joint A_i and so does ${}^{v}\mathbf{b}_i$ for joint B_i, and $\xi_{i,\max}$ and $\delta_{i,\max}$ define the maximum ranges of motion of the passive joints A_i and B_i, respectively.

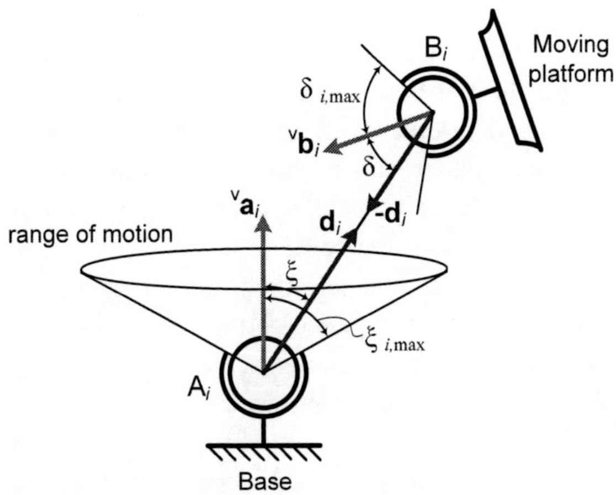

Figure 6. The limited range of motion of the passive joint.

Leg interference

Each leg including the strut cannot interfere with any other leg during manipulation. Assume that the strut of each leg is composed of a cylinder with diameter D. To avoid the leg interference, the minimum allowable distance between any two legs should be the sum of the radii of the two leg struts. This can be determined as the distance between two lines associated to any two leg struts, i.e.,

$$\begin{cases} \text{distance}(OB_1, A_iB_i) \ge D, & i=1,\ 2,\ 3 \\ \text{distance}(A_jB_j, A_kB_k) \ge D, & j=1,\ 2,\ 3,\ k=j+1,..,3 \end{cases} \tag{54}$$

It has been proposed that there are four possible situations of calculating the distance between two legs [22], Fig. 7. These four conditions can be treated as the problem of finding the minimum distance between two “line segments” (not two lines). The solving technique for this problem is addressed in Appendix.

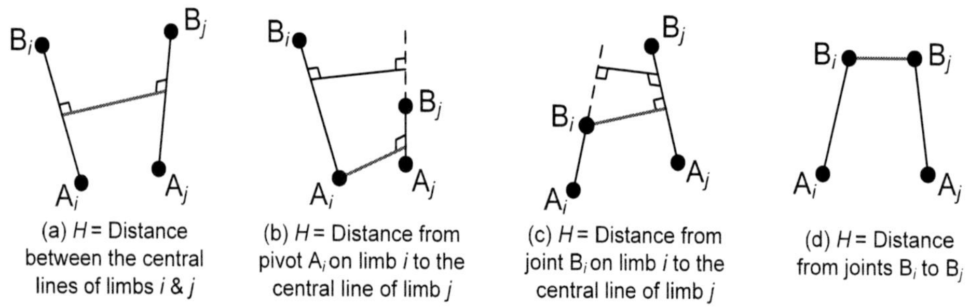

Figure 7. Four situations of calculating the distance H between two legs.

Approach

The workspace determination of general parallel manipulators is an extraordinarily difficult problem because the translational and rotational degrees of freedom of its moving platform are normally dependent on each others. However, while we deal with such problem in the 3SPS/PS parallel manipulator, this troublesome is naturally eliminated by the structure itself. As we noted previously, this parallel manipulator has three rotational degrees of freedom and only one translational degree of freedom that is completely controlled by the strut. When the length of the strut is determined, the moving platform will become a pure orientation platform. Since the workspace determination of a pure orientation platform is quite simple, the task for evaluating the workspace of the 3SPS/PS parallel manipulator will become much easier. Hence, in what follows, the determination of workspace of this manipulator will be taken in a decomposition strategy: first, determining the workspace at different strut lengths; then, associating all workspaces obtained by different strut lengths into a whole one.

According to this strategy, the main procedure of workspace determination for the 3SPS/PS parallel manipulator is outlined as follows:

(i) Define the stretching range of the strut, i.e., $d_{\min} \le d_0 \le d_{\max}$.

(ii) Define the motion range of the Euler rotation angles (φ, θ, ψ) which can describe all possible orientations of the moving platform. The range can be set by $\varphi \in [-\pi, \pi)$, $\theta \in [0, \pi)$, and $\psi[-\pi, \pi)$ since $R(w,\varphi)R(u,-\theta)R(w,\psi) \equiv R(w,\varphi)R(u,\theta)R(w,\psi \pm \pi)$ [23].

(iii) Based on (i) and (ii), select one set of suitable values for ($d_0, \varphi, \theta, \psi$). Accordingly, solve the inverse kinematics problem so as to obtain the position parameters of the three SPS legs.

(iv) For the solution obtained in step (iii), test if any kinematic limitation to workspace is violated. If any of the three kinematic limitations is conflicted, this configuration does not exist in the solution space of workspace determination.

(v) Iterate steps (iii) and (iv) for all possible sets of ($d_0, \varphi, \theta, \psi$).

(vi) Collect all feasible solutions in step (iv). Each feasible solution corresponds to one feasible solution set of the workspace. The workspace solution is the inclusive subset of all feasible solutions.

(vii) Visualize the workspace solution.

In what follows, for helping the explanation and comprehension, we parameterize the 3SPS/PS parallel manipulator as a symmetric architecture (the base $A_1A_2A_3$ is an equilateral triangle, the moving platform $B_0B_1B_2B_3$ is an equilateral tetrahedron, and the strut is mounted at the centroid of $\Delta A_1A_2A_3$) to illustrate the solving of its orientation workspace, reachable workspace, and dexterous workspace. Table 2 gives the geometric parameters of the parallel manipulator and Table 3 lists the kinematic limitation utilized in the illustrative examples.

Table 2. Geometric parameters of the manipulator for workspace determination

g	$\mathbf{a}_1$	$\mathbf{a}_2$	$\mathbf{a}_3$	${}^v\mathbf{a}_{i,\ i=1\text{–}3}$
100	$g[1,\ 0,\ 0]$	$g\left[-1/2,\ \sqrt{3}/2,\ 0\right]$	$g[-1/2, \sqrt{3}/2,\ 0]$	$[0,\ 0,\ 1]$
h	${}^B\mathbf{b}_1$	${}^B\mathbf{b}_2$	${}^B\mathbf{b}_3$	${}^v\mathbf{b}_{i,\ i=0\text{–}3}$
100	$h[0,\ 0,\ 1]$	$h\left[0, \sqrt{3}/2,\ 1/2\right]$	$h\left[-\sqrt{6}/3, \sqrt{3}/6, 1/2\right]$	$(\mathbf{b}_3-\mathbf{b}_1)\times(\mathbf{b}_2-\mathbf{b}_1)$

Table 3. Defined values of kinematic limitation[1] for workspace determination

$d_{i,\min,\ i=1\text{–}3}$	$d_{i,\max,\ i=1\text{–}3}$	$\xi_{i,\max,\ i=1\text{–}3}$	$\delta_{i,\max,\ i=0\text{–}3}$	D
130	250	$\pi/3$	$\pi/3$	15

[1]The length limitation of the strut is not considered.

Orientation Workspace, Reachable Workspace, and Dexterous Workspace

Based on the approach introduced in section 6.2, here we study the orientation workspace, reachable workspace, and dexterous workspace of the 3SPS/PS parallel manipulator.

Orientation workspace

The orientation workspace is the set of all attainable orientations of the moving platform about a fixed point [22, 23]. The Euler angle representation is a convenient way for representing the orientation of moving platform. The three rotations φ, θ, and ψ can be plotted in a 3-D coordinate system for showing the orientation workspace. Reference [23] has discussed that among the three customary coordinate representations (Cartesian coordinate, sphere coordinate, and cylindrical coordinate), the cylindrical coordinate, Fig. 8, should be the most suitable one to represent the orientation workspace. In this chapter, we adopt this representation to show the orientation workspace.

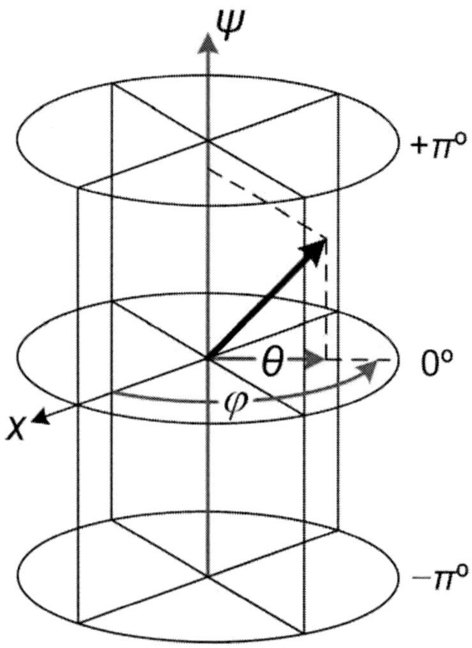

Figure 8. The cylindrical coordinate representation.

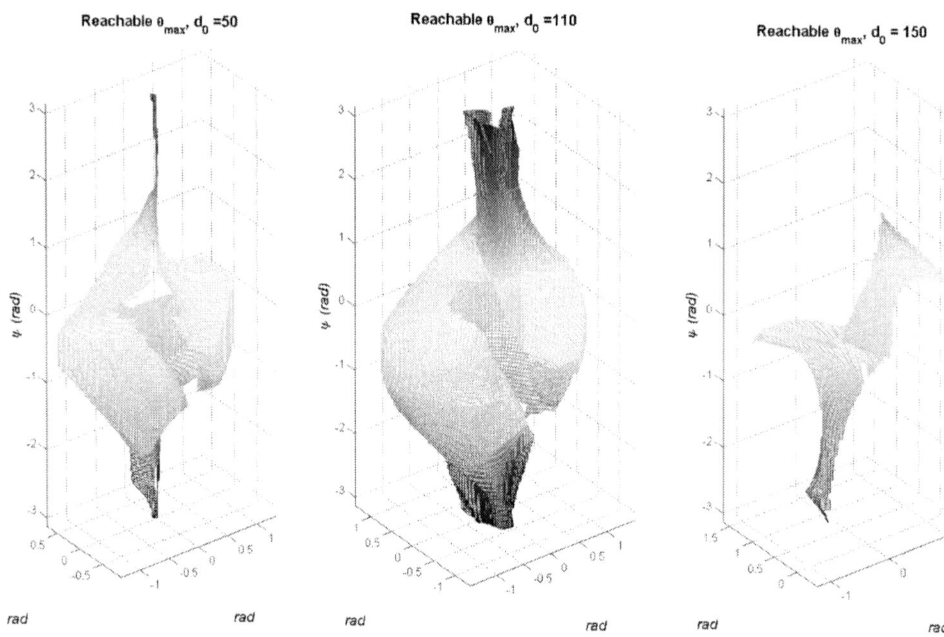

(a) $d_0 = 50$ (b) $d_0 = 110$ (c) $d_0 = 150$

Figure 9. Orientation workspace of the 3SPS/PS parallel manipulator.

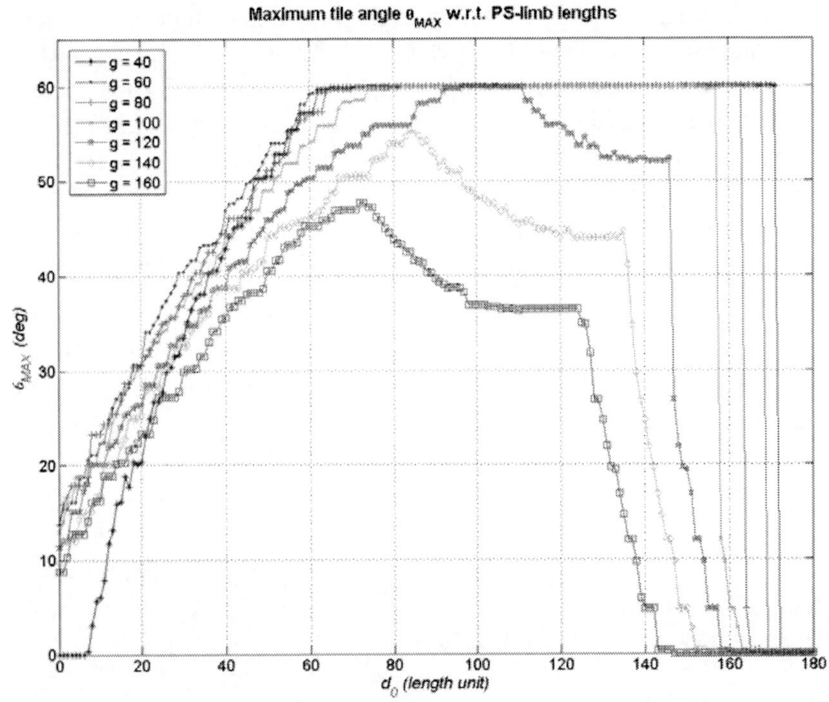

Figure 10. Maximum tile angle with respect to strut lengths under different lengths of vectors $\mathbf{a}_1$, $\mathbf{a}_2$, and $\mathbf{a}_3$ (by varying parameter g).

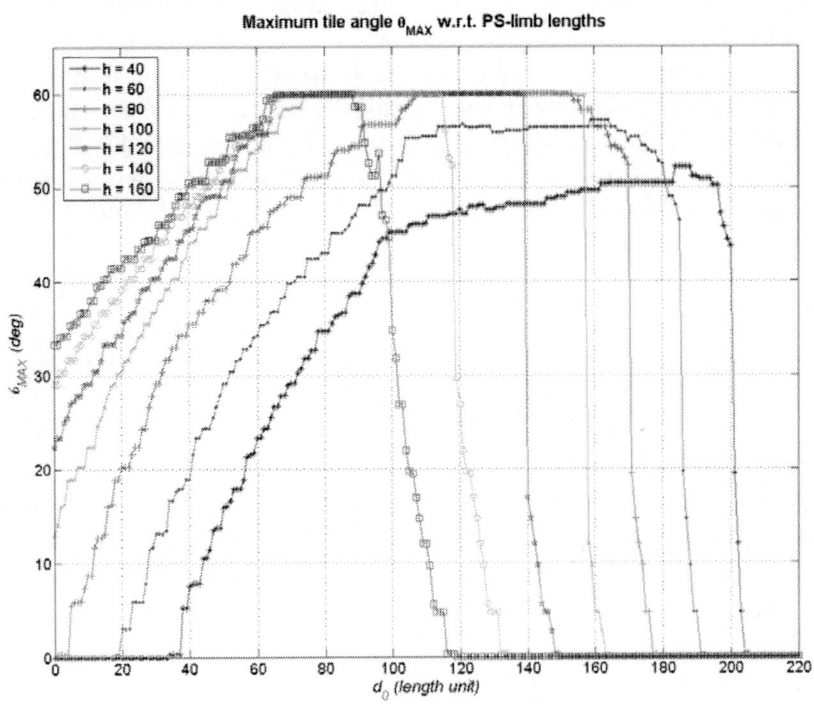

Figure 11. Maximum tile angle with respect to strut lengths under different lengths of vectors $\mathbf{b}_1$, $\mathbf{b}_2$, and $\mathbf{b}_3$ (by varying parameter h).

By following the definition of orientation workspace, it is obvious that for the 3SPS/PS parallel manipulator the fixed point about which the moving platform rotates is simply the center of joint B_0. Hence, the orientation workspace of this parallel manipulator can be obtained by evaluating the feasible Euler rotation angles, which obey all kinematic limitations for the workspace analysis, at different strut lengths. For illustration, we select d_0 = 50, 110, and 150 length units, respectively. Figure 9 shows the orientation workspace solutions of the 3SPS/PS parallel manipulator using the cylindrical coordinate as illustrated in Fig. 8. These plots are drawn by considering a given φ and ψ and it corresponding maximum reachable tile angle θ_{MAX} at a given strut length. It is obvious that, among these three conditions, the 3SPS/PS parallel manipulator possesses the largest orientation workspace while the strut is at 110 length unit. Figures 10 and 11 further explain the maximum reachable tilt angles with respect to different strut lengths under different sizes of the fixed based $A_1A_2A_3$ (by varying the geometer parameter g) and of the moving platform $B_0B_1B_2B_3$ (by varying the geometer parameter h). As a result, for the given geometric data in Table 2 (g = 100), the maximum tile angle θ_{MAX} can reach approximately 60 degrees while the strut is stretched between 75-158 length unit. On the other hand, the maximum reachable tile angle will begin to decay while the fixed base sizing parameter g is larger than 140 length unit (see Fig. 10) or the moving platform sizing parameter h is smaller than 60 length unit (see Fig. 11).

Reachable workspace

The reachable workspace is the volume of space within which every point can be reached by the moving platform in at least one orientation [21]. For the 3SPS/PS parallel manipulator, it can be first of all realized that based on a given strut length the reachable workspace will be constrained as a sphere surface centered at the joint B_0 while the three kinematic constraints are neglected. Furthermore, the complete reachable workspace of the manipulator can be basically formulated as a cylinder which is embodied by towing such sphere along a path that the center point of joint B_0 moves through. However, if the kinematic limitations are taken into account, the surface region of the reachable workspace will be constrained and will be different when different strut length is applied. Figure 12, for instance, shows the reachable workspace traced by the centroid of triangle $\Delta B_1B_2B_3$ at d_0 = 50, 110, and 150 length units, respectively. A comparison of these three reachable workspaces is given in Fig. 13. It is obvious that this parallel manipulator has a larger reachable workspace while the strut is stretched at 110 length unit. When the strut reaches 150 length unit, the reachable workspace becomes quite confined due to its kinematic constraints.

Dexterous workspace

The dexterous workspace is the volume of space within which every point can be reached by the moving platform in all possible orientations [21]. Evidently, for the 3SPS/PS parallel manipulator, while neglecting the kinematic limitations, every point in space cannot be reached by its moving platform with all possible orientations except for those points lying on the path the center point of joint B_0 passes through. Hence the dexterous workspace of this parallel manipulator is none other than a straight line traced by the center point of joint B_0. Recalling that the reachable workspace of this parallel manipulator forms a sphere-cylinder volume, the straight-line dexterous workspace associated with its reachable workspace can be

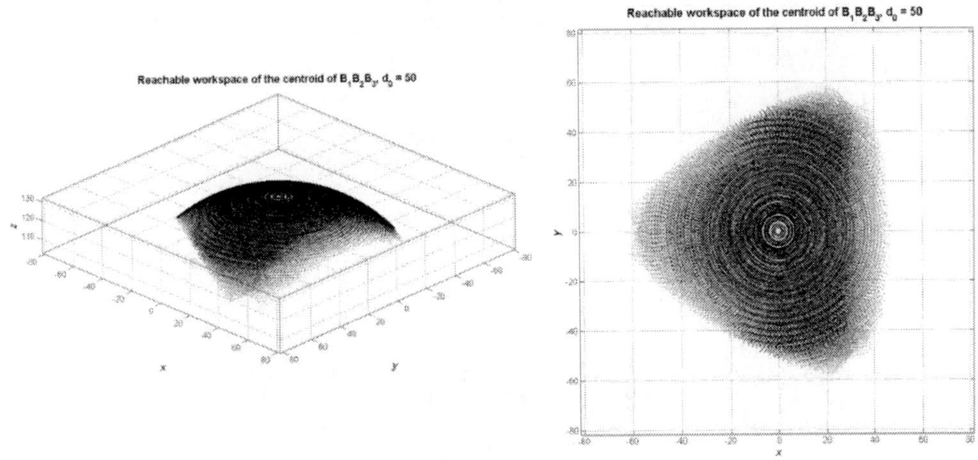

(a) d_0 = 50

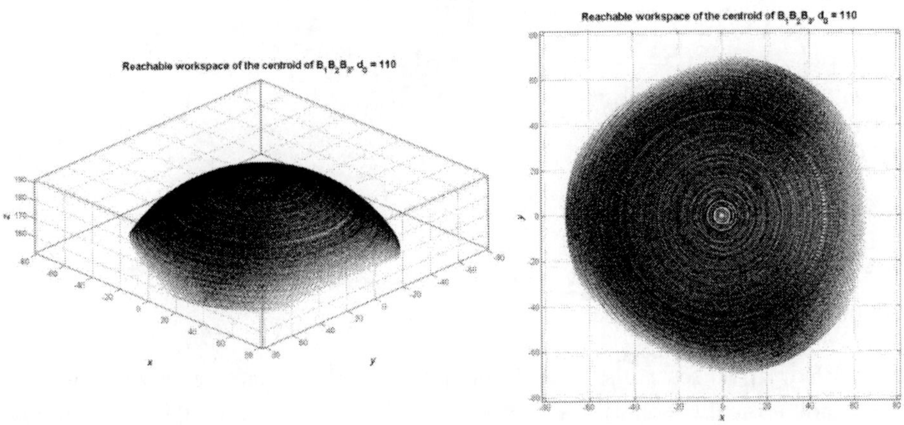

(b) d_0 = 110

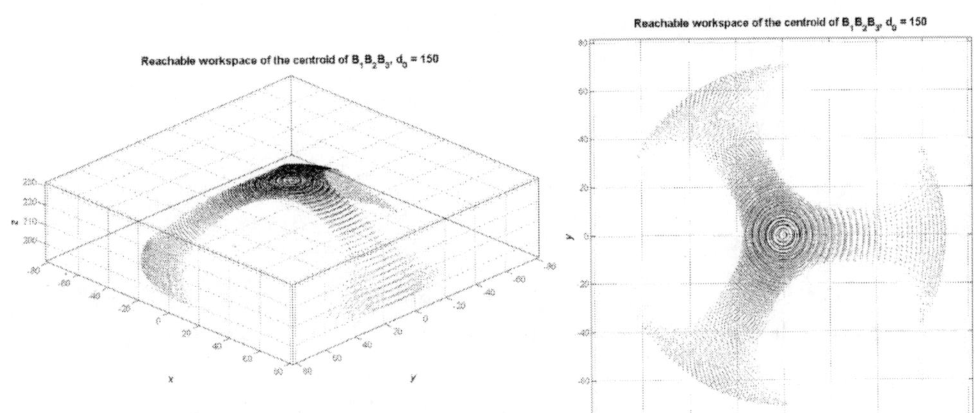

(c) d_0 = 150

Figure 12. Reachable workspace of the centroid of triangle $\Delta B_1B_2B_3$ (Left: isometric view, Right: Top view).

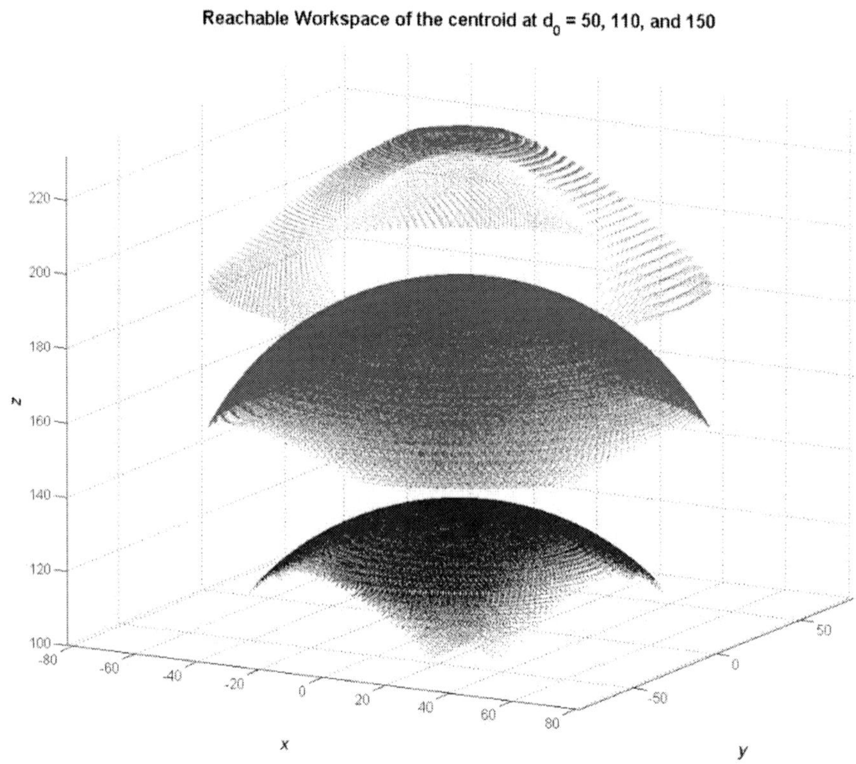

Figure 13. Comparison of the reachable workspace of the centroid of triangle $\Delta B_1B_2B_3$ at $d_0 = 50$, 110, and 150 (from bottom to top), respectively.

particularly attractive for some special applications. For example, a minimally invasive surgical robot normally controls a 4-dof motion that maneuvers the surgical instrument inserting/retracting along an entry point on patient body and rotating three-dimensionally about that point to carry out surgical operations in the instrument tip. While applying the 3SPS/PS parallel manipulator to this surgical operation, the dexterous workspace stands for the moving path of the holding instrument and the reachable workspace represents the working volume of the instrument tip. Thus the instrument can be operated in any possible orientations with a given insertion depth to the patient body.

Conclusions

This chapter examined a general type of 4-dof 3SPS/PS parallel manipulators with generally placed strut and investigated their kinematics, singularity and workspace. This general type of manipulators demonstrates a special characteristic that its only translational degree of freedom of output motion is structurally decoupled from the three rotational ones. This feature therefore dramatically simplifies its kinematic, singularity, and workspace analyses as illustrated in this chapter. As a result, the inverse and forward kinematics of the general 3SPS/PS parallel manipulator were solved, from which at most eight possible solutions can be concluded for its forward kinematics problem. The singularity of the parallel manipulator was further analyzed and represented by the inverse and direct singular configurations, and the workspace analysis was carried out via it orientation workspace and reachable workspace. The chapter further presents the dexterous workspace and provides a clear and comprehensive study of the kinematics, singularity, and workspace for the partially decoupled 4-dof general 3SPS/PS parallel manipulator.

The studied parallel manipulator is particularly applicable to those situations that require the end-effector always rotating about a fixed point but translating along any arbitrary direction in any moment. For example, this kind of special motion is required by robotically-assisted minimally invasive surgery (MIS). For an MIS robot, it needs to manipulate the surgical tool, which is held by the end-effector, to penetrate patient's abdominal wall followed by performing a series of surgical operation. In such a way, the surgical tool is performing a 4-dof, 3R1T motion where the center of rotation is fixed at some point on patient's body. Besides, for making the tool to easily adapt with various MIS procedures, the translational dof of the tool is expected to be uncoupled with the rotational ones. Obviously, the studied manipulator can elagently provide this special motion under in-parallel actuation. However, potential limitation in such application might be the insufficient orientation workspace for MIS operation. It is suggested that more design consideration in terms of surgical motion requirements and workspace limitation should be taken into account when introducing this manipulator into MIS application.

References

[1] Kuo, C.-H., Synthesis and Applications of Decoupled Parallel Manipulators Using Motion Constraint Generator for Minimally Invasive Surgery, *Ph.D. Thesis*, Department of Mechanical Engineering, King's College London, London, United Kingdom (2011).

[2] Fang, Y., Tsai, L.-W., "Structure Synthesis of a Class of 4-DoF and 5-DoF Parallel Manipulators with Identical Limb Structures," *The International Journal of Robotics Research*, **21**(9), pp. 799-810 (2002).

[3] Kong, X., Gosselin, C. M., "Type Synthesis of Three-Degree-of-Freedom Spherical Parallel Manipulators," *The International Journal of Robotics Research*, **23**(3), pp. 237-245 (2004).

[4] Kong, X., Gosselin, C. M., "Type Synthesis of 4-DOF SP-Equivalent Parallel Manipulators: A Virtual Chain Approach," *Mechanism and Machine Theory*, **41**(11), pp. 1306-1319 (2006).

[5] Alizade, R., Bayram, Ç., "Structural Synthesis of Parallel Manipulators," *Mechanism and Machine Theory*, **39**(8), pp. 857-870 (2004).

[6] Huang, Z., Li, Q. C., "General Methodology for Type Synthesis of Symmetrical Lower-Mobility Parallel Manipulators and Several Novel Manipulators," *The International Journal of Robotics Research*, **21**(2), pp. 131-145 (2002).

[7] Huang, Z., Li, Q. C., "Type Synthesis of Symmetrical Lower-Mobility Parallel Mechanisms Using the Constraint-Synthesis Method," *The International Journal of Robotics Research*, **22**(1), pp. 59-79 (2003).

[8] Zhao, T. S., Dai, J. S., Huang, Z., "Geometric Analysis of Overconstrained Parallel Manipulators with Three and Four Degrees of Freedom," *JSME International Journal Series C*, **45**(3), pp. 730-740 (2002).

[9] Rolland, L., "The Manta and the Kanuk: Novel 4-DOF Parallel Mechanisms for Industrial Handling," *Proceedings of ASME International Mechanical Engineering Congress and Exposition*, Nashville, Tennessee, 14-19 November 1999, pp. 831-844.

[10] Carricato, M., "Fully Isotropic Four-Degrees-of-Freedom Parallel Mechanisms for Schoenflies Motion," *The International Journal of Robotics Research*, **24**(5), pp. 397-414 (2005).

[11] Company, O., Marquet, F., Pierrot, F., "A New High-Speed 4-DOF Parallel Robot: Synthesis and Modeling Issues," *IEEE Transactions on Robotics and Automation*, **19**(3), pp. 411-420 (2003).

[12] Richard, P.-L., Gosselin, C. M., Kong, X., "Kinematic Analysis and Prototyping of a Partially Decoupled 4-DOF 3T1R Parallel Manipulator," *ASME Transactions, Journal of Mechanical Design*, **129**(6), pp. 611-616 (2007).

[13] Salgado, O., Altuzarra, O., Petuya, V., Hernández, A., "Synthesis and Design of a Novel 3T1R Fully-Parallel Manipulator," *ASME Transactions, Journal of Mechanical Design*, **130**(4), pp. 042305(1-8) (2008).

[14] Zlatanov, D., Gosselin, C. M., "A Family of New Parallel Architectures with Four Degrees of Freedom," *Proceedings of 2nd Workshop on Computational Kinematics*, Park, F. C., Iurascu, C. C., Eds., Seoul, South Korea, May 2001, pp. 57-66.

[15] Lu, Y., Hu, B., "Analyzing Kinematics and Solving Active/Constrained Forces of a 4-dof 3SPS+SP Parallel Manipulator," *Robotica*, **27**(1), pp. 29-36 (2009).

[16] Lenarčič, J., Stanišić, M. M., Parenti-Castelli, V., "Kinematic Design of a Humanoid Robotic Shoulder Complex," *Proceedings of IEEE International Conference on Robotics and Automation*, San Francisco, California, USA, 24-28 April 2000, Vol. 1, pp. 27-32.

[17] Lenarčič, J., Stanišić, M. M., Parenti-Castelli, V., 2000, "A 4-DOF Parallel Mechanism Simulating the Movement of the Human Sternum-Clavicle-Scapula Complex," in Lenarčič, J., Stanišić, M. M., Eds, *Advances in Robot Kinematics*, London, Kluwer Academic, pp. 325-332.

[18] Lenarčič, J., Stanišić, M., "A Humanoid Shoulder Complex and the Humeral Pointing Kinematics," *IEEE Transactions on Robotics and Automation*, **19**(3), pp. 499-506 (2003).

[19] Dai, J. S., Sodhi, C., Kerr, D. R., "Design and Analysis of a New Six-Component Force Transducer for Robotic Grasping," *Proceedings of the Second Biennial European Joint Conference on Engineering Systems Design and Analysis*, London, United Kingdom 1994, Vol. 64, No. 8-3, pp. 809-817.

[20] Dai, J. S., Huang, Z., Lipkin, H., "Mobility of Overconstrained Parallel Mechanisms," *ASME Transactions, Journal of Mechanical Design*, **128**(1), pp. 220-229 (2006).

[21] Tsai, L.-W., *Robot Analysis: The Mechanics of Serial and Parallel Manipulators*, John Wiley & Sons, New York (1999).

[22] Merlet, J.-P., "Determination of the Orientation Workspace of Parallel Manipulators," *Journal of Intelligent and Robotic Systems*, **13**(2), pp. 143-160 (1995).

[23] Bonev, I. A., Ryu, J., "A New Approach to Orientation Workspace Analysis of 6-DOF Parallel Manipulators," *Mechanism and Machine Theory*, **36**(1), pp. 15-28 (2001).

In: Robotics: State of the Art and Future Trends
Editors: G. Legnani and I. Fassi

ISBN: 978-1-62100-403-5

Chapter 3

Manipulators Workspace Analysis as Based on a Numerical Approach: Theory and Applications

***Gianni Castelli*[*] *and Erika Ottaviano*[†]**
DiMSAT, University of Cassino, Via Di Biasio 43, 03043 Cassino (Fr), Italy

Abstract

This chapter presents an overall evaluation of a numerical general procedure for the determination and evaluation of the workspace for serial and parallel manipulators by using a discretization of the operational space. Workspace determination is usually an intermediate but critical step in analyzing manipulators, therefore it is very important to have a powerful tool to provide its estimation for a given architecture. In the literature several methods have been proposed to determine the workspace of manipulators by using analytical or numerical approaches for serial or parallel architectures. Most of the proposed methods for workspace analysis and determination have been defined and are useful only for serial or for parallel architectures. The proposed procedure can be applied to serial and parallel mechanisms and allows for the creation of a cloud of points representing the workspace. The workspace investigation is carry out by using a suitable formulation based on Direct or Inverse Kinematics, depending on the manipulator architecture. Furthermore, numerical data is further processed in a CAD environment in order to create an useful representation of manipulators' workspace, which can be implemented in virtual representation of an industrial application for a 3D layout optimization. Numerical investigations are reported as related to serial and parallel industrial robots, and regarding to new prototypes of cable-based parallel manipulators.

Introduction

In the last decades several methods have been developed for determining the manipulators workspace which is generally an intermediate but critical step in analyzing and

[*] E-mail address: g.castelli@unicas.it
[†] E-mail address: ottaviano@unicas.it

synthesizing manipulators. Therefore, it is very important to have a theory to safeguard its estimation and for the conceptual design. The manipulator workspace is the set of poses, referred to the base coordinate system, that can be reached by a reference point of the manipulator wrist [1]. Furthermore, different types of workspaces have been proposed in literature. Those can be grouped as [2], [3]:

- position workspace (or reachable workspace), is the set of positions that may be reached by the end-effector reference point with at least one orientation;
- orientation workspace, all possible orientations that can be reached while the end-effector reference point is in a fixed location;
- total orientation workspace, all the locations of end-effector reference point that may be reached with all the orientations among a set defined by ranges on the orientation angles;
- constant orientation workspace, is defined as the set of all possible locations that can be reached by the end-effector reference point with a specified orientation.

The position workspace and its boundary are important characteristics for industrial manipulators which are often used in a workcell layout. In particular, the position workspace boundary is a curve (in plane) or a surface (in space) and defines the extent of reach of the wrist [4].

Fundamental characteristics of the manipulator position workspace can be recognized as:

- the shape and the volume,
  - for a serial manipulator the shape is a solid of revolution when the first joint is of revolute type, as shown in Figure 1a), and it is a solid of extrusion when the first joint is of translational type;
  - for parallel architectures the shape is a solid that is in general bounded by convex/concave surfaces, as shown in Figure 1b).
- values of distance reaches and reach ranges, which allow to evaluate the workspace limits to place the robot in a workcell layout;
- the shape and dimension of holes, whose existence is determined by a region of unreachable points that can individuate a straight-line surrounded by the workspace;
- the shape and extension of voids, which are regions of unreachable points that are buried within the workspace points.

There are different methods that can be used to identify the workspace or its boundary, and generally they are chosen as function of manipulator architecture. They can be classified as:

- analytical methods;
- numerical methods.

Analytical methods allow to identify the workspace of a manipulator by using a formulation in closed form, which is able to describe all feasible configurations for the manipulator. Analytical methods are often called algebraic-geometric methods, because they can be obtained by developing geometrical or algebraic equations [5].

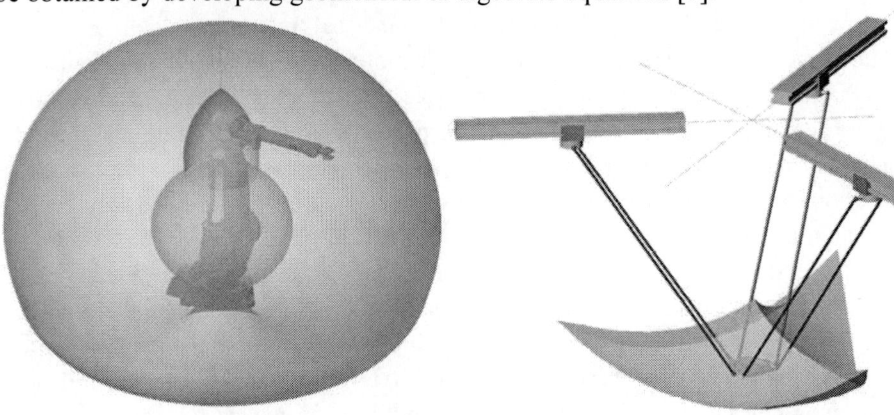

a) b)

Figure 1. Examples of position workspaces: a) for a serial manipulator; b) for a parallel manipulator.

Analytical methods have been used for the analysis of both serial and parallel manipulators, but their efficiency is related to the nature and complexity of the attached problem. These methods are usually very fast and accurate, provide an exact representation of the workspace and can be used to calculate efficiently any other characteristic of the workspace, as its volume. Main drawback of these methods concerns with the difficulty to obtain the formulation in closed form for any architecture. Furthermore, for the determination of characteristics of the workspace, in order to compare different existing manipulators or design a new one, it is almost always necessary to determine the boundary of the workspace [6].

Numerical methods are generally used to identify the workspace of a manipulator by solving formulations, which can not be expressed in closed form [6]. Problems in geometry and kinematics are often formulated using trigonometric functions and very often these can be converted to polynomials because they usually have to do with angular rotations whose main property is the preservation of length. Length relations are inherently polynomials, due to the Pythagorean Theorem. Elimination theory-based methods for constructing Grobner bases are the classical approach to solving such problems but their reliance on symbolic manipulation makes those methods seem somewhat unsuitable for all but small problems.

There are several numerical methods able to identify the workspace of manipulators. The most used are continuation methods, branch and bound algorithm and discretization methods. Continuation methods start with an initial system whose solutions are known, and then transform it gradually to the system whose solutions are sought, while tracking all solution paths along the way [7].

Branch and bound algorithms are methods for global optimization in non-convex problems and can be used to solve complex systems of equations and/or inequalities in a specific numerical space [2]. Branch and bound algorithms can be (and often are) slow, however. In the worst case they require effort that grows exponentially with problem size, but in some cases the methods converge with much less effort. The branch-and-bound algorithm is an enumerative technique, in which a solution is found based on the construction of a tree in which nodes represent the problems candidates and branches represent the new restrictions to be considered.

Through this tree, all integer solutions of the problem feasible region are listed explicitly or implicitly ensuring that all the optimal solutions will be found, as shown in Figure 2.

Discretization methods are other powerful numerical methods which consist of the discrete subdivision of the operational space in a grid of nodes. Thus, each node must satisfies the needed conditions to belong to the workspace, and the set of these nodes provides a discrete representation of the manipulator workspace. Discretization methods are generally easier than other methods, also for complex chain manipulators.

Main drawbacks of numerical methods are often related to the solution accuracy and the computational time. In particular, numerical methods provide just an approximate solution, and the accuracy level of the solution is generally inversely proportional to the computational time.

Numerical results obtained from the workspace analysis are in general not suitable to create an useful representation of the workspace. Indeed, the workspace is often represented by a cloud of points, while a representation in CAD environment can be well suited to improve the workcell layout design. A surface recognition is needed to obtain the geometric entity, which represent the workspace from this set of points.

The surface recognition problem (SRP) can be stated as follows: given a cloud of points C representing an unknown surface U, create a surface model S approximating U [8]. Furthermore, no structure or other information is assumed within the points and the surface U (assumed to be a manifold[‡]) may have arbitrary topological type[§], including boundaries, and may contain sharp features such the creases and corners.

In Figure 3 three different types of a cloud of points representation for a sphere are reported. Figure 3a) shows the cloud of points only, without any other solid geometric entity. In Figure 3b) a constant dimensions voxel (a 3D solid entity) has been used for each point. The Figure 3c) reports the mesh generation from the cloud of points.

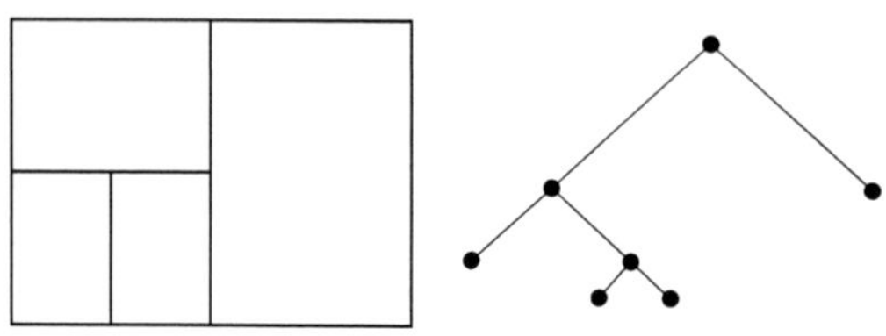

Figure 2. Branch and bound example in $\mathbf{R}^2$ after 3 iterations.

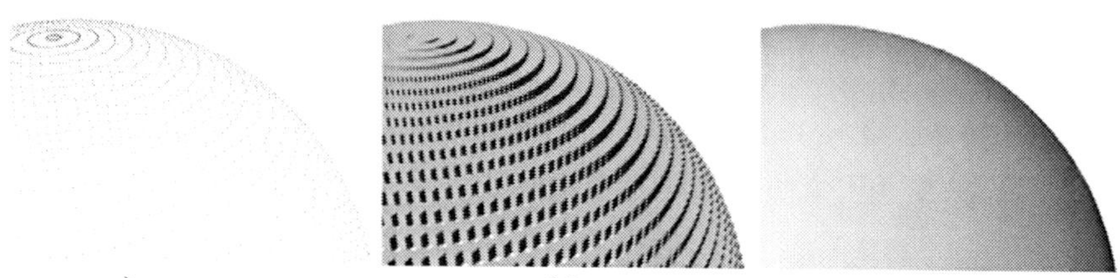

a) b) c)

Figure 3. Examples of a point cloud representation: a) point cloud; b) voxel representation; c) surface reconstruction.

‡ A manifold is a surface that does not intersect itself. More precisely, a manifold (possibly with boundary) embedded in $\mathbf{R}^3$ is a set everywhere locally homeomorphic to either a disk or a half-disk, where a homeomorphism is a continuous invertible map whose inverse is also continuous.

§ The topological type of a surface refers to its genus, the presence of boundaries, etc.

In this chapter we propose a fast and accurate computation of manipulators workspace, for serial and parallel manipulators by using the discretization of the operational space. An Introduction and State-of-art of the workspace evaluation are reported in first and second Sections, respectively. A detailed general description of the proposed algorithm and a validation test to evaluate its accuracy level have been reported in the third Section. Numerical examples have been reported in the four Section.

Related Work on Workspace Evaluation

Analytical methods are extensively used for serial manipulators, which are still the most used robots for industrial applications. Since most of the industrial manipulators are wrist-partitioned, the workspace analysis of such manipulators can be performed by considering the positioning and orienting capabilities separately [9], [10] . Many scientists have proposed analytical methods to determine the workspace of manipulators. Early studies on the workspace of serial manipulator were developed in [10]. The relationship between kinematic geometry and manipulator performances was proposed in [11], [13]; Gupta and Roth analyzed the effect of hand size on workspace in [14], and an algorithm for the workspace determination of a general N-R robot has been developed in [15]. Several authors have grouped manipulators into classes [16], [17], [18], but they have just considered special architectures, such as cuspidal or orthogonal structures, which have a simplification in the architecture. A classification of a general 3R manipulator as based on kinematic properties is proposed in [19]. As a completely new method the level-set belonging to the two-parameter set of curves, which constitutes the cross section of the workspace of the manipulator, was presented in [20].

The analytical determination of the workspace of parallel manipulators is an important issue, which has been addressed by several researchers. Procedures for workspace evaluation of parallel manipulators have been formulated by determining extreme paths [21], by computing conditions occurring at the workspace boundaries [22], [23] and by formulating ad-hoc closed-form expressions for specific architecture [24]. A geometrical algorithm to compute the parallel manipulator workspace is used in [25], [26].

A relatively new type of robots belonging to the class of parallel manipulators concerns with cable-based parallel manipulators, in which the fixed frame and platform are connected by several cables, which can be exerted or retracted by a suitable actuation system.

Several workspace classifications have been proposed for fully constrained cable robots, namely the controllable [27], wrench feasible [27], dynamic [29] and force-closure [30], [31] workspaces. They are defined as the set of poses at which the mobile platform (or end-effector) can physically reach while all the cables have positive tension, and some additional constraints are met.

Numerical methods are used to determine the workspace of a manipulator by using numerical techniques without getting to a closed form formulation. Several numerical methods can be used to identify the workspace of manipulators. Some works for serial manipulators workspace analysis are based a numerical procedures [32] and direct-kinematics scanning via probabilistic techniques [33]. A discretization algorithm has been used to identify the manipulator workspace for both serial and parallel architectures in [34].

Before the 1970's, an approach to solve multivariate polynomial systems, called polynomial continuation, was developed [35] in order to computing approximate solutions of a system of parameterized nonlinear equations, further developed by [36]. Homotopy continuation methods provide reliable and efficient numerical algorithms to compute accurate approximations to all isolated solutions of polynomial systems. During the last decades, this method has been developed into a reliable and efficient numerical algorithm for approximating all isolated zeros of polynomial systems. Modern scientific computing is marked by the advent of vector and parallel computers and the search for algorithms that are to a large extent naturally parallel. A great advantage of the homotopy continuation algorithm for solving polynomial systems is that it refers to a large degree parallel, in the sense that each isolated zero can be computed independently [36]. This natural parallelism makes the method an excellent candidate for a variety of architectures.

In a different approach, branch-and-prune methods use approximate bounds of the solution set in order to rule out portions of the search space that contain no solution [32]. They recursively reduce and bisect the initial domain until a fine-enough approximation of the solution set is derived. The convergence of this scheme is guaranteed by the fact that the bounds get tighter as the intermediate domains get smaller. Applications of such methods can be found in robot kinematics including [37], [39], [40].

The workspace analysis problem is closely related to singularity analysis. The knowledge of the number, geometry and location of kinematic singularities in an important issue for many problems in manipulator analysis, trajectory planning and control. Further, the knowledge of the relationship between singularities and kinematic parameters values can be important in manipulator design. Singularities have been extensively studied. Many authors used screw theory [41], [42], math theory exists for analyzing the singularities of smooth maps, and bifurcation analysis or geometric considerations. Basically, for serial chain manipulators the determination of kinematic singularities is equivalent to finding the determinant of the manipulator Jacobian matrix. Singularity analysis for serial manipulators has been extensively studied by several authors [16], [43], [44]. Singular configurations for parallel manipulators are particular poses of the moving platform for which the parallel robot loses its inherent infinite rigidity and in which the end-effector has uncontrollable freedom, namely, there exists an instantaneous gain or loss of DOF [45]. Singularity analyses, based on rank deficiency of the Jacobian matrix or the derivative of the input-output velocity equations of closed-loop architectures, have been addressed in the literature [45], [46], [47].

There is a variety of classifications of singularities of mechanisms from different points of view. From geometric perspective singular configurations were classified in accordance with the mechanisms screw system [41], [16]. Singularities of serial manipulators are usually referred to situations where the end-effector DOF of the mechanism decreases, the classification is more complex for parallel manipulators [48].

Therefore, even regarding to singularity analysis, serial and parallel manipulators have been treated separately. An attempt to unify the analysis is reported in [49] where a general approach to the local analysis of singular configurations of both serial and parallel manipulators is provided as based on the instantaneous screw system of the considered mechanism.

A General Numerical Algorithm for the Workspace Analysis

Several numerical procedures have been developed to evaluate the manipulators workspace. A complete representation of the workspace (orientation and position) needs a 6-dimensional space, for which is not possible a graphical and useful illustration. Thus, it is usual to represent separately position and orientation workspaces [9]. In this Chapter, the workspace (position and orientation) is identified by using the kinematics equations of manipulators in a procedure based upon the numerical discretization of the operational space, which is the 6-dimensional space where the end-effector can work.

Let us consider a multi-body system that may be either represented by a close or open kinematic chain, the reachable poses of all links are usually encoded in a n_q-dimensional vector of generalized coordinates $\mathbf{q}$, subject to a system of $n_e \leq n_q$ equations in the form

$$\Phi(\mathbf{q}) = 0 \tag{1}$$

that expresses the kinematic constraints imposed by the joints. The solution of this system of equations is, in general, a manifold C called *Configuration space*. Position analysis involves finding all possible configurations of a kinematic chain. A configuration is an assignment of a pose to each link, respecting the kinematic constraints imposed by the joints, with no regard to possible link-link interferences. It is worth noting that Eq.(1) does not consider joint limit equations, which may be introduced by considering inequalities such as

$$q_{\mathrm{i}}^{MIN} \leq q_{\mathrm{i}} \leq q_{\mathrm{i}}^{MAX} \text{ for i} = 1 \text{ to } n_q \tag{2}$$

Equation 2 refers to two types of variables q in describing joint limits: variables referring to distance and to angular position. Typically they correspond to slider and revolute joints for serial robots, respectively; but using only these two types of variables joint limits can be defined for any lower-pair joint.

In order to extend the system of Equations in (1), which refers to position analysis, to singularity analysis, one can consider some equations that represent the condition of Jacobian rank deficiency. Therefore, the solution of the obtained system of equations in (3) will contain the singular points of C with respect to some of the variables in $\mathbf{q}$. Let us considered vector $\mathbf{q}$ partitioned into three vectors $\mathbf{q}_{\mathrm{i}}$, vector of n_{i} input variables corresponding to the actuated DOFs of the multi-body system, $\mathbf{q}_{\mathrm{o}}$ is a vector of n_{o} output variables representing the pose of the end-effector, and $\mathbf{q}_{\mathrm{p}}$ is a n_{p}-dimensional vector encompassing the remaining intermediate variables. If we are interested in finding the singularities of the end-effector, Eq.(1) can be rewritten as

$$\Phi(\mathbf{q}_{\mathrm{w}}, \mathbf{q}_{\mathrm{o}}) = 0 \tag{3}$$

in which $\mathbf{q}_{\mathrm{w}} = [\mathbf{q}_{\mathrm{i}}, \mathbf{q}_{\mathrm{p}}]$. The end-effector singularities can be found by vectors $\mathbf{q}$ satisfying the system of equations

$$\Phi(\mathbf{q}_{\mathrm{w}}, \mathbf{q}_{\mathrm{o}}) = 0\,; \qquad d\Phi_z(\mathbf{q}_w, \mathbf{q}_o)^{\mathrm{T}} \xi = 0\,; \qquad \xi^{\mathrm{T}} \xi = 1 \tag{4}$$

in which ξ in an n_e-dimensional vector of unknowns. In particular, the second and third equations in (4) impose the rank deficiency of $d\Phi_z$.

The procedure proposed in this Chapter allows to compute a set of reachable poses of a manipulator which can be used to identify the workspace. The proposed algorithm can be organized in the following three main steps:

First step: Create a discretization of the 6D operational space by a grid of nodes. A 6D matrix (M) can be used to represent virtually the 6D operational space, and in particular each matrix element represents univocally a node. The value assigned to each matrix element defines if the relative node belongs to the workspace. For this algorithm a binary matrix can be used, since only a Boolean information is required (belong/do not belong), with a significant memory saving. Generally the first 3 components of the 6D matrix are related to position workspace, while the last 3 components are related to the orientation workspace. In order to create a operational space discretization, 3 parameters values must be set: operational space limits (or discretization range), discretization law, and number of nodes (and consequently the discretization steps).

The limits for position workspace must be chosen to include the whole workspace volume. Similar conditions must be set for orientation workspace, even if the range -180° to 180° can cover all possible configurations, even if there can be a loss of information.

A discretization law must define the nodes arrangement in the operational space and relationship between the continuous operational space and discretization nodes. Several discretization laws can be chosen. Besides an uniform discretization (in which discretization steps are constant), a particular law can be chosen in order to increase the nodes density in a specific region of the workspace. Moreover, the discretization grid of nodes can be arranged by using Cartesian, Cylindrical or Spherical coordinates, in order to have the best fitting according the manipulator architecture. A round function is the most used type of relationship between the continuous operational space and discretization nodes. Among several round functions, the most used are the fix, ceil and floor functions.

The number of nodes is an important parameter, which strongly influence the computational time and the accuracy: the higher the nodes number is, the higher the computational time and the accuracy level are. Consequently, the number of nodes is chosen as function of the desired level of accuracy, keeping into account a constrain due to the computational capability.

For example, in uniform Cartesian grid, the number of nodes can be related to the discretization steps by

$$D_i = \frac{L_i^{MAX} - L_i^{MIN}}{N_i} \quad \text{for i = 1 to 6} \tag{5}$$

where the D is the discretization step, L^{MAX} and L^{MIN} are the upper and lower limit values and N is the number of nodes. Coordinates of the continuous operational space can be related to the relative discretization node by

$$x_i^N = \left\lfloor \frac{x_i - L_i^{MIN}}{D_i} \right\rfloor D_i + L_i^{MIN} \quad \text{for i = 1 to 6} \tag{6}$$

where the $\lfloor\ \rfloor$ operator denotes the floor function that returns the nearest integer less than or equal to the real-valued argument inside the brackets. x^N is the coordinate of the discretization node and x is the pose coordinate of the operational space. Thus, the matrix indexes (MI) of the 6D matrix can be correlated to the operational space coordinates by the following

$$MI_i = \left\lfloor \frac{x_i - L_i^{MIN}}{D_i} \right\rfloor + 1 \text{ for i = 1 to 6} \tag{7}$$

For the evaluation of a position workspace only, the uniform Cartesian discretization grid can be thus represented thus in the scheme of Figure 4.

Values D_1, D_2 and D_3 are the discretization steps for the X_1, X_2 and X_3 axes. Integers MI_1, MI_2 and MI_3 are matrix indexes related to x_1, x_2 and x_3 coordinates respectively of the highlighted node. Each matrix entry can assume a suitable value in order to define if the relative node belongs to the manipulator workspace or not.

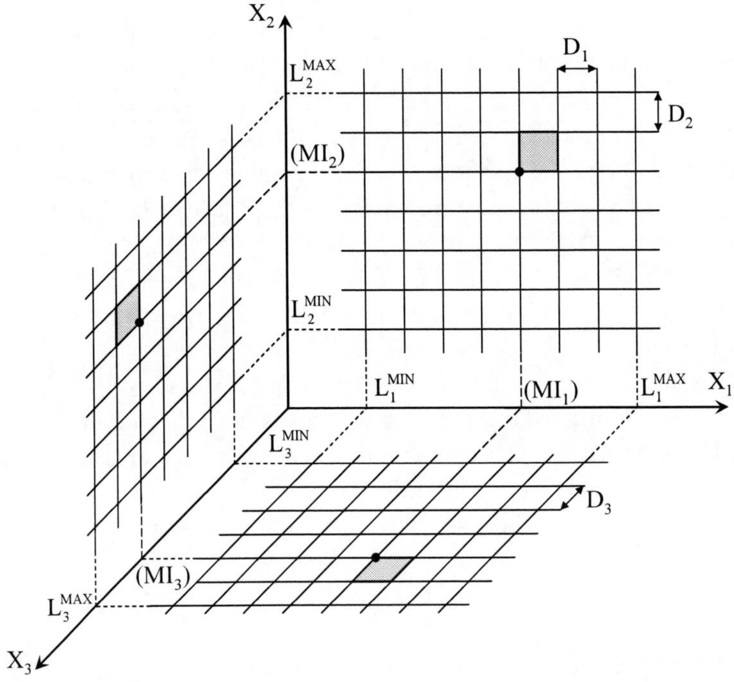

Figure 4. Cartesian uniform discretization of the 3D operational space for the position workspace evaluation.

Second step: Verify which nodes satisfy the needed conditions for belonging to the workspace by using Direct and Inverse Kinematics of the robot. For this phase the joint space must be know (e.g. minimum and maximum values for each joint), as well the constrains due to the geometric characteristics of the manipulator, in order to keep into account also singular configurations. By the kinematic analysis, the following two procedure must be used:

a. Inverse Kinematics (generally used for parallel chains). The Inverse Kinematics problem consists of solving the joint variables of a manipulator as function of manipulator configuration and particularly depending on the position of a reference point in the manipulator extremity [9]. For the proposed algorithm each node of the operational space must be analyzed. Node coordinates are the input of the Inverse Kinematics analysis and, if each output pose obtained satisfies the requested conditions (e.g. joint values, physical constrains, force conditions), then the input node belongs to the workspace.
b. Direct Kinematics (generally used for serial chains). The Direct Kinematics problem consists of solving the position and orientation of end-effector reference point as function of the values of the joint variables [9].

In this case the joints' values are the input of the kinematic analysis and a discretization of the joint space is needed, similarly to the discretization of the operational space. By using the Direct Kinematics analysis to the set of joints' coordinates, a set of output poses is obtained. An output pose belongs to the workspace if the required conditions (e.g. end-effector coordinates, physical constrains, force conditions) are satisfied. By using the chosen discretization law (applied to the operational space to create the grid of nodes), the set of output poses provides the set of nodes which belong to the workspace.

Third step: The set of nodes that verify the required conditions is a discrete representation of the manipulator workspace. This data can be used to extract some useful indexes in order to describe the characteristics of a manipulators workspace, such as volume and cross-section areas. Moreover, the boundary workspace can be evaluated as the set of nodes that have at least a neighbor node that does not belong to the workspace.

In order to evaluate the performance of the proposed procedure, a test case has been developed as based on a five bar parallel mechanism, for which the kinematic scheme is reported in Figure 5.

The mechanism dimensions are: l_0 = 800 mm and $l_1 = l_2 = l_3 = l_4$ = 300 mm; the reference point H is constrained to have positive values of the y coordinate only. By using a suitable analytical formulation, the reachable position workspace area by the reference point H has been evaluated and it is equal to 123899.28 mm^2.

Applying above-mentioned the discretization method, following steps are developed:

First step:

Discretization range: 0 mm $\geq x_1 \geq$ 800 mm; 0 mm $\geq x_2 \geq$ 600 mm

Number of nodes: (15000 x 15000) = 225E+10^6

Discretization steps: $D_1 = \dfrac{800-0}{15000} = 0.053$ mm ; $D_2 = \dfrac{600-0}{15000} = 0.04$ mm

$$\text{Discretization law:}\begin{cases}\text{Coordinates of nodes:}\begin{cases}x_1^N = \left\lfloor \dfrac{x}{15000} \right\rfloor 15000\\ x_2^N = \left\lfloor \dfrac{y}{15000} \right\rfloor 15000\end{cases}\\ \text{Matrix indexes:}\begin{cases}MI_1 = \left\lfloor \dfrac{x}{15000} \right\rfloor\\ MI_2 = \left\lfloor \dfrac{y}{15000} \right\rfloor\end{cases}\end{cases}$$

Second step:

Due to the manipulator architecture, Inverse Kinematics formulation has been used, taking into account the position limits of the reference point H ($x_2 \geq 0$ mm).

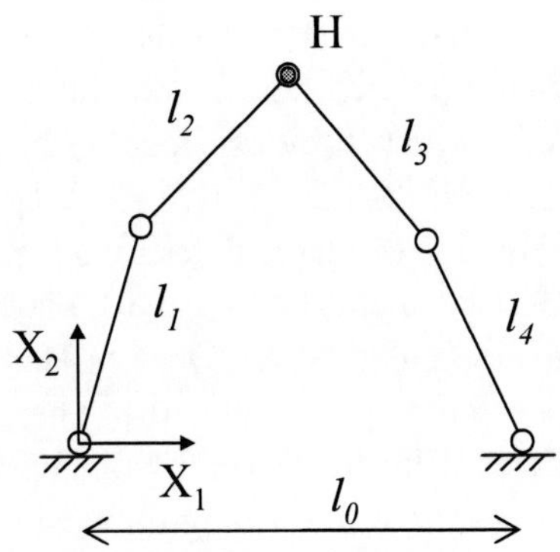

Figure 5. Kinematic scheme of a five bar parallel mechanism.

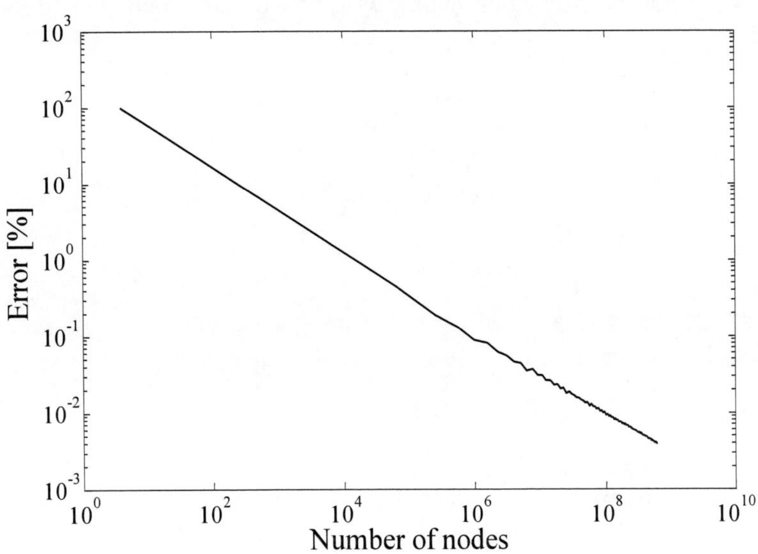

Figure 6. Workspace area error versus number of nodes.

Third step:

The workspace computation has been completed after 36.1 s by a PC with an Intel Pentium 4 processor with a frequency clock of 3 GHz. 58073778 nodes belong to the position workspace, and consequently, the workspace area is equal to 123907.24 mm^2. The workspace area error between the numerical and analytical computation is less than 8 mm^2 and equal to the 0.0064 % of the value obtained by the analytical formulation. Furthermore, in order to evaluate the accuracy level of the proposed procedure, a set of computations has been carried out for several numbers of nodes. The workspace area error obtained from this set of computation has been reported in Figure 6. In particular, with a number of nodes of 1E+6 it is possible to get an error less than 0.1 %.

Workspace Analysis for Serial Chain Manipulators

In the following the proposed algorithm is applied to evaluate the workspace for serial manipulators. Computer aided design (CAD) systems are used to design and create digital models in order to have a virtual representation of the real objects, and it is thus possible to create a virtual representation of a workcell by considering the manipulator and its workspace. Therefore, the obtained numerical results have been moreover processed in CAD software in order to create an useful representation of manipulator's workspace, which can be implemented in virtual environment of an industrial application for the 3D layout optimization.

When dealing with serial manipulators the Denavit-Hartenberg notation is commonly used to describe the kinematic chain of the robot [6]. Therefore, the workspace region is computed by scanning joint angles within the mobility range. When the position workspace is considered, one has to compute vector $\mathbf{x}_0$ describing the position of the operation point in the fixed frame as function of the first three joints' mobility.

The transformation matrices bring the position vector from one reference frame to another one. A general transformation matrix $\mathbf{T}_{j+1}$, which maps frame j+1 to frame j can be expressed as

$$\mathbf{T}_{j+1} = \begin{bmatrix} c\theta_{j+1} & -s\theta_{j+1} & 0 & a_j \\ s\theta_{j+1} c\alpha_j & c\theta_{j+1} c\alpha_j & -s\alpha_j & -d_{j+1}\, s\alpha_j \\ s\theta_{j+1} s\alpha_j & c\theta_{j+1} s\alpha_j & c\alpha_j & d_{j+1}\, c\alpha_j \\ 0 & 0 & 0 & 1 \end{bmatrix} \tag{8}$$

where $c\theta_{j+1}$ stands for $\cos\theta_{j+1}$, and so on. Consequently, the position vector $\mathbf{x}_0$ of the operation point H with respect to the base frame can be computed as

$$\mathbf{x}_0 = \mathbf{T}_1\, \mathbf{T}_2\, \mathbf{T}_3\, \mathbf{x}_3 \tag{9}$$

As an illustrative example, an industrial manipulator COMAU NH3 220-2.7 has been considered [50]. This manipulator has an architecture with six degrees of freedom, and kinematic parameters values according to the Denavit-Hartenberg notation can be found in the data sheet available in the data sheet [50]. Thus, the manipulator workspace in Figure 7 can be determined by computing the reachable points as functions of joint angles when the manipulator architecture is given. The operation point H is considered on the wrist of the manipulator in the last reference frame being used to compute the position vector $\mathbf{x}_3$ with a homogenous-coordinate transformation.

The point cloud representing the manipulator workspace is further processed in order to recover the workspace surface, as shown in Figure 7.

The obtained numerical result can be further processed and included in an industrial environment for the mechanical design and optimization of a robotized workcell. In particular, Figure 8 shows a workcell for automotive industry with 2 COMAU NH3 220-2.7 robots in a real industrial environment.

As another case of study for serial manipulators, the ADEPT Cobra i800 manipulator is considered. According to the formulation reported in Eqs. (8) and (9) and kinematic parameters given in [51] the position workspace of the robot has been obtained and reported in Figure 9 and further included in a real industrial environment, as shown in Figure 10.

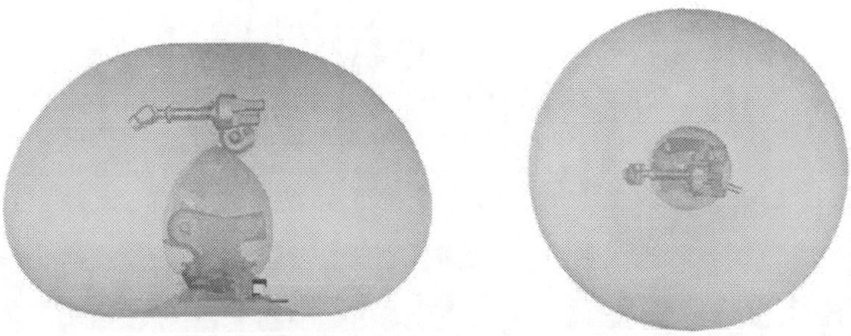

a) b)

Figure 7. CAD reconstruction model of the position workspace for the serial manipulator COMAU NH3 220-2.7: a) lateral view; b) top view.

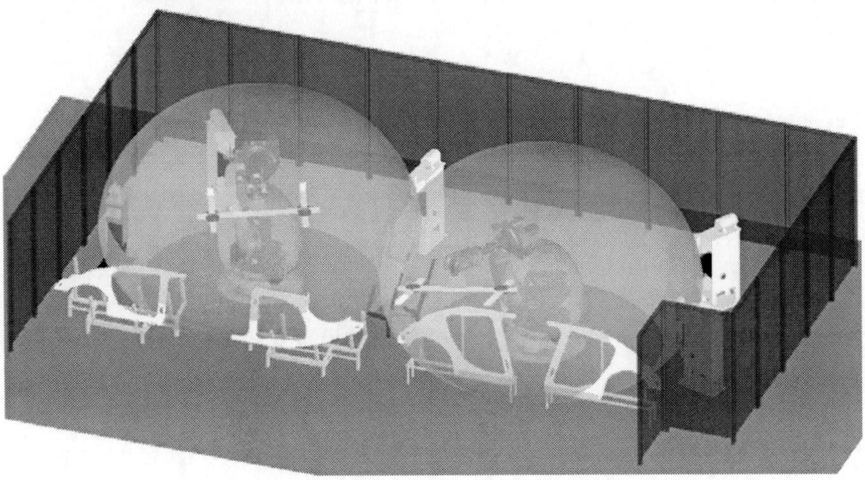

Figure 8. 3D view of the CAD model of two serial manipulators COMAU NH3 220-2.7 installed in an automotive industry workcell.

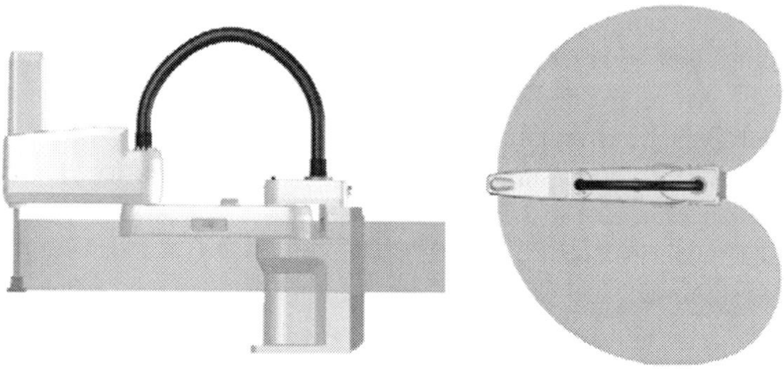

a) b)

Figure 9. CAD reconstruction model of the position workspace for the serial manipulator ADEPT Cobra i800: a) lateral view; b) top view.

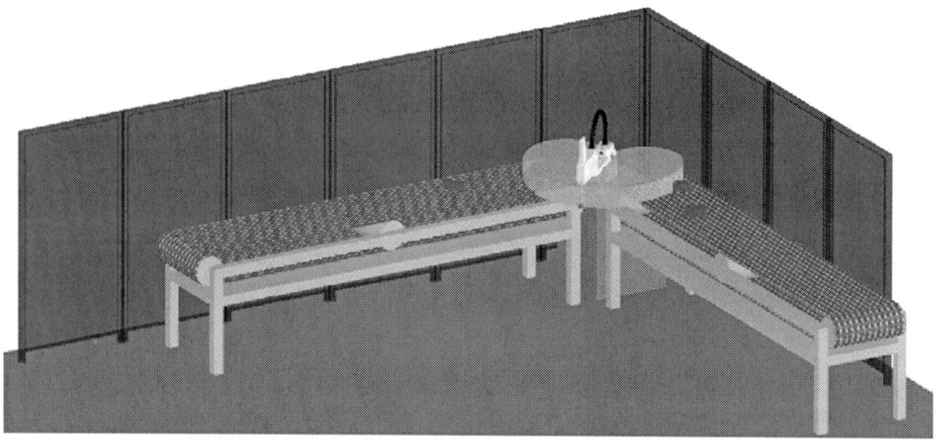

Figure 10. 3D view of the CAD model of a serial manipulator ADEPT Cobra i800 installed in a workcell for pick and place application.

Workspace Analysis for Parallel Chain Manipulators

Parallel manipulators consist of a fixed and moving platforms, which are connected by several legs constituting closed kinematic chains. Given the geometry of the manipulator, the position analysis can be formulated by writing a vector loop equation involving the position vectors of fixed origin **O** and reference point H on the moving platform, leg attachment points in the base ($\mathbf{A}_i$) and moving platform ($\mathbf{B}_i$) and the i-th leg $\mathbf{l}_i$. Then the Inverse Kinematics problem for a parallel manipulator can be formulated as

$$\mathbf{l}_i = \mathbf{A}_i - \mathbf{r} - \mathbf{R}\mathbf{B}_i \tag{10}$$

in which $\mathbf{l}_i$ is the leg length and **R** is the matrix that expresses the relative orientation of the moving frame with respect to the fixed reference frame.

According to the Inverse Kinematics formulation in Eq.(10) the industrial parallel manipulator Quattro s800H commercialized by ADEPT is considered as a case of study. Given the geometric parameters of the robot, which can be found in the data sheet [52] the Inverse Kinematic problem can be solved by using Eq. (10) in order to obtain the workspace of the manipulator, which is shown in Figure 11. Figure 12 shows a workcell for a pick-and-place application of the ADEPT Quattro s800H parallel manipulator in industrial environment.

Cable-based parallel manipulators consist of a fixed base (or frame) and mobile platform, which are connected by several cables. Therefore, they are structurally similar to the classical parallel ones, in which legs are replaced by cables. Due to the nature of cables, they posses in general good characteristics such as: good inertial properties, their actuator-transmission systems can be conveniently fixed on the base and cables are lighter and thus, they can have higher payload-to-weight ratio. Moreover, they can be relatively low-cost, modular, and easy to reconfigure according to their design. Regarding to the workspace analysis, due to the nature of cables, which can only pull but do not push the end-effector, some additional constraints to position analysis are required.

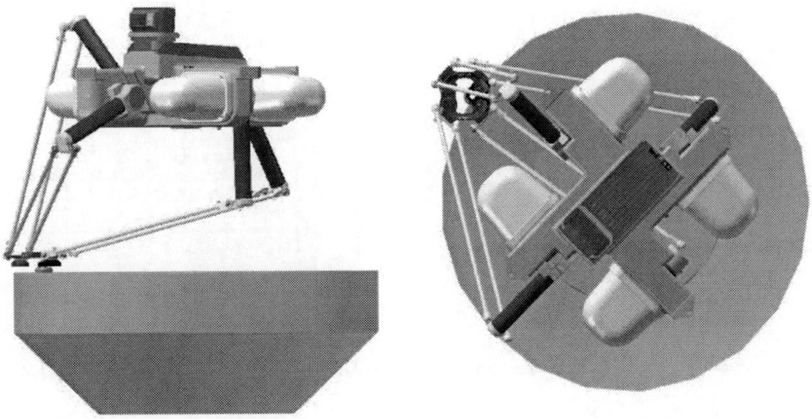

a) b)

Figure 11. CAD reconstruction model of the position workspace for the parallel manipulator ADEPT Quattro s800H: a) lateral view; b) top view.

In particular, static or dynamic equilibrium of the end-effector should be ensured for a given pose of the manipulator. If the static equilibrium is taken into account, in addition to Eq. (10) one as to consider

$$-\mathrm{J}^{\mathrm{T}}\mathbf{F} = \mathbf{W} \tag{11}$$

in which $\mathbf{F} = \begin{bmatrix} F_1 & \cdots & F_n \end{bmatrix}^T$ represents the vector of scalar n cable forces; $\mathbf{W}$ is the resultant external end-effector wrench vector expressed in the fixed frame; J is the Jacobian matrix. If Dynamics is required, then in Eq. (11) $\mathbf{W}$ must be replaced with the resultant of inertial effects vector expressed in the fixed frame.

In the following a Cartesian Cable Suspended Robot (CCSR) is considered as an illustrative example. It has been developed for industrial application [53]. Figure 13 shows the CAD reconstruction of the position workspace of the manipulator and Figure 14 shows its placement in an industrial environment.

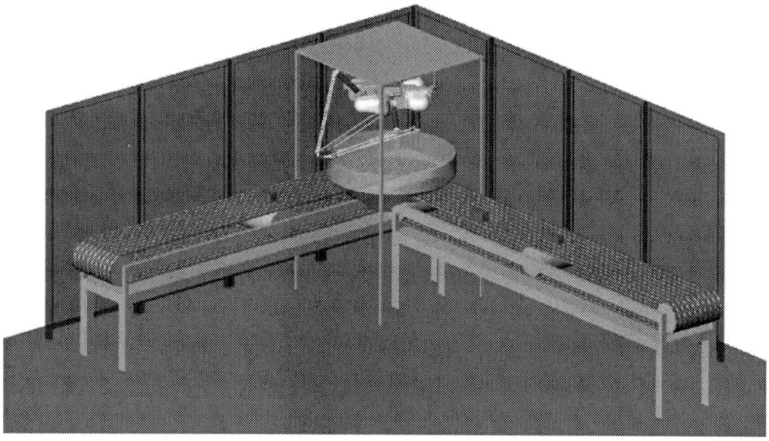

Figure 12. 3D view of the CAD model of a parallel manipulator ADEPT Quattro s800H installed in a workcell for pick and place application.

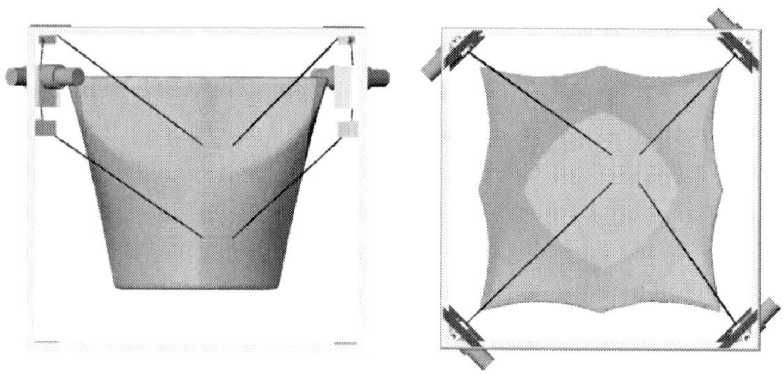

a) b)

Figure 13. CAD reconstruction model of the position workspace for the cable parallel manipulator CCSR: a) lateral view; b) top view.

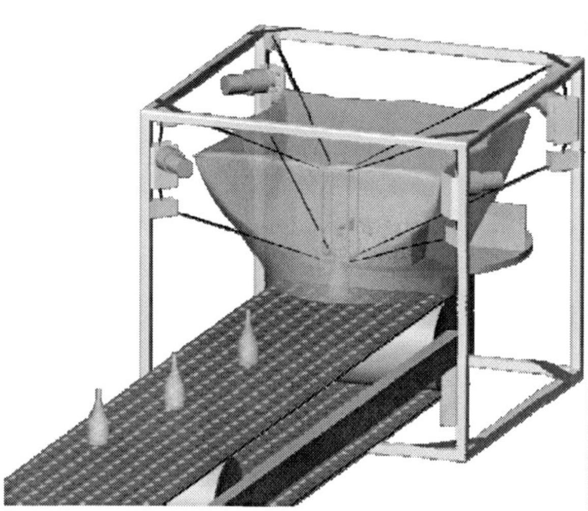

Figure 14. 3D view of the CAD model of a cable parallel manipulator CCSR installed in a workcell for bottles packaging.

Conclusions

In this chapter an overall evaluation of a numerical algorithm for the determination and analysis of workspace characteristics for serial and parallel manipulators has been presented. The proposed procedure allows to evaluate main position and orientation workspace characteristics (such as workspace volume and boundary position workspace) by using a numerical procedure concerning the discretization of the operational space. Examples are reported for serial and parallel manipulator architectures in order to show the capability of the proposed procedure. In addition, as practical cases of study, the proposed manipulators have been processed in CAD systems in order to provide a real scenario of industrial environment, by using the numerical data obtain from the position workspace investigation.

References

[1] ISO Standard 8373:1994, *Manipulating Industrial Robots – Vocabulary.*

[2] J.P. Merlet, *Parallel Robots*, 2nd ed., Springer, Dordrecht, (2006).

[3] T.W. Lee, D.C.H. Yang, "On the Evaluation of Manipulator Workspace", *ASME Journal of Mechanisms, Transmissions, and Automation in Design*, **105**:70-77, (1982).

[4] S. Dibakar, T.S. Mruthyunjaya, "A computational geometry approach for determination of boundary of workspaces of planar manipulators with arbitrary topology", *Mechanism and Machine Theory* **34**:149-169, (1999).

[5] D. Cox, J. Little, D. O'Shea, "*An Introduction to Computational Algebraic Geometry and Commutative Algebra*", 2nd ed., Springer, Berlin, (1997).

[6] J. Angeles, "*Fundamentals of Robotic Mechanical Systems*", 2nd ed., Springer, New York, (2003).

[7] A.J. Sommese, C.W. Wampler, "*Numerical solution of systems of polynomials arising in engineering and science*", World Scientific Press, (2005).

[8] S. Azernikov, A. Fischer, "Efficient surface reconstruction method for distributed CAD", *Computer-Aided Design*, **36**:799-808, (2004).

[9] J.J. Craig, "*Introduction to Robotics: Mechanics and Control*", Addison-Wesley, New York, (1989).

[10] L. RuiQin, D. Jian, "Orientation angle workspaces of planar serial three-link manipulators", *Science in China Series E: Technological Sciences*, **52**(4): 975-985, (2009).

[11] B. Roth, "Performance Evaluation of Manipulators from a Kinematic Viewpoint", *National Bureau of Standards Special Publication*, **459**:39-61, (1975).

[12] A. Kumar, K.J. Waldron, "The Workspace of a Mechanical Manipulator" *ASME Journal of Mechanical Design* **103**:665-672, (1981).

[13] F. Freudenstein, E.J.F. Primrose, "On the Analysis and Synthesis of the Workspace of a Three-Link Turning-Pair Connected Robot Arm", *ASME Journal of Mechanisms, Transmission and Automation in Design* **106**:365-370, (1984).

[14] K.C. Gupta, B. Roth, "Design Considerations for Manipulator Workspace", *ASME Journal of Mechanical Design* **104**:704-711, (1982).

[15] Y.C. Tsai, A.H. Soni, "An Algorithm for the Workspace of a General n-R Robot", *ASME Journal of Mechanisms Transmissions and Automation in Design* **105**:52-57, (1983).

[16] J.W. Burdick, "A Classification of 3R Regional Manipulator Singularities and Geometries", *Mechanism and Machine Theory*, **30**(1):71-89, (1995).

[17] P. Wenger, "Some guidelines for the kinematic design of new manipulators", *Mechanism and Machine Theory* **35**(3):437-449, (2000).

[18] M. Zein, P. Wenger, D. Chablat, "An Exhaustive Study of the Workspaces Topologies of All 3R Orthogonal Manipulators with Geometrical Simplifications", *Proceedings of the International Workshop on Computational Kinematics* CK2005, Cassino, (2005), paper 34.

[19] E. Ottaviano, M. Husty, M. Ceccarelli, "Identification of the workspace boundary of a general 3-R manipulator", *ASME Journal of Mechanical Design* **128**(1):236-242 (2006).

[20] E. Ottaviano, M. Husty, M. Ceccarelli, "A Level-Set Method for Workspace Analysis of Serial Manipulators", *10th International Symposium on Advances in Robot Kinematics*, (2006), pp. 307-314.

[21] C.M. Gosselin, "Determination of the Workspace of 6 d.o.f. Parallel Manipulators", *ASME Journal of Mechanical Design* **112**:331-336, (1990).

[22] J.P. Merlet, "Geometrical Determination of the Workspace of a Constrained Parallel Manipulator" *Proceedings of the 3rd International Workshop on Advances in Robotics Kinematics*, Ferrara, (1992), pp. 326-329.

[23] K. Abdel-Malek, J. Yang, "Workspace boundaries of serial manipulators using manifold stratification", *International Journal of Advanced Manufacturing Technology* **28**:1211-1229, (2006).

[24] R. Clavel, "DELTA, a fast robot with parallel geometry", *18th International Symposium on Industrial Robot*, (1988), Lausanne, pp. 91-100.

[25] D.I. Kim, W.K. Ching, Y. Youm "Geometrical approach for the workspace of 6-dof parallel manipulators" *IEEE International Conference on Robotics and Automation*, Albuquerque, (1997), pp. 2986-2991.

[26] I.A. Bonev, J. Ryu, "A Geometrical Method for Computing the Constant-Orientation Workspace of 6 PRRS Parallel Manipulators", *Mechanism and Machine Theory* **36**: 1-13, (2001).

[27] T. Bruckmann, L. Mikelsons, T. Brandt, M. Hiller, D. Schramm, "Wire Robots Part I: Kinematics, Analysis & Design" in Ryu J.H. "*Parallel Manipulators, New Developments*", I-Tech Education and Publishing, Vienna, 109-132, (2008).

[28] A. Riechel, I. Ebert-Uphoff, "Wrench-Based Analysis of Cable-Driven Robots", Proceedings of IEEE International Conference on Robotics and Automation, (2004), pp. 4950-4955.

[29] G. Barrette, C.M. Gosselin, "Determination of the Dynamic Workspace of Cable-Driven Planar Parallel Mechanisms", *ASME Journal of Mechanical Design*, **127**(2):242-248, (2005).

[30] C.B. Pham, S.H. Yeo, G. Yang, M.S. Kurbanhusen, C. I-Ming, "Force-Closure Workspace Analysis of Cable-Driven Parallel Mechanisms", *Mechanism and Machine Theory*, **41**(1):53-69, (2006).

[31] X. Diao, O. Ma, "A Method of Verifying Force-Closure Condition for General Cable Manipulators with Seven Cables", *Mechanism and Machine Theory*, **42**:1563-1576, (2007).

[32] E. Hansen, "*Global Optimization Using Interval Analisis*", Marcel Dekker Inc., New York, (1992).

[33] J. Rastegar, B. Fardanesh, "Manipulator Workspace Analysis Using the Monte Carlo Method", *Mechanism and Machine Theory* **25**(2):233-239, (1990).

[34] G. Castelli, E. Ottaviano, M. Ceccarelli, "A Fairly General Algorithm to Evaluate Workspace characteristics of Serial and Parallel Manipulators", *International Journal of Mechanics Based Design of Structures and Machines*, **36**(1); 14-33, (2008).

[35] B. Roth, F. Freudenstein, "Synthesis of path-generating mechanisms by numerical methods," *ASME Journal of Engineering for Industry*, **85**:298-307, (1963).

[36] A.P. Morgan, "*Solving Polynomial Systems Using Continuation for Engineering and Scientific Problems*", Prentice-Hall, (1987).

[37] T.Y. Li, "*Numerical solution of multivariate polynomial systems by homotopy continuation methods*", Acta Numerica Cambridge University Press, pp. 399-436, (1997).

[38] A. Castellet, F. Thomas, "An algorithm for the solution of inverse kinematics problems based on an interval method," *Advances in Robot Kinematics*, (1998), pp. 393-403.

[39] J.P. Merlet, "Solving the forward kinematics of a Gough-type parallel manipulator with interval analysis," *International Journal of Robotics Research*, **23**(3):221-236, (2004).

[40] J.M. Porta, L. Ros, F. Thomas, "A Linear Relaxation Technique for the Position Analysis of Multi-Loop Linkages," *IEEE Transactions on Robotics*, **25**(2):225-140, (2009).

[41] K.H. Hunt, "Special configurations of robot-arms via screw theory", *Robotica*, **4**:171-179, (1986).

[42] J.M. McCarthy, "*Geometric Design of Linkages*", Springer, New York, (2000).

[43] K.Y. Tsai, J. Arnold, D. Kohli, "Generic maps of mechanical manipulators", *Mechanism and Machine Theory* **28**(1):53-64, (1993).

[44] M. Husty, E. Ottaviano, M. Ceccarelli, "A Geometrical Characterization of Workspace Singularities in 3R Manipulators", *10th International Symposium on Advances in Robot Kinematics*, Batz-sur-Mer, (2008), pp. 411-418.

[45] C.M. Gosselin, J. Angeles, "Singular analysis of closed-loop kinematic chains", *IEEE Transaction of Robotics & Automation* **6**(3):281-290, (1990).

[46] O. Ma, J. Angeles, "Architecture singularities of platform manipulators", *IEEE International Conference on Robotics and Automation*, Sacramento, (1991), pp. 1542-1547.

[47] H.R.M. Daniali, P. J. Zsombor-Murray, J. Angeles, "Singularity analysis of a general class of planar parallel manipulators", *IEEE International Conference on Robotics and Automation*, Nagoya, (1995), pp. 1547-1552.

[48] D. Zlatanov, R.G. Fenton, B. Benhabib, "Identification and classification of the singular configurations of mechanisms", *Proceedings of the International Workshop on Computational Kinematics* CK1995, Sophia Antipolis, (1995), pp. 63-172.

[49] A. Muller, "Local Analysis of Singular Configurations of Open and Closed Loop Manipulators", *Multibody System Dynamics*, **8**:299-328, (2002).

[50] COMAU website: *Smart NH 3* page: http://www.comau.com/index.jsp?ixPageId=286&ixMenuId=152.

[51] ADEPT website: *SCARA Robot - Adept Cobra i800* page http://www.adept.com/products/robots/scara/cobra-i800/general.
[52] ADEPT website: *Parallel Robot (Delta Robot): Adept Quattro s800H* page http://www.adept.com/products/robots/parallel/quattro-s800h/general.
[53] G. Castelli, E. Ottaviano, A. Gonzalez, "Analysis and Simulation of a New Cartesian Cable-Suspended Robot", *Journal of Mechanical Engineering Science*, DOI: 10.1243/09544062JMES1976, (2009).

Reviewed by

Nestor Eduardo Nava Rodríguez, Robotics Lab, Universidad Carlos III, Avda. Universidad 30, 28911 Leganés, Madrid (Spain).

Antonio González, Dpto. Ingeniería Mecánica de la Universidad de Castilla la Mancha, Avda. de Camilo José Cela s/n, Ciudad Real (Spain).

In: Robotics: State of the Art and Future Trends
Editors: G. Legnani and I. Fassi
ISBN: 978-1-62100-403-5

Chapter 4

ON THE MOBILITY OF 3-DOF PARALLEL MANIPULATORS VIA SCREW THEORY

***E. Rodriguez-Leal*[1,*], *J. S. Dai*[2] *and G. R. Pennock*[3]**

[1]Department of Mechatronics and Automation, Tecnológico de Monterrey, Monterrey, Mexico

[2]Department of Mechanical Engineering, King's College London, University of London, Strand, London, UK

[3]School of Mechanical Engineering, Purdue University, West Lafayette, IN, U. S.

Abstract

Screw theory was proposed by Sir Robert Ball towards the end of the nineteenth century but laid dormant until the work of Brand in 1947. Then Dimentberg in 1965 and Philips and Hunt in 1967 illustrated the elegance of this theory in spatial kinematics. Hunt in particular used the theory of screws to reveal geometrical insight into the analysis of closed kinematic chains and spatial mechanisms. The duality of statics and kinematics offered by screw theory has opened a new horizon in parallel robot design as researchers recur to this mathematical tool for the calculation of the instantaneous kinematics and the determination of the mobility of a given robotic architecture. This chapter presents the mobility analysis of two families of 3-DOF parallel manipulators using screw theory. The first family consists of the 2R1T 3-PSP, 3-PPS, 3-PCU, and a new 3-CUP parallel manipulator, while the second family is formed by the fully translational 3-RPC-Y, 3-RCC, and the novel 3-RPC-T parallel mechanism. The analysis obtains the branch motion-screws for the abovementioned architectures and determines the sets of platform constraint-screws. The mobility of each manipulator platform is thus obtained by determining the reciprocal screws to the platform constraint-screw sets and the platforms are identified to have three instantaneous independent degrees of freedom which are: (i) a translation along an axis perpendicular to the base; and (ii) two rotations about two skew axes for the first family of manipulators, while the second family has fully translational mobility. The mobility analysis performed in this chapter is validated using motion simulations.

[*] E-mail address: ernesto.rodriguez@itesm.mx

Introduction

Most of the early designs of parallel manipulators had platforms with full mobility, i.e. six degrees-of-freedom and the kinematic analysis of these manipulators is available in the literature [1-4]. The current trend, however, is for the platform of a parallel manipulator to have less than 6-DOF (i.e., limited or lower mobility). These limited mobility parallel manipulators are designed primarily for specific tasks as opposed to general purpose tasks. One important feature of these limited mobility manipulators is the reduced manufacturing cost when compared to a 6-DOF parallel mechanism (supposing that the development and design costs are equal for both types of mechanisms), since fewer legs, joints, and actuators are required. The major disadvantages include the loss of flexibility for a variety of applications, and the reduction of workspace and stiffness. Limited mobility parallel manipulators can be divided into three groups: (i) translational manipulators [5-11], which consist of platforms that can translate along three mutually orthogonal axes with a constant Jacobian matrix; (ii) wrist manipulators [12-17], which exhibit rotations about three mutually orthogonal axes; and (iii) complex motion manipulators [16-40], which combine translational and rotational degrees-of-freedom. Two main categories are well accepted for the classification of 5 DOF parallel mechanisms: the (i) 3R2T (three rotations and two translations), and (ii) 3T2R (three translations and two rotations). Few examples of (i) are the contributions by Huang and Li [18-21], which include the 3-*RR*(*RRR*) parallel manipulator, consisting of three kinematic chains based on two parallel active *R* joints and a *S* joint per leg [21]. Huang and Li also studied a micro-motional 3T2R parallel mechanism with 5-*UPU* architecture [21].

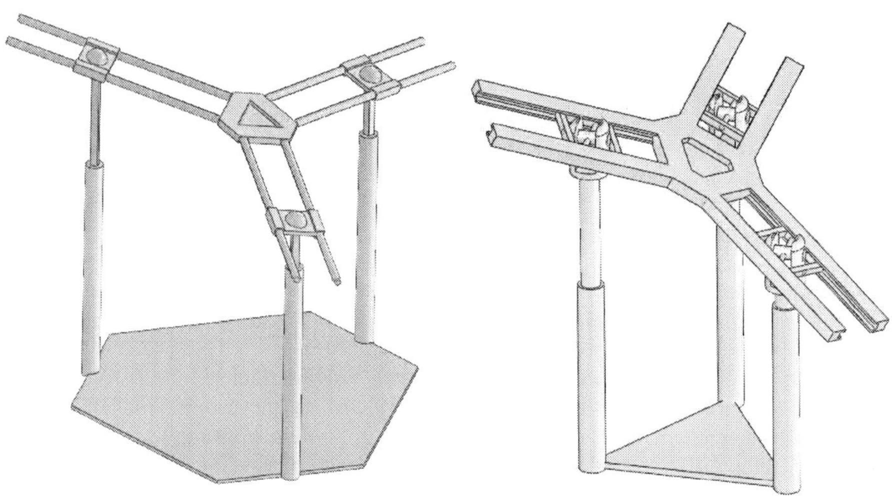

Figure 1. The (a) 3-*PSP*, and (b) 3-*CUP* parallel manipulators.

4 DOF parallel mechanisms can be also divided into three categories: (i) 3R1T (three rotations and one translation), (ii) 3T1R (three translations and one rotation), and (iii) 2R2T (two rotations and two translations). Few examples of 3R1T architectures are the three legged 5*R*-chain parallel manipulator proposed by Zlatanov and Gosselin [22], the designs by Huang

and Li [20-21], or the three legged *CPS-UPS-PS* parallel mechanism studied by Gallardo et al [23]. Kong and Gosselin proposed a general procedure for the type synthesis of 3T1R manipulators using screw theory [24], this study was followed by Richard et al [25] resulting in the 1-*CRR*+3-*CRRR Quadrupteron* parallel mechanism. Further contributions are the designs studied by Angeles [26], Huang and Li [27] and Yang et al [28]. Finally, an example of a 2R2T parallel mechanism is given by Li et al [29].

The following categories apply for the 3 DOF parallel mechanisms: (i) 2R1T (two rotations and one translation), and (ii) 2T1R (two translations and one rotation). The 3-*RPS* parallel manipulator proposed in 1983 by Hunt [3] is one of the architectures with 2R1T motion. Further examples are the 3-*PSP* [30] and the 3-*PPS* [31] architectures (see Figs. 1a and 2a, respectively), which were studied by Tischler et al [30-31], and belong to a family of parallel manipulators inspired on the Type-I Tripode joint [41] when the *P* joints on the socket are replaced by actuators. Liu and Wang proposed a 3-*PCU* manipulator (see Fig. 2b) with 2R1T motion [33]. An example of parallel mechanisms with 2T1R motion is the *HALF* robot designed by Liu et al [34], which has 2-*PRU* 1-*PR(Pa)R* kinematic chains and was further developed into a family of parallel mechanisms, such as the *HANA* and the *HALF*-II [35].

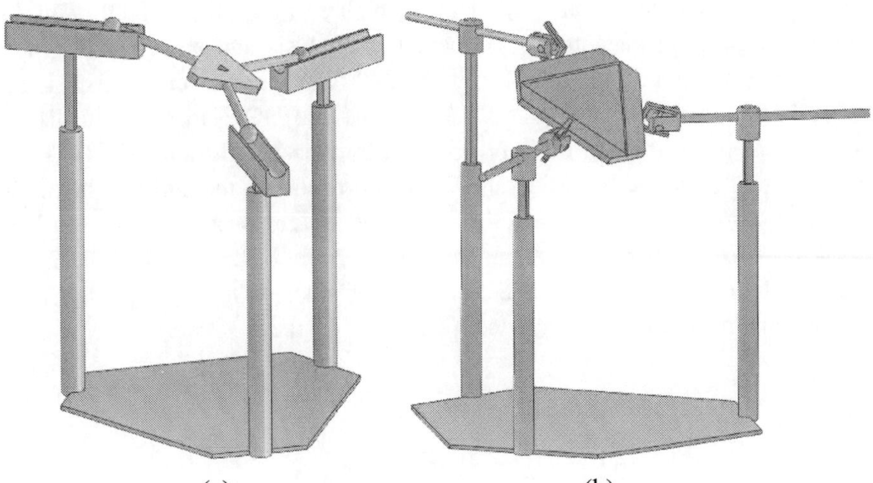

(a) (b)

Figure 2. The (a) 3-*PPS*, and (b) 3-*PCU* parallel manipulators.

This paper presents the mobility analysis of two families of limited mobility parallel manipulators (i) the four 2R1T parallel manipulators shown in Figs 1-2, and (ii) the three fully translational parallel mechanisms shown in Figs. 3-4. The family of fully translational parallel mechanisms consists of the 3-*RPC*-Y manipulator developed by Callegari and Tarantini [42], the 3-*RCC*-Y (or 3-*RPRC*-Y) mechanism proposed by Callegari and Marzetti [7], and a new 3-*RPC*-T parallel mechanism (see Fig. 4). The family of 2R1T parallel manipulators comprises the 3-*PSP*, the 3-*PPS*, the 3-*PCU*, and a new 3-*CUP* parallel manipulator (also referred to as the 3-*PRUP* parallel manipulator). A schematic drawing of this manipulator is shown in Fig. 1b. The mobility analysis for all architectures is performed using the screw-based method proposed by Dai et al [43].

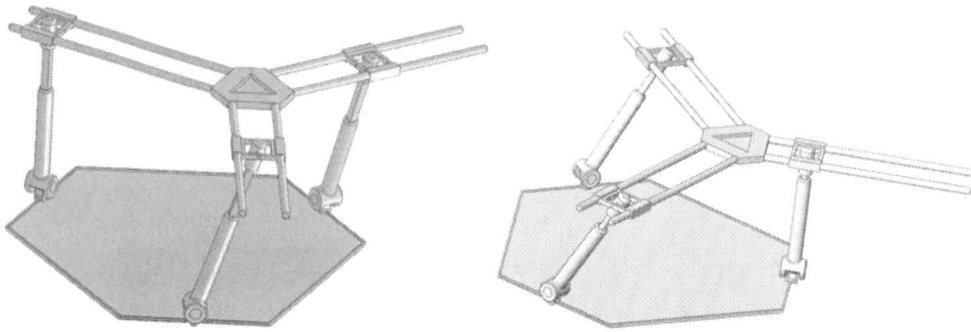

(a) (b)

Figure 3. The (a) 3-*RPC*-Y, and (b) 3-*RCC*-Y parallel manipulators.

The mobility of the 3-*RPC*-Y parallel mechanism is approached at first instance by Callegari and Tarantini [42] with the application of the general Grübler-Kutzbach mobility criterion, which shows that the platform is overconstrained. Since the topology of the parallel mechanism complies with the conditions for full Cartesian motion proposed by Hervé and Sparacino [44], the platform is identified to have three translational DOF. This is further proved by performing an instantaneous kinematics analysis and showing that the platform cannot experience rotation. The mobility of the 3-*RCC*-Y parallel mechanism is studied by Callegari and Marzetti [7], resulting in three translational DOF since the mechanism meets the conditions for full Cartesian motion proposed by Hervé and Sparacino [44]. This paper follows the method proposed by Dai et al [43] and corroborates the mobility identified for the 3-*RPC*-Y, and 3-*RCC*-Y mechanisms in [42], and [7] respectively, and determines the mobility of the 3-*RPC*-T parallel mechanism. The mobility of the 3-*PSP*, 3-*PPS*, and 3-*PCU* parallel manipulators has been the subject of previous studies [30-31, 33, 38]. The results are further validated with simulations using commercial kinematics software.

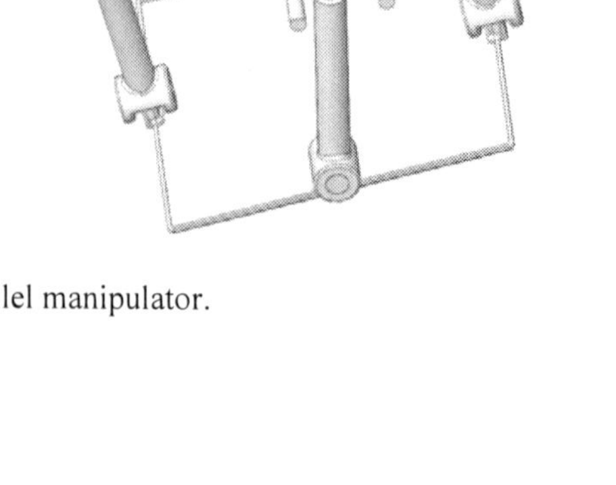

Figure 4. The 3-*RPC*-T parallel manipulator.

Mobility Analysis

The mobility is one of the most fundamental studies for mechanism kinematics and synthesis. The term degree-of-freedom is known as the number of independent coordinates needed to define the configuration of a kinematic chain or mechanism [45]. Mobility is thus, the determination of the number of degrees-of-freedom in the kinematic and dynamic models. In this book, the task of mobility analysis includes the geometrical interpretation of the degrees-of-freedom presented by the platforms of the parallel mechanisms. The first contribution to mobility analysis was made by Chebychev in 1869 with a formula for calculating the independent variables in a mechanism [46]. A comprehensive historical review of the calculation of the mobility of mechanisms is made by Gogu [47], where various methods used in the literature for the past 150 years are presented. The most accepted formula for calculating the number of degrees-of-freedom of a mechanism is the Grübler-Kutzbach mobility criterion [48], which is written as follows

$$F = \lambda(l - j - 1) + \sum_{i=1}^{j} f_i \tag{1}$$

where F is the number of DOF of the mechanism; λ is the DOF of the task space; l is the total number of links; j is the total number of joints; and f_i is the DOF permitted by joint i. It is well known that the Grübler-Kutzbach mobility criterion does not provide the correct result for the mobility of certain mechanisms (referred to as overconstrained mechanisms). The reason is that the Grübler-Kutzbach mobility criterion does not consider the geometry of the mechanism [15]. An additional limitation is that the criterion does not indicate if the identified DOF is a translational or a rotational DOF. The mobility of overconstrained mechanisms has been studied by means of alternative methods, for example, Angeles and Gosselin [49] used the dimension of the nullspace of the Jacobian matrix for finding the mobility of overconstrained mechanisms. Authors such as Angeles [26], Fanghella and Galletti [50,51], and Hervé [52] used group theory for analyzing the mobility properties of single-loop kinematic chains, this work is extended by Rico and Ravani [53], and Rico et al [54] translated the mobility criterion presented in [53] from finite kinematics based on group theory into infinitesimal kinematics based on Lie algebra. Screw theory is another approach that has been used by several researchers for investigating the mobility of mechanisms, in particular parallel mechanisms. Huang and Li [19] used screw theory for determining the mobility and kinematic properties of lower mobility parallel mechanisms by the constraint system of the mechanism. Zhao et al [55] proposed in 2004 the concept of configuration degrees-of-freedom (CDOF) for the mobility of spatial mechanisms. In 2006 Zhao et al [56] proposed a programmable algorithm that investigates the CDOF of a spatial parallel mechanism with reciprocal-screw theory. Dai et al [43] made a remarkable contribution by revealing the screw systems of parallel mechanisms and their inter-relationships, relating the screw systems to motion and constraints and by identifying the constraints as (i) platform constraints, (ii) mechanism constraints, and (iii) redundant constraints. These concepts result in a new approach for performing the mobility analysis based on the decompositions of motion- and constraint-screw systems.

Screw Theory

A unit screw $\boldsymbol{\$}$, see Fig. 5, is defined by the following operation

$$\boldsymbol{\$} = \begin{bmatrix} \boldsymbol{s} \\ \boldsymbol{s} \times \boldsymbol{r} + h\boldsymbol{s} \end{bmatrix} \tag{2}$$

where $\boldsymbol{s}$ is a unit vector pointing along the direction of the screw axis, $\boldsymbol{r}$ is the position vector to any point on the screw axis to the origin, and h is the pitch of the screw.

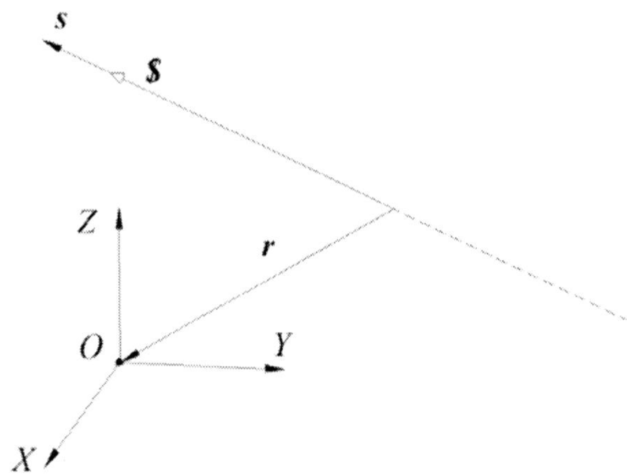

Figure 5. A screw in a tridimensional space.

Considering that the screw is a six-dimensional vector, the general screw notation in Plücker coordinates (also known as Ray coordinates) is

$$\boldsymbol{\$} = \begin{bmatrix} \boldsymbol{s} \\ \boldsymbol{s}_o \end{bmatrix} = \begin{bmatrix} \boldsymbol{s} \\ \boldsymbol{s} \times \boldsymbol{r} + h\boldsymbol{s} \end{bmatrix} = \begin{bmatrix} L, & M, & N; & P, & Q, & R \end{bmatrix}^{\mathrm{T}} \tag{3}$$

where $\boldsymbol{s}$ is the so called primary part and has L, M, and N components, while $\boldsymbol{s}_o$ that is also called secondary part has P, Q, and R components. Hence, a screw multiplied by a magnitude can be used to express a twist of freedom in a six-dimensional space. The primary part of a twist expresses the angular velocity, while the secondary part is the linear velocity. Furthermore, a screw can also be used to represent a system of forces and couples as a wrench when multiplied by a given magnitude. For a wrench, the primary part of the screw expresses a force, while the secondary part is a couple.

For a revolute joint $h = 0$, resulting in the following unit screw

$$\boldsymbol{\$}_0 = \begin{bmatrix} \boldsymbol{s} \\ \boldsymbol{s} \times \boldsymbol{r} \end{bmatrix} \tag{4}$$

The pitch of a prismatic joint is infinite, resulting in

$$\boldsymbol{\$}_{\infty} = \begin{bmatrix} \boldsymbol{0} \\ \boldsymbol{s} \end{bmatrix} \tag{5}$$

where $\boldsymbol{\$}_0$ and $\boldsymbol{\$}_{\infty}$ denote zero and infinite pitch screws, respectively.

When a wrench acts on a rigid body in such a manner that it produces no virtual work while the body is undergoing an infinitesimal twist, the two screws that carry the wrench and the twist are said to be reciprocal. In other words, two screws (say $\boldsymbol{\$}_1$ and $\boldsymbol{\$}_2$) are reciprocal if their orthogonal product $\boldsymbol{\$}_1 \circ \boldsymbol{\$}_2 = 0$. In coordinate form this is equivalent to $\boldsymbol{\$}_1^{\mathrm{T}}\ \boldsymbol{\Delta}\ \boldsymbol{\$}_2 = 0$ where

$$\boldsymbol{\Delta} = \begin{bmatrix} \mathbf{0} & \mathbf{I} \\ \mathbf{I} & \mathbf{0} \end{bmatrix} \tag{6}$$

is the interchange operator for the orthogonal product [57]. This operator is a six-by-six matrix in which the submatrix **I** is the three-by-three identity matrix, and the submatrix **0** is the three-by-three zero matrix. The above operation leads to the following equation:

$$\boldsymbol{v} \cdot \boldsymbol{f} + \boldsymbol{\omega} \cdot \boldsymbol{m} = 0 \tag{7}$$

where $\boldsymbol{\omega}$ and $\boldsymbol{v}$ are the primary and secondary parts of the twist, and $\boldsymbol{f}$ and $\boldsymbol{m}$ are the primary and secondary parts of the wrench. The reciprocity conditions between $\boldsymbol{\$}_1$ and $\boldsymbol{\$}_2$ are listed in Table 1. Note that σ and r_{12} according to Fig. 6 represent the angle between the axes of $\boldsymbol{\$}_1$ and $\boldsymbol{\$}_2$, and the distance along the perpendicular line from $\boldsymbol{\$}_1$ and $\boldsymbol{\$}_2$, respectively.

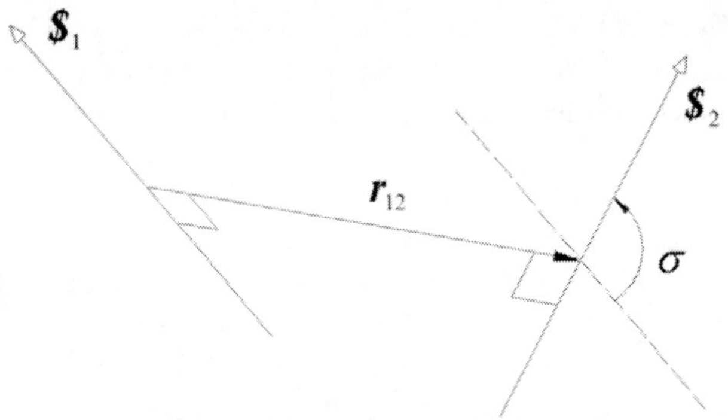

Figure 6. Reciprocal screws $\$_1$ and $\$_2$.

Table 1. Reciprocity conditions between $\$_1$ and $\$_2$

(i)	No condition	when $h_1 = h_2 = \infty$
(ii)	$\cos\sigma = 0$	when h_1 or $h_2 = \infty$
(iii)	$(h_1 + h_2) \cos\sigma - r_{12} \sin\sigma = 0$	when h_1 nor $h_2 = \infty$

Note from the reciprocity conditions in Table 1 that (i) two infinite-pitch screws are always reciprocal to each other, (ii) A $\$_0$ is reciprocal to a $\$_\infty$ when their axes are perpendicular, and (iii) two zero-pitch screws are reciprocal when their axes are coplanar.

A screw system of order n, also called an n-system (where $0 \le n \le 6$), comprises all the screws that are linearly dependent on n given linearly independent screws. It is well known that given an n-system, there is a unique reciprocal screw system of order 6-n, which comprises all the screws that are reciprocal to the correspondent n-system.

As revealed by Dai et al [43] and assuming that the screws of each leg are linearly independent, there are two systems of screws for the ith leg: (i) the motion-screw system, and (ii) the constraint-screw system; which span the base-to-platform motion and constraint, respectively. Further, each system is reciprocal to the other for each ith leg. The intersection and unions of these systems lead to four basic screw systems for the entire manipulator. Two of these systems concern the platform: (i) the platform motion-screw system, and (ii) the platform constraint-screw system; which are the intersection of all leg motion-screw systems and the union of all legs constraint-screw systems, respectively. The two remaining screw systems describe the motion and constraint of the entire manipulator: (iii) the manipulator motion-screw system; and (iv) the manipulator constraint-screw system, which are the union of all leg motion-screw systems and the intersection of all leg constraint-screw systems, respectively.

Mobility Analysis of the 3-*R<u>P</u>C*-Y and 3-*RPC*-T Parallel Mechanisms

This parallel mechanism connects the platform with the base by means of three *RPC* kinematic chains, see Fig. 7. The subscript ij denotes joint j (= 1, 2, 3, 4) on leg i. Hence, on leg i: the R joint connected to the base is denoted as j_{i1}; the P joint is denoted as j_{i2}, and the C joint is composed by the R and P joints j_{i3}, and j_{i4}, respectively. Consider that the R joint axis $\boldsymbol{s}_{i1}$ is set to be collinear with vector $\boldsymbol{a}_i$ (representing the position of the R joint), the P joint axis $\boldsymbol{s}_{i2}$ is collinear with vector $\boldsymbol{d}_i$ (denoting the position of the leg i), and the C joint axes $\boldsymbol{s}_{i3}$ and $\boldsymbol{s}_{i4}$ are set to be collinear with vector $\boldsymbol{b}_i$ (which represents the stroke of the C joint). If the platform is initially assembled parallel to the base, then the joint axes $\boldsymbol{s}_{i1}$, $\boldsymbol{s}_{i3}$, and $\boldsymbol{s}_{i4}$ are invariantly parallel to each other and perpendicular to $\boldsymbol{s}_{i2}$.

The 3-*R<u>P</u>C*-Y parallel mechanism is then obtained when three identical legs are equally distributed on the base ($\psi_1 = 0°$, $\psi_2 = 120°$, and $\psi_3 = 240°$), resulting in a star configuration (recall star is denoted as Y) of the R joints on the base. Likewise, as shown in Fig. 7, the C joints are equally distributed on the platform (where $\chi_1 = 0°$, $\chi_2 = 120°$, and $\chi_3 = 240°$). The P joints j_{i2} are selected as the active joints for this parallel mechanism. Note that each leg is comprised by a system of four motion-screws, namely

$$\$_{i1} = \begin{bmatrix} \boldsymbol{s}_{i1} \\ \boldsymbol{0} \end{bmatrix}, \; \$_{i2} = \begin{bmatrix} \boldsymbol{0} \\ \boldsymbol{s}_{i2} \end{bmatrix}, \; \$_{i3} = \begin{bmatrix} \boldsymbol{s}_{i3} \\ (\boldsymbol{a}_i + \boldsymbol{d}_i) \times \boldsymbol{s}_{i3} \end{bmatrix}, \text{ and } \$_{i4} = \begin{bmatrix} \boldsymbol{0} \\ \boldsymbol{s}_{i4} \end{bmatrix} \tag{8}$$

where

$$\boldsymbol{d}_i = \left[-d_i \mathrm{c}\theta_{i1}\mathrm{s}\psi_i,\ d_i\mathrm{c}\theta_{i1}\mathrm{c}\psi_i,\ d_i\mathrm{s}\theta_{i1}\right]^{\mathrm{T}},\ \boldsymbol{s}_{i1} = \boldsymbol{s}_{i3} = \boldsymbol{s}_{i4} = \left[\mathrm{c}\psi_i,\ \mathrm{s}\psi_i,\ 0\right]^{\mathrm{T}},$$

$$\text{and } \boldsymbol{s}_{i2} = \left[-\mathrm{c}\theta_{i1}\mathrm{s}\psi_i,\ \mathrm{c}\theta_{i1}\mathrm{c}\psi_i,\ \mathrm{s}\theta_{i1}\right]^{\mathrm{T}} \tag{9}$$

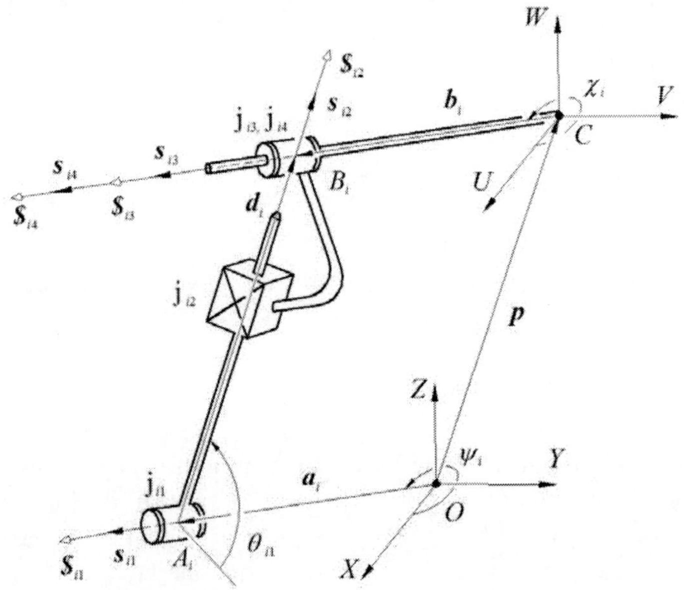

Figure 7. Leg *i* of the 3-*RPC* parallel mechanism.

Performing the vector operations in Eqs. (8)-(9), results in the branch motion-screw systems for leg *i*

$$\{\$_i\} = \left\{\begin{array}{l} \$_{i1} = \left[\mathrm{c}\psi_i,\ \mathrm{s}\psi_i,\ 0;\ 0,\ 0,\ 0\right]^{\mathrm{T}}, \\ \$_{i2} = \left[0,\ 0,\ 0;\ -\mathrm{c}\theta_{i1}\mathrm{s}\psi_i,\ \mathrm{c}\theta_{i1}\mathrm{c}\psi_i,\ \mathrm{s}\theta_{i1}\right]^{\mathrm{T}}, \\ \$_{i3} = \left[\mathrm{c}\psi_i,\ \mathrm{s}\psi_i,\ 0;\ -d_i\mathrm{s}\theta_{i1}\mathrm{s}\psi_i,\ d_i\mathrm{s}\theta_{i1}\mathrm{c}\psi_i,\ -d_i\mathrm{c}\theta_{i1}\right]^{\mathrm{T}}, \\ \$_{i4} = \left[0,\ 0,\ 0;\ \mathrm{c}\chi_i,\ \mathrm{s}\chi_i,\ 0\right]^{\mathrm{T}} \end{array}\right\} \tag{10}$$

The branch motion screws $\{\$_1\}$, $\{\$_2\}$, and $\{\$_3\}$ for the 3-$R\underline{P}C$-Y parallel mechanism are determined when $\psi_1 = 0°$, $\psi_2 = 120°$, and $\psi_3 = 240°$ are substituted into the general branch motion-screw system $\{\$_i\}$ given by Eq. (10), i.e.

$$\{\$_1\} = \left\{\begin{array}{l} \$_{11} = \left[1,\ 0,\ 0;\ 0,\ 0,\ 0\right]^{\mathrm{T}}, \\ \$_{12} = \left[0,\ 0,\ 0;\ 0,\ \mathrm{c}\theta_{11},\ \mathrm{s}\theta_{11}\right]^{\mathrm{T}}, \\ \$_{13} = \left[1,\ 0,\ 0;\ 0,\ d_1\mathrm{s}\theta_{11},\ -d_1\mathrm{c}\theta_{11}\right]^{\mathrm{T}}, \\ \$_{14} = \left[0,\ 0,\ 0;\ 1,\ 0,\ 0\right]^{\mathrm{T}} \end{array}\right\} \tag{11}$$

$$\{\$_2\} = \begin{Bmatrix} \$_{21} = \left[-1,\ \sqrt{3},\ 0;\ 0,\ 0,\ 0\right]^{\mathrm{T}}, \\ \$_{22} = \left[0,\ 0,\ 0;\ -\sqrt{3}\mathrm{c}\theta_{21},\ -\mathrm{c}\theta_{21},\ 2\mathrm{s}\theta_{21}\right]^{\mathrm{T}}, \\ \$_{23} = \left[-1,\ \sqrt{3},\ 0;\ -\sqrt{3}d_2\mathrm{s}\theta_{21},\ -d_2\mathrm{s}\theta_{21},\ -2d_2\mathrm{c}\theta_{21}\right]^{\mathrm{T}}, \\ \$_{24} = \left[0,\ 0,\ 0;\ -1,\ \sqrt{3},\ 0\right]^{\mathrm{T}} \end{Bmatrix} \tag{12}$$

$$\{\$_3\} = \begin{Bmatrix} \$_{31} = \left[-1,\ -\sqrt{3},\ 0;\ 0,\ 0,\ 0\right]^{\mathrm{T}}, \\ \$_{32} = \left[0,\ 0,\ 0;\ \sqrt{3}\mathrm{c}\theta_{31},\ -\mathrm{c}\theta_{31},\ 2\mathrm{s}\theta_{31}\right]^{\mathrm{T}}, \\ \$_{33} = \left[-1,\ -\sqrt{3},\ 0;\ \sqrt{3}d_3\mathrm{s}\theta_{31},\ -d_3\mathrm{s}\theta_{31},\ -2d_3\mathrm{c}\theta_{31}\right]^{\mathrm{T}}, \\ \$_{34} = \left[0,\ 0,\ 0;\ -1,\ -\sqrt{3},\ 0\right]^{\mathrm{T}} \end{Bmatrix} \tag{13}$$

The new 3-*RPC*-T parallel mechanism from Fig. 4 is produced, when three identical *RPC* kinematic chains are distributed on the base by setting $\psi_1 = 0°$, $\psi_2 = 90°$, and $\psi_3 = 180°$, resulting in a T configuration of the j_{i1} joints. In a similar manner, the joints j_{i4} are distributed on the platform by setting $\chi_1 = 0°$, $\chi_2 = 90°$, and $\chi_3 = 180°$. Note the joint axis $\boldsymbol{s}_{11}$ is collinear with $\boldsymbol{s}_{31}$ and they are coplanar with the *XY* plane, while the $\boldsymbol{s}_{13}$, $\boldsymbol{s}_{33}$, $\boldsymbol{s}_{14}$, $\boldsymbol{s}_{34}$ joint axes are collinear and coplanar to the *UV* plane. Due to the geometry of the abovementioned joint axes, the legs 1 and 3 present a synchronous motion. This paper considers the selection of joints j_{11}, j_{21}, and j_{32} as a nonunique combination of active joints defined for this mechanism.

Note that since the kinematic chain of this mechanism is identical to the *R<u>P</u>C*-Y parallel mechanism, the branch motion-screws correspond to the system expressed by Eq. (8), when $\psi_1 = 0°$, $\psi_2 = 90°$, and $\psi_3 = 180°$ are substituted into Eq. (10). The branch motion-screw systems $\{\$_1\}$, $\{\$_2\}$, and $\{\$_3\}$ for the 3-*RPC*-T parallel mechanism are written as

$$\{\$_1\} = \begin{Bmatrix} \$_{11} = [1,\ 0,\ 0;\ 0,\ 0,\ 0]^{\mathrm{T}}, \\ \$_{12} = [0,\ 0,\ 0;\ 0,\ \mathrm{c}\theta_{11},\ \mathrm{s}\theta_{11}]^{\mathrm{T}}, \\ \$_{13} = [1,\ 0,\ 0;\ 0,\ d_1\mathrm{s}\theta_{11},\ -d_1\mathrm{c}\theta_{11}]^{\mathrm{T}}, \\ \$_{14} = [0,\ 0,\ 0;\ 1,\ 0,\ 0]^{\mathrm{T}} \end{Bmatrix} \tag{14}$$

$$\{\$_2\} = \begin{Bmatrix} \$_{21} = [0,\ 1,\ 0;\ 0,\ 0,\ 0]^{\mathrm{T}}, \\ \$_{22} = [0,\ 0,\ 0;\ -\mathrm{c}\theta_{21},\ 0,\ \mathrm{s}\theta_{21}]^{\mathrm{T}}, \\ \$_{23} = [0,\ 1,\ 0;\ -d_2\mathrm{s}\theta_{21},\ 0,\ -d_2\mathrm{c}\theta_{21}]^{\mathrm{T}}, \\ \$_{24} = [0,\ 0,\ 0;\ 0,\ 1,\ 0]^{\mathrm{T}} \end{Bmatrix} \tag{15}$$

$$\{\$_3\} = \begin{Bmatrix} \$_{31} = [-1,\ 0,\ 0;\ 0,\ 0,\ 0]^{\mathrm{T}}, \\ \$_{32} = [0,\ 0,\ 0;\ 0,\ -\mathrm{c}\theta_{31},\ \mathrm{s}\theta_{31}]^{\mathrm{T}}, \\ \$_{33} = [-1,\ 0,\ 0;\ 0,\ -d_3\mathrm{s}\theta_{31},\ -d_3\mathrm{c}\theta_{31}]^{\mathrm{T}}, \\ \$_{34} = [0,\ 0,\ 0;\ -1,\ 0,\ 0]^{\mathrm{T}} \end{Bmatrix} \tag{16}$$

Since the branch motion-screw system comprised in leg i $\{\$_i\}$ is a four-screw system and their screws are independent, see Eq. (10), the branch constraint-screw system $\{\$_i^r\}$ is formed by the following two independent reciprocal screws $\$_{ij}^r$

$$\{\$_i^r\} = \begin{Bmatrix} \$_{i1}^r = [0,\ 0,\ 0;\ -\mathrm{s}\psi_i,\ \mathrm{c}\psi_i,\ 0]^{\mathrm{T}}, \\ \$_{i2}^r = [0,\ 0,\ 0;\ 0,\ 0,\ 1]^{\mathrm{T}} \end{Bmatrix} \tag{17}$$

Note that $\$_{i1}^r$ represent couples whose axes are coplanar and perpendicular to $\boldsymbol{s}_{i1}$, while $\$_{i2}^r$ are couples acting about the Z-axis. The branch constraint-screw systems $\{\$_1^r\}$, $\{\$_2^r\}$ and $\{\$_3^r\}$ for the 3-*R<u>P</u>C*-Y parallel mechanism can be written as

$$\{\$_1^r\} = \begin{Bmatrix} \$_{11}^r = [0,\ 0,\ 0;\ 0,\ 1,\ 0]^{\mathrm{T}}, \\ \$_{12}^r = [0,\ 0,\ 0;\ 0,\ 0,\ 1]^{\mathrm{T}} \end{Bmatrix} \tag{18}$$

$$\{\$_2^r\} = \begin{Bmatrix} \$_{21}^r = [0,\ 0,\ 0;\ \sqrt{3},\ 1,\ 0]^{\mathrm{T}}, \\ \$_{22}^r = [0,\ 0,\ 0;\ 0,\ 0,\ 1]^{\mathrm{T}} \end{Bmatrix} \tag{19}$$

$$\{\$_3^r\} = \begin{Bmatrix} \$_{31}^r = [0,\ 0,\ 0;\ -\sqrt{3},\ 1,\ 0]^{\mathrm{T}}, \\ \$_{32}^r = [0,\ 0,\ 0;\ 0,\ 0,\ 1]^{\mathrm{T}} \end{Bmatrix} \tag{20}$$

Note that $\{\$_1^r\}$, $\{\$_2^r\}$, and $\{\$_3^r\}$ provide six constraints to the platform, leading to the assumption that the mechanism will have zero DOF. This is consistent with the observation made by Callegari and Tarantini [42], and Callegari and Marzetti [7] concerning the overconstrained condition of the parallel mechanism. In addition, note that $\$_{12}^r$, $\$_{22}^r$, and $\$_{32}^r$ are common constraints. The platform constraint-screw system $\{\$^r\}$ is given by the following nonunique set of independent reciprocal screws $\$_{11}^r$, $\$_{21}^r$, and $\$_{12}^r$

$$\{\$^r\} = \left\{ \begin{array}{l} \$^r_{11} = [0,\ 0,\ 0;\ 0,\ 1,\ 0]^{\mathrm{T}}, \\ \$^r_{21} = \left[0,\ 0,\ 0;\ \sqrt{3},\ 1,\ 0\right]^{\mathrm{T}}, \\ \$^r_{12} = [0,\ 0,\ 0;\ 0,\ 0,\ 1]^{\mathrm{T}} \end{array} \right\} \tag{21}$$

As for the 3-*RPC*-T parallel mechanism, the branch constraint-screw system $\{\$^r_i\}$ is formed by the two independent reciprocal screws $\$^r_{ij}$ expressed in Eq. (17) and evaluating $\psi_1 = 0°$, $\psi_2 = 90°$, and $\psi_3 = 180°$, result in the following screw systems

$$\left\{\$^r_1\right\} = \left\{ \begin{array}{l} \$^r_{11} = [0,\ 0,\ 0;\ 0,\ 1,\ 0]^{\mathrm{T}}, \\ \$^r_{12} = [0,\ 0,\ 0;\ 0,\ 0,\ 1]^{\mathrm{T}} \end{array} \right\} \tag{22}$$

$$\left\{\$^r_2\right\} = \left\{ \begin{array}{l} \$^r_{21} = [0,\ 0,\ 0;\ 1,\ 0,\ 0]^{\mathrm{T}}, \\ \$^r_{22} = [0,\ 0,\ 0;\ 0,\ 0,\ 1]^{\mathrm{T}} \end{array} \right\} \tag{23}$$

$$\left\{\$^r_3\right\} = \left\{ \begin{array}{l} \$^r_{31} = [0,\ 0,\ 0;\ 0,\ 1,\ 0]^{\mathrm{T}}, \\ \$^r_{32} = [0,\ 0,\ 0;\ 0,\ 0,\ 1]^{\mathrm{T}} \end{array} \right\} \tag{24}$$

The reciprocal screws expressed by Eqs. (22)-(24) represent physical constraints exerted by each kinematic chain on the platform. Note that $\$^r_{12}$, $\$^r_{22}$, and $\$^r_{32}$ are common constraints and any two of these can be considered redundant constraints, while screws $\$^r_{11}$, and $\$^r_{31}$ represent an additional common constraint. Therefore the platform constraint-screw system $\{\$^r\}$ is given by the following nonunique set of independent reciprocal screws $\$^r_{11}$, $\$^r_{21}$, and $\$^r_{12}$

$$\{\$^r\} = \left\{ \begin{array}{l} \$^r_{11} = [0,\ 0,\ 0;\ 0,\ 1,\ 0]^{\mathrm{T}}, \\ \$^r_{21} = [0,\ 0,\ 0;\ 1,\ 0,\ 0]^{\mathrm{T}}, \\ \$^r_{12} = [0,\ 0,\ 0;\ 0,\ 0,\ 1]^{\mathrm{T}} \end{array} \right\} \tag{25}$$

The reciprocal screws to $\{\$^r\}$ from Eqs. (21) and (25) give the platform motion-screw system $\{\$^f\}$ of the 3-*R<u>P</u>C*-Y, and the 3-*RPC*-T parallel mechanisms

$$\{\$^f\} = \left\{ \begin{array}{l} \$^f_1 = [0,\ 0,\ 0;\ 1,\ 0,\ 0]^{\mathrm{T}}, \\ \$^f_2 = [0,\ 0,\ 0;\ 0,\ 1,\ 0]^{\mathrm{T}}, \\ \$^f_3 = [0,\ 0,\ 0;\ 0,\ 0,\ 1]^{\mathrm{T}} \end{array} \right\} \tag{26}$$

This shows that the 3-*RPC* parallel mechanism has three DOF, which are translations along the *X*-, *Y*-, and *Z*-axes. This result is consistent with the mobility predicted for the 3-*RPC*-Y parallel manipulator by Callegari and Marzetti in [7].

Mobility Analysis of the 3-*RCC*-Y Parallel Mechanism

When an additional rotational DOF (denoted as joint j_{i5}) is given to the *P* joints j_{i2} of the *RPC*-Y parallel mechanism shown in Fig. 7, results in a *RPRRP* kinematic chain. For this mechanism, consider now that the joint axis $\boldsymbol{s}_{i1}$ is set to be collinear with vector $\boldsymbol{a}_i$, the $\boldsymbol{s}_{i3}$ and $\boldsymbol{s}_{i4}$ joint axes are collinear with vector $\boldsymbol{b}_i$, and the $\boldsymbol{s}_{i2}$ and $\boldsymbol{s}_{i5}$ joint axes are collinear with vector $\boldsymbol{d}_i$, see Fig. 8. The joints j_{i2} are selected as active joints. For this architecture, three identical legs are equally distributed on the base ($\psi_1 = 0°$, $\psi_2 = 120°$, and $\psi_3 = 240°$), while the *C* joints (i.e. joints j_{i3} and j_{i4}) are equally distributed on the platform ($\chi_1 = 0°$, $\chi_2 = 120°$, and $\chi_3 = 240°$). The resulting parallel mechanism is a 3-*RPRC*-Y, also denoted as a 3-*RCC*-Y architecture.

As shown in Fig. 8, each leg is comprised by a system of five screws. For convenience, the joint screws are expressed in the global frame G (*X*, *Y*, *Z*) and calculated relative to *O*. In addition, consider that the platform is initially assembled to be parallel to the base. Note that for this architecture, the joint axes $\boldsymbol{s}_{i1}$, $\boldsymbol{s}_{i2}$, $\boldsymbol{s}_{i3}$, and $\boldsymbol{s}_{i4}$, are identical to their corresponding joint axes of the 3-*RPC*-Y parallel mechanism (see Figs. 7 and 8). Therefore, the joint screws $\boldsymbol{\$}_{i1}$, $\boldsymbol{\$}_{i2}$, $\boldsymbol{\$}_{i3}$, and $\boldsymbol{\$}_{i4}$ for the 3-*RCC*-Y parallel mechanism are given by Eq. (8), while the joint screw $\boldsymbol{\$}_{i5}$ is written as

$$\boldsymbol{\$}_{i5} = \begin{bmatrix} \boldsymbol{s}_{i5} \\ \boldsymbol{a}_i \times \boldsymbol{s}_{i5} \end{bmatrix} \tag{27}$$

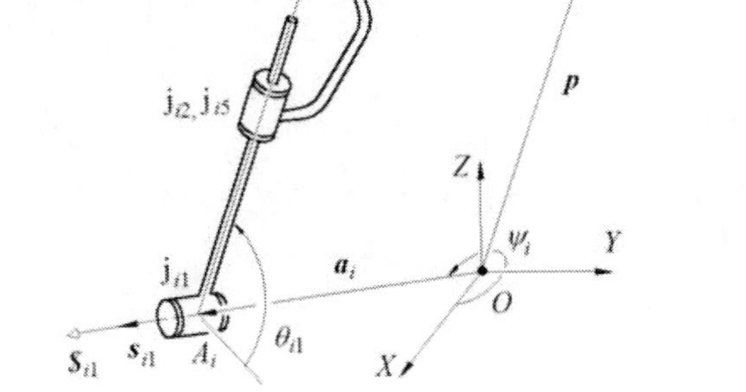

Figure 8. Leg *i* of the 3-*RCC*-Y (3-*RPRC*-Y) parallel mechanism.

Note that $\boldsymbol{s}_{i5} = \boldsymbol{s}_{i2}$. The joint screw $\boldsymbol{\$}_{i5}$ is obtained when the vector operations from Eqs. (9) and (27) are performed

$$\boldsymbol{\$}_{i5} = \left[-c\theta_{i1}s\psi_i,\ c\theta_{i1}c\psi_i,\ s\theta_{i1};\ a_i s\theta_{i1}s\psi_i,\ -a_i s\theta_{i1}c\psi_i,\ a_i c\theta_{i1}\right]^{\mathrm{T}} \tag{28}$$

The branch motion-screw systems $\{\boldsymbol{\$}_1\}$, $\{\boldsymbol{\$}_2\}$, and $\{\boldsymbol{\$}_3\}$ of the *RCC*-Y parallel mechanism result in

$$\{\boldsymbol{\$}_1\} = \left\{\begin{array}{l} \boldsymbol{\$}_{11} = [1,\ 0,\ 0;\ 0,\ 0,\ 0]^{\mathrm{T}}, \\ \boldsymbol{\$}_{12} = [0,\ 0,\ 0;\ 0,\ c\theta_{11},\ s\theta_{11}]^{\mathrm{T}}, \\ \boldsymbol{\$}_{13} = [1,\ 0,\ 0;\ 0,\ d_1 s\theta_{11},\ -d_1 c\theta_{11}]^{\mathrm{T}}, \\ \boldsymbol{\$}_{14} = [0,\ 0,\ 0;\ 1,\ 0,\ 0]^{\mathrm{T}}, \\ \boldsymbol{\$}_{15} = [0,\ c\theta_{11},\ s\theta_{11};\ 0,\ -as\theta_{11},\ ac\theta_{11}]^{\mathrm{T}} \end{array}\right\} \tag{29}$$

$$\{\boldsymbol{\$}_2\} = \left\{\begin{array}{l} \boldsymbol{\$}_{21} = \left[-1,\ \sqrt{3},\ 0;\ 0,\ 0,\ 0\right]^{\mathrm{T}}, \\ \boldsymbol{\$}_{22} = \left[0,\ 0,\ 0;\ -\sqrt{3}c\theta_{21},\ -c\theta_{21},\ 2s\theta_{21}\right]^{\mathrm{T}}, \\ \boldsymbol{\$}_{23} = \left[-1,\ \sqrt{3},\ 0;\ -\sqrt{3}d_2 s\theta_{21},\ -d_2 s\theta_{21},\ -2d_2 c\theta_{21}\right]^{\mathrm{T}}, \\ \boldsymbol{\$}_{24} = \left[0,\ 0,\ 0;\ -1,\ \sqrt{3},\ 0\right]^{\mathrm{T}}, \\ \boldsymbol{\$}_{25} = \left[-\sqrt{3}c\theta_{21},\ -c\theta_{21},\ 2s\theta_{21};\ \sqrt{3}as\theta_{21},\ as\theta_{21},\ 2ac\theta_{21}\right]^{\mathrm{T}} \end{array}\right\} \tag{30}$$

$$\{\boldsymbol{\$}_3\} = \left\{\begin{array}{l} \boldsymbol{\$}_{31} = \left[-1,\ -\sqrt{3},\ 0;\ 0,\ 0,\ 0\right]^{\mathrm{T}}, \\ \boldsymbol{\$}_{32} = \left[0,\ 0,\ 0;\ \sqrt{3}c\theta_{31},\ -c\theta_{31},\ 2s\theta_{31}\right]^{\mathrm{T}}, \\ \boldsymbol{\$}_{33} = \left[-1,\ -\sqrt{3},\ 0;\ \sqrt{3}d_3 s\theta_{31},\ -d_3 s\theta_{31},\ -2d_3 c\theta_{31}\right]^{\mathrm{T}}, \\ \boldsymbol{\$}_{34} = \left[0,\ 0,\ 0;\ -1,\ -\sqrt{3},\ 0\right]^{\mathrm{T}}, \\ \boldsymbol{\$}_{35} = \left[\sqrt{3}c\theta_{31},\ -c\theta_{31},\ 2s\theta_{31};\ -\sqrt{3}as\theta_{31},\ as\theta_{31},\ 2ac\theta_{31}\right]^{\mathrm{T}} \end{array}\right\} \tag{31}$$

Since each branch motion-screw system is composed by five independent screws, see Eqs. (29)-(31), the branch constraint-screw is given by the reciprocal screw $\boldsymbol{\$}_{i1}^r$

$$\boldsymbol{\$}_{i1}^r = \left[0,\ 0,\ 0;\ s\psi_i s\theta_{i1},\ -c\psi_i s\theta_{i1},\ c\theta_{i1}\right]^{\mathrm{T}} \tag{32}$$

The constraints exerted by the reciprocal screws $\boldsymbol{\$}_{i1}^r$ represent couples whose axes are orthogonal to vectors $\boldsymbol{d}_i$, and $\boldsymbol{a}_i$. The platform constraint-screw system $\{\boldsymbol{\$}^r\}$ consist of the following set of three independent reciprocal screws $\boldsymbol{\$}_{11}^r$, $\boldsymbol{\$}_{21}^r$, and $\boldsymbol{\$}_{31}^r$

$$\{\$^r\} = \begin{Bmatrix} \$^r_{11} = [0,\ 0,\ 0;\ 0,\ -\mathrm{s}\theta_{11},\ \mathrm{c}\theta_{11}]^{\mathrm{T}}, \\ \$^r_{21} = \left[0,\ 0,\ 0;\ \sqrt{3}\mathrm{s}\theta_{21},\ \mathrm{s}\theta_{21},\ 2\mathrm{c}\theta_{21}\right]^{\mathrm{T}}, \\ \$^r_{31} = \left[0,\ 0,\ 0;\ -\sqrt{3}\mathrm{s}\theta_{31},\ \mathrm{s}\theta_{31},\ 2\mathrm{c}\theta_{31}\right]^{\mathrm{T}} \end{Bmatrix} \tag{33}$$

Since $\{\$^r\}$ from Eq. (33) is a three screw system and the constraint-screws $\$^r_{i1}$ are independent, it is expected that the parallel mechanism has 3-DOF, therefore the platform is not overconstrained. This remark is consistent with the observations of Callegari and Marzetti [7] on the 3-*RCC*-Y parallel mechanism.

The platform motion-screw system $\{\$^f\}$ is obtained by determining the reciprocal screws to $\{\$^r\}$ from Eq. (33); resulting in a screw system identical to Eq. (26). This indicates that the platform of the 3-*RCC*-Y (3-*R<u>P</u>RC*-Y) parallel mechanism has three DOF, which are translations along the *X*-, *Y*-, and *Z*-axes. The resulting mobility is consistent with the observations of Callegari and Marzetti in [7].

Mobility Analysis of the 3-*<u>P</u>SP* and 3-*CUP* Parallel Manipulators

Schematic drawings of the kinematic chains of the 3-*<u>P</u>SP* and 3-*CUP* parallel manipulators are shown in Figs. 9-10, respectively. Note that the axes of joints j_{i5} intersect at point *C*, their strokes are denoted by vectors $\boldsymbol{b}_i$ and the magnitude of the stroke is denoted as b_i. The prismatic joints located on the platform are distributed by an angular position given by the angle χ_i, which is a rotation about the *W*-axis. For this paper consider that j_{i5} are equally distributed on the platform, i.e., $\chi_1 = 0°$, $\chi_2 = 120°$, $\chi_3 = 240°$.

Note that the branch motion-screw set $\{\$_i\}$ is a system comprised of five independent joint motion-screws $\$_{ij}$. For convenience, the following joint screws will be expressed in the global frame G and calculated relative to *O*

$$\$_{i1} = \begin{bmatrix} \boldsymbol{0} \\ \boldsymbol{s}_{i1} \end{bmatrix}, \quad \$_{i2} = \begin{bmatrix} \boldsymbol{s}_{i2} \\ (\boldsymbol{a}_i + \boldsymbol{d}_i) \times \boldsymbol{s}_{i2} \end{bmatrix}, \quad \$_{i3} = \begin{bmatrix} \boldsymbol{s}_{i3} \\ (\boldsymbol{a}_i + \boldsymbol{d}_i) \times \boldsymbol{s}_{i3} \end{bmatrix},$$
$$\$_{i4} = \begin{bmatrix} \boldsymbol{s}_{i4} \\ (\boldsymbol{a}_i + \boldsymbol{d}_i) \times \boldsymbol{s}_{i4} \end{bmatrix}, \text{ and } \$_{i5} = \begin{bmatrix} \boldsymbol{0} \\ \boldsymbol{s}_{i5} \end{bmatrix} \tag{34}$$

where

$$\boldsymbol{a}_i = [a\mathrm{c}\psi_i,\ a\mathrm{s}\psi_i,\ 0]^{\mathrm{T}},\ \boldsymbol{d}_i = [0,\ 0,\ d_i]^{\mathrm{T}},\ \boldsymbol{s}_{i1} = [0,\ 0,\ 1]^{\mathrm{T}},$$
$$\text{and } \boldsymbol{s}_{i5} = \left[r_x,\ r_y,\ r_z\right]^{\mathrm{T}} \tag{35}$$

where r_x, r_y, and r_z represent the components on the *X*, *Y*, and *Z*-axes of the unit vector $\boldsymbol{b}_i$. The transformation which expresses this unit vector in the global frame G can be written as

$$\begin{bmatrix} r_x \\ r_y \\ r_z \end{bmatrix} = \begin{bmatrix} u_x & v_x & w_x \\ u_y & v_y & w_y \\ u_z & v_z & w_z \end{bmatrix} \cdot \left[c\chi_i, \ s\chi_i, \ 0 \right]^{T} = \begin{bmatrix} u_x c\chi_i + v_x s\chi_i \\ u_y c\chi_i + v_y s\chi_i \\ u_z c\chi_i + v_z s\chi_i \end{bmatrix} \quad (36)$$

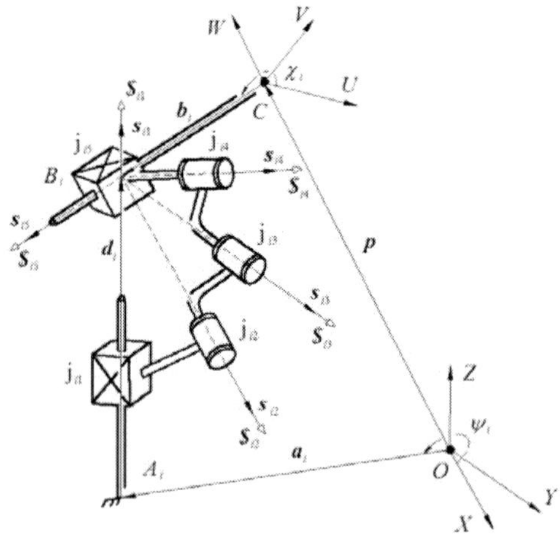

Figure 9. Leg i of a 3-*<u>P</u>SP* manipulator.

The elements of the rotation matrix in Eq. (36) are the direction cosines of $\boldsymbol{u}$, $\boldsymbol{v}$ and $\boldsymbol{w}$, which are defined parallel to the *U*, *V* and *W* axes.

Assume that the joint axes $\boldsymbol{s}_{i2}$, $\boldsymbol{s}_{i3}$, and $\boldsymbol{s}_{i4}$ of the 3-*<u>P</u>SP* parallel manipulator are parallel to the *X*, *Y*, and *Z* axes, respectively. In this case, the three *R* joints represent a canonical *S* joint, and the joint axis can be written as

$$\boldsymbol{s}_{i2} = [1, \ 0, \ 0]^{T}, \quad \boldsymbol{s}_{i3} = [0, \ 1, \ 0]^{T}, \quad \text{and} \quad \boldsymbol{s}_{i4} = [0, \ 0, \ 1]^{T} \quad (37)$$

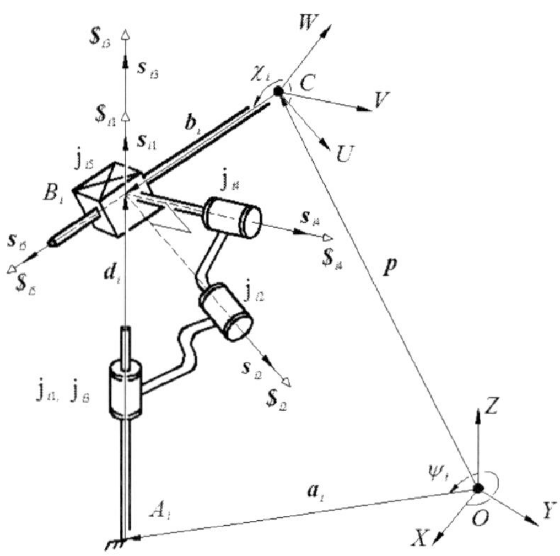

Figure 10. Leg i of the 3-*CUP* (3-*<u>P</u>RUP*) manipulator.

On the other hand, the 3-*CUP* parallel manipulator can be considered as a non-canonical 3-*PSP* parallel mechanism if the joint axis s_{i2} is coaxial to s_{i1}, the joint axis s_{i4} is parallel to the *V*-axis, while the joint axis s_{i3} is parallel to the *XY* plane and is perpendicular to s_{i4}. Note that s_{i2}, s_{i3}, and s_{i4} intersect at B_i and form a non-canonical *S* joint. In this case, the joints j_{i1} and j_{i2} behave as *C* joints, while the joints j_{i2} and j_{i4} can be represented as *U* joints. The parallel manipulator is, therefore, a 3-*PRUP* also denoted as 3-*CUP* architecture. For this mechanism, the vectors s_{i2}, s_{i3}, and s_{i4} can be written as

$$s_{i2} = [0,\ 0,\ 1]^{\mathrm{T}},\quad s_{i3} = [-v_y,\ v_x,\ 0]^{\mathrm{T}},\quad \text{and}\quad s_{i4} = [v_x,\ v_y,\ v_z]^{\mathrm{T}} \tag{38}$$

The branch motion-screw set $\{\$_i\}$ of the 3-*PSP* parallel manipulator is obtained by substituting Eqs. (35) and (37) into the screws of Eqs. (34). After performing the respective cross-vector operations results in

$$\{\$_i\} = \left\{\begin{array}{l} \$_{i1} = [0,\ 0,\ 0;\ 0,\ 0,\ 1]^{\mathrm{T}}, \\ \$_{i2} = [1,\ 0,\ 0;\ 0,\ d_i,\ -as\psi_i]^{\mathrm{T}}, \\ \$_{i3} = [0,\ 1,\ 0;\ -d_i,\ 0,\ ac\psi_i]^{\mathrm{T}}, \\ \$_{i4} = [0,\ 0,\ 1;\ as\psi_i,\ -ac\psi_i,\ 0]^{\mathrm{T}}, \\ \$_{i5} = [0,\ 0,\ 0;\ r_x,\ r_y,\ r_z]^{\mathrm{T}} \end{array}\right\} \tag{39}$$

where the branch motion-screw sets for each leg $\{\$_1\}$, $\{\$_2\}$, and $\{\$_3\}$ of the 3-*PSP* parallel manipulator can be written as

$$\{\$_1\} = \left\{\begin{array}{l} \$_{11} = [0,\ 0,\ 0;\ 0,\ 0,\ 1]^{\mathrm{T}}, \\ \$_{12} = [1,\ 0,\ 0;\ 0,\ d_1,\ 0]^{\mathrm{T}}, \\ \$_{13} = [0,\ 1,\ 0;\ -d_1,\ 0,\ a]^{\mathrm{T}}, \\ \$_{14} = [0,\ 0,\ 1;\ 0,\ -a,\ 0]^{\mathrm{T}}, \\ \$_{15} = [0,\ 0,\ 0;\ u_x,\ u_y,\ u_z]^{\mathrm{T}} \end{array}\right\} \tag{40}$$

$$\{\$_2\} = \left\{\begin{array}{l} \$_{21} = [0,\ 0,\ 0;\ 0,\ 0,\ 1]^{\mathrm{T}}, \\ \$_{22} = [2,\ 0,\ 0;\ 0,\ 2d_2,\ -\sqrt{3}a]^{\mathrm{T}}, \\ \$_{23} = [0,\ 2,\ 0;\ -2d_2,\ 0,\ -a]^{\mathrm{T}}, \\ \$_{24} = [0,\ 0,\ 2;\ \sqrt{3}a,\ a,\ 0]^{\mathrm{T}}, \\ \$_{25} = [0,\ 0,\ 0;\ \sqrt{3}v_x - u_x,\ \sqrt{3}v_y - u_y,\ \sqrt{3}v_z - u_z]^{\mathrm{T}} \end{array}\right\} \tag{41}$$

$$\{\$_3\} = \left\{ \begin{array}{l} \$_{31} = \left[0,\ 0,\ 0;\ 0,\ 0,\ 1\right]^{\mathrm{T}}, \\ \$_{32} = \left[2,\ 0,\ 0;\ 0,\ 2d_3,\ \sqrt{3}a\right]^{\mathrm{T}}, \\ \$_{33} = \left[0,\ 2,\ 0;\ -2d_3,\ 0,\ -a\right]^{\mathrm{T}}, \\ \$_{34} = \left[0,\ 0,\ 2;\ -\sqrt{3}a,\ a,\ 0\right]^{\mathrm{T}}, \\ \$_{35} = \left[0,\ 0,\ 0;\ -\sqrt{3}v_x - u_x,\ -\sqrt{3}v_y - u_y,\ -\sqrt{3}v_z - u_z\right]^{\mathrm{T}} \end{array} \right\} \tag{42}$$

The branch motion-screw set $\{\$_i\}$ of the 3-*CUP* parallel manipulator can be obtained by substituting Eqs. (35) and (38) into the screws of Eqs. (34) and performing the respective cross-vector operations, resulting in

$$\{\$_i\} = \left\{ \begin{array}{l} \$_{i1} = \left[0,\ 0,\ 0;\ 0,\ 0,\ 1\right]^{\mathrm{T}}, \\ \$_{i2} = \left[0,\ 0,\ 1;\ a\mathrm{s}\psi_i,\ -a\mathrm{c}\psi_i,\ 0\right]^{\mathrm{T}}, \\ \$_{i3} = \left[-v_y,\ v_x,\ 0;\ -d_i v_x,\ -d_i v_y,\ ar_y\right]^{\mathrm{T}}, \\ \$_{i4} = \left[v_x,\ v_y,\ v_z;\ a\mathrm{s}\psi_i v_z - d_i v_y,\ d_i v_x - a\mathrm{c}\psi_i v_z,\ a(\mathrm{c}\psi_i v_y - \mathrm{s}\psi_i v_x)\right]^{\mathrm{T}}, \\ \$_{i5} = \left[0,\ 0,\ 0;\ r_x,\ r_y,\ r_z\right]^{\mathrm{T}} \end{array} \right\} \tag{43}$$

where $\{\$_1\}$, $\{\$_2\}$, and $\{\$_3\}$ for the 3-*CUP* parallel manipulator can be written as

$$\{\$_1\} = \left\{ \begin{array}{l} \$_{11} = \left[0,\ 0,\ 0;\ 0,\ 0,\ 1\right]^{\mathrm{T}}, \\ \$_{12} = \left[0,\ 0,\ 1;\ 0,\ -a,\ 0\right]^{\mathrm{T}}, \\ \$_{13} = \left[-v_y,\ v_x,\ 0;\ -d_1 v_x,\ -d_1 v_y,\ av_x\right]^{\mathrm{T}}, \\ \$_{14} = \left[v_x,\ v_y,\ v_z;\ -d_1 v_y,\ d_1 v_x - av_z,\ av_y\right]^{\mathrm{T}}, \\ \$_{15} = \left[0,\ 0,\ 0;\ u_x,\ u_y,\ u_z\right]^{\mathrm{T}} \end{array} \right\} \tag{44}$$

$$\{\$_2\} = \left\{ \begin{array}{l} \$_{21} = \left[0,\ 0,\ 0;\ 0,\ 0,\ 1\right]^{\mathrm{T}}, \\ \$_{22} = \left[0,\ 0,\ 2;\ \sqrt{3}a,\ a,\ 0\right]^{\mathrm{T}}, \\ \$_{23} = \left[-2v_y,\ 2v_x,\ 0;\ -2d_2 v_x,\ -2d_2 v_y,\ a(\sqrt{3}v_y - v_x)\right]^{\mathrm{T}}, \\ \$_{24} = \left[v_x,\ v_y,\ v_z;\ \frac{\sqrt{3}}{2}av_z - d_2 v_y,\ d_2 v_x + \frac{av_z}{2},\ -\frac{a}{2}(\sqrt{3}v_x + v_y)\right]^{\mathrm{T}}, \\ \$_{25} = \left[0,\ 0,\ 0;\ \sqrt{3}v_x - u_x,\ \sqrt{3}v_y - u_y,\ \sqrt{3}v_z - u_z\right]^{\mathrm{T}} \end{array} \right\} \tag{45}$$

$$\{\$_3\} = \begin{Bmatrix} \$_{31} = \left[0,\ 0,\ 0;\ 0,\ 0,\ 1\right]^{\mathrm{T}}, \\ \$_{32} = \left[0,\ 0,\ 2;\ -\sqrt{3}a,\ a,\ 0\right]^{\mathrm{T}}, \\ \$_{33} = \left[-2v_y,\ 2v_x,\ 0;\ -2d_3v_x,\ -2d_3v_y,\ -a(\sqrt{3}v_y+v_x)\right]^{\mathrm{T}}, \\ \$_{34} = \left[v_x,\ v_y,\ v_z;-\dfrac{\sqrt{3}}{2}av_z-d_3v_y,\ d_3v_x+\dfrac{av_z}{2},\ \dfrac{a}{2}(\sqrt{3}v_x-v_y)\right]^{\mathrm{T}}, \\ \$_{35} = \left[0,\ 0,\ 0;\ -\sqrt{3}v_x-u_x,\ -\sqrt{3}v_y-u_y,\ -\sqrt{3}v_z-u_z\right]^{\mathrm{T}} \end{Bmatrix} \tag{46}$$

Note that for both the 3-<u>*P*</u>*SP* and the 3-*CUP* parallel manipulators, $\{\$_i\}$ is a system of five independent motion-screws, the reciprocal screw of all branches can be written as

$$\$_{i1}^r = \left[-r_y,\ r_x,\ 0;\ -d_i r_x,\ -d_i r_y,\ a(\mathrm{c}\psi_i r_x+\mathrm{s}\psi_i r_y)\right]^{\mathrm{T}} \tag{47}$$

The reciprocal screws $\$_{11}^r$, $\$_{21}^r$, and $\$_{31}^r$ represent forces that intersect at point B_i and are perpendicular to the vector $\boldsymbol{b}_i$ and parallel to the *XY* plane. These branch constraint-screws, which apply for both the 3-<u>*P*</u>*SP* and the 3-*CUP* parallel manipulators form the following platform constraint-screw system

$$\{\$^r\} = \begin{Bmatrix} \$_{11}^r = \left[-u_y,\ u_x,\ 0;\ -d_1u_x,\ -d_1u_y,\ au_x\right]^{\mathrm{T}}, \\ \$_{21}^r = \left[\dfrac{\sqrt{3}v_y-u_y}{u_x-\sqrt{3}v_x},\ 1,\ 0;\ -d_2,\ d_2\dfrac{\sqrt{3}v_y-u_y}{u_x-\sqrt{3}v_x},\ \dfrac{a}{2}(\dfrac{\sqrt{3}u_y-3v_y}{u_x-\sqrt{3}v_x}-1)\right]^{\mathrm{T}}, \\ \$_{31}^r = \left[-\dfrac{\sqrt{3}v_y+u_y}{u_x+\sqrt{3}v_x},\ 1,\ 0;\ -d_3,\ -d_3\dfrac{\sqrt{3}v_y+u_y}{u_x+\sqrt{3}v_x},\ -\dfrac{a}{2}(\dfrac{\sqrt{3}u_y+3v_y}{u_x+\sqrt{3}v_x}+1)\right]^{\mathrm{T}} \end{Bmatrix} \tag{48}$$

the platform motion-screw system $\{\$^f\}$ is given by the reciprocal screws of $\{\$^r\}$ resulting in

$$\{\$^f\} = \begin{Bmatrix} \$_1^f = \left[L_1^f,\ 0,\ 1;\ P_1^f,\ Q_1^f,\ 0\right]^{\mathrm{T}}, \\ \$_2^f = \left[L_2^f,\ 1,\ 0;\ P_2^f,\ Q_2^f,\ 0\right]^{\mathrm{T}}, \\ \$_3^f = \left[0,\ 0,\ 0;\ 0,\ 0,\ 1\right]^{\mathrm{T}} \end{Bmatrix} \tag{49}$$

where

$$L_1^f = \frac{3a(u_x+v_y)}{2d_1u_x-(d_2+d_3)u_x+\sqrt{3}(d_2-d_3)v_x},$$

$$P_1^f = au_x \frac{d_2K_1 + d_3(u_x(\sqrt{3}u_y - 3v_y) + 3v_x(u_y + 3v_x - \sqrt{3}v_y) - 3u_x^2) - 2\sqrt{3}d_1(u_xu_y - 3v_xv_y)}{2\sqrt{3}(2d_1u_x - (d_2 + d_3)u_x + \sqrt{3}(d_2 - d_3)v_x)(u_xv_y - u_yv_x)},$$

$$K_1 = 3u_x^2 + u_x(\sqrt{3}u_y + 3v_y) - 3v_x(u_y + 3v_x + \sqrt{3}v_y),$$

$$Q_1^f = a\frac{u_y^2[\sqrt{3}u_x(d_2 - 2d_1 + d_3) + 3v_x(d_3 - d_2)] + 2u_xv_yK_2 + u_yK_3}{2\sqrt{3}(2d_1u_x - (d_2 + d_3)u_x + \sqrt{3}(d_2 - d_3)v_x)(u_xv_y - u_yv_x)},$$

$$K_2 = \sqrt{3}u_x(d_1 + d_2 + d_3) + 3v_x(d_3 - d_2) + 3\sqrt{3}d_1v_y,$$

$$K_3 = 3u_x(d_2 - d_3)(u_x + v_y) - \sqrt{3}v_x[2(d_1 + d_2 + d_3)u_x + 3v_y(d_2 + d_3)] - 3v_x^2(d_2 - d_3),$$

$$L_2^f = \frac{u_y(d_2 + d_3 - 2d_1) - \sqrt{3}v_y(d_2 - d_3)}{2d_1u_x - (d_2 + d_3)u_x + \sqrt{3}(d_2 - d_3)v_x},$$

$$P_2^f = -\frac{d_1u_x(d_2 + d_3) - 2u_xd_2d_3 + \sqrt{3}d_1(d_2 - d_3)v_x}{2d_1u_x - (d_2 + d_3)u_x + \sqrt{3}(d_2 - d_3)v_x},$$

$$Q_2^f = -\frac{d_1u_y(d_2 + d_3) - 2u_yd_2d_3 + \sqrt{3}d_1(d_2 - d_3)v_y}{2d_1u_x - (d_2 + d_3)u_x + \sqrt{3}(d_2 - d_3)v_x} \quad (50)$$

Equations (49) and (50) indicate that the platforms of the 3-*<u>P</u>SP* and the 3-*CUP* parallel manipulators have three independent degrees of freedom, namely, a translation along the *Z*-axis and two rotations about two skew axes. Note that the 3-*<u>P</u>SP* and 3-*CUP* parallel mechanisms exhibit identical motion due to the canonical and non-canonical *S* joint architectures, i.e. the *R* joints j_{i2}, j_{i3}, and j_{i4} are grouped in *U* and *C* joints (together with j_{i1}) and the axes of the *R* joints intersect invariantly at point B_i for both architectures. Hence, each branch motion-screw system for both the canonical and non-canonical *S* joint architectures is formed by the same motion-screws, exerting the same platform constraint-screws and therefore the same mobility. The following section presents a mobility analysis of the 3-*<u>P</u>PS* and the 3-*<u>P</u>CU* parallel manipulators.

Mobility Analysis of the 3-*<u>P</u>PS* and 3-*<u>P</u>CU* Parallel Manipulators

Consider now that the j_{i5} joints are decoupled from the platform and they are attached to joints j_{i1}. The joint axis $\boldsymbol{s}_{i5}$ is parallel to $\boldsymbol{a}_i$ and perpendicular to $\boldsymbol{s}_{i1}$ and intersects the Z-axis. Schematic drawings of the kinematic chains of the 3-*<u>P</u>PS* and 3-*<u>P</u>CU* parallel manipulators are shown in Figs. 11 and 12, respectively. Note that for both manipulators, the joint axes $\boldsymbol{s}_{i1}$

and $\boldsymbol{s}_{i5}$ intersect at point D_i. The stroke of joint j_{i5} is denoted by the vector $\boldsymbol{t}_i$, which is directed to the platform at point B_i where the vectors $\boldsymbol{s}_{i2}$, $\boldsymbol{s}_{i3}$, and $\boldsymbol{s}_{i4}$ intersect invariantly.

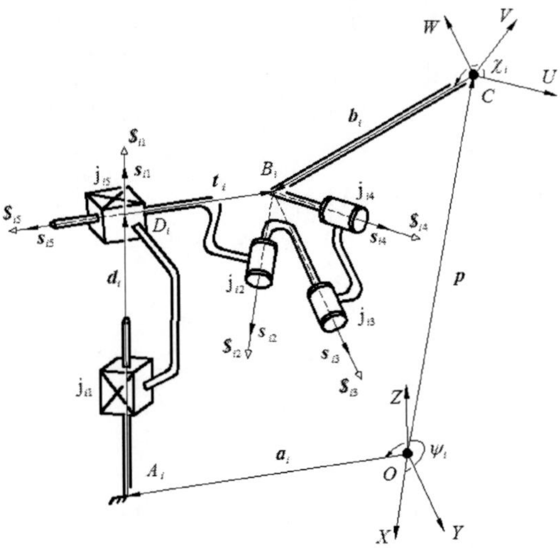

Figure 11. Leg *i* of the 3-*PPS* manipulator.

If the canonical representation of the *S* joint is considered, then $\boldsymbol{s}_{i2}$, $\boldsymbol{s}_{i3}$, and $\boldsymbol{s}_{i4}$ are orthogonal to each other and intersect at the connection point B_i. For convenience, assume that $\boldsymbol{s}_{i2}$, $\boldsymbol{s}_{i3}$, and $\boldsymbol{s}_{i4}$ are parallel to the *X*, *Y*, and *Z* axes, respectively. In this case, the architecture of the 3-*PPS* parallel manipulator is as shown in Fig. 11.

Consider a topology where the joint axis $\boldsymbol{s}_{i2}$ is coaxial with $\boldsymbol{s}_{i5}$ (i.e., parallel to $\boldsymbol{a}_i$), the joint axis $\boldsymbol{s}_{i3}$ is perpendicular to $\boldsymbol{s}_{i4}$, which is parallel to the *W*-axis. Note that $\boldsymbol{s}_{i2}$, $\boldsymbol{s}_{i3}$, and $\boldsymbol{s}_{i4}$ intersect at B_i and form a non-canonical *S* joint. In this particular topology, the joints j_{i2} and j_{i5} behave as *C* joints, while joints j_{i3} and j_{i4} can be represented as *U* joints. The resulting architecture is the 3-*PCU* parallel manipulator as shown in Fig. 12.

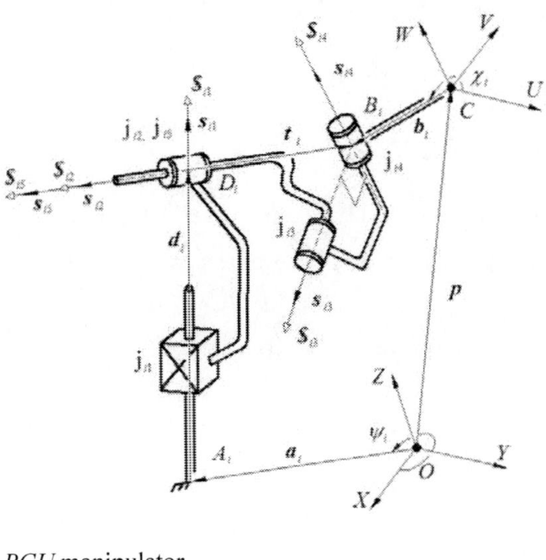

Figure 12. Leg *i* of the 3-*PCU* manipulator.

Each leg of the 3-*PPS* and the 3-*PCU* parallel manipulators is comprised of a system of five screws. For convenience, the joint screws will be expressed in the global frame G and calculated relative to *O*. The joint screws of leg *i* can be written as

$$\$_{i1}=\begin{bmatrix}\boldsymbol{0}\\ \boldsymbol{s}_{i1}\end{bmatrix},\quad \$_{i2}=\begin{bmatrix}\boldsymbol{s}_{i2}\\ (\boldsymbol{a}_i+\boldsymbol{t}_i+\boldsymbol{d}_i)\times\boldsymbol{s}_{i2}\end{bmatrix},\quad \$_{i3}=\begin{bmatrix}\boldsymbol{s}_{i3}\\ (\boldsymbol{a}_i+\boldsymbol{t}_i+\boldsymbol{d}_i)\times\boldsymbol{s}_{i3}\end{bmatrix},$$
$$\$_{i4}=\begin{bmatrix}\boldsymbol{s}_{i4}\\ (\boldsymbol{a}_i+\boldsymbol{t}_i+\boldsymbol{d}_i)\times\boldsymbol{s}_{i4}\end{bmatrix},\quad \text{and}\quad \$_{i5}=\begin{bmatrix}\boldsymbol{0}\\ \boldsymbol{s}_{i5}\end{bmatrix} \tag{51}$$

where $\boldsymbol{a}_i$, $\boldsymbol{d}_i$, and $\boldsymbol{s}_{i1}$ are given by Eq. (35) and

$$\boldsymbol{t}_i=\left[t_i c\psi_i,\ t_i s\psi_i,\ 0\right]^{\mathrm{T}},\ \text{and}\ \boldsymbol{s}_{i5}=\frac{\boldsymbol{a}_i}{a}=\left[c\psi_i,\ s\psi_i,\ 0\right]^{\mathrm{T}} \tag{52}$$

The remaining joint axes for the 3-*PPS* parallel mechanism are given by Eqs. (37), while the unit vectors $\boldsymbol{s}_{i2}$, $\boldsymbol{s}_{i3}$, and $\boldsymbol{s}_{i4}$ can be written for the 3-*PCU* parallel manipulator as

$$\boldsymbol{s}_{i2}=\left[c\psi_i,\ s\psi_i,\ 0\right]^{\mathrm{T}},\ \boldsymbol{s}_{i3}=\left[-w_z,\ 0,\ w_x\right]^{\mathrm{T}},\ \text{and}\ \boldsymbol{s}_{i4}=\left[w_x,\ w_y,\ w_z\right]^{\mathrm{T}} \tag{53}$$

Substituting Eqs. (35), (37) and (52) into Eqs. (51) and performing the cross-vector operations, results in $\{\$_i\}$ of the 3-*PPS* parallel manipulator; i.e.,

$$\{\$_i\}=\left\{\begin{array}{l}\$_{i1}=\left[0,\ 0,\ 0;\ 0,\ 0,\ 1\right]^{\mathrm{T}},\\ \$_{i2}=\left[1,\ 0,\ 0;\ 0,\ d_i,\ (t_i-a)s\psi_i\right]^{\mathrm{T}},\\ \$_{i3}=\left[0,\ 1,\ 0;\ -d_i,\ 0,\ (a-t_i)c\psi_i\right]^{\mathrm{T}},\\ \$_{i4}=\left[0,\ 0,\ 1;\ (a-t_i)s\psi_i,\ (t_i-a)c\psi_i,\ 0\right]^{\mathrm{T}},\\ \$_{i5}=\left[0,\ 0,\ 0;\ c\psi_i,\ s\psi_i,\ 0\right]^{\mathrm{T}}\end{array}\right\} \tag{54}$$

where $\{\$_1\}$, $\{\$_2\}$, and $\{\$_3\}$ can be written as

$$\{\$_1\}=\left\{\begin{array}{l}\$_{11}=\left[0,\ 0,\ 0;\ 0,\ 0,\ 1\right]^{\mathrm{T}},\\ \$_{12}=\left[1,\ 0,\ 0;\ 0,\ d_1,\ 0\right]^{\mathrm{T}},\\ \$_{13}=\left[0,\ 1,\ 0;\ -d_1,\ 0,\ a-t_1\right]^{\mathrm{T}},\\ \$_{14}=\left[0,\ 0,\ 1;\ 0,\ t_1-a,\ 0\right]^{\mathrm{T}},\\ \$_{15}=\left[0,\ 0,\ 0;\ 1,\ 0,\ 0\right]^{\mathrm{T}}\end{array}\right\} \tag{55}$$

$$\{\$_2\} = \left\{ \begin{aligned} &\$_{21} = [0,\ 0,\ 0;\ 0,\ 0,\ 1]^{\mathrm{T}}, \\ &\$_{22} = \left[2,\ 0,\ 0;\ 0,\ 2d_2,\ \sqrt{3}(t_2 - a)\right]^{\mathrm{T}}, \\ &\$_{23} = [0,\ 2,\ 0;\ -2d_2,\ 0,\ t_2 - a]^{\mathrm{T}}, \\ &\$_{24} = \left[0, 0,\ 2;\ \sqrt{3}(a - t_2),\ (a - t_2),\ 0\right]^{\mathrm{T}}, \\ &\$_{25} = \left[0,\ 0,\ 0;\ -1,\ \sqrt{3},\ 0\right]^{\mathrm{T}} \end{aligned} \right\} \tag{56}$$

$$\{\$_3\} = \left\{ \begin{aligned} &\$_{31} = [0,\ 0,\ 0;\ 0,\ 0,\ 1]^{\mathrm{T}}, \\ &\$_{32} = \left[2,\ 0,\ 0;\ 0,\ 2d_3,\ -\sqrt{3}(t_3 - a)\right]^{\mathrm{T}}, \\ &\$_{33} = [0,\ 2,\ 0;\ -2d_3,\ 0,\ t_3 - a]^{\mathrm{T}}, \\ &\$_{34} = \left[0,\ 0,\ 2;\ \sqrt{3}(t_3 - a),\ (a - t_3),\ 0\right]^{\mathrm{T}}, \\ &\$_{35} = \left[0,\ 0,\ 0;\ -1,\ -\sqrt{3},\ 0\right]^{\mathrm{T}} \end{aligned} \right\} \tag{57}$$

Substituting the unit vectors from Eqs. (34), (52), and (53) into the screws from Eqs. (51), and performing the respective cross-vector operations, result in the $\{\$_i\}$ of the 3-*PCU* parallel manipulator, i.e.,

$$\{\$_i\} = \left\{ \begin{aligned} &\$_{i1} = [0,\ 0,\ 0;\ 0,\ 0,\ 1]^{\mathrm{T}}, \\ &\$_{i2} = [\mathrm{c}\psi_i,\ \mathrm{s}\psi_i,\ 0;\ -d_i\mathrm{s}\psi_i,\ d_i\mathrm{c}\psi_i,\ 0]^{\mathrm{T}}, \\ &\$_{i3} = [-w_z,\ 0,\ w_x;\ (a - t_i)\mathrm{s}\psi_i w_x, \\ &\qquad (t_i - a)\mathrm{c}\psi_i w_x - d_i w_z,\ (a - t_i)\mathrm{s}\psi_i w_z]^{\mathrm{T}}, \\ &\$_{i4} = \left[w_x,\ w_y,\ w_z;\ (a - t_i)\mathrm{s}\psi_i w_z - d_i w_y,\right. \\ &\qquad \left.(t_i - a)\mathrm{c}\psi_i w_z + d_i w_x,\ R_{ij}\right]^{\mathrm{T}}, \\ &\$_{i5} = [0,\ 0,\ 0;\ \mathrm{c}\psi_i,\ \mathrm{s}\psi_i,\ 0]^{\mathrm{T}} \end{aligned} \right\} \tag{58}$$

where

$$R_{ij} = (a - t_i)(\mathrm{c}\psi_i w_y - \mathrm{s}\psi_i w_x) \tag{59}$$

The branch motion-screw sets $\{\$_1\}$, $\{\$_2\}$, and $\{\$_3\}$ result in

$$\{\$_1\} = \left\{ \begin{array}{l} \$_{11} = [0,\ 0,\ 0;\ 0,\ 0,\ 1]^{\mathrm{T}}, \\ \$_{12} = [1,\ 0,\ 0;\ 0,\ d_1,\ 0]^{\mathrm{T}}, \\ \$_{13} = [-w_z,\ 0,\ w_x; \\ \qquad 0,\ (t_1 - a)w_x - d_1 w_z,\ 0]^{\mathrm{T}}, \\ \$_{14} = [w_x,\ w_y,\ w_z; \\ \quad -d_1 w_y,\ (t_1 - a)w_z + d_1 w_x,\ (a - t_1)w_y]^{\mathrm{T}}, \\ \$_{15} = [0,\ 0,\ 0;\ 1,\ 0,\ 0]^{\mathrm{T}} \end{array} \right\} \tag{60}$$

$$\{\$_2\} = \left\{ \begin{array}{l} \$_{21} = [0,\ 0,\ 0;\ 0,\ 0,\ 1]^{\mathrm{T}}, \\ \$_{22} = \left[-1,\ \sqrt{3},\ 0;\ -\sqrt{3}d_2,\ -d_2,\ 0\right]^{\mathrm{T}}, \\ \$_{23} = \left[-2w_z,\ 0,\ 2w_x;\ \sqrt{3}(a - t_2)w_x, \right. \\ \qquad \left. (a - t_2)w_x - 2d_2 w_z,\ \sqrt{3}(a - t_2)w_z\right]^{\mathrm{T}}, \\ \$_{24} = \left[w_x,\ w_y,\ w_z;\ \dfrac{\sqrt{3}}{2}(a - t_2)w_z - d_2 w_y, \right. \\ \qquad \left. \dfrac{(a - t_2)}{2} w_z + d_2 w_x,\ R_{25} \right]^{\mathrm{T}}, \\ \$_{25} = \left[0,\ 0,\ 0;\ -1,\ \sqrt{3},\ 0\right]^{\mathrm{T}} \end{array} \right\} \tag{61}$$

$$\{\$_3\} = \left\{ \begin{array}{l} \$_{31} = [0,\ 0,\ 0;\ 0,\ 0,\ 1]^{\mathrm{T}}, \\ \$_{32} = \left[-1,\ -\sqrt{3},\ 0;\ \sqrt{3}d_3,\ -d_3,\ 0\right]^{\mathrm{T}}, \\ \$_{33} = \left[-2w_z,\ 0,\ 2w_x;\ \sqrt{3}(t_3 - a)w_x, \right. \\ \qquad \left. (a - t_3)w_x - 2d_3 w_z,\ \sqrt{3}(t_3 - a)w_z\right]^{\mathrm{T}}, \\ \$_{34} = \left[w_x,\ w_y,\ w_z;\ \dfrac{\sqrt{3}}{2}(t_3 - a)w_z - d_3 w_y, \right. \\ \qquad \left. \dfrac{(a - t_3)}{2} w_z + d_3 w_x,\ R_{35} \right]^{\mathrm{T}}, \\ \$_{35} = \left[0,\ 0,\ 0;\ -1,\ -\sqrt{3},\ 0\right]^{\mathrm{T}} \end{array} \right\} \tag{62}$$

where

$$R_{25} = -\frac{(a-t_2)}{2}(w_y + \sqrt{3}w_x) \text{ , and } R_{35} = \frac{(a-t_3)}{2}(\sqrt{3}w_x - w_y) \quad (63)$$

Note that for the 3-*PPS* and 3-*PCU* parallel manipulators, $\{\$_i\}$ contains five independent screws, each leg provides the following constraint-screw that is reciprocal to $\{\$_i\}$

$$\$_{i1}^r = \left[-s\psi_i,\ c\psi_i,\ 0;\ -d_i c\psi_i,\ -d_i s\psi_i,\ a-t_i\right]^{\mathrm{T}} \quad (64)$$

The reciprocal screws $\$_{11}^r$, $\$_{21}^r$, and $\$_{31}^r$ are forces that intersect point B_i and are perpendicular to the plane defined by each leg and the *Z*-axis. The platform constraint-screw system $\{\$^r\}$ for both architectures is

$$\{\$^r\} = \left\{\begin{array}{l} \$_{11}^r = \left[0,\ 1,\ 0;\ -d_1,\ 0,\ a-t_1\right]^{\mathrm{T}}, \\ \$_{21}^r = \left[\sqrt{3},\ 1,\ 0;\ -d_2,\ \sqrt{3}d_2,\ -2(a-t_2)\right]^{\mathrm{T}}, \\ \$_{31}^r = \left[-\sqrt{3},\ 1,\ 0;\ -d_3,\ -\sqrt{3}d_3,\ -2(a-t_3)\right]^{\mathrm{T}} \end{array}\right\} \quad (65)$$

The platform motion-screw system $\{\$^f\}$ for the 3-*PPS* and the 3-*PCU* parallel manipulators is given by the reciprocal screws of $\{\$^r\}$ in Eq. (65), that is

$$\{\$^f\} = \left\{\begin{array}{l} \$_1^f = \left[L_1^f,\ 0,\ 1;\ P_1^f,\ Q_1^f,\ 0\right]^{\mathrm{T}}, \\ \$_2^f = \left[\dfrac{\sqrt{3}(d_2-d_3)}{-2d_1+d_2+d_3},\ 1,\ 0;\right. \\ \qquad \left.\dfrac{2d_2d_3-d_1(d_2+d_3)}{-2d_1+d_2+d_3},\ \dfrac{\sqrt{3}d_1(d_2-d_3)}{-2d_1+d_2+d_3},0\right]^{\mathrm{T}}, \\ \$_3^f = \left[0,\ 0,\ 0;\ 0,\ 0,\ 1\right]^{\mathrm{T}} \end{array}\right\} \quad (66)$$

where

$$L_1^f = -\frac{6a-2(t_1+t_2+t_3)}{-2d_1+d_2+d_3},\quad P_1^f = \sqrt{3}\frac{3a(d_2-d_3)+d_3t_1-2d_1t_2+2d_3t_2+2d_1t_3-d_2(t_1+2t_3)}{6d_1-3(d_2+d_3)},$$
$$Q_1^f = t_1 - \frac{a(4d_1+d_2+d_3)-2d_1(t_1+t_2+t_3)}{-2d_1+d_2+d_3} \quad (67)$$

The platform motion-screw system $\{\$^f\}$ given by Eqs. (66) indicate that the platforms of the 3-*PPS* and 3-*PCU* parallel manipulators have three independent degrees-of-freedom, which are a translation along the *Z*-axis and two rotations about two skew axes. In a similar manner to the 3-*PSP* and 3-*CUP* architectures, the 3-*PPS* and 3-*PCU* parallel mechanisms exhibit identical motion due to the constraints exerted by the canonical and non-canonical *S* joint architectures on their respective platforms.

Motion Simulations

In order to verify the plausibility of the screw-based mobility analysis performed in this paper, CAD models of the 3-$R\underline{P}C$-Y, 3-*RCC*-Y, 3-$R\underline{P}C$-T, 3-$\underline{P}SP$, 3-*CUP*, 3-$\underline{P}PS$, and 3-$\underline{P}CU$ parallel mechanisms are generated and subjected to motion simulations. Tables 2-6 contain the parameters used in the simulations such as the total simulation time, the velocity of the active joints and the dimensional parameters of the abovementioned parallel mechanisms.

In addition, Tables 2-6 present the simulation results of the initial and final poses of the platforms, where p_x, p_y, and p_z are the Cartesian coordinates of the platform centroid *C*; while α, β, and γ are the projection angles about the *X*-, *Y*-, and *Z*-axes that express the orientation of the platform and displayed in solid, dashed and dotted plots, respectively (see Figs. 13-16).

The platform of the 3-$R\underline{P}C$-Y parallel mechanism has three degrees-of-freedom of translational motion according to Eq. (26). This is corroborated by the simulation results, where the position of *C*, its linear velocity and linear acceleration show in Figs. 13(a), 13(c), and 13(e) that the platform is able to position itself along any axis of the global frame. The results of the projected angles of the platform indicate that there is no rotation about any axis of the global frame, see Fig. 14(d).

Table 2. Simulation parameters and results for the 3-$R\underline{P}C$-Y parallel mechanism

Simulation Parameters		Initial Configuration	Final Configuration
a = 433.01 mm	d_1 [mm]	446.29	519.80
Time = 5 s	d_2 [mm]	422.69	472.14
$\dot{d}_1$ = 10 mm/s	d_3 [mm]	470.34	456.30
$\dot{d}_2$ = 3 mm/s	θ_{11} [deg]	115.01	121.51
$\dot{d}_3$ = -5 mm/s	θ_{12} [deg]	88.73	69.82
	θ_{13} [deg]	63.96	76.21
	p_x [mm]	124.64	-31.26
	p_y [mm]	-197.10	-271.67
	p_z [mm]	422.58	443.15
	α, β, γ [deg]	0	0

Likewise, it is concluded in previous sections that the platform of the 3-*RCC*-Y parallel mechanism has fully Cartesian motion. Table 3 presents the simulation parameters used for obtaining the positioning profile of the platform as shown in Fig. 13(b). Furthermore, the plots of the linear velocity and linear acceleration of the platform are presented in Figs. 13(d) and 13(f). Note that the projected angles of the platforms of the 3-$R\underline{P}C$-Y and 3-*RCC*-Y parallel mechanisms are identical, see Fig. 14(d). The simulations prove that the manipulator has fully translational motion.

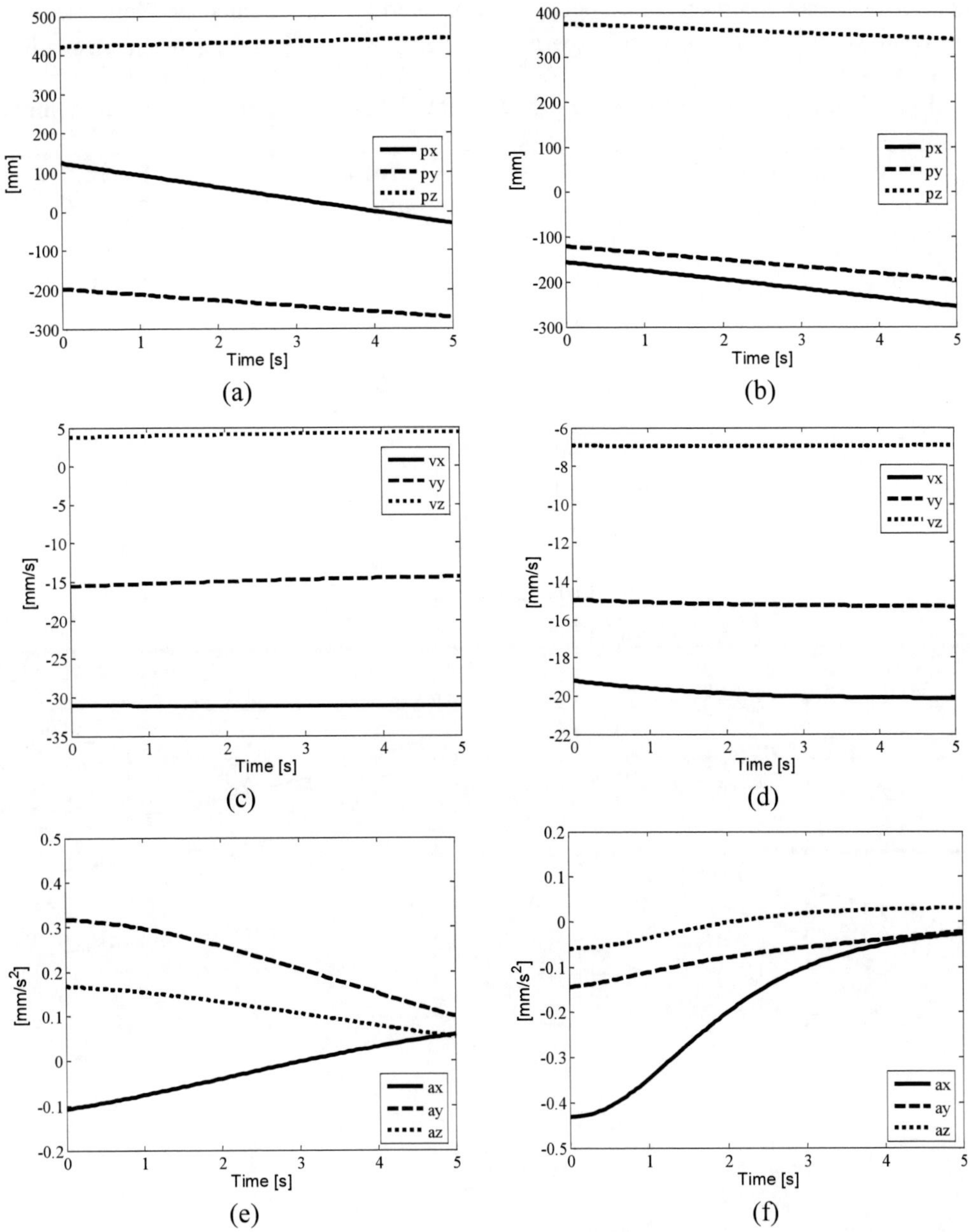

Figure 13. Simulation plots for the 3-*RPC*-Y and 3-*RCC*-Y parallel mechanisms.

The 3-*RPC*-T parallel mechanism studied previously has three translational degrees-of-freedom, according to the platform motion-screw system from Eq. (26). The mobility of this parallel mechanism is verified with a motion simulation using the parameters listed in Table 4. Recall that for this architecture, the joints j_{11}, j_{21}, and j_{32} are selected as actuators.

According to the simulation plots for the position, linear velocity and linear acceleration shown in Figs. 14(a)-14(c), the platform of the 3-*RPC*-T parallel mechanism is able to position along any of the three Cartesian axes of the global frame, while Fig. 14(d) is the simulation result for the projection angles, angular velocity and angular acceleration of the platform and shows that the platform is parallel to the base for all positions of the mechanism.

Table 3. Simulation parameters and results for the 3-*RCC*-Y parallel mechanism

Simulation Parameters		Initial Configuration	Final Configuration
$a = 433.01$ mm	d_1 [mm]	394.00	393.01
Time = 5 s	d_2 [mm]	422.81	466.56
$\dot{d}_1 = -2$ mm/s	d_3 [mm]	382.41	361.67
$\dot{d}_2 = 5$ mm/s	θ_{11} [deg]	107.83	119.75
$\dot{d}_3 = -5$ mm/s	θ_{12} [deg]	62.51	46.64
	θ_{13} [deg]	101.24	109.75
	p_x [mm]	-155.69	-255.04
	p_y [mm]	-120.64	-196.67
	p_z [mm]	375.08	340.27
	α, β, γ [deg]	0	0

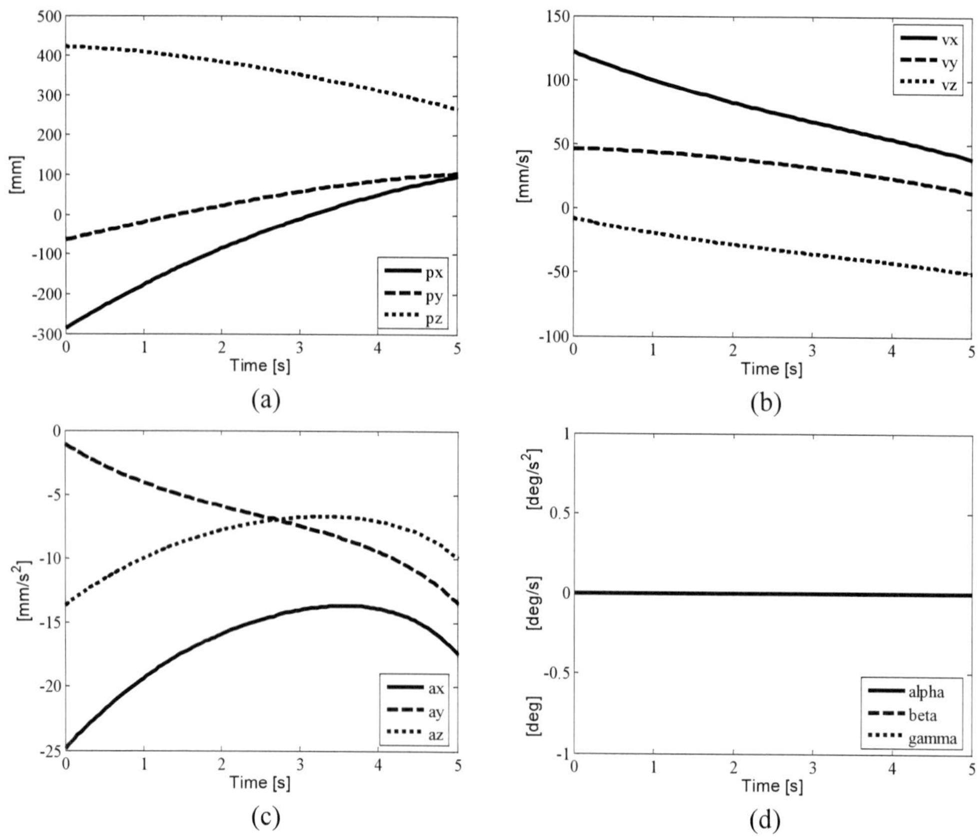

Figure 14. Simulation plots for the 3-*RPC*-T parallel mechanism.

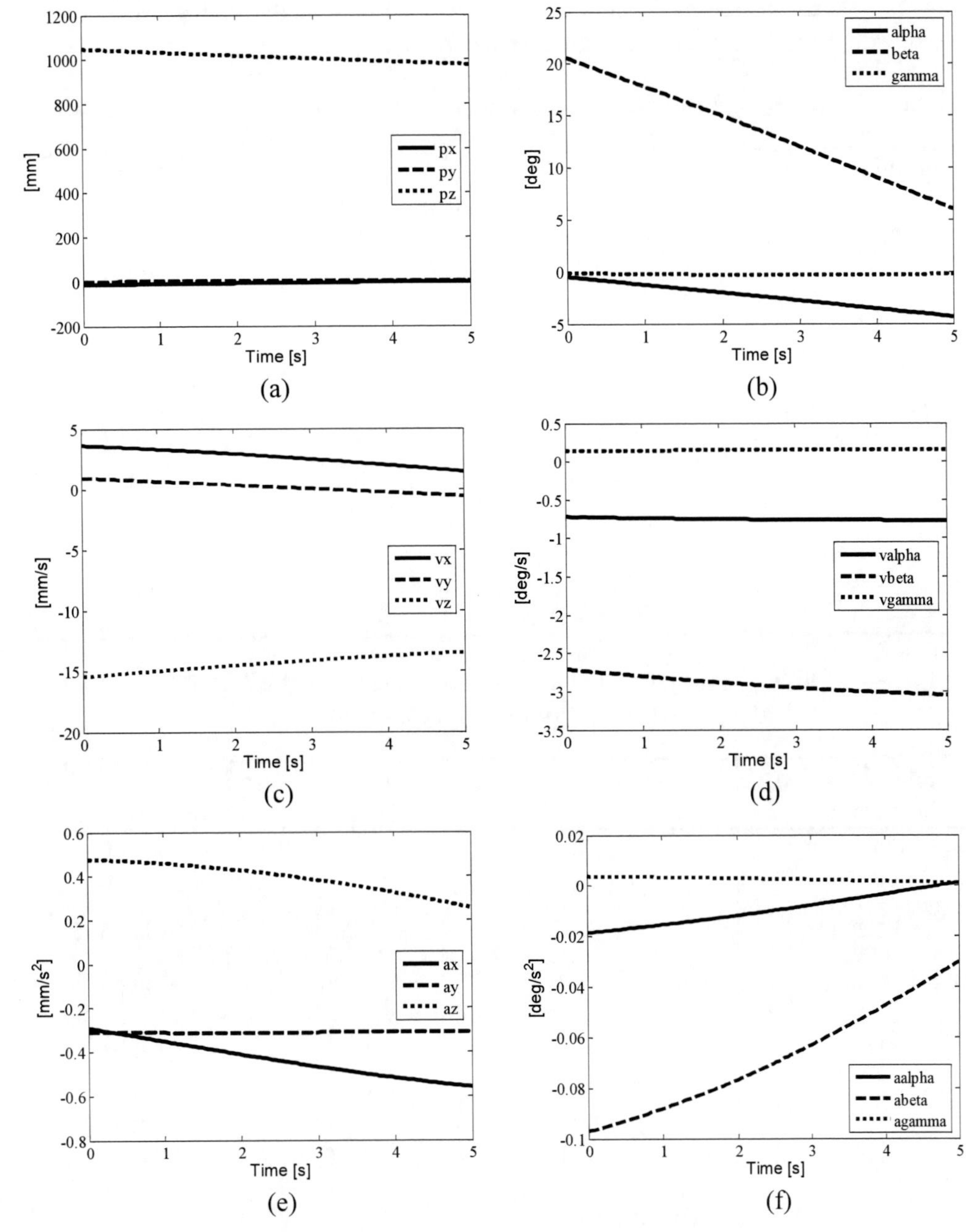

Figure 15. Simulation plots for the 3-*<u>P</u>SP* and 3-*CUP* parallel mechanisms.

As presented previously, the platforms of the 3-*<u>P</u>SP* and 3-*CUP* parallel mechanisms have three independent DOF, which according to Eq. (49) is rotation about two skew axes and translation along the *Z*-axis. Furthermore, it is determined that both mechanisms present an identical motion due to the constraints imposed by the canonical and non-canonical *S* joint configurations. The simulation parameters and the results of the position and orientation simulations are contained in Table 5. The behavior of the position of the centroid *C*, its linear velocity and acceleration are presented in Figs. 15(a), 15(c), and 15(e) and show that the platform is able to change its position along the three axes of the global frame. This is consistent with the platform screw-motion system from Eq. (49).

Table 4. Simulation parameters and results for the 3-*RPC*-T parallel mechanism

Simulation Parameters		Initial Configuration	Final Configuration
a = 225 mm	d_1 [mm]	427.64	282.51
Time = 5 s	d_2 [mm]	511.03	280.55
$\dot{\theta}_{11}$ = -1.8 rpm	d_3 [mm]	511.03	280.55
$\dot{d}_3$ = -75 mm/s	θ_{11} [deg]	98.69	68.09
$\dot{\theta}_{21}$ = -1 rpm	θ_{12} [deg]	55.81	110.89
	θ_{13} [deg]	110.89	55.81
	p_x [mm]	-287.15	100.05
	p_y [mm]	-64.63	105.41
	p_z [mm]	422.73	262.11
	α, β, γ [deg]	0	0

Table 5. Simulation parameters and results for the 3-*PSP* and 3-*CUP* parallel mechanisms

Simulation Parameters		Initial Configuration	Final Configuration
a = 433.01 mm	d_1 [mm]	881.42	931.42
Time = 5 s	d_2 [mm]	1121.39	971.39
$\dot{d}_1$ = 10 mm/s	d_3 [mm]	1128.56	1028.56
$\dot{d}_2$ = -30 mm/s	θ_{11}, θ_{12}, θ_{13} [deg]	90.00	90.00
$\dot{d}_3$ = -20 mm/s	p_x [mm]	-13.54	-0.33
	p_y [mm]	0.77	1.73
	p_z [mm]	1048.85	977.02
	α [deg]	-0.60	-4.3
	β [deg]	21.00	6.00
	γ [deg]	-0.10	-0.23

Equation (66) shows the results of the mobility analysis for the 3-*PPS* and 3-*PCU* parallel mechanisms. For both manipulators, the platform has three DOF, namely translation along the *Z*-axis and rotations about two axes. Note that these architectures exhibit an identical motion, where a similar case is presented by the 3-*PSP* and 3-*CUP* parallel mechanisms.

The simulation and dimensional parameters for the 3-*PPS* and 3-*PCU* parallel mechanisms are listed in Table 6. Figures 16(a), 16(c), and 16(e) indicate that the platform of the mechanism is able to position along any of the three Cartesian axes of the global frame, while Fig. 16(b), 16(d), and 16(f) show that the projected angles of rotation, angular velocity and angular acceleration of the platform are given about all axes of frame G, according to Eq. (66). The simulations show that the motion for both architectures is identical, as presented in the previous section.

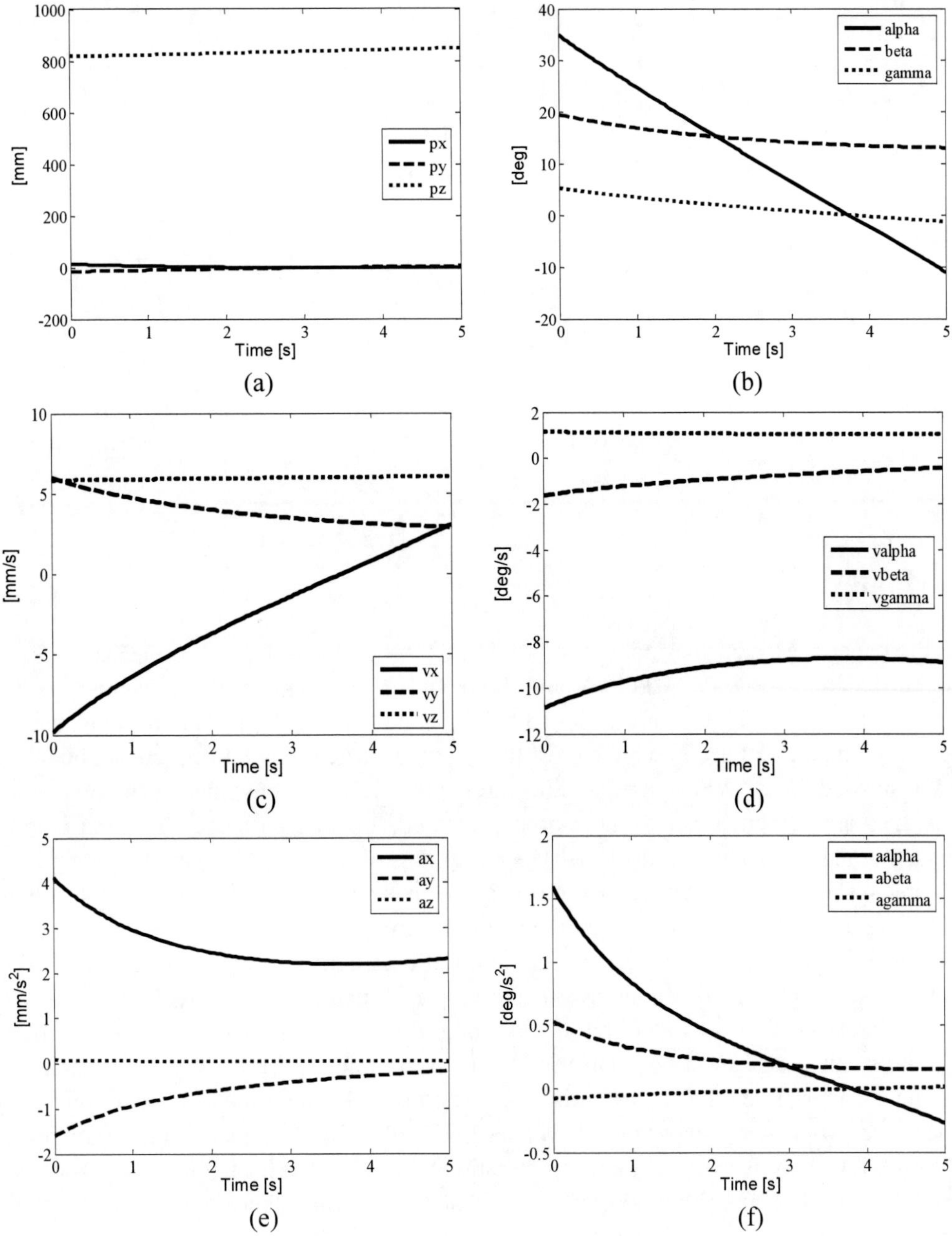

Figure 16. Simulation plots for the 3-*PPS* and 3-*PCU* parallel mechanisms.

Table 6. Simulation parameters and results for the 3-$\underline{P}PS$ and 3-$\underline{P}CU$ parallel mechanisms

Simulation Parameters		Initial Configuration	Final Configuration
$a = 433.13$ mm	d_1 [mm]	714.17	814.17
$b = 189.34$ mm	d_2 [mm]	719.69	919.69
Time = 5 s	d_3 [mm]	817.14	767.14
$\dot{d}_1 = 20$ mm/s	θ_{11} [deg]	90.00	90.00
$\dot{d}_2 = 40$ mm/s	θ_{12} [deg]	90.00	90.00
$\dot{d}_3 = -10$ mm/s	θ_{13} [deg]	90.00	90.00
	p_x [mm]	2.55	10.45
	p_y [mm]	5.59	-4.87
	p_z [mm]	750.33	833.67
	α [deg]	-17.00	28.00
	β [deg]	12.00	7.00
	γ [deg]	-1.70	1.50

Conclusions

This paper presented the mobility analysis of two families of parallel mechanisms: (i) fully translational 3-*R$\underline{P}$C*-Y, 3-*RCC*-Y and the new 3-*RPC*-T parallel manipulators; and (ii) 2R1T 3-*$\underline{P}$SP*, 3-*$\underline{P}$PS*, 3-*$\underline{P}$CU* and a new 3-*CUP* parallel manipulator. Having obtained the branch motion-screws for all architectures, the paper identified the platform constraint-screw sets and showed that all *RPC*- based architectures show the same constraints, namely: couples whose axes are perpendicular to the *R* joint axes and coplanar with the base, and couples acting about the normal axis to the base. The constraints of the 3-*RCC*-Y parallel manipulator are couples whose axes are orthogonal to the *R* joint axis and the leg of each kinematic chain. For the remainder architectures, the constraints are identified to be forces that are perpendicular to the vector defining the position of the leg acting on the *S* and *U* joints on the platform, and couples acting about an axis that is perpendicular to the base. The mobility of the platform was then obtained by determining the reciprocal screws to the platform constraint-screw sets and the platforms were identified to have three instantaneous independent translational DOF for family (i), and translation along the axis perpendicular to the base and two rotations about two skew axes for family (ii). The paper validated the results of the mobility analysis by using motion simulations, showing that for family (i) the platforms are able to translate along any axis of the global frame, while there is no rotation of the platform. The paper concluded that for family (ii), a non-canonical architecture *S* joint presents the same mobility independently of the assembly configuration of the joints. Future research will focus on the kinematics of the new 3-*RPC*-T, and 3-*CUP* parallel manipulators.

References

[1] Mohamed M.G., and Duffy J., 1985, "A Direct Determination of the Instantaneous Kinematics of Fully Parallel Robot Manipulators", *ASME Journal of Mechanism, Transmissions and Automation in Design,* **107**(2), pp. 226-229.

[2] Innocenti, C., and Parenti-Castelli, V., 1990, "Direct position analysis of the Stewart platform mechanism", *Mechanism and Machine Theory*, **25**(6), pp. 611–621.

[3] Hunt, K.H., 1983, "Structural Kinematics of In-Parallel-Actuated Robot-Arms", *ASME Journal of Mechanism, Transmissions and Automation in Design,* **105**(4), pp. 705-712.

[4] Waldron, K.J., Raghavan, M, and Roth, B., 1989, "Kinematic of a Hybrid Series-Parallel Manipulation System", *ASME Journal of Dynamic Systems, Measurement and Control,* **111**(2), pp. 211-221.

[5] Frisoli A., Checcacci D., Salsedo F. and Bergamasco M., 2000, "Synthesis by screw algebra of translating in-parallel actuated mechanisms", *Advances in Robot Kinematics*, J. Lenarcic and M. M. Stanisic (eds.), Kluwer Academatic Publishers, pp. 433–440.

[6] Kong, X., and Gosselin, C.M., 2002, "Kinematics and singularity analysis of a novel type of 3-CRR 3-DOF translational parallel manipulator", *The International Journal of Robotics Research*, **21**(9), pp. 791–798.

[7] Callegari M, and Marzetti P., 2003, "Kinematics of a family of parallel translating mechanisms", Proc. of the RAAD'03, *12th Int. Workshop on Robotics in Alpe-Adria-Danube region*, May 7-10, Cassino, Italy.

[8] Callegari, M., Palpacelli, M.C., and Scarponi, M., 2005, "Kinematics of the 3-CPU Parallel Manipulator Assembled for Motions of Pure Translation", *IEEE Proc. Intl. Conf. Robotics and Automation ICRA*, Barcelona, April 18-22, pp.4031-4036.

[9] Kong, X., and Gosselin, C.M., 2004, "Type Synthesis of 3-DOF Translational Parallel Manipulators Based on Screw Theory", *ASME Journal of Mechanical Design*, **126**(1), pp. 83-92.

[10] Rodriguez Velazco, E., and Rodriguez Leal, E., 2011, "Kinematic Analysis of the 3-RPC-T Fully Translational Mechanism", IFToMM World Congress, June 19-25, Guanajuato, Mexico.

[11] Rodriguez Leal, E., and Dai, J.S., 2010, "Mobility Analysis of a Family of Fully Translational Parallel Mechanisms using Screw Theory", IFToMM Symposium on Mechanism Design for Robotics, September 28-30, Mexico City, Mexico.

[12] Gosselin, C.M., and Angeles, J., 1989, "The optimum kinematic design of a spherical three-degree-of-freedom parallel manipulator", *ASME Journal of Mechanisms, Transmissions and Automation in Design*, **111**(2), pp. 202-207.

[13] Innocenti, C., and Parenti-Castelli, V., 1993, "Echelon form solution of direct kinematics for the general fully- parallel spherical wrist", *Mechanism and Machine Theory*, **28**(4), pp. 553-561.

[14] Alizade, R.I., Tagiyev, N.R., and Duffy, J., 1994, "A forward and reverse displacement analysis of an in-parallel spherical manipulator", *Mechanism and Machine Theory*, **29**(1), pp. 125-137.

[15] Kong, X., and Gosselin, C.M., 2004, "Type Synthesis of 3-DOF spherical parallel manipulators based on screw theory", *ASME Journal of Mechanical Design*, **126**(1), pp. 101-108.

[16] Fang, Y, and Tsai, L.W., 2004, "Structure Synthesis of a class of 3-DOF rotational parallel manipulators", *IEEE Transactions on Robotics and Automation*, **20**(1), 117-121.

[17] Di Gregorio, R., 2001, "Kinematics of a new spherical parallel manipulator with three equal legs: the 3-URC wrist", *Journal of Robotic Systems*, **18**(5), pp. 213-219.

[18] Huang, Z., and Li, Q.C., 2001, "Two novel symmetrical 5-DOF parallel mechanisms" (in Chinese), *J. Yanshan Univ.*, **25**(4), pp. 283–286.

[19] Huang, Z., and Li, Q.C., 2002, "General methodology for type synthesis of lower mobility symmetrical parallel manipulators and several novel manipulators", *The International Journal of Robotics Research*, **21**(2), pp. 131–146.

[20] Huang, Z., and Li, Q.C., 2002, "Some novel lower mobility parallel mechanisms", *Proc. ASME Design Eng. Tech. Conf. and Computers and Information in Eng. Conf. IDETC/CIE 2002,* DETC2002-34299., Sept. 29–Oct. 2, Montreal, QC, Canada.

[21] Huang, Z., and Li, Q.C., 2002, "On the type synthesis of lower mobility parallel manipulators", *Proc. Workshop on Fundamental Issues and Future Research Directions for Parallel Mechanisms and Manipulators*, October 3-4, Quebec, QC, Canada, pp. 272–283.

[22] Zlatanov, D., and Gosselin, C.M., 2001, "A new parallel architecture with four degrees of freedom", *Proc. 2nd Workshop on Computational Kinematics*, May 19–22, Seoul, Korea, pp. 57–66.

[23] Gallardo-Alvarado, J., Rico-Martinez, J.M., and Alici, G., 2004, "Kinematics and singularity analysis of a 4-dof parallel manipulator using screw theory", *Mechanism and Machine Theory*, **41**(9), pp. 1048-1061.

[24] Kong, X., and Gosselin, C.M., 2004, "Type Synthesis of 3T1R 4-DOF Parallel Manipulators Based on Screw Theory", *IEEE Trans. on Robotics and Automation*, **20**(2), pp. 181-190.

[25] Richard, P.C., Gosselin, C.M., Kong, X., 2007, "Kinematic Analysis and Prototyping of a Partially Decoupled 4-DOF 3T1R Parallel Manipulator", *ASME Journal of Mechanical Design*, **129**(6), pp. 611-616.

[26] Angeles J., 2002, "The qualitative synthesis of parallel manipulators", *Proc. Workshop on Fundamental Issues and Future Research Directions for Parallel Mechanisms and Manipulators*, October 3-4, Quebec, QC, Canada, pp. 160–169.

[27] Huang, Z., and Li, Q.C., 2002, "Construction and kinematic properties of 3-UPU parallel mechanism", *Proc. ASME Design Eng. Tech. Conf. and Computers and Information in Eng. Conf. IDETC/CIE 2002,* DETC2002-34321, Sept. 29–Oct. 2, Montreal, QC, Canada.

[28] Yang T.L., Jin Q., Liu A.X. Yao F.H., and Luo Y., 2001, "Structure synthesis of 4-DOF (3-translation and 1-rotation) parallel robot mechanisms based on the units of single-opened-chain", *Proc. ASME Design Eng. Tech. Conf. and Computers and Information in Eng. Conf. IDETC/CIE 2001*, DETC2001/DAC-21152, September 9-12, Pittsburgh, PA, USA.

[29] Li, Q., Wu, C., Hu, X., 2008, "Type Synthesis of 2R2T 4-DOF Parallel Mechanisms Using Constraint Synthesis Method", *Parallel Robotics: Recent Advances in Research and Application,* J. Wang and X.L. Liu (eds), Nova Publishers, New York, USA.

[30] Tischler, C.R., Hunt, K.H., and Samuel A.E., 1998, "On Optimizing the Kinematic Geometry of a Dextrous Robot Finger", *The International Journal of Robotics Research*, **17**(10), pp. 1055-1067.

[31] Tischler, C.R., Hunt, K.H., and Samuel A.E., 1998, "A spatial extension of Cardanic movement: its geometry and some derived mechanisms", *Mechanism and Machine Theory*, **33**(5), pp. 1249-1276.

[32] Liu, X.J., and Wang, J., 2005, "Some new spatial 3-DOF fully-parallel robotic mechanisms", *Robot Manipulators: new research, Nova Science*, New York, USA.

[33] Liu, X.J., Wang, J, Gao, F., Wang, L., 2001, "On the analysis of a new spatial three-degrees-of-freedom parallel manipulator", *IEEE Transactions on Robotics and Automation*, **17**(6) pp. 959–968.

[34] Liu, X.J., Wang, J., and Pritschow, G., 2005, "A New Family of Spatial 3-DoF Fully-Parallel Manipulators with high rotational capability", *Mechanism and Machine Theory*, **40**(4), pp. 475-494.

[35] Rodriguez Leal, E., Dai, J.S., and Pennock, G.R., 2011, "Kinematic Analysis of a 5-RSP Parallel Mechanism with Centralized Motion", Special Issue: Fundamental issues and new trends of parallel mechanisms and manipulators, Meccanica, Springer Verlag, 46(1), pp. 221-237.

[36] Rodriguez Leal, E., and Dai, J.S., 2010, "Evolutionary Design of Parallel Mechanisms", Lambert Academic Publications, Saarbrücken, Germany. ISBN 9783838378763

[37] Rodriguez Leal, E., Dai, J.S., and Pennock, G.R., 2009, "Inverse Kinematics and Motion Simulation of a 2-DOF Parallel Manipulator with 3-PUP Legs", 5th International Conference on Computational Kinematics, Springer Verlag, ISBN 9783642019463 Duisburg, Germany.

[38] Rodriguez Leal, E., Dai, J.S., and Pennock, G.R., 2009, "A Study of the Mobility of a Family of 3-DOF Parallel Manipulators", Proc. ASME IDETC/CIE 2009, San Diego, CA, USA.

[39] Rodriguez Leal, E., and Dai, J.S., 2010, "Inverse and Forward Instantaneous Kinematics of a 5-RSP Parallel Mechanism", IFToMM Symposium on Mechanism Design for Robotics, September 28-30, Mexico City, Mexico.

[40] Cuan, E., and Rodriguez Leal, E., 2011, "Mobility Analysis and Inverse Kinematics of a Novel 2R1T Parallel Manipulator", Proc. ASME IDETC/CIE 2011, August 28-31, Washington, CA, USA.

[41] Durum, M.M, 1975, "Kinematic Properties of Tripode (Tri-pot) Joints", *ASME Transactions: Journal of Engineering for Industry,* **97**(May), pp. 708-713.

[42] Callegari, M., and Tarantini, M., 2003, "Kinematic analysis of a novel translational platform", *ASME Journal of Mechanical Design*, **125**(2), pp. 308–315.

[43] Dai, J.S., Huang, Z., and Lipkin, H., 2006, "Mobility of Overconstrained Parallel Mechanisms," *ASME Journal of Mechanical Design*, Vol. **128**(1), pp. 220-229.

[44] Hervé, J. M., and Sparacino, F., 1991, ''*Structural Synthesis of Parallel Robots Generating Spatial Translation*", Proc. of the 5th Int. Conf. on Advanced Robotics, Pisa, Italy, Vol. 1, pp. 808–813.

[45] Ionescu, T.G., 2003, "IFToMM terminology/English", *Mech. Mach. Theory,* **38**(7-10), pp. 607-682.

[46] Chebychev, P.A., 1869, "Theorie des mecanismes connus sous le nom de parallelogrammes, 2eme partie" (in French), *Memoires presentes al_Academie imperiale des sciences de Saint-Petersbourg par divers savants*.

[47] Gogu, G., 2005, "Mobility of mechanisms: a critical review", *Mechanism and Machine Theory,* **40**(9), pp. 1068–1097.

[48] Uicker, J.J., Jr., Pennock, G.R., and Shigley, J.E., 2003, "Theory of Machines and Mechanisms", *Third Edition, Oxford University Press*, New York, pp. 734.

[49] Angeles, J., and Gosselin, C.M., 1988, "Determination du degre de liberte des chaines cinematiques" (in French), *Trans. CSME*, **12**(4), pp. 219–226.

[50] Fanghella, P., 1998, "Kinematics of spatial linkages by group algebra: a structure-based approach", *Mechanism and Machine Theory*, **23**(3), pp. 171–183.

[51] Fanghella, P., and Galletti, C., 1994, "Mobility analysis of single-loop kinematic chains: an algorithmic approach based on displacement groups", *Mechanism and Machine Theory*, **29**(8) 1187–1204.

[52] Hervé, J.M., 1999, "The Lie group of rigid body displacements, a fundamental tool for mechanism design", *Mechanism and Machine Theory*, **34**(5), pp. 719–730.

[53] Rico-Martinez, J.M., and Ravani, B., 2003, "On mobility analysis of linkages using group theory", *ASME Journal of Mechanical Design*, **125**(1) pp. 70–80.

[54] Rico, J.M., Gallardo, J., and Ravani, B., 2003, "Lie algebra and the mobility of kinematic chains", *Journal of Robotic Systems*, **20**(8), pp. 477–499.

[55] Zhao, J.S., Zhou, K., and Feng, Z.J., 2004, "A theory of degrees of freedom for mechanisms", *Mechanism and Machine Theory*, **39**(6), pp. 621–643.

[56] Zhao, J.S., Feng, Z.J., and Dong, J.X., 2006, "Computation of the configuration degree of freedom of a spatial parallel mechanism by using reciprocal screw theory", *Mechanism and Machine Theory*, **41**(12), pp. 1486-1504.

[57] Davidson, J.K., and Hunt, K.H., 2004, "Robots and Screw Theory: applications of kinematics and statics to robotics", *Oxford University Press*, New York.

Related papers by the authors Rodriguez-Leal, E., Dai, J.S., and Pennock, G.R.:

1. "Kinematic Analysis of a 5-RSP Parallel Mechanism with Centralized Motion," *Meccanica*, An International Journal of Theoretical and Applied Mechanics, Springer Netherlands, Springer Science+Business Media, **46**(1), February 2011, pp. 221-237.

2. "A Centralized 5-RSP Parallel Mechanism," *Proceedings of the 2008 ASME International Design Engineering Technical Conferences and Computers and Information in Engineering Conference*, The 32nd ASME Mechanisms and Robotics Conference, CD-ROM Format, New York City, New York, August 3-6, 2008.

3. "The Duality of Biomimetics and Artiomimetics in the Creative Process of Design," *Proceedings of the 2008 ASME International Design Engineering Technical Conferences and Computers and Information in Engineering Conference*, The 32nd ASME Mechanisms and Robotics Conference, CD-ROM Format, New York City, New York, August 3-6, 2008.

4. "Kinematic Analysis of a New Family of Centralized Parallel Mechanisms," *Proceedings of the Second International Workshop on Fundamental Issues and Future Research Directions for Parallel Mechanisms and Manipulators*, Montpellier, France, September 21 and 22, 2008, pp. 57-65.

5. "Inverse Kinematics and Motion Simulation of a 2-DOF Parallel Manipulator with 3-PUP Legs," *Proceedings of the 4th International Workshop on Robotics*, Duisburg, Germany, May 4-8, 2009.

6. "Using Artiomimetics to Develop a New Family of Centralized Single-DOF Parallel Manipulators," *Proceedings of the IEEE/ASME International Conference on Reconfigurable Mechanisms and Robots (ReMAR 2009)*, London, England, June 22-24, 2009.

7. "Kinematics of a Centralized 3-DOF Parallel Manipulator with 3-PUP Legs," *Proceedings of the 2009 ASME International Design Engineering Technical Conferences and Computers and Information in Engineering Conference*, The 33rd ASME Mechanisms and Robotics Conference, CD-ROM Format, San Diego, California, August 30-September 2, 2009.

8. "Constraint Redundancy in Mobility of Parallel Manipulators,"
(Malik, A.N., Dai, J.S., and Pennock, G.R.), *Proceedings of the 2008 ASME International Design Engineering Technical Conferences and Computers and Information in Engineering Conference*, The 32nd ASME Mechanisms and Robotics Conference, CD-ROM Format, New York City, New York, August 3-6, 2008.

In: Robotics: State of the Art and Future Trends
Editors: G. Legnani and I. Fassi
ISBN: 978-1-62100-403-5

Chapter 5

An Application of Screw Algebra to the Jerk Analysis of a Class of Six-Degrees-of-Freedom Three-Legged Parallel Manipulators

Jaime Gallardo-Alvarado*[*], *Karla A. Camarillo-Gómez
and Ramón Rodríguez-Castro
Department of Mechanical Engineering,
Instituto Tecnológico de Celaya, México

Abstract

The jerk analysis is a topic of growing interest in quite different scientific communities. In fact, the jerk can be a powerful mathematical tool to understand human motions, characterize singularities, compute impact forces, avoid the wearing of mechanical components like cams, smooth path planning trajectories of rigid bodies in robot manipulators and so on. In this chapter the jerk analysis of a class of three–legged parallel manipulators designed to realize six degrees of freedom is approached by means of the theory of screws. Simple and compact expressions involving active and passive joint rates are obtained by applying reciprocal–screw theory. As an intermediate step, semi–closed form expressions to solve the displacement analysis are easily derived owing the compact topology of the manipulators under study. A numerical example, which is verified with the aid of commercially available software, is included in order to show the application of the chosen methodology of kinematic analysis.

Keywords: parallel manipulator, jerk analysis, Klein form, reciprocal screw theory, Kinematics

---

[*]E-mail address: gjaime@itc.mx

1. Notation

In this section most of the notation used throughout the contribution is listed.

$\boldsymbol{a}$: linear acceleration vector

$\boldsymbol{A}$: reduced acceleration state

A, B: nominal positions of revolute joints

C: center of a spherical joint

e: distance between the centers of two spherical joints

E: the physical space

h: pitch of a screw

$\mathbf{I}$: identity matrix

ISA: instantaneous screw axis

$\boldsymbol{j}$: linear jerk vector

$\boldsymbol{J}$: reduced jerk state

$\mathbf{J}_*$: screw-coordinate Jacobian matrix of a limb

$\mathbf{J}$: active screw-coordinate Jacobian matrix of the manipulator

$\boldsymbol{p}, \boldsymbol{q}, \boldsymbol{r}$: position vectors

P: prismatic joint

q, $\dot{q}$, $\ddot{q}$, $\dddot{q}$: generalized coordinate and its time derivatives

R: revolute joint

$\mathbf{R}$: rotation matrix

$\hat{\boldsymbol{s}}$, $\boldsymbol{s}_*$: primal and dual parts of a screw

S: spherical joint

$\mathbf{T}$: 4×4 homogeneous transformation matrix

$\hat{\boldsymbol{u}}$: unit vector

$\boldsymbol{v}$: linear velocity vector

$\boldsymbol{V}$: velocity state

xyz: reference frame attached to the moving platform

XYZ: fixed reference frame

$\boldsymbol{\alpha}$: angular acceleration vector

$\alpha = \dot{}$:ωangular acceleration

$\boldsymbol{\beta}$: free or bound vector

Δ: operator of polarity

$\mathbf{\Gamma}_a$: instantaneous acceleration center

$\mathbf{\Gamma}_j$: instantaneous jerk center

$\mathbf{\Gamma}_{P/Q}$: a point of the instantaneous screw axis

λ: arbitrary real number

ω: angular velocity

$\boldsymbol{\omega}$: angular velocity vector

Ψa: acceleration matrix

Ψj: jerk matrix

$\rho = \ddot{\omega}$: angular jerk

$\boldsymbol{\rho}$: angular jerk vector

$\boldsymbol{\mathcal{A}}$: Lie screw of acceleration

$\boldsymbol{\mathcal{J}}$: Lie screw of jerk

$\{\boldsymbol{b}_1, \boldsymbol{b}_2, \boldsymbol{b}_3\}$: right-handed orthonormal basis

$\Re$: the field of real numbers

$\Re^3$: three-dimensional vector space over $\Re$

$(\bullet)$: inner product of three-dimensional vector algebra

$\$$: infinitesimal screw

$\{*; *\}$: the Klein form

$[* \quad *]$: the Lie product

$\mathbf{0}$: zero matrix

On the other hand, subscripts and superscripts are used to indicate the involved bodies or reference frames, e.g. ${}_{\#}\omega_{\%}$ denotes the joint rate velocity of body $\%$ with respect to body $\#$, while ${}^{\#}\boldsymbol{v}^{\%}_{\&}$ denotes the linear velocity vector of point $\&$ fixed in body $\%$ as it is observed from body or reference frame $\#$.

2. Introduction

The weaknesses and strengths of the Gough–Stewart platform [1, 2] have been extensively discussed, sometimes bravely, in the kinematician community. This hexapod is affected by a higher non–linear and complex coupled finite kinematics which, among other disadvantages, difficults its possible commercialization due to the necessity to include expensive and sophisticated control systems. Remarkable efforts have been devoted to solve the forward displacement analysis of a Gough–Stewart platform, see for instance [3, 4, 5], yielding elegant solutions mainly in the form of univariate polynomial equations capable to compute each possible pose, position and orientation, that the moving platform can reach with respect to the fixed platform, however with such methods it is not possible to determine, without using sensors, the actual configuration of the parallel manipulator. A closed–form solution to solve this formidable problem, or in other words a linear input–output relationship involving the actuated joint displacement vector and the task–space displacement vector, called by Freudenstein as "The Everest of modern kinematics" is without doubt the most challenging goal for any kinematician, maybe an unrealistic task. On the other hand, it is straightforward to show that due to the closed–loop nature of the mechanism, the number of limbs is a crucial factor limiting the motions of the moving platform and creating complex singularities in its workspace. With these considerations in mind and considering that a rigid body is statically balanced when it is supported on three points, reducing the number of limbs of the Gough–Stewart platform one can increase the workspace, improve the manipulability and diminish the presence of singularities of the manipulator. Certainly, the development of three–legged parallel manipulators, preserving the six degrees of freedom, is a viable option to alleviate the drawbacks of the Gough–Stewart platform. In [6, 7] the singularity analysis of a 3–RRRS parallel manipulator to realize six degrees of freedom, where each limb has two active revolute joints, is investigated. Three–legged parallel manipulators with PRPS and PPSP type limbs were proposed, respectively, in Behi [8] and Byun and Cho [9]. Simpler both forward kinematics and control are some relevant benefits when parallel manipulators with fewer than six limbs are used, for details the reader is referred to [10]-[15].

On the other hand, according to Viviani et al [16, 17] it is well–known that human motions are realized by controlling the jerk. Without doubt, parallel manipulators are a potential option in biomechanics and medical applications, see for instance [18]–[22], and therefore the development of efficient jerk analysis methods available for parallel manipulators become a necessity. The jerk is the time rate of change acceleration and is related to the rate of change of force, namely an impulse which is considered as a hammer blow force. The third–time derivative is of interest in quite different scientific communities. Particularly, among the analysis and control of human body movements, the jerk has relevant applications in the motion planning of robots and machine tools as well as in the higher order analyses of serial and parallel manipulators. Suh [23] and Schot [24] investigated the torsion of spatial curves using the jerk. Žefran et al [25] presented a coordinate–free formulation in order to generate shortest paths, and minimum acceleration and jerk properties of rigid bodies undergoing spherical motions, which correspond to the subgroup $SO(3)$ of rotational motions of the Euclidean group $SE(3)$, and general motions of $SE(3)$. Cheng [26] reported a procedure to obtain jerk curves which are expressed in explicit parametric forms. Moreover, it is well known that there is a direct relationship between the jerk and the movements of the human body [27]– [29]. In this way, Crossman and Goodeve [30] have shown that when high spatial precision is required, many movements related with the hands present irregularities and multiple velocity peaks. Gielen et al [31] noted that the charactecteristic pattern of cerebellar ataxia, related with the jerk and submovements, is contained in the trajectory of the hand during repeated arm movements. Goldvasser et al [32] investigated high curvature analysis and integrated absolute jerk for differentiating healthy and cerebellopathy patients performing pointing tasks.With the purpose to understand how the central nervous system controls the kinematics of rapid finger and hand movement, Novak et al [33] proposed an objective algorithm to identify overlapping submovements, detecting appreciable inflections in the acceleration traces which are related, of course, with the jerk. Viviani and coworkers [16, 17] proved that there is a correlation between perception, motion planning, and jerk in human motions. Furthermore, interesting mechanical applications of the jerk are, among many others, as follows: (i) the classical design of cam-follower systems; (ii) the correct characterization of singularities in closed kinematic chains; (iii) the synthesis of spherical and in general spatial linkages; (iv) the computation of impact forces due to rapid changes of the acceleration; (v) the improvement of path planning trajectories in the neighborhood of singular regions. Following that trend, Zahraee and Neff [34] reported the jerk analysis of some commercial cams, whereas Nguyen and Kim [35] introduced a method to smooth cam profiles by means of modified spline curves. Lee and Lin [36] have shown that, in milling machining operations, undesirable chattering effects can be reduced by controlling the continuity of jerks associated with the trajectory of the cutter.Finally, as shown by Erkorkmaz and Altintas [37] and Macfarlane and Croft [38], it is possible to control vibrations within kinematic structures by minimizing the jerk. Sparis and Moroutsos [39] applied the jerk analysis to improve and control the position analysis of planar mechanisms, when the mechanism is near to a singular configuration. Urbinati and Pennestri [40] developed a computer program for determining the jerk analysis of a rigid body with respect to another body or reference frame using matrix methods. Gasparetto and Zanotto [41] resort to fifth–order B–splines to compose smooth path planning trajectories in robot manipulators. Mandal and Nazkar [42] designed cam displacement functions based upon the jerk analysis.

In this contribution, the kinematics, including the jerk analysis, of a class of six–degrees–of–freedom three–legged parallel manipulators is addressed by means of the theory of screws. The rest of the chapter is organized as follows: section 3 provides a little information concerning with the kinematic properties of rigid bodies as well as an introductory subsection dealing with fundamental concepts of screw theory commonly used in the kinematic analysis of robot manipulators, section 4 gives a description of the family of parallel manipulators under study, in sections 5 and 6 the finite and infinitesimal kinematics of a member of the considered class of parallel manipulators are carried–out systematically, finally, in section 7 a numerical example is provided in order to show the application of the chosen method to solve the kinematics of this class of parallel manipulators.

3. Preliminary Concepts

In this section a brief review of well known concepts dealing with the kinematics of a rigid body and the theory of screws are given in order to provide a proper foundation for the subsequent sections.

3.1. Motion Properties of a Rigid Body

3.1.1. Velocity, Acceleration and Jerk Vectors

It is well known that the angular velocity cannot be obtained as the time derivative of any vector function. A robust definition of angular velocity is that provided by Kane and Levinson [43]. With reference to Fig. 1, let $\{\boldsymbol{b}_1, \boldsymbol{b}_2, \boldsymbol{b}_3\}$ be a right-handed set of mutually

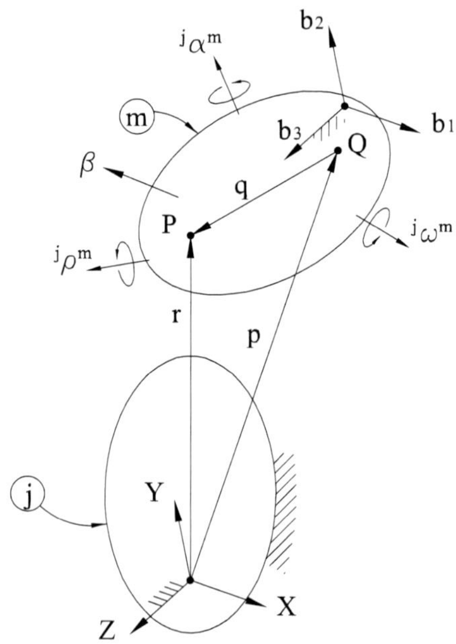

Figure 1. A rigid body m in general spatial motion.

perpendicular normalized vectors embedded on rigid body m, which is observed from body j. Then the angular velocity of body m with respect to body j, vector ${}^j\boldsymbol{\omega}^m$, is defined as follows

$$ {}^j\boldsymbol{\omega}^m \equiv \boldsymbol{b}_1 \frac{{}^j d\boldsymbol{b}_2}{dt} \bullet \boldsymbol{b}_3 + \boldsymbol{b}_2 \frac{{}^j d\boldsymbol{b}_3}{dt} \bullet \boldsymbol{b}_1 + \boldsymbol{b}_3 \frac{{}^j d\boldsymbol{b}_1}{dt} \bullet \boldsymbol{b}_2 \tag{1} $$

where the dot ($\bullet$) denotes the usual inner product of three-dimensional vector algebra. Expression 1 allows elegant formulation of a kinematic analogy between angular and linear kinematic quantities [44]. Further, assuming that $\boldsymbol{\beta} = \beta_1\boldsymbol{b}_1 + \beta_2\boldsymbol{b}_2 + \beta_3\boldsymbol{b}_3$ is a bound vector on m, it can be shown, for details the reader is referred to Rico and Duffy [45], that the angular velocity ${}^j\boldsymbol{\omega}^m$, as given in Eq. (1), satisfies the relationship

$$ \frac{{}^j d}{dt}\boldsymbol{\beta} = {}^j\boldsymbol{\omega}^m \times \boldsymbol{\beta}. \tag{2} $$

Whereas, if $\boldsymbol{\beta}$ is a free vector on m then it follows that

$$ \frac{{}^j d}{dt}\boldsymbol{\beta} = {}^j\boldsymbol{\omega}^m \times \boldsymbol{\beta} + \frac{{}^m d}{dt}\boldsymbol{\beta} \tag{3} $$

The time derivatives of vectors according to different frames are essential for obtaining the linear and angular properties of rigid bodies, in that way Eqs. (2) and (3) play a central role. Consider for instance that in Fig. 1 body j is a fixed reference frame (Earth). The position vector of point P with respect to point Q, vector $\boldsymbol{q}$, can be written as

$$ \boldsymbol{q} = \boldsymbol{r} - \boldsymbol{p} \tag{4} $$

Taking into account, see Eq. (2), that ${}^j d\boldsymbol{q}/dt = {}^j\boldsymbol{\omega}^m \times \boldsymbol{r}_{P/Q}$ is the linear velocity of P as observed from Q with respect to frame j, while ${}^j d\boldsymbol{r}/dt = {}^j\boldsymbol{v}_P^m$ is the linear velocity of point P with respect to j, and ${}^j d\boldsymbol{p}/dt = {}^j\boldsymbol{v}_Q^m$ is the linear velocity of point Q with respect to j. Then differentiating Eq. (4) and substituting yields

$$ {}^j\boldsymbol{\omega}^m \times \boldsymbol{r}_{P/Q} = {}^j\boldsymbol{v}_P^m - {}^j\boldsymbol{v}_Q^m \tag{5} $$

Assuming that the instantaneous velocity of point P vanishes then ${}^j\boldsymbol{v}_P^m = \boldsymbol{0}$, $\boldsymbol{r}_{P/Q} = \boldsymbol{\Gamma}_{P/Q}$ and Eq. (5) reduces to

$$ {}^j\boldsymbol{\omega}^m \times \boldsymbol{\Gamma}_{P/Q} = -{}^j\boldsymbol{v}_Q^m \tag{6} $$

which gives the expression for finding the instantaneous screw axis, ISA for brevity. Please note that an infinite number of points P satisfies Eq. (6), however only one point P is sufficient to characterize the ISA. Furthermore as it is shown in Legnani et al [46] the ISA maybe represented by a matrix meeting the conditions of Plücker coordinates. Premultiplying Eq. (6) by ${}^j\boldsymbol{\omega}^m$ and considering, see Wylie and Barret [47], that the vector triple product can be decomposed as

$$ {}^j\boldsymbol{\omega}^m \times ({}^j\boldsymbol{\omega}^m \times \boldsymbol{\Gamma}_{P/Q}) = ({}^j\boldsymbol{\omega}^m \bullet \boldsymbol{\Gamma}_{P/Q}){}^j\boldsymbol{\omega}^m - {}_j\omega_m^2 \boldsymbol{\Gamma}_{P/Q} \tag{7} $$

where ${}_j\omega_m$ is the magnitude of the angular velocity of body m with respect to body j. Furthermore, taking into account that the Chasles' theorem establishes, see Angeles [48], that

Given a rigid body moving with a general displacement, a set of its points, located on a line L, undergo identical displacements of minimum magnitude. Moreover, line L and the minimum magnitude displacement are parallel to the axis of the rotation involved.

Then, there is a point P fixed to body m whose linear velocity ${}^j\boldsymbol{v}_P^m$ is parallel to vector ${}^j\boldsymbol{\omega}^m$. In other words, the perpendicular component of the linear velocity of that point vanishes. Clearly, such a point is on the ISA and can be chosen in such a way that

$$ {}^j\boldsymbol{\omega}^m \bullet \boldsymbol{\Gamma}_{P/Q} = 0 \tag{8} $$

Hence, one point P on the ISA can be obtained using the equation

$$ \boldsymbol{\Gamma}_{P/Q} = {}^j\boldsymbol{\omega}^m \times {}^j\boldsymbol{v}_Q^m / {}_j\omega_m^2 \tag{9} $$

Concluding the velocity analysis, the ISA is parallel to the angular velocity ${}^j\boldsymbol{\omega}^m$ and passes through the point P computed with Eq. (9).

In order to approach the acceleration analysis, the time derivative of Eq. (5) leads to

$$ {}^j\boldsymbol{\alpha}^m \times \boldsymbol{r}_{P/Q} = {}^j\boldsymbol{a}_P^m - {}^j\boldsymbol{a}_Q^m - {}^j\boldsymbol{\omega}^m \times \left({}^j\boldsymbol{\omega}^m \times \boldsymbol{r}_{P/Q}\right) \tag{10} $$

where ${}^j\boldsymbol{\alpha}^m \equiv {}^j d({}^j\boldsymbol{\omega}^m)/dt$ is the angular acceleration of body m with respect to body j, ${}^j\boldsymbol{a}_P^m \equiv {}^j d({}^j\boldsymbol{v}_P^m)/dt$ is the linear acceleration of point P and ${}^j\boldsymbol{a}_Q^m \equiv {}^j d({}^j\boldsymbol{v}_Q^m)/dt$ is the linear acceleration of point Q.

Assuming that $\boldsymbol{r}_{P/Q} = \boldsymbol{\Gamma}_a$ is a position vector of P with respect to Q in which the instantaneous linear acceleration ${}^j\boldsymbol{a}_P^m$ vanishes and considering that ${}^j\boldsymbol{\omega}^m = \omega_X\hat{\imath} + \omega_Y\hat{\jmath} + \omega_Z\hat{k}$ and ${}^j\boldsymbol{\alpha}^m = \alpha_X\hat{\imath} + \alpha_Y\hat{\jmath} + \alpha_Z\hat{k}$. Then Eq. (10), provided that ${}^j\boldsymbol{a}_P^m = \boldsymbol{0}$, allows the instantaneous acceleration center $\mathbf{\Gamma}_a$ to be found. In fact, under these considerations, Eq. (10) can be rewritten in a matrix–vector form as follows

$$\Psi a \mathbf{\Gamma}_a = -{}^j\boldsymbol{a}_Q^m \tag{11}$$

where the *acceleration matrix* Ψa, see Mohamed [49], is given by

$$\Psi a = \begin{bmatrix} -\omega_Y^2 - \omega_Z^2 & -\alpha_Z + \omega_Y\omega_X & \alpha_Y + \omega_Z\omega_X \\ \alpha_Z + \omega_Y\omega_X & -\omega_X^2 - \omega_Z^2 & -\alpha_X + \omega_Z\omega_Y \\ \omega_Z\omega_X - \alpha_Y & \alpha_X + \omega_Z\omega_Y & -\omega_X^2 - \omega_Y^2 \end{bmatrix} \tag{12}$$

As pointed out by Mohamed [49], the instantaneous acceleration center $\mathbf{\Gamma}_a$ can be interpreted geometrically as the intersection point of the planes formed by the parametric components of the linear acceleration ${}^j\boldsymbol{a}_Q^m$ involved in Eq. (11). Furthermore, the instantaneous center of acceleration is a unique point if, and only if, the matrix Ψa is non–singular otherwise the rigid body either undergoes translational displacements or the analysis yields a *line of minimal acceleration*, for details the reader is referred to Legnani [50]. On the other hand, the acceleration analysis of rigid bodies can be approached also successfully by using methods involving homogeneous matrices, for details see Legnani et al [46, 51].

With the purpose to approach the jerk analysis, differentiating Eq. (10), and using Eq. (2), the reduction of terms yields

$$\begin{aligned} {}^j\boldsymbol{\rho}^m \times \boldsymbol{r}_{P/Q} = {}^j\boldsymbol{j}_P^m - {}^j\boldsymbol{j}_Q^m - 2{}^j\boldsymbol{\alpha}^m \times \left({}^j\boldsymbol{\omega}^m \times \boldsymbol{r}_{P/Q}\right) - {}^j\boldsymbol{\omega}^m \times \left({}^j\boldsymbol{\alpha}^m \times \boldsymbol{r}_{P/Q}\right) - \\ {}^j\boldsymbol{\omega}^m \times \left[{}^j\boldsymbol{\omega}^m \times \left({}^j\boldsymbol{\omega}^m \times \boldsymbol{r}_{P/Q}\right)\right] \end{aligned} \tag{13}$$

where ${}^j\boldsymbol{\rho}^m \equiv {}^jd({}^j\boldsymbol{\alpha}^m)/dt = {}^j\dot{\boldsymbol{\alpha}}^m$ is the angular jerk of body m with respect to body j, ${}^j\boldsymbol{j}_P^m \equiv {}^jd({}^j\boldsymbol{a}_P^m)/dt$ is the linear jerk of point P, and ${}^j\boldsymbol{j}_Q^m \equiv {}^jd({}^j\boldsymbol{a}_Q^m)/dt$ is the linear jerk of point Q.

Following the trend of the computation of the instantaneous acceleration center, the instantaneous jerk center $\mathbf{\Gamma}_j$ is calculated assuming that there is a point of the body m in which the instantaneous linear jerk vanishes, in other words ${}^j\boldsymbol{j}_P^m = \mathbf{0}$. Under this assumption and considering that ${}^j\boldsymbol{\rho}^m = \dot{\alpha}_X\hat{\imath} + \dot{\alpha}_Y\hat{\jmath} + \dot{\alpha}_Z\hat{k}$, then Eq. (13) can be rewritten in a matrix–vector form as follows

$$\Psi j \mathbf{\Gamma}_j = -{}^j\boldsymbol{j}_Q^m \tag{14}$$

where $\Psi j = [\mathrm{C}_1 \quad \mathrm{C}_2 \quad \mathrm{C}_3]$ is the *jerk matrix* whose columns C_i are given by

$$\mathrm{C}_1 = \begin{bmatrix} -3\alpha_Y\omega_Y - 3\alpha_Z\omega_Z \\ 2\omega_Y\alpha_X + \alpha_Y\omega_X - \omega_Z^3 - \omega_Z\omega_X^2 - \omega_Y^2\omega_Z + \dot{\alpha}_Z \\ 2\omega_Z\alpha_X + \alpha_Z\omega_X - \dot{\alpha}_Y + \omega_Y\omega_X^2 + \omega_Y^3 + \omega_Z^2\omega_Y \end{bmatrix} \tag{15}$$

$$\mathrm{C}_2 = \begin{bmatrix} 2\alpha_Y\omega_X + \omega_Y\alpha_X + \omega_Z^3 - \dot{\alpha}_Z + \omega_Y^2\omega_Z + \omega_Z\omega_X^2 \\ -3\alpha_X\omega_X - 3\alpha_Z\omega_Z \\ \dot{\alpha}_X + 2\omega_Z\alpha_Y - \omega_X^3 + \alpha_Z\omega_Y - \omega_Z^2\omega_X - \omega_X\omega_Y^2 \end{bmatrix} \tag{16}$$

and

$$\mathrm{C}_3 = \begin{bmatrix} \dot{\alpha}_Y + 2\alpha_Z\omega_X + \omega_Z\alpha_X - \omega_Y\omega_X^2 - \omega_Y^3 - \omega_Z^2\omega_Y \\ \omega_X\omega_Y^2 - \dot{\alpha}_X + 2\alpha_Z\omega_Y + \omega_Z^2\omega_X + \omega_Z\alpha_Y + \omega_X^3 \\ -3\alpha_X\omega_X - 3\alpha_Y\omega_Y \end{bmatrix} \tag{17}$$

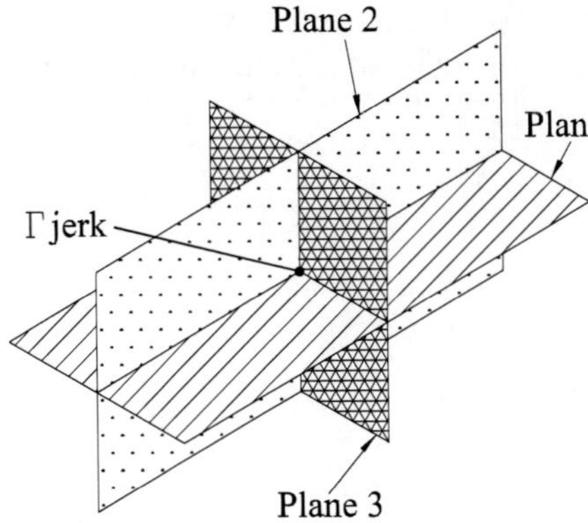

Figure 2. Geometric interpretation of the jerk center.

Equation (14) indicates that, if the matrix Ψj is non–singular then there is a unique point of body m with null linear jerk. However, if Ψj is singular, i.e. $\det(\Psi j) = 0$, then a geometrical interpretation of the jerk center, if any, is required. For example, if body m is undergoing purely translational motions, then the corresponding angular jerk vanishes, i.e. ${}^j\boldsymbol{\rho}^m = {}^j\boldsymbol{\alpha}^m = {}^j\boldsymbol{\omega}^m = \mathbf{0}$, causing the singularity of Ψj, a particular case. Further, according to Eq. (14) the jerk center can be interpreted geometrically, provided $\det(\Psi j) \neq 0$, as the point of intersection of the three planes, see Fig. 2, representing the three components of the linear jerk ${}^j\boldsymbol{j}_Q^m = j_X\hat{\imath} + j_Y\hat{\jmath} + j_Z\hat{k}$ of the point Q fixed to the moving body m. Indeed, these three planes are defined as follows

$$\text{Plane 1}:$$
$$(-3\alpha_Y\omega_Y - 3\alpha_Z\omega_Z)X + (2\alpha_Y\omega_X + \omega_Y\alpha_X + {\omega_Z}^3 - \dot{\alpha}_Z + {\omega_Y}^2\omega_Z + \omega_Z{\omega_X}^2)Y + (\dot{\alpha}_Y + 2\alpha_Z\omega_X + \omega_Z\alpha_X - \omega_Y{\omega_X}^2 - {\omega_Y}^3 - {\omega_Z}^2\omega_Y)Z = j_X \tag{18}$$

$$\text{Plane 2}:$$
$$(2\omega_Y\alpha_X + \alpha_Y\omega_X - {\omega_Z}^3 - \omega_Z{\omega_X}^2 - {\omega_Y}^2\omega_Z + \dot{\alpha}_Z)X + (-3\alpha_X\omega_X - 3\alpha_Z\omega_Z)Y + (\omega_X{\omega_Y}^2 - \dot{\alpha}_X + 2\alpha_Z\omega_Y + {\omega_Z}^2\omega_X + \omega_Z\alpha_Y + {\omega_X}^3)Z = j_Y \tag{19}$$

and

$$\text{Plane 3}:$$
$$(2\omega_Z\alpha_X + \alpha_Z\omega_X - \dot{\alpha}_Y + \omega_Y{\omega_X}^2 + {\omega_Y}^3 + {\omega_Z}^2\omega_Y)X + (\dot{\alpha}_X + 2\omega_Z\alpha_Y - {\omega_X}^3 + \alpha_Z\omega_Y - {\omega_Z}^2\omega_X - \omega_X{\omega_Y}^2)Y + (-3\alpha_X\omega_X - 3\alpha_Y\omega_Y)Z = j_Z \tag{20}$$

Finally, Eq. (14) is compact, simple and allows computation of the jerk ellipsoid for a given linear jerk ${}^j\boldsymbol{j}_Q^m$.

3.1.2. Kinematic States of a Rigid Body

The kinematic state of a rigid body is to be defined by a set of vectors from which the linear kinematic properties of any point in the rigid body may be determined at the level of the required analysis, in this work up to the jerk analysis.

A kinematic state is composed by a primal or angular part and a dual or linear part and must satisfy the conditions of helicoidal vector fields. Let E be the physical space, a helicoidal vector field is defined as follows

$$X : E^3 \rightarrow \Re^3 \quad X(P) = X(P) \quad \forall P \in E \tag{21}$$

where $X(P) \in \Re^3$. In addition, there is a vector $\boldsymbol{h} \in \Re^3$, called the vector of the helicoidal field, such that for any $Q \in E$ it satisfies

$$X(Q) = X(P) + \boldsymbol{h} \times \boldsymbol{r}_{Q/P} \tag{22}$$

Let O be an arbitrary point embedded in body m, see Fig. 1. The twist about a screw of body m with respect to body j, nowadays recalled as the velocity state, was introduced by Ball [52] at the beginning of the twentieth century as a six–dimensional vector ${}^j\boldsymbol{V}_O^m$ given by

$$ {}^j\boldsymbol{V}_O^m \equiv \begin{bmatrix} {}^j\boldsymbol{\omega}^m \\ {}^j\boldsymbol{v}_O^m \end{bmatrix} \tag{23}$$

where ${}^j\boldsymbol{\omega}^m$ and ${}^j\boldsymbol{v}_O^m$, namely the primal and dual parts, are the angular and linear velocities of body m taking point O as the reference pole. Unlike the dual part, the primal part ${}^j\boldsymbol{\omega}^m$ does not depend on point O and can be considered as a property of the rigid bodies. Furthermore, the velocity state ${}^j\boldsymbol{V}_O^m$ satisfies the conditions of helicoidal vector fields. In fact, assuming that P is an arbitrary point of body m and considering that ${}^j\boldsymbol{\omega}^m$ is the vector of the helicoidal field then it is possible to write according to Eq. (22),

$$ {}^j\boldsymbol{V}_Q^m = \begin{bmatrix} {}^j\boldsymbol{\omega}^m \\ {}^j\boldsymbol{v}_O^m + {}^j\boldsymbol{\omega}^m \times \boldsymbol{r}_{Q/O} \end{bmatrix} = \begin{bmatrix} {}^j\boldsymbol{\omega}^m \\ {}^j\boldsymbol{v}_Q^m \end{bmatrix} \tag{24}$$

The reduced acceleration state of body m with respect to body j, six–dimensional vector ${}^j\boldsymbol{A}_O^m$, is obtained as the time derivative, via a helicoidal field, of the velocity state ${}^j\boldsymbol{V}_O^m$. In reality such is not so clear than it appears. The first definition of reduced acceleration state, or accelerator for brevity, is credited to Brand [53], however its derivation using the concept of helicoidal vector field is due to Rico and Duffy [45], who concluded that the accelerator is given by

$$ {}^j\boldsymbol{A}_O^m = \begin{bmatrix} {}^j\boldsymbol{\alpha}^m \\ {}^j\boldsymbol{a}_O^m - {}^j\boldsymbol{\omega}^m \times {}^j\boldsymbol{v}_O^m \end{bmatrix} \tag{25}$$

To elucidate quickly that expression (25) satisfies the conditions of helicoidal vector fields consider that upon the accelerator ${}^j\boldsymbol{A}_O^m$ it follows that

$$\begin{aligned} {}^j\boldsymbol{A}_Q^m &= \begin{bmatrix} {}^j\boldsymbol{\alpha}^m \\ {}^j\boldsymbol{a}_O^m - {}^j\boldsymbol{\omega}^m \times {}^j\boldsymbol{v}_O^m + {}^j\boldsymbol{\alpha}^m \times \boldsymbol{r}_{Q/O} \end{bmatrix} \\ &= \begin{bmatrix} {}^j\boldsymbol{\alpha}^m \\ {}^j\boldsymbol{a}_O^m - {}^j\boldsymbol{\omega}^m \times ({}^j\boldsymbol{v}_Q^m - {}^j\boldsymbol{\omega}^m \times \boldsymbol{r}_{Q/O}) + {}^j\boldsymbol{\alpha}^m \times \boldsymbol{r}_{Q/O} \end{bmatrix} \\ &= \begin{bmatrix} {}^j\boldsymbol{\alpha}^m \\ {}^j\boldsymbol{a}_O^m + {}^j\boldsymbol{\omega}^m \times ({}^j\boldsymbol{\omega}^m \times \boldsymbol{r}_{Q/O}) + {}^j\boldsymbol{\alpha}^m \times \boldsymbol{r}_{Q/O} - {}^j\boldsymbol{\omega}^m \times {}^j\boldsymbol{v}_Q^m \end{bmatrix} \\ &= \begin{bmatrix} {}^j\boldsymbol{\alpha}^m \\ {}^j\boldsymbol{a}_Q^m - {}^j\boldsymbol{\omega}^m \times {}^j\boldsymbol{v}_Q^m \end{bmatrix} \end{aligned} \tag{26}$$

In order to find the *reduced jerk state or jerkor for brevity*, six–dimensional vector ${}^j\boldsymbol{J}_O^m$, of body m with respect to body j taking point O as the reference pole, consider that a general representation of it can be written as

$$ {}^j\boldsymbol{J}_O^m = \begin{bmatrix} {}^j\boldsymbol{\rho}^m \\ {}^j\boldsymbol{j}_O^m + \boldsymbol{c} \end{bmatrix} \tag{27}$$

where $\boldsymbol{c}$ is the complementary vector of the dual part of the jerkor. Henceforth, if P and Q are two points of the rigid body m, then their corresponding jerkors are given by

$$ {}^j\boldsymbol{J}_P^m = \begin{bmatrix} {}^j\boldsymbol{\rho}^m \\ {}^j\boldsymbol{j}_P^m + \boldsymbol{c} \end{bmatrix} \tag{28}$$

and

$$ {}^j\boldsymbol{J}_Q^m = \begin{bmatrix} {}^j\boldsymbol{\rho}^m \\ {}^j\boldsymbol{j}_Q^m + \boldsymbol{c} \end{bmatrix} \tag{29}$$

Assuming that the jerkor given in Eq. (22) satisfies the conditions of a helicoidal vector field, then it follows that

$$ {}^j\boldsymbol{j}_P^m + F(P) = {}^j\boldsymbol{j}_Q^m + F(Q) + {}^j\boldsymbol{\rho}^m \times \boldsymbol{r}_{P/Q} \tag{30}$$

where according to Eq. (13)

$$ {}^j\boldsymbol{j}_P^m = {}^j\boldsymbol{\rho}^m \times \boldsymbol{r}_{P/Q} + {}^j\boldsymbol{j}_Q^m + 2{}^j\boldsymbol{\alpha}^m \times \left({}^j\boldsymbol{\omega}^m \times \boldsymbol{r}_{P/Q}\right) + {}^j\boldsymbol{\omega}^m \times \left({}^j\boldsymbol{\alpha}^m \times \boldsymbol{r}_{P/Q}\right) + {}^j\boldsymbol{\omega}^m \times \left[{}^j\boldsymbol{\omega}^m \times \left({}^j\boldsymbol{\omega}^m \times \boldsymbol{r}_{P/Q}\right)\right] \tag{31}$$

Hence, see Eqs. (5) and (10), one obtains

$$ F(P) + 2{}^j\boldsymbol{\alpha}^m \times {}^j\boldsymbol{v}_P^m + {}^j\boldsymbol{\omega}^m \times {}^j\boldsymbol{a}_P^m = F(Q) + 2{}^j\boldsymbol{\alpha}^m \times {}^j\boldsymbol{v}_Q^m + {}^j\boldsymbol{\omega}^m \times {}^j\boldsymbol{a}_Q^m \tag{32}$$

Please note that Eq. (32) is in reality a constant vector which can be precisely the zero vector and therefore is evident that

$$ \left. \begin{aligned} F(P) &= -2{}^j\boldsymbol{\alpha}^m \times {}^j\boldsymbol{v}_P^m - {}^j\boldsymbol{\omega}^m \times {}^j\boldsymbol{a}_P^m \\ F(Q) &= -2{}^j\boldsymbol{\alpha}^m \times {}^j\boldsymbol{v}_Q^m - {}^j\boldsymbol{\omega}^m \times {}^j\boldsymbol{a}_Q^m \end{aligned} \right\} \tag{33}$$

Afterwards, a robust definition of jerkor results in

$$ {}^j\boldsymbol{J}_O^m = \begin{bmatrix} {}^j\boldsymbol{\rho}^m \\ {}^j\boldsymbol{j}_O^m - 2{}^j\boldsymbol{\alpha}^m \times {}^j\boldsymbol{v}_O^m - {}^j\boldsymbol{\omega}^m \times {}^j\boldsymbol{a}_O^m \end{bmatrix} \tag{34}$$

without doubt a remarkable and elegant advance in the higher order kinematic analysis of rigid bodies.

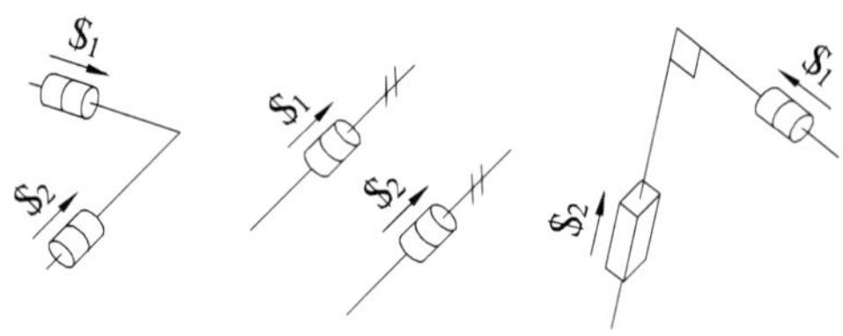

Figure 3. Typical systems of reciprocal screws.

3.2. Fundamentals of Screw Theory. Not All the Screws are in the Hardware

A screw is a straight line with which a definite linear magnitude termed the pitch is associated, Ball [52]. Furthermore, a screw $\$ = (\hat{s}, s_O)$ can be considered as a six–dimensional vector composed of a primal part and a dual part, where the primal part $\hat{s}$ is a unit vector along the screw axis, while the dual part s_O is the moment produced by the normalized vector $\hat{s}$ about a selected point O, namely the pole, which is instantaneously coincident with a point of a reference frame, usually its origin. The moment s_O is calculated as follows

$$s_O = h\hat{s} + \hat{s} \times r_{O/P} \tag{35}$$

where $r_{O/P}$ is a vector pointed from a point P on the screw axis to point O. Any lower kinematic pair can be represented either by a screw or a group of screws. A particular case is the prismatic pair. In this joint the pitch of the screw goes to infinity and therefore the corresponding primal part is the zero vector while the dual part is precisely the direction of the line.

The screw algebra is the set of elements of the form $\$ = (\hat{s}, s_O)$ with the following operations: $\forall \$_1 = (\hat{s}_1, s_{O1}), \$_2 = (\hat{s}_2, s_{O2})$ and $\lambda \in \Re$:

- Addition, $\$_1 + \$_2 = (\hat{s}_1 + \hat{s}_2, s_{O1} + s_{O2})$
- Multiplication by a scalar, $\lambda \$_1 = (\lambda \hat{s}_1, \lambda s_{O1})$
- The Lie product, $[\$_1 \quad \$_2] = (\hat{s}_1 \times \hat{s}_2, \hat{s}_1 \times s_{O2} - \hat{s}_2 \times s_{O1})$

Furthermore, the Lie algebra is endowed with two bilinear symmetric forms or inner products:

- The Killing form, $< \$_1; \$_2 >= \hat{s}_1 \bullet \hat{s}_2$
- The Klein form, $\{\$_1; \$_2\} = \hat{s}_1 \bullet s_{O2} + \hat{s}_2 \bullet s_{O1})$

It is said that the screws $\$_1$ and $\$_2$ are reciprocal if $\{\$_1; \$_2\} = 0$. While the Killing form is not used in this work, the Klein form plays a central role.

With reference to Fig. 3, if the screws $\$_1$ and $\$_2$ represent revolute joints, then these screws are reciprocal when their primal parts intersect or are parallel. Other possibility of reciprocal screws immediately emerges if the primal part of $\$_1$, representing a revolute joint, is perpendicular to the dual part of the screw $\$_2$ denoting a prismatic joint.

Infinitesimal screws and kinematic states form inseparable kinematic entities. In a serial manipulator, the velocity state of the end–effector, body m, with respect to the base link, body 0, can be written as a linear combination of the joint–rate velocities of the open kinematic chain, see Sugimoto and Duffy [54], as

$$ {}_0\omega_1\ {}^0\$^1 + {}_1\omega_2\ {}^1\$^2 + \ldots + {}_{m-1}\omega_{m-1}\ {}^{m-1}\$^m = {}^0\boldsymbol{V}_O^m \tag{36} $$

while the corresponding accelerator can be written, see Rico and Duffy [45], in screw form as

$$ {}_0\dot{\omega}_1\ {}^0\$^1 + {}_1\dot{\omega}_2\ {}^1\$^2 + \ldots + {}_{m-1}\dot{\omega}_{m-1}\ {}^{m-1}\$^m + \boldsymbol{\mathcal{A}} = {}^0\boldsymbol{A}_O^m \tag{37} $$

where $\boldsymbol{\mathcal{A}}$ is the Lie screw of acceleration which is computed as,

$$ \begin{aligned} \boldsymbol{\mathcal{A}} = & \left[{}_0\omega_1\ {}^0\$^1 \quad {}_1\omega_2\ {}^1\$^2 + \ldots + {}_{m-1}\omega_m\ {}^{m-1}\$^m\right] + \\ & \left[{}_1\omega_2\ {}^1\$^2 \quad {}_2\omega_3\ {}^2\$^3 + \ldots + {}_{m-1}\omega_m\ {}^{m-1}\$^m\right] + \ldots + \\ & \left[{}_{m-2}\omega_{m-1}\ {}^{m-2}\$^{m-1} \quad {}_{m-1}\omega_m\ {}^{m-1}\$^m\right] \end{aligned} \tag{38} $$

Finally, according to Rico et al [55] the jerkor in screw form is given by

$$ {}_0\ddot{\omega}_1\ {}^0\$^1 + {}_1\ddot{\omega}_2\ {}^1\$^2 + \ldots + {}_{m-1}\ddot{\omega}_{m-1}\ {}^{m-1}\$^m + \boldsymbol{\mathcal{J}} = {}^0\boldsymbol{J}_C^m \tag{39} $$

where $\boldsymbol{\mathcal{J}}$ is the i-th Lie screw of jerk which is computed as,

$$ \begin{aligned} \boldsymbol{\mathcal{J}} = 2\Big[& {}_0\omega_1{}^0\$^1 \quad {}_1\dot{\omega}_2{}^1\$^2 + \ldots + {}_{m-1}\dot{\omega}_m\ {}^m\$^m \\ & + \left[{}_1\omega_2{}^1\$^2 \quad {}_2\omega_3{}^2\$^3 + \ldots + {}_{m-1}\omega_m\ {}^{m-1}\$^m\right] \\ & + \left[{}_2\omega_3{}^2\$^3 \quad {}_3\omega_4{}^3\$^4 + \ldots + {}_{m-1}\omega_m\ {}^{m-1}\$^m\right] \\ & + \ldots + \left[{}_{m-2}\omega_{m-1}\ {}^{m-2}\$^{m-1} \quad {}_{m-1}\omega_m\ {}^{m-1}\$^m\right]\Big] + \\ 2\Big[& {}_1\omega_2{}^1\$^2 \quad {}_2\dot{\omega}_3{}^2\$^3 + \ldots + {}_{m-1}\dot{\omega}_m\ {}^m\$^m \\ & + \left[{}_2\omega_3{}^2\$^3 \quad {}_3\omega_4{}^3\$^4 + \ldots + {}_{m-1}\omega_m\ {}^{m-1}\$^m\right] \\ & + \left[{}_3\omega_4{}^3\$^4 \quad {}_4\omega_5{}^4\$^5 + \ldots + {}_{m-1}\omega_m\ {}^{m-1}\$^m\right] \\ & + \ldots + \left[{}_{m-2}\omega_{m-1}\ {}^{m-2}\$^{m-1} \quad {}_{m-1}\omega_m\ {}^{m-1}\$^m\right]\Big] + \\ & + \ldots + 2\left[{}_{m-2}\omega_{m-1}{}^{m-2}\$^{m-1} \quad {}_{m-1}\dot{\omega}_m{}^{m-1}\$^m\right] \\ & + \left[{}_0\dot{\omega}_1{}^0\$^1 \quad {}_1\omega_2{}^1\$^2 + \ldots + {}_{m-1}\omega_m\ {}^{m-1}\$^m\right] \\ & + \left[{}_1\dot{\omega}_2{}^1\$^2 \quad {}_2\omega_3{}^2\$^3 + \ldots + {}_{m-1}\omega_m\ {}^{m-1}\$^m\right] \\ & + \ldots + \left[{}_{m-2}\dot{\omega}_{m-1}{}^{m-2}\$^{m-1} \quad {}_{m-1}\omega_m{}^{m-1}\$^m\right] \\ & + \Big[{}_0\omega_1{}^0\$^1 \left[{}_0\omega_1{}^0\$^1 \quad {}_1\omega_2{}^1\$^2 + \ldots + {}_{m-1}\omega_m\ {}^{m-1}\$^m\right]\Big] \\ & + \Big[{}_1\omega_2{}^1\$^2 \left[{}_1\omega_2{}^1\$^2 \quad {}_2\omega_3{}^2\$^3 + \ldots + {}_{m-1}\omega_m\ {}^{m-1}\$^m\right]\Big] \\ & + \ldots + \Big[{}_{m-2}\omega_{m-1}{}^{m-2}\$^{m-1} \left[{}_{m-2}\omega_{m-1}{}^{m-2}\$^{m-1} \quad {}_{m-1}\omega_m{}^{m-1}\$^m\right]\Big] \end{aligned} \tag{40} $$

It is straightforward to demonstrate that expressions (36), (37) and (39) can be applied to parallel manipulators, the inclusion of subscripts and superscripts make the difference between the distinct limbs.

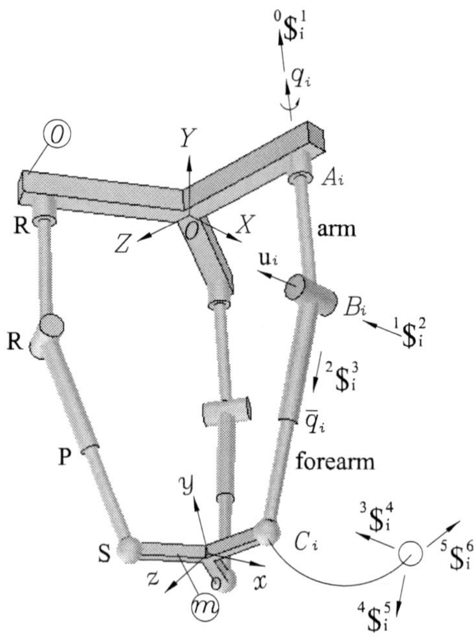

Figure 4. The 3–RRPS parallel manipulator.

4. Description of the Parallel Manipulators under Study

The class of parallel manipulator considered in this study has the following features:

- Six degrees of freedom are available
- The moving and fixed platforms are connected each other by means of three kinematically identical limbs
- Each limb is provided with two active joints
- The limbs are connected to the moving and fixed platforms by means, respectively, of a spherical joint and a revolute joint
- Due to the spherical joints, the proposed class of parallel manipulators does not require of additional conditions of manufacture like intersection of screws or similar restrictions

Some parallel manipulators meeting such characteristics are specified as 3–RRPS, 3–RRRS, 3–RPRS and 3–RPPS where R P S stand for revolute joint, prismatic joint and spherical joint. The 3–RRPS parallel manipulator is depicted in Fig. 4.

Without loss of generality, hereafter the analysis is focused on the 3–RRPS parallel manipulator. The chosen parallel manipulator for the analysis is composed by a moving platform m, a fixed platform 0, and three kinematically identical limbs. Each limb of the manipulator is connected to the fixed and moving platforms by means, respectively, of one active revolute joint q_i, in the remainder of this work $i = 1, 2, 3$, with nominal position

A_i located by vector $\boldsymbol{A}_i$, and one passive spherical joint with center C_i located by vector $\boldsymbol{C}_i$, whereas a passive revolute joint with nominal position B_i located by vector $\boldsymbol{B}_i$ connects the arm and the forearm in the same limb. The length of the forearm is variable and it is represented by an active prismatic joint $\overline{q}_i$ which completes the required two degrees of freedom for the limb. The axes of the revolute joints in the same limb are perpendicular each other and therefore the spherical joints move on concentric spheres around the corresponding point B_i.

5. Finite Kinematics

5.1. Forward Displacement Analysis

Let XYZ be a reference frame attached at the fixed platform and let xyz be a reference frame attached at the moving platform. The forward position analysis is formulated as follows: given the input generalized coordinates $\{\overline{q}_i, q_i\}$, compute the pose, position and orientation, of the moving platform with respect to the fixed platform. To this end, it is evident that the coupler revolute joints $B_i(i = 1, 2, 3)$ restrict the motion of the spherical joints $C_i(i = 1, 2, 3)$ in such a way that,

$$(\boldsymbol{C}_i - \boldsymbol{B}_i) \bullet \hat{\boldsymbol{u}}_i = 0 \tag{41}$$

where $\hat{\boldsymbol{u}}_i = \mathbf{R}_i \hat{\boldsymbol{w}}_i$ is a unit vector along the axis of the i–th revolute joint according to the rotation matrix $\mathbf{R}_i$ given by the motive angle q_i along the Y axis and the same vector $\hat{\boldsymbol{u}}_i$ but expressed in the moving reference frame xyz, vector $\hat{\boldsymbol{w}}_i$, whereas the bullet $(\bullet)$ denotes the inner product of the usual three–dimensional vectorial algebra. Moreover, the variable–length rods $\overline{q}_i$ provide three closure equations as follows,

$$(\boldsymbol{C}_i - \boldsymbol{B}_i) \bullet (\boldsymbol{C}_i - \boldsymbol{B}_i) - \overline{q}_i^2 = 0 \tag{42}$$

Furthermore, from the equilateral triangle $\Delta C_1 C_2 C_3$ of side e, three compatibility equations are given by,

$$(\boldsymbol{C}_i - \boldsymbol{C}_j) \bullet (\boldsymbol{C}_i - \boldsymbol{C}_j) - e^2 = 0 \quad i, j = 1, 2, 3 \,\mathrm{mod}(3) \tag{43}$$

Expressions (41)–(43) can be reduced systematically into a 8–th univariate polynomial equation by applying recursively the Sylvester dialytic elimination method, for details the reader is referred to [56, 57, 58], and therefore there are eight feasible poses that the moving platform can reach with respect to the fixed platform. Clearly, each center of a spherical joint admits at most two distinct positions. Once the coordinates of the spherical joints are computed, the forward position analysis can be considered as the well–known 4×4 homogeneous matrix ${}^0\mathbf{T}^m$ given by,

$${}^0\mathbf{T}^m = \begin{bmatrix} {}^0\mathbf{R}^m & \boldsymbol{r}_{o/O} \\ \mathbf{0} & 1 \end{bmatrix} \tag{44}$$

where ${}^0\mathbf{R}^m$ is the rotation matrix between the moving and fixed platforms that can be calculated using the method introduced in Gallardo–Alvarado et al [59] whereas $\boldsymbol{r}_{o/O} =$

$(\boldsymbol{C}_1+\boldsymbol{C}_2+\boldsymbol{C}_2)/3$ is the position vector of the center o of the moving platform with respect to the center O of the fixed platform.

It is worth to note how easy is the solution of the forward position analysis of this manipulator in comparison to the original Stewart platform. Certainly, from a computational point of view is cheaper the solution of a 8-order polynomial equation instead of a hazardous 40-order polynomial equation. Furthermore, since the determination of the actual configuration of most parallel manipulators are achieved using sensors, then semi–closed form solutions, like the reported here minimize the number of such elements.

5.2. Inverse Displacement Analysis

The inverse position analysis consists of finding the input joint angles $\{\overline{q}_i, q_i\}$ given the pose, matrix ${}^0\mathbf{R}^m$ and vector $\boldsymbol{r}_{o/O}$, of the moving platform with respect to the fixed platform. It is straightforward to demonstrate that,

$$\boldsymbol{C}_i = {}^0\mathbf{R}^m \boldsymbol{D}_i + \boldsymbol{r}_{o/O} \tag{45}$$

where $\boldsymbol{D}_i$ is the vector $\boldsymbol{C}_i$ expressed in the moving reference frame xyz. After, the generalized coordinates are calculated by means of Eqs. (41) and (42).

6. Infinitesimal Kinematics

Screw theory has been proved to be an excellent medium to investigate the kinematics of parallel manipulators and therefore it is chosen as the mathematical tool to approach the jerk analysis of the parallel manipulator under study. To this end, naturally, it is required to approach, in this order, the velocity and acceleration analyses. The notation of the screws, which will be written using Plcker coordinates, is depicted in Fig. 4.

6.1. Velocity Analysis

Let ${}^0\boldsymbol{\omega}^m$ and ${}^0\boldsymbol{v}_C^m$ be the angular and linear velocities of the center C of the moving platform taking point C as the reference pole. The velocity state of the moving platform with respect to the fixed platform, six–dimensional vector ${}^0\boldsymbol{V}_C^m = ({}^0\boldsymbol{\omega}^m, {}^0\boldsymbol{v}_C^m)$, can be written in screw form through any of the connector chains as,

$$ {}_0\omega_1^i\ {}^0\$_i^1 + {}_1\omega_2^i\ {}^1\$_i^2 + {}_2\omega_3^i\ {}^2\$_i^3 + {}_3\omega_4^i\ {}^3\$_i^4 + {}_4\omega_5^i\ {}^4\$_i^5 + {}_5\omega_6^i\ {}^5\$_i^6 = {}^0\boldsymbol{V}_C^m \tag{46}$$

The inverse velocity analysis is concerned with the computation of the joint–rate velocities of the manipulator for a prescribed velocity state ${}^0\boldsymbol{V}_C^m$. Clearly, the inverse velocity analysis is obtained via expression (46) as,

$$\boldsymbol{\omega}_i = \mathbf{J}_i^{-1}\ {}^0\boldsymbol{V}_C^m \tag{47}$$

where $\boldsymbol{\omega}_i = [{}_0\omega_1^i\ {}_1\omega_2^i\ {}_2\omega_3^i\ {}_3\omega_4^i\ {}_4\omega_5^i\ {}_5\omega_6^i]^T$ is a square matrix containing the first order joint rates of the i–th limb and $\mathbf{J}_i = [{}^0\$_i^1\ {}^1\$_i^2\ {}^2\$_i^3\ {}^3\$_i^4\ {}^4\$_i^5\ {}^5\$_i^6]$ is the screw–coordinate Jacobian matrix in the same limb.

In what follows the forward velocity analysis is simplified considerably using reciprocal–screw theory. To this aim, please note that the screw ${}^{4}\$_{i}^{5}$ is reciprocal to all the screws representing revolute joints in the same limb. Therefore, with the application of the Klein form of this screw to both sides of expression (46), the reduction of terms leads to,

$$\{{}^{4}\$_{i}^{5};{}^{0}\mathbf{V}_{C}^{m}\} = \dot{\bar{q}}_{i} \tag{48}$$

Similarly, since the screw ${}^{3}\$_{i}^{4}$ is reciprocal to all the screws in the same limb, excluding screw ${}^{0}\$_{i}^{1}$, then with the reduction of terms it is possible to write,

$$\{{}^{3}\$_{i}^{4};{}^{0}\mathbf{V}_{C}^{m}\} = \dot{q}_{i}\{{}^{3}\$_{i}^{4};{}^{0}\$_{i}^{1}\} \tag{49}$$

Casting into a matrix–vector form expressions (48) and (49) it follows that the input–output velocity equation of the manipulator is given by,

$$\mathbf{J}^{T}\,\boldsymbol{\Delta}\,{}^{0}\mathbf{V}_{C}^{m} = [\dot{\bar{q}}_{1}\;\dot{\bar{q}}_{2}\;\dot{\bar{q}}_{3}\;\dot{q}_{1}\{{}^{3}\$_{i}^{4};{}^{0}\$_{i}^{1}\}\;\dot{q}_{2}\{{}^{3}\$_{i}^{4};{}^{0}\$_{i}^{1}\}\;\dot{q}_{3}\{{}^{3}\$_{i}^{4};{}^{0}\$_{i}^{1}\}]^{T} \tag{50}$$

where $\mathbf{J} = [{}^{4}\$_{1}^{5}\;{}^{4}\$_{2}^{5}\;{}^{4}\$_{3}^{5}\;{}^{3}\$_{1}^{4}\;{}^{3}\$_{2}^{4}\;{}^{3}\$_{3}^{4}]$ is the active screw–coordinate Jacobian matrix of the manipulator and $\boldsymbol{\Delta} = \begin{bmatrix} \mathbf{0} & \mathbf{I} \\ \mathbf{I} & \mathbf{0} \end{bmatrix}$ is a partitioned 6×6 matrix according to the identity matrix $\mathbf{I}$ and the zero matrix $\mathbf{0}$.

6.2. Acceleration Analysis

Let ${}^{0}\dot{\boldsymbol{\omega}}^{m}$ and ${}^{0}\boldsymbol{a}_{C}^{m}$ be the angular and linear accelerations of the center of the moving platform. The reduced acceleration state of the moving platform with respect to the fixed platform, vector ${}^{0}\boldsymbol{A}_{C}^{m} = ({}^{0}\dot{\boldsymbol{\omega}}^{m}, {}^{0}\boldsymbol{a}_{C}^{m} - {}^{0}\boldsymbol{\omega}^{m} \times {}^{0}\boldsymbol{v}_{C}^{m})$, can be written in screw form through any of the limbs as,

$${}_{0}\dot{\omega}_{1}^{i}\,{}^{0}\$_{i}^{1} + {}_{1}\dot{\omega}_{2}^{i}\,{}^{1}\$_{i}^{2} + {}_{2}\dot{\omega}_{3}^{i}\,{}^{2}\$_{i}^{3} + {}_{3}\dot{\omega}_{4}^{i}\,{}^{3}\$_{i}^{4} + {}_{4}\dot{\omega}_{5}^{i}\,{}^{4}\$_{i}^{5} + {}_{5}\dot{\omega}_{6}^{i}\,{}^{5}\$_{i}^{6} + \boldsymbol{\mathcal{A}}_{i} = {}^{0}\boldsymbol{A}_{C}^{m} \tag{51}$$

where $\boldsymbol{\mathcal{A}}_{i}$ is the i-th Lie screw of acceleration which is calculated as,

$$\begin{aligned} \boldsymbol{\mathcal{A}}_{i} = {} & [{}_{0}\omega_{1}^{i}\,{}^{0}\$_{i}^{1}\quad {}_{1}\omega_{2}^{i}\,{}^{1}\$_{i}^{2} + \ldots + {}_{5}\omega_{6}^{i}\,{}^{5}\$_{i}^{6}] + \\ & [{}_{1}\omega_{2}^{i}\,{}^{1}\$_{i}^{2}\quad {}_{2}\omega_{3}^{i}\,{}^{2}\$_{i}^{3} + \ldots + {}_{5}\omega_{6}^{i}\,{}^{5}\$_{i}^{6}] + \\ & [{}_{2}\omega_{3}^{i}\,{}^{2}\$_{i}^{3}\quad {}_{3}\omega_{4}^{i}\,{}^{3}\$_{i}^{4} + \ldots + {}_{5}\omega_{6}^{i}\,{}^{5}\$_{i}^{6}] + \\ [{}_{3}\omega_{4}^{i}\,{}^{3}\$_{i}^{4}\quad & {}_{4}\omega_{5}^{i}\,{}^{4}\$_{i}^{5} + {}_{5}\omega_{6}^{i}\,{}^{5}\$_{i}^{6}] + [{}_{4}\omega_{5}^{i}\,{}^{4}\$_{i}^{5}\quad {}_{5}\omega_{6}^{i}\,{}^{4}\$_{i}^{5}] \end{aligned} \tag{52}$$

Therefore, given the accelerator ${}^{0}\boldsymbol{A}_{C}^{m}$, the solution of the inverse acceleration analysis is carried–out by means of expression (51). In fact

$$\dot{\boldsymbol{\omega}}_{i} = \mathbf{J}_{i}^{-1}({}^{0}\boldsymbol{A}_{C}^{m} - \boldsymbol{\mathcal{A}}_{i}) \tag{53}$$

where $\dot{\boldsymbol{\omega}}_{i} = [{}_{0}\dot{\omega}_{1}^{i}\;{}_{1}\dot{\omega}_{2}^{i}\;{}_{2}\dot{\omega}_{3}^{i}\;{}_{3}\dot{\omega}_{4}^{i}\;{}_{4}\dot{\omega}_{5}^{i}\;{}_{5}\dot{\omega}_{6}^{i}]^{T}$ is a matrix containing the second order joint rates of the i–th limb. On the other hand, following a similar procedure to the presented to solve the forward velocity analysis, the input–output acceleration expression to solve the forward acceleration analysis results in,

$$\mathbf{J}^{T}\,\boldsymbol{\Delta}\,{}^{0}\boldsymbol{A}_{C}^{m} = \begin{bmatrix} \ddot{\bar{q}}_1 + \{{}^{4}\$_{1}^{5};\boldsymbol{\mathcal{A}}_1\} \\ \ddot{\bar{q}}_2 + \{{}^{4}\$_{2}^{5};\boldsymbol{\mathcal{A}}_2\} \\ \ddot{\bar{q}}_3 + \{{}^{4}\$_{3}^{5};\boldsymbol{\mathcal{A}}_3\} \\ \ddot{q}_1\{{}^{3}\$_{1}^{4};{}^{0}\$_{1}^{1}\} + \{{}^{3}\$_{1}^{4};\boldsymbol{\mathcal{A}}_1\} \\ \ddot{q}_2\{{}^{3}\$_{1}^{4};{}^{0}\$_{2}^{1}\} + \{{}^{3}\$_{2}^{4};\boldsymbol{\mathcal{A}}_2\} \\ \ddot{q}_3\{{}^{3}\$_{1}^{4};{}^{0}\$_{3}^{1}\} + \{{}^{3}\$_{3}^{4};\boldsymbol{\mathcal{A}}_3\} \end{bmatrix} \tag{54}$$

where some elements like the Jacobian $\mathbf{J}$ and the operator of polarity $\boldsymbol{\Delta}$ are the same of the velocity analysis, which simplifies the implementation of such expressions in computer codes.

6.3. Jerk Analysis

Let ${}^{0}\ddot{\boldsymbol{\omega}}^{m}$ and ${}^{0}\boldsymbol{j}_{C}^{m}$ be the angular and linear jerks of the center of the moving platform. The reduced jerk state of the moving platform with respect to the fixed platform, vector ${}^{0}\boldsymbol{J}_{C}^{m} = ({}^{0}\ddot{\boldsymbol{\omega}}^{m}, {}^{0}\boldsymbol{j}_{C}^{m} - 2\,{}^{0}\dot{\boldsymbol{\omega}}^{m} \times {}^{0}\boldsymbol{v}_{C}^{m} - {}^{0}\boldsymbol{\omega}^{m} \times {}^{0}\boldsymbol{a}_{C}^{m})$, can be written in screw form as,

$${}_{0}\ddot{\omega}_{1}^{i}\;{}^{0}\$_{i}^{1} + {}_{1}\ddot{\omega}_{2}^{i}\;{}^{1}\$_{i}^{2} + {}_{2}\ddot{\omega}_{3}^{i}\;{}^{2}\$_{i}^{3} + {}_{3}\ddot{\omega}_{4}^{i}\;{}^{3}\$_{i}^{4} + {}_{4}\ddot{\omega}_{5}^{i}\;{}^{4}\$_{i}^{5} + {}_{5}\ddot{\omega}_{6}^{i}\;{}^{5}\$_{i}^{6} + \boldsymbol{\mathcal{J}}_{i} = {}^{0}\boldsymbol{J}_{C}^{m} \tag{55}$$

where $\boldsymbol{\mathcal{J}}_{i}$ is the i-th Lie screw of jerk which is computed as,

$$\begin{aligned} \boldsymbol{\mathcal{J}}_{i} = 2\Big[& {}_{0}\omega_{1}{}^{0}\$^{1} \quad {}_{1}\dot{\omega}_{2}{}^{1}\$^{2} + \ldots + {}_{4}\dot{\omega}_{5}{}^{4}\$^{5} + {}_{5}\dot{\omega}_{6}{}^{5}\$^{6} \\ & + [{}_{1}\omega_{2}{}^{1}\$^{2} \quad {}_{2}\omega_{3}{}^{2}\$^{3} + \ldots + {}_{4}\omega_{5}{}^{4}\$^{5} + {}_{5}\omega_{6}{}^{5}\$^{6}] \\ & + [{}_{2}\omega_{3}{}^{2}\$^{3} \quad {}_{3}\omega_{4}{}^{3}\$^{4} + \ldots + {}_{4}\omega_{5}{}^{4}\$^{5} + {}_{5}\omega_{6}{}^{5}\$^{6}] \\ & + \ldots + [{}_{4}\omega_{5}{}^{4}\$^{5} \quad {}_{5}\omega_{6}{}^{5}\$^{6}] \Big] \\ +2\Big[& {}_{1}\omega_{2}{}^{1}\$^{2} \quad {}_{2}\dot{\omega}_{3}{}^{2}\$^{3} + \ldots + {}_{4}\dot{\omega}_{5}{}^{4}\$^{5} + {}_{5}\dot{\omega}_{6}{}^{5}\$^{6} \\ & + [{}_{2}\omega_{3}{}^{2}\$^{3} \quad {}_{3}\omega_{4}{}^{3}\$^{4} + \ldots + {}_{4}\omega_{5}{}^{4}\$^{5} + {}_{5}\omega_{6}{}^{5}\$^{6}] \\ & + [{}_{3}\omega_{4}{}^{3}\$^{4} \quad {}_{4}\omega_{5}{}^{4}\$^{5} + \ldots + {}_{4}\omega_{5}{}^{4}\$^{5} + {}_{5}\omega_{6}{}^{5}\$^{6}] \\ & + \ldots + [{}_{4}\omega_{5}{}^{4}\$^{5} \quad {}_{5}\omega_{6}{}^{5}\$^{6}] \Big] \\ & + \ldots + 2\,[{}_{4}\omega_{5}{}^{4}\$^{5} \quad {}_{5}\dot{\omega}_{6}{}^{5}\$^{6}] \\ & + [{}_{0}\dot{\omega}_{1}{}^{0}\$^{1} \quad {}_{1}\omega_{2}{}^{1}\$^{2} + \ldots + {}_{4}\omega_{5}{}^{4}\$^{5} + {}_{5}\omega_{6}{}^{5}\$^{6}] \\ & + [{}_{1}\dot{\omega}_{2}{}^{1}\$^{2} \quad {}_{2}\omega_{3}{}^{2}\$^{3} + \ldots + {}_{4}\omega_{5}{}^{4}\$^{5} + {}_{5}\omega_{6}{}^{5}\$^{6}] \\ & + \ldots + [{}_{4}\dot{\omega}_{5}{}^{4}\$^{5} \quad {}_{5}\omega_{6}{}^{5}\$^{6}] \\ + & \Big[{}_{0}\omega_{1}{}^{0}\$^{1}\,[{}_{0}\omega_{1}{}^{0}\$^{1} \quad {}_{1}\omega_{2}{}^{1}\$^{2} + \ldots + {}_{4}\omega_{5}{}^{4}\$^{5} + {}_{5}\omega_{6}{}^{5}\$^{6}]\Big] \\ & + \Big[{}_{1}\omega_{2}{}^{1}\$^{2}\,[{}_{1}\omega_{2}{}^{1}\$^{2} \quad {}_{2}\omega_{3}{}^{2}\$^{3} + \ldots + {}_{5}\omega_{6}{}^{5}\$^{6}]\Big] \\ & + \ldots + \Big[{}_{4}\omega_{5}{}^{4}\$^{5}\,[{}_{4}\omega_{5}{}^{4}\$^{5} \quad {}_{5}\omega_{6}{}^{5}\$^{6}]\Big] \end{aligned} \tag{56}$$

Therefore, the solution of the inverse jerk analysis of the manipulator is carried–out upon expression (55) as follows,

$$\dddot{\boldsymbol{\omega}}_i = \mathbf{J}_i^{-1}({}^0\boldsymbol{J}_C^m - \boldsymbol{\mathcal{J}}_i) \tag{57}$$

where $\dddot{\boldsymbol{\omega}}_i = [{}_0\dddot{\omega}_1^i \; {}_1\dddot{\omega}_2^i \; {}_2\dddot{\omega}_3^i \; {}_3\dddot{\omega}_4^i \; {}_4\dddot{\omega}_5^i \; {}_5\dddot{\omega}_6^i]^T$ is a matrix containing the third order joint rates of the i–th limb.

Finally, following the trend of the velocity and acceleration analyses, the input–output jerk expression to solve the forward jerk analysis results in,

$$\mathbf{J}^T \, \boldsymbol{\Delta} \, {}^0\boldsymbol{J}_C^m = \begin{bmatrix} \dddot{\overline{q}}_1 + \{{}^4\$_1^5; \boldsymbol{\mathcal{J}}_1\} \\ \dddot{\overline{q}}_2 + \{{}^4\$_2^5; \boldsymbol{\mathcal{J}}_2\} \\ \dddot{\overline{q}}_3 + \{{}^4\$_3^5; \boldsymbol{\mathcal{J}}_3\} \\ \dddot{q}_1\{{}^3\$_1^4; {}^0\$_1^1\} + \{{}^3\$_1^4; \boldsymbol{\mathcal{J}}_1\} \\ \dddot{q}_2\{{}^3\$_1^4; {}^0\$_2^1\} + \{{}^3\$_2^4; \boldsymbol{\mathcal{J}}_2\} \\ \dddot{q}_3\{{}^3\$_1^4; {}^0\$_3^1\} + \{{}^3\$_3^4; \boldsymbol{\mathcal{J}}_3\} \end{bmatrix} \tag{58}$$

Note that expressions (50), (54) and (58) are compact, can be easily translated into computer codes and do not require the calculation of the corresponding passive joint–rates of the manipulator. The benefits of using screw theory in the kinematic analysis of parallel manipulators cannot be denied when expressions like these are systematically obtained.

7. Numerical Example

In this section an application of the proposed method is presented. The numerical example consists of solving the workspace and forward kinematics of a 3–RRPS parallel manipulator where the input–joint angles are associated to the revolute joints mounted on the fixed platform.

To this aim, consider that the nominal coordinates of the revolute joints A_i, using SI units thorough the numerical example, connecting the limbs to the fixed platform form an equilateral triangle $\Delta A_1 A_2 A_3$ with corners expressed in the reference frame XYZ chosen as,

$$A_1 = (0.25, 0, 0), A_2 = (-0.125, 0, -0.2165), A_3 = -0.125, 0, 0.2165)$$

Whereas the coordinates of the coupler revolute joints B_i form an equilateral triangle $\Delta B_1 B_2 B_3$ with corners given by,

$$B_1 = (0.25, -0.25, 0), B_2 = (-0.125, -0.25, -0.2165), B_3 = -0.125, -0.25, 0.2165)$$

Furthermore, consider $e = 0.2165$ as the side of the equilateral triangle $\Delta C_1 C_2 C_3$ associated to the moving platform.

On the other hand, the example concern a movement for which the generalized coordinates are assigned as follows

$$\left.\begin{array}{l} q_1 = 0.5\sin(t), q_2 = 7\pi/6 + 0.2\sin(t), q_3 = \pi/3 + 0.5\sin(t) \\ \overline{q}_1 = 0.371 + 0.14\sin(t), \overline{q}_2 = 0.371 + 0.15\sin(t), \overline{q}_3 = 0.371 + 0.12\sin(t) \end{array}\right\}$$

where the interval of the time t is selected as $0 \leq t \leq 2\pi$. Of course, at the time $t = 0$ the parallel manipulator is at his home position.

With the above given information, the first part of the numerical example consists of determining the home position of the robot. To this end, it is necessary to solve the forward displacement analysis considering $t = 0$. According to expressions (41)–(43), the eight real solutions of the forward position analysis result in,

$$\left.\begin{array}{l} (0.0, -0.6, 0.0) \\ (0.00000118500, -0.2786094246, 0.0) \\ (0.00830142581, -0.3145891810, 0.00000128525) \\ (-0.00415310759, -0.3146131463, -0.00719136267) \\ (-0.00415398605, -0.3146245895, 0.00719558092) \\ (0.03477387012, -0.5378997733, -0.06023357070) \\ (0.03477428560, -0.5378999747, 0.06023442521) \\ (-0.06955293399, -0.5379014094, 0.00000250518) \end{array}\right\} \quad (59)$$

The user has the liberty to select any of the eight feasible solutions as the home position of the robot. Taking the first solution as the home position of the robot, and considering variations of the input joint angles as,

$$-0.0174 < \Delta q_i < 0.0174, \quad -0.25 < \Delta \overline{q}_i < 0.25$$

the workspace of the parallel manipulator is depicted in Fig. 5. Due to the lack of a finite closed–form solution, only representative points are computed here.

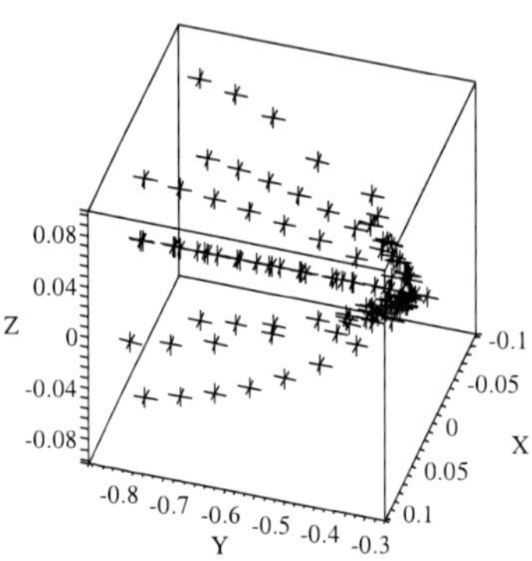

Figure 5. Boundary of the workspace of the center of the moving platform.

As it was expected, the workspace generated numerically is a symmetric volume which could give some insight about the acceptable manipulability and isotropy of the robot at hand, although these topics beyond the scope of this work it is interesting to include a simple comment about it.

The next part of the exercise is concerned with the forward velocity and acceleration analyses. The resulting angular and linear velocities and accelerations of the center of the moving platform, given the instantaneous generalized velocities and accelerations of the manipulator, are shown in Fig. 6. Furthermore, these results are compared in the same figure with simulations generated with the aid of commercially available software like ADAMS ©.

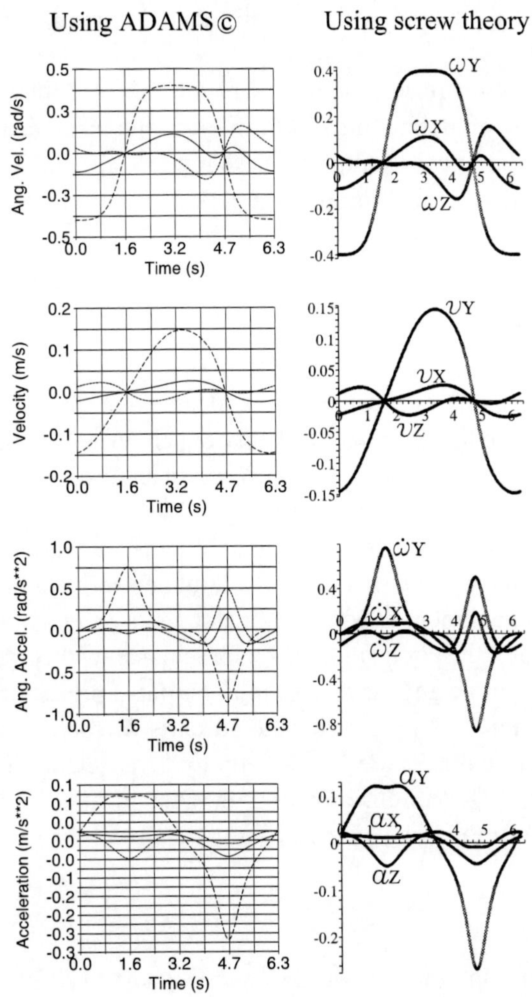

Figure 6. Time history of the angular and linear velocities and accelerations of the center of the moving platform.

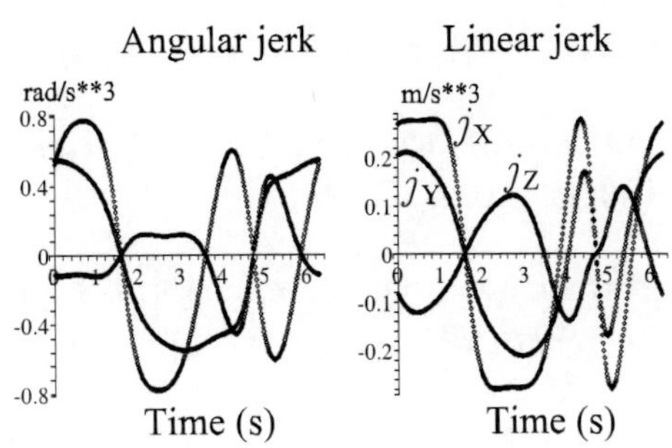

Figure 7. Time history of the angular and linear jerks of the center of the moving platform.

Please note that the results obtained at this point using screw theory are in excellent agreement when it are compared with a totally different approach such as the software ADAMS©.

The final part of the exercise is concerned with the computation of the angular and linear jerks of the center of the moving platform given the instantaneous generalized jerks, the main goal of this numerical example. These numerical results are provided in Fig. 7. In that way it is important to mention that the jerk analysis of mechanical systems is not available in ADAMS©.

8. Conclusions

In this work the jerk analysis of a class of six–degrees–of–freedom three–legged parallel manipulator is approached by means of the theory of screws. As an intermediate step, semi–closed form solutions concerned with the displacement analysis are easily derived owing the compact topology of the manipulator under study, while in a classical hexapod the solution of the forward position analysis is based upon a 40–order univariate polynomial equation, all solutions of this analysis for the parallel manipulator here studied stem from the roots of an 8-order univariate polynomial equation, the benefits of a computational point of view are evident. Furthermore, the application of reciprocal–screw theory, the main contribution of this work, leads to simple and compact input–output expressions to solve the velocity, acceleration and jerk analyses of the considered class of robot manipulators. Finally, a numerical example whose results are verified, up to the acceleration analysis, with the aid of special software like ADAMS©is provided. In that way, it is worth to mention that such software does not include the jerk analysis.

Acknowledgments

This work has been supported by DGEST and Conacyt of México.

References

[1] Gough, V.E., Whitehall, S.G., Universal tire test machine, *Proceedings Institution of Technical Congress FISITA,* 1961, pp. 117.

[2] Stewart, D., A platform with six degrees of freedom, *Proceedings Institution of Mechanical Engineers,* **180**(1)(1965), 371-386.

[3] Innocenti, C., Forward kinematics in polynomial form of the general Stewart platform, *ASME Journal of Mechanical Design*, **123**(2)(2001), 254-260.

[4] Rolland, L., Certified solving of the forward kinematics problem with an exact algebraic method for the general parallel manipulator, *Advanced Robotics,* **19**(2005), 9951025.

[5] Raghavan, M., The Stewart platform of general geometry has 40 configurations, *ASME Journal of Mechanical Design,* **115**(1993), 277-282.

[6] Guilin, Y., Chen, I–M., Lin, W., Angeles, J., Singularity analysis of three-legged parallel robots based on passive–joint velocities, *Proceedings 2001 IEEE International Conference on Robotics & Automation,* 2001, pp. 2407–2412.

[7] Angeles, J., Yang, G., Chen, I-M., Singularity analysis of three-legged, six-dof platform manipulators with RRRS legs, *Proceedings 2001 IEEE/ASME International Conference on Advanced Intelligent Mechatronics Proceedings,* 2001, pp. 32–36.

[8] Behi, F., Kinematic analysis for a six-degree-of-freedom 3-prps parallel mechanism, *IEEE Journal of Robotics and Innovation,* **4**(5)(1988), 561565.

[9] Byun, Y., Cho, H.S., Analysis of a novel 6-dof, 3-ppsp parallel manipulator, *The International Journal of Robotics Research,* **16**(6)(1997), 859-872.

[10] Zlatanov, D., Dai, M.Q., Fenton, R.G., Benhabib, B., Mechanical design and kinematic analysis of a three–legged six–degree-of-freedom parallel manipulator, *Proceedings 22nd Biennial Mechanisms Conference,* 1992, pp. 529–536.

[11] Geng, Z., Haynes, L.S., A 3-2-1 kinematic configuration of a Stewart platform and its application to six–degree–of–freedom pose measurements, *Robotics & Computer–Integrated Manufacturing,* **11**(1)(1994), 23–34.

[12] Portman, T.V., Sandler, B.Z., Tripod robot with cylindrically actuated limbs: structure and kinematics, *Mechanism and Machine Theory,* **37**(12)(2002) 1447-1463.

[13] Zabalza, I., Ros, J., Gil, J.J., Pintor, J.M., Jiménez, J.M., TRI-SCOTT. A new kinematic structure for a 6-DOF decoupled parallel manipulator, *Proceedings WORKSHOP on Fundamental Issues and Future Research Directions for Parallel Mechanisms and Manipulators,* 2002, pp. 1-4.

[14] Goldsmith, P.B., Design and kinematics of a three-legged parallel manipulator, *IEEE Transactions on Robotics and Automation,* **19**(4)(2003), 726-731.

[15] Jin, Y., Chen, I-M., Yang, G., Kinematic design of a family of 6-DOF partially decoupled parallel manipulators, *Mechanism and Machine Theory,* **44**(5)(2009), 912-922.

[16] Viviani, P., Schneider,R., A development study of the relationship between geometry and kinematics in drawing movements, *Journal of Experimental Psychology in Human Perception and Performance,* **17**(1991), 198-218.

[17] Viviani, P., Flash, T., Minimum-jerk, two-thirds power law, and isochrony: converging approaches to movement planning, *Journal of Experimental Psychology in Human Perception and Performance,* **21**(1995), 32-53.

[18] Howard, R.A., Rosvold, J.M., Darcy, S.P., Corr, D.T., Shrive, N.G., Tapper, J.E., Ronsky, J.L., Reproduction of in vivo motion using a parallel robot, *ASME Journal of Biomechanical Engineering,* **129**(5)(2007), 743(7 pages).

[19] Lessard, S., Bigras, P., Bonev, I.A., A new medical parallel robot and its static balancing optimization, *ASME Journal of Medical Devices,* **1**(2007), 272–278.

[20] Li, Y., Xu, Q., Design and development of a medical parallel robot for cardiopulmonary resuscitation, *IEEE/ASME Transactions on Mechatronics,* **12**(3)(2007), 265-273.

[21] Stan, S.–D., Bălan, R., Mătie, V., Rad, C., Kinematics and fuzzy control of ISOGLIDE3 medical parallel robot, *MECHANIKA,* **1**(75)(2009), 62-66.

[22] Tsoi, Y.H., Xie, S.Q., Design and control of a parallel robot for ankle rehabilitation, *International Journal of Intelligent Systems Technologies and Applications,* **8**(1/2/3/4)(2010), 100-113.

[23] Suh, C., Higher order analysis of spatial coupler curves, *Journal of Mechanisms,* **6**(1971), 81-95.

[24] Schot, S.H., Jerk: the time rate of change of acceleration, *American Journal of Physics,* **46**(1978), 10901094.

[25] Žefran, M., Kumar, V., Croke, C., On the generation of smooth three-dimensional rigid body motions, *IEEE Transactions on Robotics and Automation,* **14**(1995), 579-589.

[26] Cheng, W.-T. Synthesis of universal motion curves ingeneralized model, *ASME Journal of Mechanical Design,* **124**(2002), 284293.

[27] Morasso, P., Spatial control arm movements, *Experimental Brain Reserach,* **42**(1981), 223-227.

[28] Flash, T., Hogan, N., The coordination of arm movements: an experimentally confirmed mathematical model, *Journal of Neuroscience,* **5**(1985), 16881703.

[29] Uno, Y., Kawato, M., Suzuki, R., Formation and control of optimal trajectory in human multijoint arm movements, *Biological Cybernetics*, **61**(1989), 89-101.

[30] Crossman, E.R.F.W., Goodeve, P.J., Feedback control of handmovements and Fitts law, *Quarterly Journal of Experimental Psychology,* **A35**(1983), 251-278.

[31] Gielen, C.C., Vrijenhock, E.J., Neggers, S.F., Arm position constraints during pointing and reaching in 3-D space, *Journal of Neurophysiology,* **78**(1997), 660673.

[32] Goldvasser, D., McGibbon, C.A., Krebs, D.E., High curvature and jerk analyses of arm ataxia, *Biological Cybernetics,* **84**(2001), 84–90.

[33] Novak, K.E., Miller, L.E., Houk, J.C., Kinematic properties of rapid hand movements in a knob turning task, *Experimental Brain Research,* **132**(2000), 419433.

[34] Zahraee, M.A., Neff, G.P., Computer–aided analysis of cam mechanisms: a reverse engineering approach, *Proceedings of the Second International Mechanical Engineering Conference, Shiraz,* 1996, pp. 905912.

[35] Nguyen, V.U., Kim, D., Cam profile smoothing by modified spline curves, *In Proceedings of the 12th IFToMM World Congress,* 2007.

[36] Lee, T.S., Lin,Y.J., An improved sculptured part surface design method with jerk continuity consideration for smooth machining, *International Journal of Advanced Manufacturing Technology,* **15**(1999), 640-648.

[37] Erkorkmaz, K., Altintas, Y., High speed CNC system desing, Part I. Jerk limited trajectory generation and quintic spline interpolation, *International Journal of Machine Tools and Manufacture,* **41**(2001), 1323-1345.

[38] Macfarlane, S., Croft, E.A., Jerk-bounded manipulator trajectory planning: desing for real-time applications, *IEEE Transactions on Robotics and Automation,* **19**(2003), 42-52.

[39] Sparis, P.D., Moroutsos, S.G., A new matrix method for the kinematic analysis and motion simulation of planar mechanisms with lower kinematic pairs, *ASME Journal of Mechanisms Transmission and Automation in Design,* **106**(1984), 429436.

[40] Urbinati, F., Pennestri, E., A tool for third order kinematic analysis of spatial mechanisms, *Proceedings of the 25th Biennial Mechanisms Conference,* Atlanta, paper DETC/MECH–5901, 1998.

[41] Gasparetto, A., Zanotto, V., A new method for smooth trajectory planning of robot manipulators, *Mechanism and Machine Theory*, **42**(4)(2007), 455-471.

[42] Mandal, M., Nazcar, T.K., Introduction of control points in splines for synthesis of optimized cam motion program, *Mechanism and Machine Theory*, **44**(1)(2009), 255-271.

[43] Kane, T.R., Levinson, D.A., *Dynamics: Theory and Applications,* (McGraw Hill, NewYork), 1985.

[44] Rico-Martínez, J.M., Gallardo–Alvarado, J., A complete kinematic analogy, *International Journal of Mechanical Engineering Education,* , **30**(2002), 8-16.

[45] Rico, J.M., Duffy, J., An application of screw algebra to the acceleration analysis of serial chains, *Mechanism and Machine Theory,* **31**(5), 1996, 445-457.

[46] Legnani, G., Casolo, F., Righettini, P., Zappa, B., A homogeneous matrix approach to 3D kinematics and dynamics. Part 1: theory, *Mechanism and Machine Theory,* **31**(5) (1996), 573-587.

[47] Wylie, C.R., Barret, L.C., *Advanced Engineering Mathematics.* (McGraw-Hill Book Company, fifth edition, New York), 1982.

[48] Angeles, J., *Fundamentals of Robotic Mechanical Systems,* (Springer Science-Business Media, NewYork), 2007.

[49] Mohamed, M.G., Kinematics of rigid bodies in general spatial motion: second-order motion properties, *Appled Mathematical Modelling,* **21**(1997), 471479.

[50] Legnani G., Una semplice formulazione matriciale per la determinazione del punti di accelerazione minima nel moto rigido spaziale, *Proceedings Congresso Nazionale AIMETA88,* Bari, 1988, pp. 353-356.

[51] Legnani, G., Casolo, F., Righettini, P., Zappa, B., A homogeneous matrix approach to 3D kinematics and dynamics. Part 2: applications, *Mechanism and Machine Theory,* **31**(5) (1996), 589-605.

[52] Ball, R.S., *A Treatise on the Theory of Screws,* (Cambridge University Press, Cambridge, UK), 1900 (reprint 1998).

[53] Brand, L., *Vector and tensor analysis*, (John Wiley, NewYork), 1947.

[54] Sugimoto, K., Duffy, J., Application of linear algebra to screw systems, *Mechanism and Machine Theory,* **17**(1)(1982), 73-83.

[55] Rico, J.M., Gallardo, J., Duffy, J., Screw theory and higher order analysis of open serial and closed chains, *Mechanism and Machine Theory,* **34**(4)(1999), 559-586.

[56] Merlet, J.P., Manipulateurs paralleles, 4eme partie: mode d'assemblage et cinematique directe sous forme polynomiale, *INRIA Research Report no. 1135,* 1989, Centre de Sophia Antipolis, Valbonne.

[57] Tsai, L.W., *Robot Analysis,* (Wiley, New York: N.Y.), 1999.

[58] J. Gallardo-Alvarado, R. Rodríguez-Castro, Islam, Md N., Analytical solution of the forward position analysis of parallel manipulators that generate 3-RS structures, *Advanced Robotics*, **22**(2008), 215–234.

[59] Gallardo-Alvarado, J., Aguilar-Nájera, C.R., Casique-Rosas, L., Pérez-González, L., Rico-Martínez, J.M., Solving the kinematics and dynamics of a modular spatial hyper-redundant manipulator by means of screw theory, *Multibody System dynamics,* **20**(2008), 307325.

In: Robotics: State of the Art and Future Trends
Editors: G. Legnani and I. Fassi
ISBN: 978-1-62100-403-5

Chapter 6

DYNAMICS AND SYSTEM IDENTIFICATION OF PARALLEL ROBOTS

Gloria J. Wiens[*] and Young-Hoon Oh
Department of Mechanical and Aerospace Engineering,
University of Florida, Gainesville, FL, U. S.

Abstract

This past decade of rapidly expanding application opportunities for parallel kinematic mechanisms/machines (PKMs) has created an increasing awareness of their dynamic needs. Due to little research on the dynamics of PKMs or lack thereof, the intent of this chapter is to provide the fundamentals for dynamic modeling, analysis and system identification of parallel robots, also known as PKMs. Through both simulated and experimental implementation, the challenges in the determination of parameter values necessary for modeling the dynamics of parallel kinematic machines, e.g., mass, inertia, and friction, is also demonstrated. This is done using the unscented Kalman filter (UKF), a model based identification method. Implementing both a force based model and an energy based model, the use of the UKF method for PKMs is evaluated in terms of accuracy, convergence, ability to handle noisy data, and ease of implementation.

Introduction

Although due in part to their anthropomorphic nature, serial manipulators comprised of open chain linkages are commonly used in industry. However, since the science of robotics has developed, interest in parallel kinematic mechanisms has grown such that many of the serial robots' actuation transmissions are actually comprised of closed chain linkages (the general class in which parallel kinematic mechanisms belong). The Stewart platform [1] was one of the first implementations of the parallel kinematic mechanism and has been widely adapted as an aircraft simulator.

[*] E-mail address: gwiens@ufl.edu

As of today, the limited usage of parallel kinematic mechanisms (PKMs) is largely due to small workspace, small regions of dexterity, variable end-effectors stiffness, non-unique solutions of forward kinematics, workspace singularities, self-motions and incomprehension of dynamic performances. In the last two decades, there have been many debates on the 'true' attributes of PKMs and their applications, especially since their 1994 début as a new machine tool paradigm. Their subsequent failure to be widely adopted by the machine tool community can be attributed to the following [2].

> "The PKM suffers from a lack of comprehensive calibration techniques, from weak or not well-defined dynamics behavior, and from the need for complex control algorithms for the compensation of stiffness variations in the workspace."

Although it has been claimed by researchers that parallel kinematic mechanisms/machines (also known as parallel robots) have higher structural stiffness, and proportionally distribute the payload to the links, these attributes are highly dependent on their configuration design (e.g., Stewart-Gough, Delta-like, Triaglide, Linapod-like). Using application driven design methods, suitable PKM configurations have been realized that exhibit sufficient structural stiffness and high speed motions for a wide variety of applications (e.g., motion simulations, assembly and packaging end-effectors, surgical, MEMS devices and entertainment).

To date, research on the dynamics of the PKM has been somewhat limited. The dynamic equations of motion have been derived for both serial robots and the generalized Stewart platforms using Lagrangian, Newton-Euler, and Kane's method. Of the work available in the literature, a sampling of researchers who investigated modeling, dynamics and/or parameter identification in the robotics community are: [3-17]. Fregolent and Sestieri [15] derived a series of simple linear problems from the non-linear problem, and solved them with a frequency-response function (FRF). Tinone and Aoshima [16] used a repetitive controller in order to reduce bias from measurement error. Excluding the friction parameters, Lin [11] used minimal linear combinations (MLC) of inertial parameters for avoiding the difficulties to estimate the parameters. The MLC was used for estimating the parameters in the serial manipulators, but it can also be used in the parallel manipulator. Further insight into PKM and serial robot dynamics and identification has been provided in the following: vibration analysis [18], trajectory analysis [19, 20], force analysis [21], accuracy analysis [22], forward analysis [23], stiffness analysis [24], and parameter estimation [25-28]. In reference [28-32], the authors attempted parameter identification of PKM's rigid body dynamics. For the work presented in this chapter, the unscented Kalman filter method will be used for demonstrating identification of the inertial parameters for the PKM's rigid-body model.

Focusing on the dynamics aspects of PKMs, Wiens *et al.* [29-31, 33-40] and Oh [32] established metrics and methodologies for characterizing and identifying various dynamic parameters of parallel kinematic mechanisms. Hardage and Wiens [30, 34, 36, 40] developed a methodology for identifying the structural dynamic parameters of PKMs. Component-mode synthesis and Lagrangian techniques were used to derive the equations of motion capturing the structural vibrations. For experimental validation of these equations, modal testing was performed to measure the stiffness and frequency of the Sandia National Laboratories' Hexel PKM. The results demonstrated that the Hexel machine could not be expected to achieve high structural performance nor be widely adopted as a machine tool.

Using Kane's method for deriving the equations of motion of the rigid-body dynamics, dynamic performance indices for parallel manipulators were developed by Johnson and Wiens [35]. To eliminate mixed units in this metric, the energy expression was modified to be in terms of linear velocities of the platform only. The performance indices clearly demonstrated that to fully characterize the dynamic performance of the machine, kinematics based indices are not sufficient. This was further substantiated by Walker and Wiens' [33] research on studying the effects of kinematic errors on dynamic performance.

Register *et al.* [29] characterized the effects of friction using a parameter estimation algorithm (extended Kalman filter). Friction forces were shown to be dominant over inertial and gravitational forces in linear and angular motions of the struts. The friction parameter for the single joint strut was estimated to within 10 percent using the estimation algorithm; however the identification of friction parameters in multiple joint types was not addressed. Shamblin *et al.* [31, 37, 38] conducted further analysis and experimental identification of the individual struts and the assembled PKM, introducing an identification methodology for extracting inertial parameters when faced with the challenges of high friction.

For the modeling of robotic systems, an important characteristic of the inertial parameters is that their values are typically invariant in time. In other words, the inertial parameters will stay the same regardless of the input and output values to the robotic system. However, such parameters are generally not provided by the manufacturer and due to the complexity and multiple moving components they are difficult to identify by simple physical measurements of the components. In such situations, a parameter estimation method can be used. Thus, in order to obtain a high fidelity dynamics model, one must resort to the use of system identification and parameter estimation techniques.

System identification methods are well known and utilized by researchers working in the areas of dynamics, controls, and estimation theory. System identification can be categorized into state, parameter and dual estimation. The parameter estimation method is used for approximating the unknown parameters of the system from measured data which contains the known system input and output variables. There are many methods existing in parameter estimation, however due to its mathematical simplicity, a commonly used method is least squares. The least squares method is widely used to find system parameter and state estimation. Least squares methods, however, exhibits convergence problems in the presence of noise and lacks the ability to effectively handle nonlinearities in system identification. This has led to the development of various mutations of least squares [3]. Of these mutations, a method that has gained popularity is the extended Kalman filter. A more efficient way to identify nonlinear system parameters, however, is the unscented Kalman filter method (UKF) [29]. Therefore, the unscented Kalman filter method is selected as a method of choice for dealing with the dynamic complexity of PKMs. A major advantage for the UKF is that it does not require the calculation of Jacobian and Hessian matrices. This approach has been found to yield faster and more convenient solutions for the nonlinear equations of the PKM.

To address the dynamic modeling needs of PKMs, this chapter first establishes a set of fixed and moving frames necessary for describing the kinematics of the PKM. Using a lumped-mass modeling approach, the system's equation of motion are derived that captures not only the inertial affects of the (platform end-effector) but those of the struts (parallel mechanisms) on the PKM dynamics. This is followed by implementation of parameter identification method for determining the invariant parameters of the PKM dynamics. The parameter identification methods considered in this chapter are model-based, requiring a theoretical model of the expected system dynamics. In the derivation of the theoretical model, two different approaches were taken in formulating a model-base. One approach uses the system's equations of motion, referred to as the force-based model. The other approach uses the expressions from the system's energy equations, referred to as the energy-based model. The same force-based model also serves as a theoretical representation of the physical system in the simulated identification implementation studies. In the following sections, the theoretical modeling of the general class of Stewart-Gough parallel kinematic mechanisms (PKM) is presented for which supplemented details can be found in reference [32]. The chapter continues with theoretical presentation of unscented Kalman filtering (UKF) parameter identification method followed by both a simulated and experimental implementation. While the results are limited to the Stewart-Gough class of PKMs, it should be noted that the methodology presented in this chapter and the issues addressed are directly applicable to PKMs, in general.

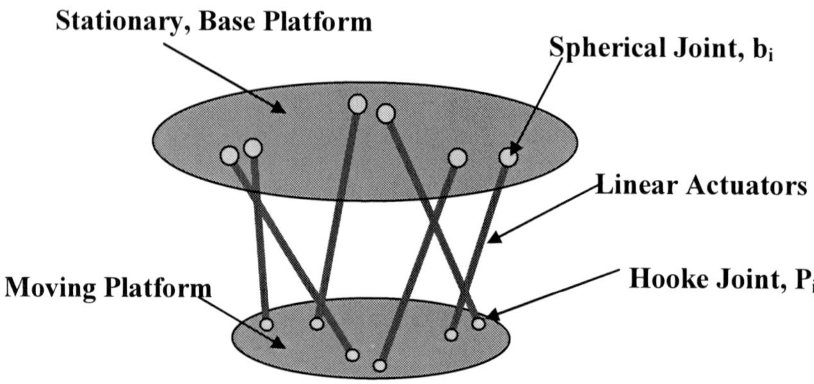

Figure 6.1. Schematic of a Stewart-Gough PKM.

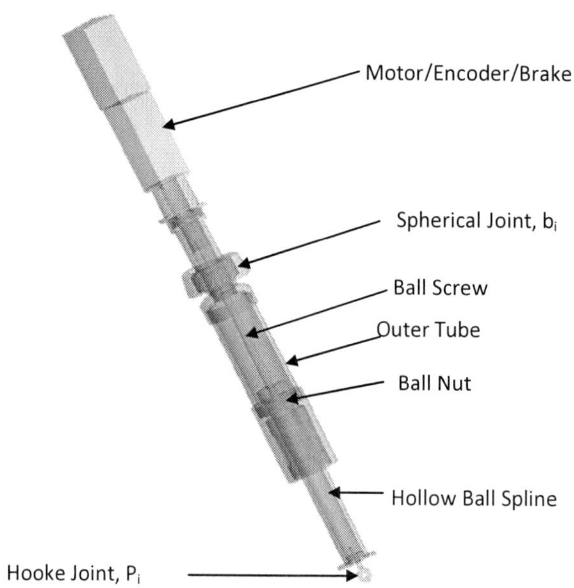

Figure 6.2. Physical Diagram of Strut i.

Dynamic Modeling and Analysis of Parallel Robots

In this section, the dynamic modeling of the Stewart-Gough type PKM is detailed. This type of PKM has six extendable struts (linear actuators) that provide 6-degrees of freedom (d.o.f.) to a moving platform relative to a base platform, refer to schematic in Figure 6.1. Figure 6.2 shows a physical diagram of a representative strut, and Figure 6.3 is its corresponding lumped mass model where details are provided in the section on equations of motion.

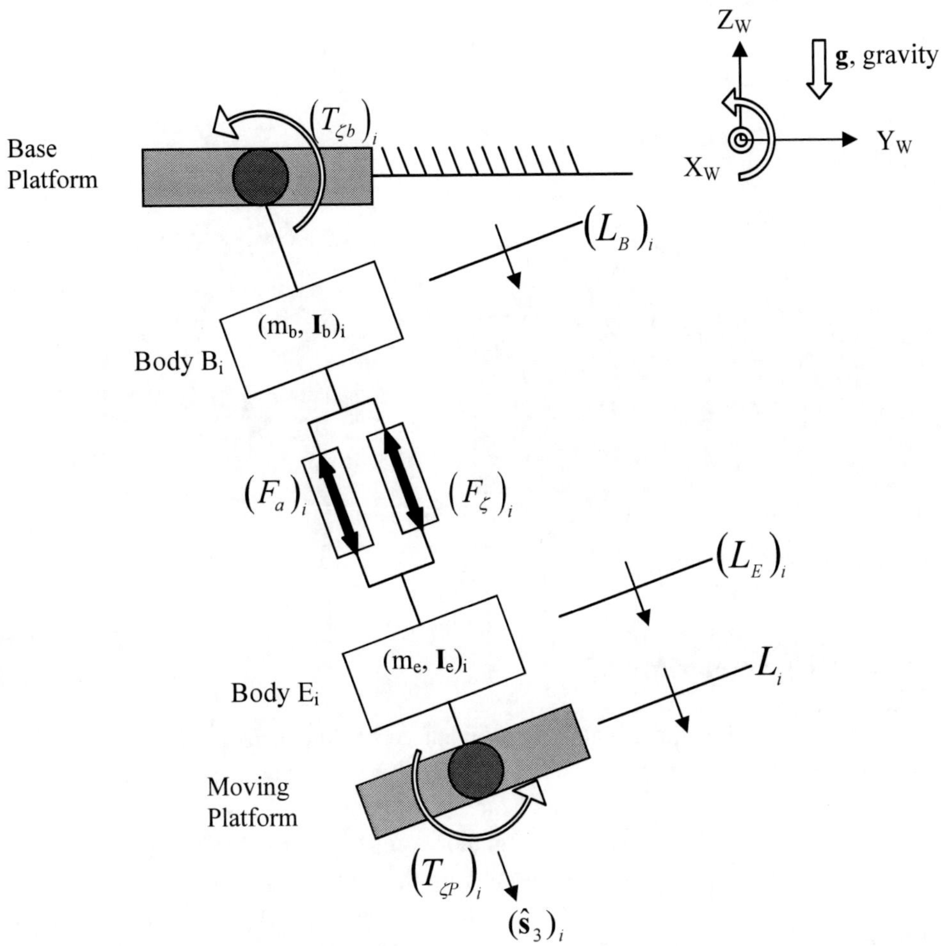

Figure 6.3. Lumped Mass Parameter Diagram for Strut i.

System Description – Special 6-6 PKM

Figure 6.4 shows the University of Florida – Space, Automation and Manufacturing Mechanisms (UF-SAMM) Laboratory's Special 6-6 PKM [42], a subset of Stewart-Gough class of PKMs. The special 6-6 PKM consists of a moving platform with its 6 degrees of motion actuated via six extendable struts. The special 6-6 PKM has a unique telescoping strut arrangement such that both the platform and base consist of three sets of three joints lying on lines that form the sides of a triangle. It is this joint configuration that yields reduced computation requirement for the forward kinematic solutions compared to that required of the general Stewart-Gough PKM as well as allows for simpler physical joint designs. As shown in Figure 6.2, each strut consists of a motor, encoder, brake, spherical joint, ball screw, Hooke joint (also known as universal joint or cardan joint), hollow ball spline/nut, and outer tube. Each of the struts (linear actuators) is connected to the stationary base platform through a large hollow spherical joint. This connection allows the strut to rotate through space as its length is varied. A servomotor, whose armature is connected to a ball screw that passes through the center of the hollow spherical joint, drives the strut. To prevent rotation about the servomotor axis, the combination of a hollow ball spline and a Hooke joint is used as the connection between the moving platform and the strut.

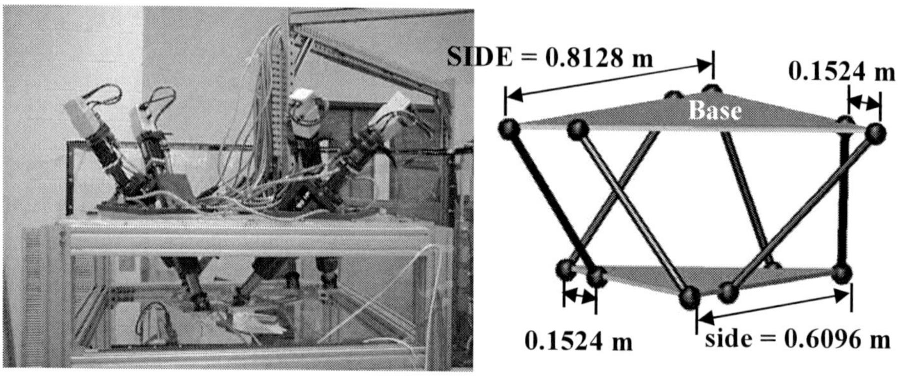

(side:SIDE=0.75)

Figure 6.4. UF-SAMM, Special 6-6 PKM Configuration.

Lumped Mass Parameter Model

The physical system has three major components to be categorized: the stationary base, the strut and the moving platform. The stationary base serves as the inertial reference frame for the system. The moving platform has six degrees of freedom and consists of a single mass. It has a mass m_P and inertia tensor $\mathbf{I}_P$. The last major components are the struts. Since each of the actuators only produce translational motion along the longitudinal axis of their respective strut, each actuating component is modeled as a linear force applied across a prismatic joint. For the UF-PKM, the struts are connected between the stationary base platform and a moving platform with a spherical joint and a Hooke joint, respectively. However, without loss of generality and for modeling purposes, each strut can be theoretical modeled as forming a kinematic chain (H-P-S) which consists of a Hooke joint, a prismatic joint (representing the linear actuation), and a spherical joint. Both kinematic chains have the same kinematic degrees of freedom. Figure 6.2 and Figure 6.3 show the physical diagram of a single strut i for the UF PKM and the corresponding lumped mass model, respectively. For Body B_i, the $(m_b)_i$ and $(\mathbf{I}_b)_i$ are respectively the composite mass and inertia tensor for all the components in the physical model that do not extend (or translate along strut axis $(\hat{S}_3)_i$) as strut i (linear actuator) changes in length. For the UF-PKM, this consists of an outer tube, ball screw, ball of the spherical joint, motor, encoder and brake. The strut length is defined by the distance between center of the moving platform-strut joint (P_i) and base platform-strut joint (b_i). For Body E_i, the $(m_e)_i$ and $(\mathbf{I}_e)_i$ are respectively the composite mass and inertia for the remaining components that exhibit a translational motion along the $(\hat{s}_3)_i$ axis. For the UF-PKM, this consists of a ball nut, hollow ball spline, and Hooke joint. The actuator's force $(F_a)_i$ can be modeled as a force exerted between these two lumped mass entities. The joint friction within the kinematic chain can be modeled as both translational and torsional friction, $(F_\varsigma)_i$, $(T_{\varsigma P})_i$, and $(T_{\varsigma b})_i$, respectively. The quantities $(L_B)_i$ and $(L_E)_i$ denote respectively the instantaneous center of mass positions of the lumped masses, $(m_e)_i$ and $(m_b)_i$, defined along strut axis $(\hat{s}_3)_i$ relative to point (b_i). The L_i is the instantaneous length of strut i. In reiterating the above and illustrated in Figure 6.3, the strut masses are lumped into two parts, Body E_i $(m_e, I_e)_i$ and Body B_i $(m_b, I_b)_i$, which both experience rotational motion relative to the base joint, i.e., rotate with $(\hat{s}_3)_i$ axis. In addition, Body E_i experiences translational motion along $(\hat{s}_3)_i$ where Body B_i does not. Since Body E_i translates freely in the direction of the strut axis, the location of its center of mass is located at a variable distance measured from the base platform-strut joint location (b_i) along the strut axis $(\hat{s}_3)_i$. As stated earlier, this distance is denoted by $(L_E)_i$. The velocity and acceleration of Body E_i can be expressed in terms of the linear velocity and acceleration of the extendable part of the strut in the direction of the strut axis and the strut angular velocity and acceleration about axes defined in the base joint. Body B_i has a center of mass that is located a fixed distance $(L_B)_i$ from the base joint (b_i) and is measured in the direction of the strut axis. Since this distance is fixed, the velocity and acceleration of Body B_i can be expressed in terms of the strut's angular velocity and acceleration about the axes defined in the base joint (b_i). To summarize, the total number of the lumped masses in the Stewart-Gough PKM model are thirteen: two for each strut and one for the moving platform.

Coordinate System Assignment

In the analysis of kinematics and dynamics, it is typical that a number of coordinate systems are defined to describe the motion of bodies. To formulate the dynamic equations, three types of coordinate systems are defined: global (the inertial reference frame), strut (S-coordinate system), and moving platform coordinate systems (translational and rotational coordinate system).

Global and Moving Platform Coordinate Systems

The global (inertial) coordinate system is denoted as {W} coordinate system. It is a fixed, Cartesian inertial reference frame. This coordinate system can be located arbitrarily, however, for the PKMs considered, the {W} coordinate system is fixed at the geometric center location of the stationary base platform as shown in Figure 6.5. The moving platform coordinate system consists of two frames: a translational and rotational coordinate system. The translation frame is called the {F} coordinate system and the rotation frame is called the {P} coordinate system. As in Figure 6.5, the origin of the {F} coordinate system moves with the center of mass location of the moving platform but the orientation of the {F} frame remains parallel with the {W} frame for all time. It shares the same origin with the {P} coordinate system. The {P} frame is affixed to the moving platform, where its absolute position and orientation is used to describe the motion of the moving platform relative to the {W} frame. The {F} frame is used as an intermediate step for obtaining the {P} frame's absolute position and orientation. The {P} frame orientation is arbitrary, but is typically selected to have its axes aligned with the principal axes of inertia. To avoid singularities in the computation, the axes are also selected to have specific orientations when the platform is in its 'home' position.

The naming convention used for denoting the center location of the strut joint on the base and moving platform as well as the center of mass location of the moving platform are shown in Figure 6.6. All joints on the stationary base are labeled point b_i where i=1,6 for six struts. Their locations are defined in the {W} coordinates (e.g., ${}^{W}b_i$). On the moving platform, the locations of the joints are denoted as P_i. The moving platform joint (P_i)'s position measured from the origin of the {F} or {P} coordinate system to point P_i is denoted by vector $({}^{W}\mathrm{r}_{pc})_i$, i.e., location of P_i relative to the platform's center of mass, C, expressed in {W} coordinates.

Strut Coordinate System

The strut coordinate system is denoted as the $\{S\}_i$ coordinate system (refer to Figure 6.7 which shows the theoretical Hooke joint model for the base joint b_i). The $(\hat{\mathbf{s}}_1)_i$ unit vector is used to denote the first axis in the Hooke joint. The angular rate about this axis is defined as $(\omega_{s_{s_1}})_i$. Used as a reference to define other vectors and without loss of generality, the direction of the $(\hat{\mathbf{s}}_1)_i$ vector is arbitrarily defined to be parallel to the X_W-direction of {W} coordinate system. I.e., the X-axis of the X-Y-Z coordinate frame in Figure 6.7 theoretically remains parallel to the {W} coordinate system for all time. The $(\hat{\mathbf{s}}_3)_i$ unit vector is defined as the unit vector along the line of action between joint P_i and b_i, which is calculated as

$$(\hat{\mathbf{s}}_3)_i = \frac{\left({}^{W}\mathbf{r}_P\right)_i - \left({}^{W}\mathbf{r}_b\right)_i}{\left\|\left({}^{W}\mathbf{r}_P\right)_i - \left({}^{W}\mathbf{r}_b\right)_i\right\|} \tag{6.1}$$

Note, the subscripts i denote i^{th} strut where i=1,6 for the PKM's under consideration.

The $(\hat{\mathbf{s}}_2)_i$ unit vector is formed as the cross product of $(\hat{\mathbf{s}}_1)_i$ and $(\hat{\mathbf{s}}_3)_i$ divided by the magnitude of the cross product of $(\hat{\mathbf{s}}_1)_i$ and $(\hat{\mathbf{s}}_3)_i$. The $(\hat{\mathbf{s}}_{23})_i$ unit vector is defined as the cross product of $(\hat{\mathbf{s}}_2)_i$ and $(\hat{\mathbf{s}}_3)_i$. Thus, the three orthogonal unit vectors form the strut coordinate system expressed by equations (6.1) through (6.3) in {W} frame coordinates.

$$(\hat{\mathbf{s}}_1)_i = [1 \quad 0 \quad 0]^T \text{ and } (\hat{\mathbf{s}}_2)_i = \frac{(\hat{\mathbf{s}}_1 \times \hat{\mathbf{s}}_3)_i}{\|\hat{\mathbf{s}}_1 \times \hat{\mathbf{s}}_3\|_i} \tag{6.2}$$

$$(\hat{\mathbf{s}}_{23})_i = (\hat{\mathbf{s}}_2 \times \hat{\mathbf{s}}_3)_i \tag{6.3}$$

where the strut coordinate system $\{S\}_i = \{\hat{\mathbf{s}}_2, \hat{\mathbf{s}}_3, \hat{\mathbf{s}}_{23}\}_i$. This frame definition is used for describing the strut motion and to facilitate the derivation of the strut dynamic terms in the equations of motion.

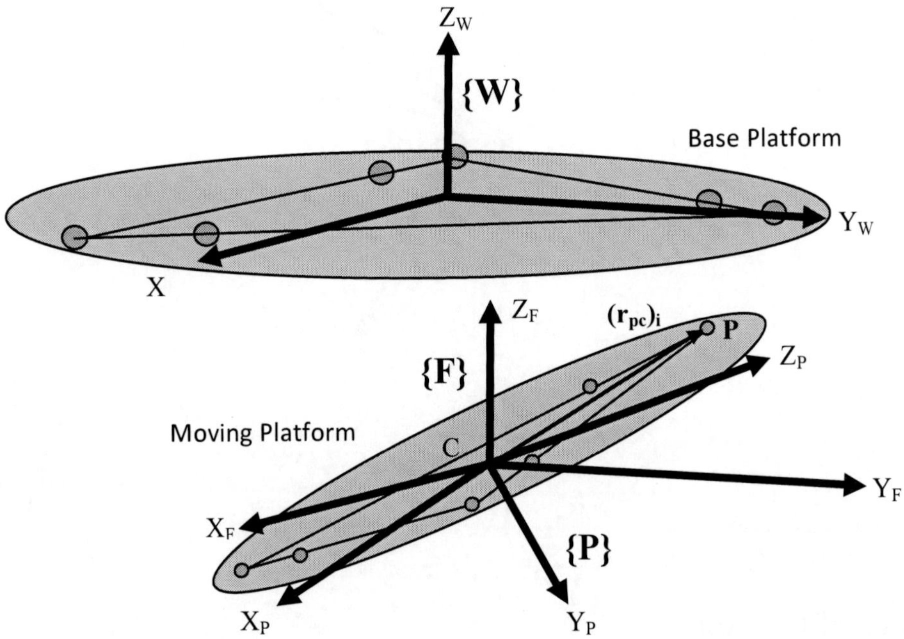

Figure 6.5. Coordinate System Assignment.

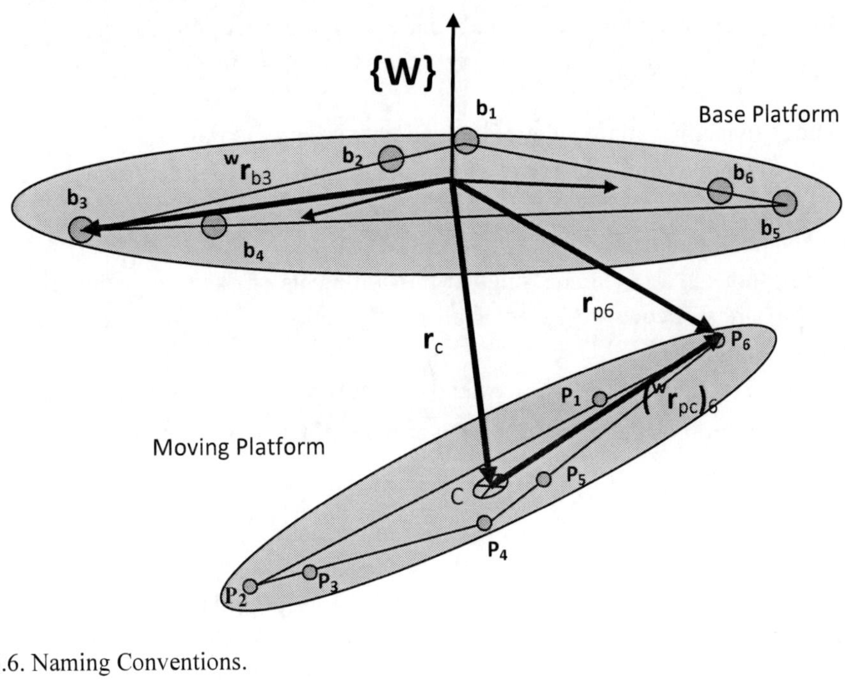

Figure 6.6. Naming Conventions.

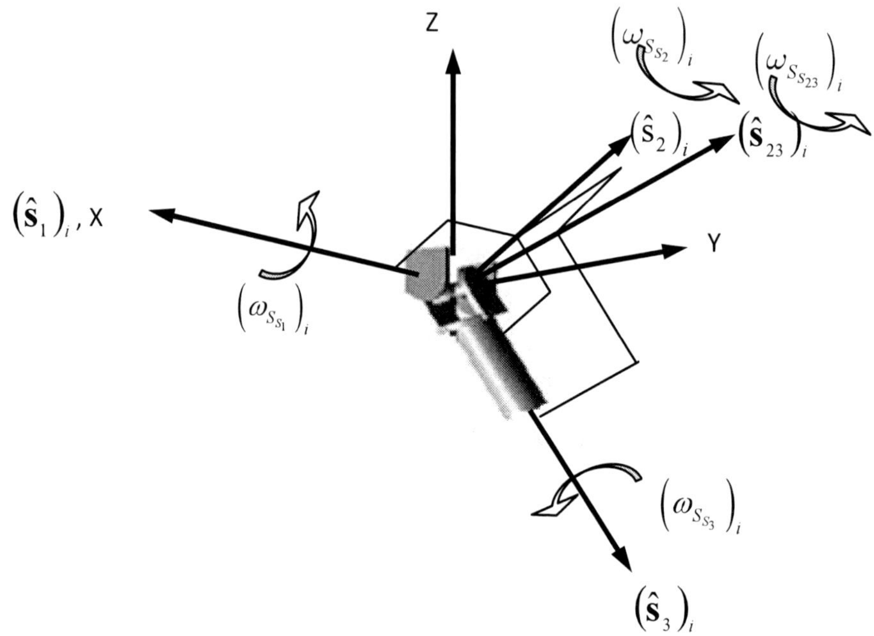

Figure 6.7. Strut Coordinate System.

Equations of Motion

To conduct model-based dynamic parameter identification, one must first formulate the equations of motion used to represent the expected dynamic model of the system. The equations used in this chapter are based on prior theoretical modeling by references [32, 43-47]. Assuming a lumped-mass rigid body model, Kane's method is selected for deriving the dynamics equations of motion. The following summarizes the assumptions made.

- Rigid body dynamics of a system of lumped masses and inertias.
- Strut inertias are symmetric about strut longitudinal axis defined along line of action between the strut base and platform joint centers.
- No kinematic joint errors.
- PKM does not have kinematic singularities within its workspace (volume), e.g., det[J] and det[U] are not equal to zero for all configurations (these terms are defined in the following sections).
- Center of mass locations of the extendable portion of the strut (Body E)$_i$ and non-extendable portion of the strut (Body B)$_i$ remain along $(\hat{\mathbf{s}}_3)_i$. Both $(m_e)_i$ and $(m_b)_i$ rotate with $(\hat{\mathbf{s}}_3)_i$ axis (Refer to Figure 6.2 and 6.3).
- Unmodeled dynamics are attributed to joint imperfections, mass variations, actuator and controller dynamics and the inertial effect of rotational motion of struts about the struts longitudinal axis. These are generally assumed negligible but are retained in the theoretical model to denote sources of error in parameter identification.

- Each strut is designed to restrict free spin rotation about its longitudinal axis ($\hat{\mathbf{s}}_3$)$_i$.
- Modified joint friction model (F_ζ)$_i$ that includes both viscous and Coulomb friction is assumed [47]. In the analysis presented, the torsional friction within the joints between the base and moving platforms and the struts are assumed negligible (i.e., $(T_{\zeta b})_i = (T_{\zeta P})_i = 0$, i=1,6).

Generalized Coordinates

To describe the motion of the 6 degrees of freedom PKM system, a set of 6 generalized coordinates must be chosen. In this system, the generalized coordinates are chosen to be the center of mass location of the moving platform (x_c, y_c, z_c) with reference to the {W} coordinate system. For the moving platform's absolute orientation, the z-x-z Euler angles are chosen to avoid computational difficulties and to allow direct integration of the angular velocity. The z-x-z Euler angles represents, sequentially, a rotation ϕ about the z-axis of the {W} coordinate system, a rotation θ about the x-axis of the rotated coordinate system, and a rotation ψ about the z-axis of the rotated coordinate system. For this system, to avoid the mathematical singularity, the z-x-z Euler angles are set to (ϕ, θ, ψ) = (0°, -90°, 0°) when the platform orientation is in its 'home' position (parallel to the base frame {W} and at a specified (x_c, y_c, z_c)). Thus, at 'home' position, the {P} frames are expressed as {Y_p= -Z_F, X_p= X_F, Z_p= Y_F}.

The symbol $\mathbf{q}$ is used to express the (6x1) vector of the generalized coordinates. For Kane's equations, generalized speeds are also required. The generalized speeds ($\dot{\mathbf{q}}$) correspond to the time derivative of the generalized coordinates. The generalized coordinates and the generalized speeds can be expressed as follows, respectively.

$$\mathbf{q} = [x_c, y_c, z_c, \phi, \theta, \psi]^T \tag{6.4}$$

$$\dot{\mathbf{q}} = [\dot{x}_c, \dot{y}_c, \dot{z}_c, \dot{\phi}, \dot{\theta}, \dot{\psi}]^T \tag{6.5}$$

where $\mathbf{q}$ is a (6x1) vector of the absolute position and orientation coordinates of the platform at its center of mass denoted by point C in Figure 6.6. The $[\ ]^T$ denotes the transpose of the vector [ ]. In equation (6.5), the ($\dot{}$) denotes the time derivative of ( ). The generalized coordinates can be further expressed as $q = [{}^W r_c^T, \beta^T]^T$ where the platform's absolute position (${}^W\mathbf{r}_c$) referenced in {W} is

$${}^W r_c = [x_C \quad y_C \quad z_C]^T \tag{6.6}$$

The orientation coordinates of the moving platform can also be expressed by a vector of the Euler angles, $\boldsymbol{\beta}$, where

$$\boldsymbol{\beta} = [\phi \quad \theta \quad \psi]^T \tag{6.7}$$

Similarly, generalized speed can be expressed as

$$\dot{\mathbf{q}} = [\mathbf{v}_c{}^T, \dot{\boldsymbol{\beta}}^{\mathrm{T}}]^T \tag{6.8}$$

where $\mathbf{v}_c$ is the absolute linear velocity of the platform's center of mass.

Kinematics

In order to obtain the dynamic response of the PKM system, the position, velocity, and acceleration of all bodies must be expressed in terms of the generalized coordinates. However, to obtain the position, velocity and acceleration of Bodies B_i and E_i in terms of generalized coordinates, the inverse kinematics must be solved. The inverse kinematics is that given position and orientation data of the moving platform, one can determine the length (L_i) and orientation $(\hat{\mathbf{s}}_3)_i$ of the struts. The subscript i in these and the following expressions and equations denote that there is a corresponding expression and set of numerical values for each strut i for i=1,6. Without loss of generality, the i subscript will be dropped where it is understood that for all joints, there are i=1,6 set of expressions. Given the orientation of the moving platform, a standard (3x3) rotation matrix is defined to map the orientation from the {P} coordinate system to {F} or {W} coordinate systems per the following.

$$^{W}\mathbf{R}_{\mathbf{p}} = {}^{W}\mathbf{R}_{F}\,{}^{F}\mathbf{R}_{\mathbf{p}} = {}^{F}\mathbf{R}_{\mathbf{p}} = \begin{bmatrix} c\psi c\varphi\text{-}c\theta s\varphi s\psi & -s\psi c\varphi - c\theta s\varphi c\psi & s\theta s\varphi \\ c\psi s\varphi + c\theta c\varphi s\psi & \text{-}s\psi s\varphi + c\theta c\varphi c\psi & -s\theta c\varphi \\ s\psi s\theta & c\psi s\theta & c\theta \end{bmatrix} \tag{6.9}$$

where s and c denote the sine and cosine functions, respectively, and $^{W}\mathbf{R}_{F} = [\mathrm{I}]$ since {F} and {W} are parallel frames. The matrix $[\mathrm{I}]$ is a (3x3) identity matrix.

Since the position of the strut joint with the moving platform joint (P_i) is denoted by $(\mathbf{r}_{pc})_i$ with reference to frame {P}, pre-multiplying $(\mathbf{r}_{pc})_i$ by $^{W}\mathbf{R}_{\mathbf{p}}$ yields the value of P_i position $(^{W}\mathrm{r}_{P})_i$ in the {F} coordinate system. Then to complete the mapping to {W} frame, the translation from the origin of {F} frame to {W} frame ($^{\mathrm{W}}\mathbf{r}_c$) must be done. The total mapping of P_i is shown in equation (6.10).

$$(^{W}\mathrm{r}_{P})_i = {}^{W}\mathrm{r}_c + {}^{W}\mathbf{R}_{\mathbf{p}}\,{}^{P}(\mathbf{r}_{\mathbf{pc}})_i \tag{6.10}$$

Once the location of the position P_i is defined, then the strut length and orientation can be determined. Subtracting the location of point b_i from the location of P_i gives vectorially the length and orientation of the strut.

$$(L\hat{\mathbf{s}}_3)_i = ({}^W\mathrm{r}_P - {}^W\mathrm{r}_b)_i \tag{6.11}$$

Since the unit vector of the strut $(\hat{\mathbf{s}}_3)_i$ expressed in {W} coordinates is aligned to the individual strut i, it can be found with the following equation.

$$(\hat{\mathbf{s}}_3)_i = \frac{({}^W\mathbf{r}_P - {}^W\mathbf{r}_b)_i}{\left\| {}^W\mathbf{r}_P - {}^W\mathbf{r}_b \right\|_i} \tag{6.12}$$

The velocity of point P_i ($(\mathbf{v}_p)_i$) can be calculated by differentiating $({}^W\mathrm{r}_P)_i$ in equation (6.10) with respect to time.

$$(\mathrm{v}_P)_i = \mathrm{v}_c + (\boldsymbol{\omega}_c \times {}^W\mathbf{R}_\mathbf{p}\mathbf{r}_\mathbf{pc})_i = \mathrm{v}_c + (\boldsymbol{\omega}_c \times {}^W\mathbf{r}_\mathbf{pc})_i \tag{6.13}$$

where $\boldsymbol{\omega}_c$ is the moving platform's absolute angular velocity defined in the inertial frame {W} coordinates.

Since the velocity of point P must be expressed in terms of the generalized coordinates, an expression for $\boldsymbol{\omega}_c$ in terms of the generalized coordinates and generalized speeds must be found. That is,

$$\boldsymbol{\omega}_c = \begin{bmatrix} \mathbf{0} & \mathbf{c\phi} & \mathbf{s\phi s\theta} \\ \mathbf{0} & \mathbf{s\phi} & -\mathbf{c\phi s\theta} \\ \mathbf{1} & \mathbf{0} & \mathbf{c\theta} \end{bmatrix} \begin{bmatrix} \dot{\phi} \\ \dot{\theta} \\ \dot{\psi} \end{bmatrix} = [\mathbf{U}]\dot{\boldsymbol{\beta}} \tag{6.14}$$

where $[\mathbf{U}]$ is the matrix transformation from the Euler angular rates to the absolute angular velocity vector for the platform, $\boldsymbol{\omega}_c$. The derivation of matrix $[\mathbf{U}]$ is provided in Oh [32].

When substituting equation (6.14) into (6.13), the equation yields the velocity of point P_i in the {W} coordinate system as a function of the generalized coordinates and generalized speeds. The component of this velocity in the $(\hat{\mathbf{s}}_3)_i$ direction is the strut i extension rate, which can be obtained by dotting the velocity of point P_i with the $(\hat{\mathbf{s}}_3)_i$ vector as shown in equation (6.15). This operation results in the magnitude and sense of the strut extension rate.

$$\dot{L}_i = (\mathrm{v}_P \cdot \hat{\mathbf{s}}_3)_i \tag{6.15}$$

Alternatively to equation (6.10), the location of point P_i can be obtained from equation (6.11).

$$({}^W\mathrm{r}_P)_i = (L\hat{\mathbf{s}}_3 + {}^W\mathrm{r}_b)_i \tag{6.16}$$

Differentiating equation (2.16) yields the velocity of point P_i.

$$(\mathrm{v}_P)_i = (\dot{L}\hat{\mathbf{s}}_3 + \tilde{\boldsymbol{\omega}}_s \times L\hat{\mathbf{s}}_3)_i \tag{6.17}$$

where the total strut angular velocity in {W} coordinates is specified as

$$(\tilde{\boldsymbol{\omega}}_S)_i = (\omega_{S_{S_2}})_i(\hat{\mathbf{s}}_2)_i + (\omega_{S_{S_3}})_i(\hat{\mathbf{s}}_3)_i + (\omega_{S_{S_{23}}})_i(\hat{\mathbf{s}}_{23})_i \tag{6.18}$$

Since $(\hat{\mathbf{s}}_3 \times \hat{\mathbf{s}}_3)_i = 0$, $(\mathrm{v}_P)_i$ can also be written as

$$(\mathrm{v}_P)_i = \dot{L}\hat{\mathbf{s}}_3 + \boldsymbol{\omega}_s \times L\hat{\mathbf{s}}_3 \tag{6.19}$$

and the strut angular velocity component $(\boldsymbol{\omega}_s)_i$ without the angular velocity about $\hat{\mathbf{S}}_3$ axis is

$$(\boldsymbol{\omega}_S)_i = (\omega_{S_{S_2}}\hat{\mathbf{s}}_2 + \omega_{S_{S_{23}}}\hat{\mathbf{s}}_{23})_i \tag{6.20}$$

which can also be expressed as

$$(\boldsymbol{\omega}_S)_i = \left(\left(\frac{1}{L}\right)(\hat{\mathbf{s}}_3 \times \mathbf{v}_P)\right)_i \tag{6.21}$$

It should be noted

$$\boldsymbol{\omega}_S \cdot \hat{\mathbf{s}}_3 = 0 \tag{6.22}$$

Hence, in a physical setup, each strut is typically designed to restrict free spin rotation about its longitudinal axis $(\hat{\mathbf{s}}_3)_i$, therefore, the dot product of $(\tilde{\boldsymbol{\omega}}_S)_i$ and $(\hat{\mathbf{s}}_3)_i$ are zero ($(\tilde{\boldsymbol{\omega}}_S \cdot \hat{\mathbf{s}}_3)_i = 0$). Thus, the strut angular velocity about $(\hat{\mathbf{s}}_3)_i$ axis is $(\boldsymbol{\omega}_S)_i$ and the strut entities Body B_i and Body E_i have the same angular velocities, $(\boldsymbol{\omega}_E)_i = (\boldsymbol{\omega}_B)_i = (\tilde{\boldsymbol{\omega}}_S)_i$. In actuality, there are small rotations about $(\hat{\mathbf{s}}_3)_i$, depending on joint designs for joint b_i and P_i. However, the dynamics due to $\tilde{\boldsymbol{\omega}}_S$ in the $\hat{\mathbf{s}}_3$ can still be considered negligible. Hence, from this point forward, $\boldsymbol{\omega}_E$ and $\boldsymbol{\omega}_B$ are assumed equal to $\boldsymbol{\omega}_S$.

The notation for the absolute locations and velocities of center of mass of Bodies E_i and B_i and the platform joint P_i can be expressed as:

$$({}^W\mathrm{r}_E)_i = (L_E\hat{\mathbf{s}}_3 + {}^W\mathrm{r}_b)_i \tag{6.23}$$

$$({}^W\mathrm{r}_B)_i = (L_B\hat{\mathbf{s}}_3 + {}^W\mathrm{r}_b)_i \tag{6.24}$$

$$(\mathbf{v}_E)_i = (\dot{L}_E\hat{\mathbf{s}}_3 + \boldsymbol{\omega}_E \times (L_E\hat{\mathbf{s}}_3))_i \tag{6.25}$$

$$\left(\mathbf{v}_B\right)_i = \left(\boldsymbol{\omega}_B \times \left(L_B\hat{\mathbf{s}}_3\right)\right)_i \tag{6.26}$$

$$\left(\mathbf{v}_P\right)_i = \mathbf{v}_c + \left(\boldsymbol{\omega}_c \times {}^W\mathbf{r}_{pc}\right)_i = \left(\dot{L}\hat{\mathbf{s}}_3 + \boldsymbol{\omega}_S \times \left(L\hat{\mathbf{s}}_3\right)\right)_i \tag{6.27}$$

Strut angular accelerations are determined from the linear velocities and accelerations. To derive the equations of motion in terms of only the platform's motion, the velocities and accelerations of the length of the strut variable (L) can be defined as follows:

From equation (6.15),

$$\dot{L}_i = \dot{L}_{E_i} = \left(\mathbf{v}_P \cdot \hat{\mathbf{s}}_3\right)_i \tag{6.28}$$

$$\ddot{L}_i = \ddot{L}_{E_i} = \frac{d}{dt}\left(\mathbf{v}_P \cdot \hat{\mathbf{s}}_3\right)_i = \frac{d}{dt}\left(\mathbf{v}_P\right)_i \cdot \left(\hat{\mathbf{s}}_3\right)_i + \left(\mathbf{v}_P\right)_i \cdot \frac{d}{dt}\left(\hat{\mathbf{s}}_3\right)_i \tag{6.29}$$

$$= \frac{d}{dt}\left(\mathbf{v}_c + \left(\omega_c \times {}^W\mathbf{r}_{pc}\right)_i\right) \cdot \left(\hat{\mathbf{s}}_3\right)_i + \left(\mathbf{v}_P\right)_i \cdot \left(\omega_s \times \hat{\mathbf{s}}_3\right)_i$$

$$= \left[\mathbf{a}_c + \left(\boldsymbol{\alpha}_c \times \left({}^W\mathbf{r}_{pc}\right)_i\right) + \boldsymbol{\omega}_c \times \left(\boldsymbol{\omega}_c \times \left({}^W\mathbf{r}_{pc}\right)_i\right)\right] \cdot \left(\hat{\mathbf{s}}_3\right)_i + \left(\dot{L}\hat{\mathbf{s}}_3 + \left(\boldsymbol{\omega}_s \times L\hat{\mathbf{s}}_3\right)\right)_i \cdot \left(\omega_s \times \hat{\mathbf{s}}_3\right)_i$$

$$= \left[\mathbf{a}_c + \left(\boldsymbol{\alpha}_c \times \left({}^W\mathbf{r}_{pc}\right)_i\right) + \boldsymbol{\omega}_c \times \left(\boldsymbol{\omega}_c \times \left({}^W\mathbf{r}_{pc}\right)_i\right)\right] \cdot \left(\hat{\mathbf{s}}_3\right)_i + L_i\left(\left(\omega_s \times \hat{\mathbf{s}}_3\right) \cdot \left(\omega_s \times \hat{\mathbf{s}}_3\right)\right)_i$$

because

$$\left(\omega_s \times \hat{\mathbf{s}}_3\right)_i = \left[\left(\omega_{s_{s2}}\hat{\mathbf{s}}_2 + \omega_{s_{s23}}\hat{\mathbf{s}}_{23}\right) \times \hat{\mathbf{s}}_3\right]_i = \left(\omega_{s_{s2}} \times \hat{\mathbf{s}}_{23} + \omega_{s_{s23}} \times \left(-\hat{\mathbf{s}}_2\right)\right)_i \tag{6.30}$$

and

$$\left(\left(\omega_s \times \hat{\mathbf{s}}_3\right) \cdot \left(\omega_s \times \hat{\mathbf{s}}_3\right)\right)_i = \left(\omega_{s_{s2}}{}^2 + \omega_{s_{s23}}{}^2\right)_i = \left\|\omega_s\right\|_i^2 = \left(\omega_s\right)_i^2 \tag{6.31}$$

where $\left(\omega_s\right)_i = \left\|\boldsymbol{\omega}_s\right\|_i = \left(\sqrt{\boldsymbol{\omega}_{s_{s2}}{}^2 + \boldsymbol{\omega}_{s_{s23}}{}^2}\right)_i$. Therefore,

$$\ddot{L}_i = \ddot{L}_{E_i} = \left[\mathbf{a}_c + \left(\boldsymbol{\alpha}_c \times \left({}^W\mathbf{r}_{pc}\right)_i\right) + \boldsymbol{\omega}_c \times \left(\boldsymbol{\omega}_c \times \left({}^W\mathbf{r}_{pc}\right)_i\right)\right] \cdot \left(\hat{\mathbf{s}}_3\right)_i + \left(\omega_s^2 L\right)_i \tag{6.32}$$

The linear acceleration of the center of mass of Bodies E and B and the moving platform can be determined with simply differentiating equations (6.25) to (6.27). I.e.,

$$\left(\mathbf{a}_P\right)_i = \mathbf{a}_c + \left(\boldsymbol{\alpha}_c \times^W \mathbf{r}_{pc}\right)_i + \boldsymbol{\omega}_c \times \left(\boldsymbol{\omega}_c \times \left({}^W \mathbf{r}_{pc}\right)_i\right)$$

$$= \frac{d}{dt}\left(\dot{L}\hat{\mathbf{s}}_3 + \boldsymbol{\omega}_s \times L\hat{\mathbf{s}}_3\right)_i \quad (6.33)$$

$$= \left(\ddot{L}\hat{\mathbf{s}}_3 + 2\dot{L}\left(\boldsymbol{\omega}_s \times \hat{\mathbf{s}}_3\right) + \boldsymbol{\alpha}_s \times \left(L\hat{\mathbf{s}}_3\right) + \boldsymbol{\omega}_s \times \left(\boldsymbol{\omega}_s \times \left(L\hat{\mathbf{s}}_3\right)\right)\right)_i$$

$$\begin{aligned} \left(\mathbf{a}_E\right)_i &= \frac{d}{dt}\left(\dot{L}_E\hat{\mathbf{s}}_3 + \boldsymbol{\omega}_E \times L_E\hat{\mathbf{s}}_3\right)_i \\ &= \left(\ddot{L}_{E_i}\hat{\mathbf{s}}_3 + 2\dot{L}_{E_i}(\boldsymbol{\omega}_S \times \hat{\mathbf{s}}_3) + \boldsymbol{\alpha}_s \times L_{E_i}\hat{\mathbf{s}}_3 + \boldsymbol{\omega}_S \times (\boldsymbol{\omega}_S \times L_{E_i}\hat{\mathbf{s}}_3)\right)_i \end{aligned} \quad (6.34)$$

$$\left(\mathbf{a}_B\right)_i = \left(\boldsymbol{\alpha}_B \times L_{B_i}\hat{\mathbf{s}}_3 + \boldsymbol{\omega}_B \times (\boldsymbol{\omega}_B \times L_{B_i}\hat{\mathbf{s}}_3)\right)_i \quad (6.35)$$

Since for each strut the corresponding angular strut velocities of the Bodies E and B are the same, the angular accelerations for these bodies are the same and can be derived by crossing $\left(\mathbf{a}_P\right)_i$ with $\left(\hat{\mathbf{s}}_3\right)_i$, i.e.,

$$\begin{aligned} \left(\hat{\mathbf{s}}_3 \times \mathbf{a}_P\right)_i &= \left(\hat{\mathbf{s}}_3\right)_i \times \left(\ddot{L}\hat{\mathbf{s}}_3 + 2\dot{L}\left(\boldsymbol{\omega}_s \times \hat{\mathbf{s}}_3\right) + \boldsymbol{\alpha}_s \times \left(L\hat{\mathbf{s}}_3\right) + \boldsymbol{\omega}_s \times \left(\boldsymbol{\omega}_s \times \left(L\hat{\mathbf{s}}_3\right)\right)\right)_i \\ &= \left(2\dot{L}\boldsymbol{\omega}_s + L\boldsymbol{\alpha}_s\right)_i \end{aligned} \quad (6.36)$$

Therefore,

$$\left(\boldsymbol{\alpha}_E\right)_i = \left(\boldsymbol{\alpha}_B\right)_i = \left(\boldsymbol{\alpha}_S\right)_i = \left(\frac{1}{L}\right)\left(\left(\hat{\mathbf{s}}_3 \times \mathbf{a}_P\right) - 2\dot{L}\boldsymbol{\omega}_S\right)_i \quad (6.37)$$

The angular acceleration of the moving platform can be obtained by differentiating $\boldsymbol{\omega}_c$ given in equation (6.14).

$$\boldsymbol{\alpha}_c = [\mathbf{U}]\ddot{\boldsymbol{\beta}} + [\dot{\mathbf{U}}]\dot{\boldsymbol{\beta}} \quad (6.38)$$

where

$$[\dot{\mathbf{U}}] = \begin{bmatrix} 0 & -s\phi\dot{\phi} & (c\phi s\theta\dot{\phi} + s\phi c\theta\dot{\theta}) \\ 0 & c\phi\dot{\phi} & (s\phi s\theta\dot{\phi} - c\phi c\theta\dot{\theta}) \\ 0 & 0 & -s\theta\dot{\theta} \end{bmatrix} \quad (6.39)$$

Kane's Equation

Using the generalized coordinates defined in equation (6.1), Kane's method is now used to derive the equations of motion. Its generic form is:

$$\sum_{j=1}^{w}\left(\frac{\partial \mathbf{v}_j}{\partial \dot{q}_k}\cdot\left\{\mathbf{F}_j-\mathbf{F}_j^*\right\}\right)+\sum_{j=1}^{w}\left(\frac{\partial \boldsymbol{\omega}_j}{\partial \dot{q}_k}\cdot\left\{\mathbf{T}_j-\mathbf{T}_j^*\right\}\right)=0 \quad k=1,2,\ldots,n \tag{6.40}$$

where the $\frac{\partial \mathbf{v}_j}{\partial \dot{q}_k}$ and $\frac{\partial \boldsymbol{\omega}_j}{\partial \dot{q}_k}$ are the partial derivatives of the body j's absolute linear and angular velocities with respect to the kth generalized speed $\dot{q}_k$, respectively. Here w is the number of rigid bodies and n is the number of the generalized coordinates. The $\mathbf{F}_j$ is the resultant external force acting on each body j, while $\mathbf{F}_j^*$ is the inertial force of body j. Similarly, $\mathbf{T}_j$ and $\mathbf{T}_j^*$ represent the resultant external torque/moments and inertial moment of body j about its center of mass, respectively. Due to equation (6.40) being summations of vector dot products and there are n generalized coordinates, the Kane's method results in a set of n scalar equations.

For the Stewart-Gough PKM, there exists one moving platform (plat), six-extendable strut components (Body E), and six-base strut components (Body B). Thus, rewriting Kane's equation for the Stewart-Gough PKM:

$$\begin{aligned}&\sum_{i=1}^{6}\left[\frac{\partial \mathbf{v}_E}{\partial \dot{q}_k}\cdot\left\{\mathbf{F}_E-\mathbf{F}_E^*\right\}\right]_i+\sum_{i=1}^{6}\left[\frac{\partial \boldsymbol{\omega}_E}{\partial \dot{q}_k}\cdot\left\{\mathbf{T}_E-\mathbf{T}_E^*\right\}\right]_i\\&+\sum_{i=1}^{6}\left[\frac{\partial \mathbf{v}_B}{\partial \dot{q}_k}\cdot\left\{\mathbf{F}_B-\mathbf{F}_B^*\right\}\right]_i+\sum_{i=1}^{6}\left[\frac{\partial \boldsymbol{\omega}_B}{\partial \dot{q}_k}\cdot\left\{\mathbf{T}_B-\mathbf{T}_B^*\right\}\right]_i\\&+\frac{\partial \mathbf{v}_c}{\partial \dot{q}_k}\cdot\left\{\mathbf{F}_c-\mathbf{F}_c^*\right\}+\frac{\partial \boldsymbol{\omega}_c}{\partial \dot{q}_k}\cdot\left\{\mathbf{T}_c-\mathbf{T}_c^*\right\}=0\end{aligned}\quad ,\ k=1,2,\ldots,6 \tag{6.41}$$

where $(\mathbf{v}_E, \boldsymbol{\omega}_E)$ and $(\mathbf{v}_B, \boldsymbol{\omega}_B)$ are the absolute (linear velocity, angular velocity) of Body E and Body B, respectively.

The resultant external forces for each body type are listed below.

$$\mathbf{F}_c = m_p\mathbf{g}+\mathbf{F}_{tool} \tag{6.42}$$

$$\mathbf{F}_E = m_e\mathbf{g}+(F_a-F_\zeta-F_{m/b})\hat{\mathbf{s}}_3 \tag{6.43}$$

$$\mathbf{F}_B = (F_\zeta+F_{m/b}-F_a)\hat{\mathbf{s}}_3+m_B\mathbf{g} \tag{6.44}$$

where m_p is the mass of the platform and $\mathbf{g}$ is the gravitational field vector. $\mathbf{F}_{tool}$ is the applied tool and/or payload force, such as the reaction forces generated by deburring or other cutting operations. The actuator force, F_a, acts along the vector $\hat{\mathbf{s}}_3$ with a magnitude and sense of F_a, however, the friction force ($F_\zeta \hat{\mathbf{s}}_3$) and the ball inertial screw forces ($F_{m/b} \hat{\mathbf{s}}_3$) act in the opposite direction. The $F_{m/b}$ are part of the original unmodeled set of dynamic terms but are included here for enlightment. For the external force acting on Body B ($\mathbf{F}_B$), the forces, ($F_\zeta \hat{\mathbf{s}}_3$, $F_{m/b} \hat{\mathbf{s}}_3$, and $F_a \hat{\mathbf{s}}_3$), have the same magnitude but in equal and opposite directions when compared to those acting on Body E ($\mathbf{F}_E$). For all bodies, gravity forces are defined by $m\,\mathbf{g}$ terms. In the simulation studies to be presented later in this chapter, friction and ball screw forces are assumed to be zero. Since the friction forces in the spherical and Hooke joint are assumed to be zero and the motor torques are modeled as equivalent linear actuator forces, there are no other applied external torque/moments to the system. I.e.,

$$\mathbf{T}_c = \mathbf{T}_E = \mathbf{T}_B = 0 \tag{6.45}$$

For the inertial terms ($\mathbf{F}_j^*$, $\mathbf{T}_j^*$), it can be written as:

$$\mathbf{F}_j^* = m_j \mathbf{a}_j \tag{6.46}$$

$$\mathbf{T}_j^* = \boldsymbol{\alpha}_j \circ \mathbf{I}_j'' + \boldsymbol{\omega}_j \times (\mathbf{I}_j'' \circ \boldsymbol{\omega}_j) \tag{6.47}$$

where $\mathbf{a}_j$ is the absolute linear acceleration of the center of mass of body j; $\boldsymbol{\omega}_j$ and $\boldsymbol{\alpha}_j$ are the absolute angular velocity and acceleration of body j, respectively; $\mathbf{I}_j''$ is the dyadic term for the inertia of the jth body about its center of mass; '$\circ$' denotes dyadic product of the dyadic term with a vector; and '$\times$' denotes the cross product of vectors. The reader is referred to reference [43] for detailed description of dyadics.

The partial derivative terms in the Stewart-Gough PKM Kane equation (6.41) can be written as follows using vector-matrix notation and partition matrices.

$$\frac{\partial \mathbf{v}_c}{\partial \mathbf{v}_c} = [\mathbf{I}], \qquad \frac{\partial \mathbf{v}_c}{\partial \dot{\boldsymbol{\beta}}} = [\mathbf{0}], \qquad \frac{\partial \boldsymbol{\omega}_c}{\partial \mathbf{v}_c} = [\mathbf{0}], \qquad \frac{\partial \boldsymbol{\omega}_c}{\partial \dot{\boldsymbol{\beta}}} = [\mathbf{U}] \tag{6.48}$$

where [**I**] is a (3x3) identity matrix and [**0**] is a (3x3) zero matrix. The $[\mathbf{U}]$ matrix is defined in equation (6.14). For each strut i, the following equations are stated with subscript i omitted for brevity. Note, $\hat{i}, \hat{j}, \hat{k}$ are the unit vectors of {W}, X_w, Y_w, Z_w respectively. The $\vdots$ in $[\;\;\vdots\;\;]$ are used in forming partitions in the [ ] matrix.

$$\frac{\partial \mathbf{v}_E}{\partial \mathbf{v}_c} = (\hat{\mathbf{s}}_3\hat{\mathbf{s}}_3^T) + \left(\frac{L_E}{L}\right)\left[(\hat{\mathbf{s}}_3 \times \hat{i}) \times \hat{\mathbf{s}}_3 \ \vdots \ (\hat{\mathbf{s}}_3 \times \hat{j}) \times \hat{\mathbf{s}}_3 \ \vdots \ (\hat{\mathbf{s}}_3 \times \hat{k}) \times \hat{\mathbf{s}}_3\right] \tag{6.49}$$

$$\frac{\partial \mathbf{v}_B}{\partial \mathbf{v}_c} = \left(\frac{L_B}{L}\right)\left[(\hat{\mathbf{s}}_3 \times \hat{i}) \times \hat{\mathbf{s}}_3 \ \vdots \ (\hat{\mathbf{s}}_3 \times \hat{j}) \times \hat{\mathbf{s}}_3 \ \vdots \ (\hat{\mathbf{s}}_3 \times \hat{k}) \times \hat{\mathbf{s}}_3\right] \tag{6.50}$$

$$\frac{\partial \boldsymbol{\omega}_S}{\partial \mathbf{v}_c} = \frac{\partial \boldsymbol{\omega}_B}{\partial \mathbf{v}_c} = \frac{\partial \boldsymbol{\omega}_E}{\partial \mathbf{v}_c} = \left(\frac{1}{L}\right)\left[(\hat{\mathbf{s}}_3 \times \hat{i}) \ \vdots \ (\hat{\mathbf{s}}_3 \times \hat{j}) \ \vdots \ (\hat{\mathbf{s}}_3 \times \hat{k})\right] \tag{6.51}$$

$$\frac{\partial \mathbf{v}_E}{\partial \dot{\boldsymbol{\beta}}} = \hat{\mathbf{s}}_3\left[\left({}^w\mathbf{r}_{pc} \times \hat{\mathbf{s}}_3\right)^T [\mathbf{U}]\right] + \left(\frac{L_E}{L}\right)\left[\left(\hat{\mathbf{s}}_3 \times \left([\mathbf{U}]_{Col.1} \times {}^w\mathbf{r}_{pc}\right)\right) \times \hat{\mathbf{s}}_3 \ \vdots \ \left(\hat{\mathbf{s}}_3 \times \left([\mathbf{U}]_{Col.2} \times {}^w\mathbf{r}_{pc}\right)\right) \times \hat{\mathbf{s}}_3 \ \vdots \ \left(\hat{\mathbf{s}}_3 \times \left([\mathbf{U}]_{Col.3} \times {}^w\mathbf{r}_{pc}\right)\right) \times \hat{\mathbf{s}}_3\right] \tag{6.52}$$

$$\frac{\partial \mathbf{v}_B}{\partial \dot{\boldsymbol{\beta}}} = \left(\frac{L_B}{L}\right)\left[\left(\hat{\mathbf{s}}_3 \times \left([\mathbf{U}]_{Col.1} \times {}^w\mathbf{r}_{pc}\right)\right) \times \hat{\mathbf{s}}_3 \ \vdots \ \left(\hat{\mathbf{s}}_3 \times \left([\mathbf{U}]_{Col.2} \times {}^w\mathbf{r}_{pc}\right)\right) \times \hat{\mathbf{s}}_3 \ \vdots \ \left(\hat{\mathbf{s}}_3 \times \left([\mathbf{U}]_{Col.3} \times {}^w\mathbf{r}_{pc}\right)\right) \times \hat{\mathbf{s}}_3\right] \tag{6.53}$$

$$\frac{\partial \boldsymbol{\omega}_S}{\partial \dot{\boldsymbol{\beta}}} = \frac{\partial \boldsymbol{\omega}_B}{\partial \dot{\boldsymbol{\beta}}} = \frac{\partial \boldsymbol{\omega}_E}{\partial \dot{\boldsymbol{\beta}}} = \left(\frac{1}{L}\right)\left[\left(\hat{\mathbf{s}}_3 \times \left([\mathbf{U}]_{Col.1} \times {}^w\mathbf{r}_{pc}\right)\right) \ \vdots \ \left(\hat{\mathbf{s}}_3 \times \left([\mathbf{U}]_{Col.2} \times {}^w\mathbf{r}_{pc}\right)\right) \ \vdots \ \left(\hat{\mathbf{s}}_3 \times \left([\mathbf{U}]_{Col.3} \times {}^w\mathbf{r}_{pc}\right)\right)\right] \tag{6.54}$$

where $[\mathbf{U}]_{Col.i}$ is the ith column of matrix [**U**].

Equations of Motion (Force-Based Model)

Once the above equations are substituted into the Kane's equation, a set of six scalar equations will result, where the six $(\mathbf{F}_a)_i$ terms can be factored out. The force-based model uses the resulting model expression for the vector of these six actuator force terms $(\mathbf{F}_a)_i$. When expressing these six equations of motion in an ordered, compact vector-matrix form per the defined generalized coordinates, the coefficients of these $(\mathbf{F}_a)_i$ terms yield a pseudo-Jacobian matrix

$$\mathbf{J} = \begin{bmatrix} (\hat{\mathbf{s}}_3)_1 & \vdots \cdots \vdots & (\hat{\mathbf{s}}_3)_6 \\ [\mathbf{U}]^T\left({}^w\mathbf{r}_{pc} \times \hat{\mathbf{s}}_3\right)_1 & \vdots \cdots \vdots & [\mathbf{U}]^T\left({}^w\mathbf{r}_{pc} \times \hat{\mathbf{s}}_3\right)_6 \end{bmatrix} \tag{6.55}$$

analogous to the traditional Jacobian definition for PKMs, for further derivation details refer to Oh [32]. That is, the Pseudo-Jacobian matrix can be physically interpreted as the pose of the PKM expressed in the Euler angles, especially if one takes into account that the $[\mathbf{U}]$ matrix is the transformation matrix from the Euler angular rates to the absolute angular velocity of the platform in {W} coordinates.

Solving for the actuator forces yields the following form of the equation of motion.

$$\mathbf{F}_a = \mathbf{J}^{-1}\left[\begin{bmatrix}\mathbf{F}_{pm}\\ \mathbf{T}_{pm}\end{bmatrix}+\begin{bmatrix}\mathbf{F}_{sm}\\ \mathbf{T}_{sm}\end{bmatrix}+\begin{bmatrix}\mathbf{F}_{cor}\\ \mathbf{T}_{cor}\end{bmatrix}+\begin{bmatrix}\mathbf{F}_{cen}\\ \mathbf{T}_{cen}\end{bmatrix}\right.$$
$$\left.+\begin{bmatrix}\mathbf{F}_{grp}\\ \mathbf{T}_{grp}\end{bmatrix}+\begin{bmatrix}\mathbf{F}_{grs}\\ \mathbf{T}_{grs}\end{bmatrix}+\begin{bmatrix}\mathbf{F}_{frict}\\ \mathbf{T}_{frict}\end{bmatrix}+\begin{bmatrix}\mathbf{F}_{ext}\\ \mathbf{T}_{ext}\end{bmatrix}+\begin{bmatrix}\mathbf{F}_{m/b}\\ \mathbf{T}_{m/b}\end{bmatrix}+\begin{bmatrix}\mathbf{F}_{umd}\\ \mathbf{T}_{umd}\end{bmatrix}\right] \tag{6.56}$$

where F_a denotes a vector of the magnitude and sense of the six actuator forces, each applied along their respective strut axes $(\hat{\mathbf{s}}_3)_i$. The terms on the right-side of the equation are the dynamic forces and moment contributions and the external loads grouped by their source. As expected, the resulting equations of motion are highly nonlinear.

First, the platform acceleration dependent terms (forces and moments) are grouped as $\begin{bmatrix}\mathbf{F}_{pm}^T & \mathbf{T}_{pm}^T\end{bmatrix}^T$.

$$\begin{bmatrix}\mathbf{F}_{pm}\\ \mathbf{T}_{pm}\end{bmatrix} = [\mathbf{M}_p]\begin{Bmatrix}\mathbf{a}_c\\ \ddot{\boldsymbol{\beta}}\end{Bmatrix} \tag{6.57}$$

where

$$[\mathbf{M}_p] = \begin{bmatrix} m_p[\mathbf{I}] & [0] \\ [0] & \left[\dfrac{\partial\boldsymbol{\omega}}{\partial\dot{\beta}}\right]^T [\mathbf{I}_p]\left[\dfrac{\partial\boldsymbol{\omega}}{\partial\dot{\beta}}\right]\end{bmatrix} \tag{6.58}$$

The $\begin{bmatrix}\mathbf{F}_{sm}^T & \mathbf{T}_{sm}^T\end{bmatrix}^T$ term yields the strut acceleration effects.

$$\begin{bmatrix}\mathbf{F}_{sm}\\ \mathbf{T}_{sm}\end{bmatrix} = [\mathbf{M}_{strut}]\begin{Bmatrix}\mathbf{a}_c\\ \ddot{\boldsymbol{\beta}}\end{Bmatrix} \tag{6.59}$$

where

$$[\mathbf{M}_{strut}] = \begin{bmatrix}[\mathbf{M}_{ac}] & [\mathbf{M}_{\alpha ac}]\\ [\mathbf{M}_{ac\alpha}] & [\mathbf{M}_{\alpha}]\end{bmatrix} \tag{6.60}$$

$$[\mathbf{M}_{ac}] = -\sum_{i=1}^{6}\left\{\begin{matrix}\left[\dfrac{\partial\mathbf{v}_E}{\partial\mathbf{v}_c}\right]^T(-m_E)\left[\dfrac{\partial\mathbf{v}_E}{\partial\mathbf{v}_c}\right]+\left[\dfrac{\partial\mathbf{v}_B}{\partial\mathbf{v}_c}\right]^T(-m_B)\left[\dfrac{\partial\mathbf{v}_B}{\partial\mathbf{v}_c}\right]\\ +\left[\dfrac{\partial\boldsymbol{\omega}_\mathbf{E}}{\partial\mathbf{v}_c}\right]^T(-1)[\mathbf{I}_E]\left[\dfrac{\partial\boldsymbol{\omega}_\mathbf{E}}{\partial\mathbf{v}_c}\right]+\left[\dfrac{\partial\boldsymbol{\omega}_\mathbf{B}}{\partial\mathbf{v}_c}\right]^T(-1)[\mathbf{I}_B]\left[\dfrac{\partial\boldsymbol{\omega}_\mathbf{B}}{\partial\mathbf{v}_c}\right]\end{matrix}\right\}_i \tag{6.61}$$

$$[\mathbf{M}_{\alpha ac}] = -\sum_{i=1}^{6}\left\{\begin{array}{l}\left[\frac{\partial \mathbf{v}_E}{\partial \mathbf{v}_c}\right]^T(-m_E)\left[\frac{\partial \mathbf{v}_E}{\partial \dot{\boldsymbol{\beta}}}\right]+\left[\frac{\partial \mathbf{v}_B}{\partial \mathbf{v}_c}\right]^T(-m_B)\left[\frac{\partial \mathbf{v}_B}{\partial \dot{\boldsymbol{\beta}}}\right]\\ +\left[\frac{\partial \boldsymbol{\omega}_\mathbf{E}}{\partial \mathbf{v}_c}\right]^T(-1)[\mathbf{I}_E]\left[\frac{\partial \boldsymbol{\omega}_\mathbf{E}}{\partial \dot{\boldsymbol{\beta}}}\right]+\left[\frac{\partial \boldsymbol{\omega}_\mathbf{B}}{\partial \mathbf{v}_c}\right]^T(-1)[\mathbf{I}_B]\left[\frac{\partial \boldsymbol{\omega}_\mathbf{B}}{\partial \dot{\boldsymbol{\beta}}}\right]\end{array}\right\}_i \tag{6.62}$$

$$[\mathbf{M}_{ac\alpha}] = -\sum_{i=1}^{6}\left\{\begin{array}{l}\left[\frac{\partial \mathbf{v}_E}{\partial \dot{\boldsymbol{\beta}}}\right]^T(-m_E)\left[\frac{\partial \mathbf{v}_E}{\partial \mathbf{v}_c}\right]+\left[\frac{\partial \mathbf{v}_B}{\partial \dot{\boldsymbol{\beta}}}\right]^T(-m_B)\left[\frac{\partial \mathbf{v}_B}{\partial \mathbf{v}_c}\right]\\ +\left[\frac{\partial \boldsymbol{\omega}_\mathbf{E}}{\partial \dot{\boldsymbol{\beta}}}\right]^T(-1)[\mathbf{I}_E]\left[\frac{\partial \boldsymbol{\omega}_\mathbf{E}}{\partial \mathbf{v}_c}\right]+\left[\frac{\partial \boldsymbol{\omega}_\mathbf{B}}{\partial \dot{\boldsymbol{\beta}}}\right]^T(-1)[\mathbf{I}_B]\left[\frac{\partial \boldsymbol{\omega}_\mathbf{B}}{\partial \mathbf{v}_c}\right]\end{array}\right\}_i \tag{6.63}$$

$$[\mathbf{M}_{\alpha}] = -\sum_{i=1}^{6}\left\{\begin{array}{l}\left[\frac{\partial \mathbf{v}_E}{\partial \dot{\boldsymbol{\beta}}}\right]^T(-m_E)\left[\frac{\partial \mathbf{v}_E}{\partial \dot{\boldsymbol{\beta}}}\right]+\left[\frac{\partial \mathbf{v}_B}{\partial \dot{\boldsymbol{\beta}}}\right]^T(-m_B)\left[\frac{\partial \mathbf{v}_B}{\partial \dot{\boldsymbol{\beta}}}\right]\\ +\left[\frac{\partial \boldsymbol{\omega}_\mathbf{E}}{\partial \dot{\boldsymbol{\beta}}}\right]^T(-1)[\mathbf{I}_E]\left[\frac{\partial \boldsymbol{\omega}_\mathbf{E}}{\partial \dot{\boldsymbol{\beta}}}\right]+\left[\frac{\partial \boldsymbol{\omega}_\mathbf{B}}{\partial \dot{\boldsymbol{\beta}}}\right]^T(-1)[\mathbf{I}_B]\left[\frac{\partial \boldsymbol{\omega}_\mathbf{B}}{\partial \dot{\boldsymbol{\beta}}}\right]\end{array}\right\}_i \tag{6.64}$$

Additionally,

$$\begin{bmatrix}\mathbf{F}_{pm}\\ \mathbf{T}_{pm}\end{bmatrix}+\begin{bmatrix}\mathbf{F}_{sm}\\ \mathbf{T}_{sm}\end{bmatrix}=[\mathbf{M}]\begin{Bmatrix}\mathbf{a}_c\\ \ddot{\boldsymbol{\beta}}\end{Bmatrix}=\left[[\mathbf{M}_\mathbf{p}]+[\mathbf{M}_\mathbf{strut}]\right]\begin{Bmatrix}\mathbf{a}_c\\ \ddot{\boldsymbol{\beta}}\end{Bmatrix} \tag{6.65}$$

The Coriolis and centripetal acceleration-like terms contain terms that are a function of strut and platform velocities products. The Coriolis terms are grouped in $\left[\mathbf{F}_{cor}{}^T \quad \mathbf{T}_{cor}{}^T\right]^T$.

$$\mathbf{F}_{cor} = \sum_{i=1}^{6}\left\{\begin{array}{l}\left[\frac{\partial \mathbf{v}_E}{\partial \mathbf{v}_c}\right]^T(-m_E)\left[2\dot{L}_E(\boldsymbol{\omega}_E\times\hat{\mathbf{s}}_3)-\left(\frac{2L_E\dot{L}}{L}\right)\left[\left[\hat{\mathbf{s}}_3\times\left(\mathbf{v}_c+\left(\boldsymbol{\omega}_\mathbf{c}\times{}^w\mathbf{r}_{pc}\right)\right)\right]\times\hat{\mathbf{s}}_3\right]\right]\\ +\left[\frac{\partial \mathbf{v}_B}{\partial \mathbf{v}_c}\right]^T(-m_B)\left[-\left(\frac{2L_B\dot{L}}{L}\right)\left[\left[\hat{\mathbf{s}}_3\times\left(\mathbf{v}_c+\left(\boldsymbol{\omega}_\mathbf{c}\times{}^w\mathbf{r}_{pc}\right)\right)\right]\times\hat{\mathbf{s}}_3\right]\right]\\ +\left[\frac{\partial \boldsymbol{\omega}_\mathbf{E}}{\partial \mathbf{v}_c}\right]^T\left(\frac{2\dot{L}}{L}\right)[\mathbf{I}_E]\boldsymbol{\omega}_\mathbf{E}+\left[\frac{\partial \boldsymbol{\omega}_\mathbf{B}}{\partial \mathbf{v}_c}\right]^T\left(\frac{2\dot{L}}{L}\right)[I_B]\boldsymbol{\omega}_\mathbf{B}\end{array}\right\}_i \tag{6.66}$$

$$\mathbf{T}_{cor} = \sum_{i=1}^{6}\left\{\begin{array}{l}\left[\frac{\partial \mathbf{v}_E}{\partial \dot{\boldsymbol{\beta}}}\right]^T(-m_E)\left[2\dot{L}_E(\boldsymbol{\omega}_E\times\hat{\mathbf{s}}_3)-\left(\frac{2L_E\dot{L}}{L}\right)\left[\left[\hat{\mathbf{s}}_3\times\left(\mathbf{v}_c+\left(\boldsymbol{\omega}_\mathbf{c}\times{}^w\mathbf{r}_{pc}\right)\right)\right]\times\hat{\mathbf{s}}_3\right]\right]\\ +\left[\frac{\partial \mathbf{v}_B}{\partial \dot{\boldsymbol{\beta}}}\right]^T(-m_B)\left[-\left(\frac{2L_B\dot{L}}{L}\right)\left[\left[\hat{\mathbf{s}}_3\times\left(\mathbf{v}_c+\left(\boldsymbol{\omega}_\mathbf{c}\times{}^w\mathbf{r}_{pc}\right)\right)\right]\times\hat{\mathbf{s}}_3\right]\right]\\ +\left[\frac{\partial \boldsymbol{\omega}_\mathbf{E}}{\partial \dot{\boldsymbol{\beta}}}\right]^T\left(\frac{2\dot{L}}{L}\right)[\mathbf{I}_E]\boldsymbol{\omega}_\mathbf{E}+\left[\frac{\partial \boldsymbol{\omega}_\mathbf{B}}{\partial \dot{\boldsymbol{\beta}}}\right]^T\left(\frac{2\dot{L}}{L}\right)[\mathbf{I}_B]\boldsymbol{\omega}_\mathbf{B}\end{array}\right\}_i \tag{6.67}$$

The centripetal acceleration-like terms are grouped in $\left[\mathbf{F}_{cen}{}^{T} \quad \mathbf{T}_{cen}{}^{T}\right]^{T}$.

$$\mathbf{F}_{cen}=\sum_{i=1}^{6}\left\{\begin{aligned}&\left[\frac{\partial\mathbf{v}_{E}}{\partial\mathbf{v}_{c}}\right]^{T}(-m_{E})\left[\boldsymbol{\omega}_{E}\times(\boldsymbol{\omega}_{E}\times(L_{E}\hat{\mathbf{s}}_{3}))+\left(\boldsymbol{\omega}_{E}{}^{\mathbf{T}}\boldsymbol{\omega}_{E}\right)L\hat{\mathbf{s}}_{3}\right]\\&+\left[\frac{\partial\mathbf{v}_{B}}{\partial\mathbf{v}_{c}}\right]^{T}(-m_{B})\left[\boldsymbol{\omega}_{B}\times(\boldsymbol{\omega}_{B}\times(L_{B}\hat{\mathbf{s}}_{3}))\right]+\left[\frac{\partial\boldsymbol{\omega}_{E}}{\partial\mathbf{v}_{c}}\right]^{T}(-1)\left[\boldsymbol{\omega}_{E}\times\left[\mathbf{I}_{E}\right]\boldsymbol{\omega}_{E}\right]\\&+\left[\frac{\partial\boldsymbol{\omega}_{B}}{\partial\mathbf{v}_{c}}\right]^{T}(-1)\left[\boldsymbol{\omega}_{B}\times\left[\mathbf{I}_{B}\right]\boldsymbol{\omega}_{B}\right]\\&+\left[\frac{\partial\mathbf{v}_{E}}{\partial\mathbf{v}_{c}}\right]^{T}(-m_{E})\left[\hat{\mathbf{s}}_{3}\cdot\hat{\mathbf{s}}_{3}{}^{\mathbf{T}}[\boldsymbol{\omega}_{c}\times(\boldsymbol{\omega}_{c}\times{}^{w}\mathbf{r}_{pc})]+\left(\frac{L_{E}}{L}\right)\left[\left[\hat{\mathbf{s}}_{3}\times\left(\boldsymbol{\omega}_{c}\times\left(\boldsymbol{\omega}_{c}\times{}^{w}\mathbf{r}_{pc}\right)\right)\right]\times\hat{\mathbf{s}}_{3}\right]\right]\\&+\left[\frac{\partial\mathbf{v}_{B}}{\partial\mathbf{v}_{c}}\right]^{T}(-m_{B})\left[\left(\frac{L_{B}}{L}\right)\left[\left[\hat{\mathbf{s}}_{3}\times\left(\boldsymbol{\omega}_{c}\times\left(\boldsymbol{\omega}_{c}\times{}^{w}\mathbf{r}_{pc}\right)\right)\right]\times\hat{\mathbf{s}}_{3}\right]\right]\\&+\left[\frac{\partial\boldsymbol{\omega}_{E}}{\partial\mathbf{v}_{c}}\right]^{T}\left(-\frac{1}{L}\right)\left[\mathbf{I}_{E}\right]\left[\hat{\mathbf{s}}_{3}\times\left(\boldsymbol{\omega}_{c}\times\left(\boldsymbol{\omega}_{c}\times{}^{w}\mathbf{r}_{pc}\right)\right)\right]\\&+\left[\frac{\partial\boldsymbol{\omega}_{B}}{\partial\mathbf{v}_{c}}\right]^{T}\left(-\frac{1}{L}\right)\left[\mathbf{I}_{B}\right]\left[\hat{\mathbf{s}}_{3}\times\left(\boldsymbol{\omega}_{c}\times\left(\boldsymbol{\omega}_{c}\times{}^{w}\mathbf{r}_{pc}\right)\right)\right]\end{aligned}\right\}_{i} \tag{6.68}$$

$$\begin{aligned}\mathbf{T}_{cen}=\sum_{i=1}^{6}&\left\{\begin{aligned}&\left[\frac{\partial\mathbf{v}_{E}}{\partial\dot{\boldsymbol{\beta}}}\right]^{T}(-m_{E})\left[\boldsymbol{\omega}_{E}\times(\boldsymbol{\omega}_{E}\times(L_{E}\hat{\mathbf{s}}_{3}))+\left(\boldsymbol{\omega}_{E}{}^{\mathbf{T}}\boldsymbol{\omega}_{E}\right)L\hat{\mathbf{s}}_{3}\right]+\left[\frac{\partial\mathbf{v}_{B}}{\partial\dot{\boldsymbol{\beta}}}\right]^{T}(-m_{B})\left[\boldsymbol{\omega}_{B}\times(\boldsymbol{\omega}_{B}\times(L_{B}\hat{\mathbf{s}}_{3}))\right]\\&+\left[\frac{\partial\boldsymbol{\omega}_{E}}{\partial\dot{\boldsymbol{\beta}}}\right]^{T}(-1)\left[\boldsymbol{\omega}_{E}\times\left[\mathbf{I}_{E}\right]\boldsymbol{\omega}_{E}\right]+\left[\frac{\partial\boldsymbol{\omega}_{B}}{\partial\dot{\boldsymbol{\beta}}}\right]^{T}(-1)\left[\boldsymbol{\omega}_{B}\times\left[\mathbf{I}_{B}\right]\boldsymbol{\omega}_{B}\right]\\&+\left[\frac{\partial\mathbf{v}_{E}}{\partial\dot{\boldsymbol{\beta}}}\right]^{T}(-m_{E})\left[\hat{\mathbf{s}}_{3}\cdot\hat{\mathbf{s}}_{3}{}^{\mathbf{T}}[\boldsymbol{\omega}_{c}\times(\boldsymbol{\omega}_{c}\times{}^{w}\mathbf{r}_{pc})]+\left(\frac{L_{E}}{L}\right)\left[\left[\hat{\mathbf{s}}_{3}\times\left(\boldsymbol{\omega}_{c}\times\left(\boldsymbol{\omega}_{c}\times{}^{w}\mathbf{r}_{pc}\right)\right)\right]\times\hat{\mathbf{s}}_{3}\right]\right]\\&+\left[\frac{\partial\mathbf{v}_{B}}{\partial\dot{\boldsymbol{\beta}}}\right]^{T}(-m_{B})\left[\left(\frac{L_{B}}{L}\right)\left[\left[\hat{\mathbf{s}}_{3}\times\left(\boldsymbol{\omega}_{c}\times\left(\boldsymbol{\omega}_{c}\times{}^{w}\mathbf{r}_{pc}\right)\right)\right]\times\hat{\mathbf{s}}_{3}\right]\right]\\&+\left[\frac{\partial\boldsymbol{\omega}_{E}}{\partial\dot{\boldsymbol{\beta}}}\right]^{T}\left(-\frac{1}{L}\right)\left[\mathbf{I}_{E}\right]\left[\hat{\mathbf{s}}_{3}\times\left(\boldsymbol{\omega}_{c}\times\left(\boldsymbol{\omega}_{c}\times{}^{w}\mathbf{r}_{pc}\right)\right)\right]\\&+\left[\frac{\partial\boldsymbol{\omega}_{B}}{\partial\dot{\boldsymbol{\beta}}}\right]^{T}\left(-\frac{1}{L}\right)\left[\mathbf{I}_{B}\right]\left[\hat{\mathbf{s}}_{3}\times\left(\boldsymbol{\omega}_{c}\times\left(\boldsymbol{\omega}_{c}\times{}^{w}\mathbf{r}_{pc}\right)\right)\right]\end{aligned}\right\}_{i}\\&+\left[\frac{\partial\boldsymbol{\omega}_{c}}{\partial\dot{\boldsymbol{\beta}}}\right]^{T}(-1)\left[\boldsymbol{\omega}_{c}\times\left[\mathbf{I}_{p}\right]\boldsymbol{\omega}_{c}\right]\end{aligned} \tag{6.69}$$

The gravitational dependent terms for the platform are grouped in $\left[\mathbf{F}_{grp}{}^{T} \quad \mathbf{T}_{grp}{}^{T}\right]^{T}$.

$$\mathbf{F}_{grp}=\left[\frac{\partial\mathbf{v}_{c}}{\partial\mathbf{v}_{c}}\right]^{T}m_{p}\mathbf{g} \tag{6.70}$$

$$\mathbf{T}_{grp}=0 \tag{6.71}$$

The gravitation dependent terms for the struts are grouped in $\left[\mathbf{F}_{grs}{}^T \quad \mathbf{T}_{grs}{}^T\right]^T$.

$$\mathbf{F}_{grs} = \sum_{i=1}^{6}\left\{\left[\frac{\partial \mathbf{v}_E}{\partial \mathbf{v}_c}\right]^T m_E\mathbf{g} + \left[\frac{\partial \mathbf{v}_B}{\partial \mathbf{v}_c}\right]^T m_B\mathbf{g}\right\}_i \tag{6.72}$$

$$\mathbf{T}_{grs} = \sum_{i=1}^{6}\left\{\left[\frac{\partial \mathbf{v}_E}{\partial \dot{\boldsymbol{\beta}}}\right]^T m_E\mathbf{g} + \left[\frac{\partial \mathbf{v}_B}{\partial \dot{\boldsymbol{\beta}}}\right]^T m_B\mathbf{g}\right\}_i \tag{6.73}$$

One of the sources of nonlinearities are joint frictions which are denoted by $\left[\mathrm{F}_{frict}{}^T \quad \mathrm{T}_{frict}{}^T\right]^T$ and are dependent on the PKM instantaneous configuration, velocity and friction model. E.g., the friction terms F_{ζ_i} can be modeled as modified friction.

$$\mathrm{F}_{frict} = (-1)\left[(\hat{\mathrm{s}}_3)_1 \;\vdots\; \cdots \;\vdots\; (\hat{\mathrm{s}}_3)_6\right]\begin{Bmatrix} F_{\zeta_1} \\ \vdots \\ F_{\zeta_6}\end{Bmatrix} + \sum_{i=1}^{6}\left\{\left[\frac{\partial \omega_E}{\partial \mathrm{v}_c}\right]^T \tau_{\zeta P} + \left[\frac{\partial \omega_B}{\partial \mathrm{v}_c}\right]^T \tau_{\zeta b}\right\}_i \tag{6.74}$$

$$\mathrm{T}_{frict} = \sum_{i=1}^{6}\left\{[\mathrm{U}]\left({}^{\mathrm{w}}\mathrm{r}_{pc} \times \hat{\mathrm{s}}_3\right)(-\mathrm{F}_{\zeta}) + \left[\frac{\partial \omega_E}{\partial \dot{\beta}}\right]^T \tau_{\zeta P} + \left[\frac{\partial \omega_B}{\partial \dot{\beta}}\right]^T \tau_{\zeta b} + \left[\frac{\partial \omega_c}{\partial \dot{\beta}}\right]^T \left[-\tau_{\zeta b}\right]\right\}_i \tag{6.75}$$

The $\left[\mathrm{F}_{ext}{}^T \quad \mathrm{T}_{ext}{}^T\right]^T$ consists of the external loads acting on the PKM.

$$\mathbf{F}_{ext} = \sum_{i=1}^{6}\left\{\left[\frac{\partial \mathbf{v}_E}{\partial \mathbf{v}_c}\right]^T \mathbf{F}_{ext_E} + \left[\frac{\partial \mathbf{v}_B}{\partial \mathbf{v}_c}\right]^T \mathbf{F}_{ext_B} + \left[\frac{\partial \boldsymbol{\omega}_E}{\partial \mathbf{v}_c}\right]^T \boldsymbol{\tau}_{ext_E} + \left[\frac{\partial \boldsymbol{\omega}_B}{\partial \mathbf{v}_c}\right]^T \boldsymbol{\tau}_{ext_B} + \left[\frac{\partial \mathbf{v}_c}{\partial \mathbf{v}_c}\right]^T \mathbf{F}_{ext_p}\right\}_i \tag{6.76}$$

$$\mathbf{T}_{ext} = \sum_{i=1}^{6}\left\{\left[\frac{\partial \mathbf{v}_E}{\partial \dot{\boldsymbol{\beta}}}\right]^T \mathbf{F}_{ext_E} + \left[\frac{\partial \mathbf{v}_B}{\partial \dot{\boldsymbol{\beta}}}\right]^T \mathbf{F}_{ext_B} + \left[\frac{\partial \boldsymbol{\omega}_E}{\partial \dot{\boldsymbol{\beta}}}\right]^T \boldsymbol{\tau}_{ext_E} + \left[\frac{\partial \boldsymbol{\omega}_B}{\partial \dot{\boldsymbol{\beta}}}\right]^T \boldsymbol{\tau}_{ext_B}\right\}_i + \left[\frac{\partial \boldsymbol{\omega}_c}{\partial \dot{\boldsymbol{\beta}}}\right]^T \boldsymbol{\tau}_{ext_p} \tag{6.77}$$

The $\left[\mathrm{F}_{m/b}{}^T \quad \mathrm{T}_{m/b}{}^T\right]^T$ represents the terms due to the motor and ball screw dynamics. The $\left[\mathrm{F}_{umd}{}^T \quad \mathrm{T}_{umd}{}^T\right]^T$ denotes the collection of the other unmodeled dynamics term. In the remainder of this chapter, the $\left[\mathrm{F}_{m/b}{}^T \quad \mathrm{T}_{m/b}{}^T\right]^T$ and $\left[\mathrm{F}_{umd}{}^T \quad \mathrm{T}_{umd}{}^T\right]^T$ terms are assumed to be negligible.

As expected, the resulting PKM equations of motion are highly nonlinear. For further details on generating the equations of motion, the reader is referred to Oh [32], Register [47] and Walker [45].

It should be noted that the above derivation of PKM equations of motion is procedurally the same for obtaining the equations of motions for other PKM configurations such as those with revolute joints in place of the above prismatic joints and those with fixed strut lengths with actuated translating base joints. However, the appropriate kinematic relationship of the support links (equivalent strut) with the moving platform's generalized coordinates would need to be obtained and there would be a different point of application of the actuating force or torque. Specific assumptions may also vary from model to model.

The Energy Method

For the parameter identification algorithms, the implementation of the identification algorithm with the dynamic equations of motion as the model basis involves measuring position, velocity, acceleration and actuator forces as the inputs to the identification process. Measurements of velocity and acceleration are typically prone to errors. When data is not directly measurable but must be calculated numerically from other measured data, e.g., position data, errors can rapidly grow in magnitude especially with each level of differentiation that is calculated using data containing noise. To circumvent this problem, one can instrument the setup with better quality sensors, which tends to drive up costs and require greater engineering expertise. An alternative would be to use as the model base the energy method suggested by Bhattacharya [28]. In doing so, one does not need to measure accelerations, thus eliminating one level of noise amplification in calculated inputs to the identification algorithm. Thus, an energy-based model approach was investigated as a potential alternative to the equation of motion based approach, referred to as the force-based approach in this chapter.

For the energy-based method, the PKM is assumed to move in a singular-free trajectory, permitting an energy difference determination for time step $\left(\mathrm{x}_k, \dot{\mathrm{x}}_k\right)$ through $\left(\mathrm{x}_{k+1}, \dot{\mathrm{x}}_{k+1}\right)$. The energy equations developed by Bhattacharya [28] are defined as the following scalar function.

$$h_{k+1}(x_{k+1}, \dot{x}_{k+1}, w_{k+1}) - h_k(x_k, \dot{x}_k, w_k) = \int_1^2 \mathbf{F} \cdot \dot{x}\, dt = \sum_{i=1}^{n} \left(F_{a_i}\right)_k \left(\Delta L_i\right)_k \tag{6.78}$$

where $h_k(x_k, \dot{x}_k, w_k)$ is the total energy (kinetic plus potential energy) at the kth states, $h_{k+1}(x_{k+1}, \dot{x}_{k+1}, w_{k+1})$ is total energy (kinetic plus potential energy) at the (k+1)th state, w_k and w_{k+1} are the current kth and (k+1)th inertia parameter estimates. The $\left(F_{a_i}\right)_k$ is the ith strut force at the kth state, $\left(\Delta L_i\right)_k$ is the change in the ith strut length at kth state, n is the number of struts.

For the PKMs modeled, the mass matrix is the same as that defined in equations (6.58) - (6.65). The energy at kth state is expressed as follows.

$$h_k(x_k, \dot{x}_k, w_k) = T_k + V_k \tag{6.79}$$

where the kinetic energy, $T_k = \frac{1}{2}\left[\mathbf{v}_c^{\,T}, \dot{\boldsymbol{\beta}}^T\right]\left[\mathbf{M}\right]\begin{bmatrix}\mathbf{v}_c \\ \dot{\boldsymbol{\beta}}\end{bmatrix}$ is a scalar quantity where matrix $\mathbf{M}$ is a function of the generalized coordinates. The total potential energy in the system is given by

$$V_k = (-1)\left[m_p \mathbf{g}^{T\,W}\mathbf{r}_c + \sum_{i=1}^{6}\left[m_e \mathbf{g}^{T\,W}\mathbf{r}_E\right] + \sum_{i=1}^{6}\left[m_b \mathbf{g}^{T\,W}\mathbf{r}_B\right]\right].$$

Parameter Identification Algorithms

Herein, the ultimate goal is to develop a methodology for determining accurate dynamic system parameter values for parallel kinematic machines (a highly nonlinear system) where the ease of implementing the estimation process is critical. In general, the minimal set of dynamic parameters consists of ten parameters per rigid body: the mass, inertia tensor and the locations of the center of mass per rigid body in the system. These terms can be extracted from the system's theoretical model-based equations forming the set of potentially identifiable parameters. As the system equations become more complex and nonlinear, the extraction of the parameters in forming the traditional vector of identifiable parameters (w_k), as required by least squares (LS) and extended Kalman filter (EKF) methods, becomes insurmountable. For this reason, the unscented Kalman filter methodology is presented and implemented as a method of choice in identifying the dynamic system parameters of PKMs.

The EKF method has widely been used because of its benefits [47]. For the PKM system, the EKF's ability to handle noisy data and highly non-linear, multi-input/multi-output systems is attractive. However recently, Wan and Van der Merwe [41] demonstrated the unscented Kalman filter (UKF) as an alternative to EKF, pointing out the flaws and shortcomings of the EKF. The EKF involves a linearization approximation within its algorithm when applied to nonlinear systems. This causes an elimination of the expectations for the optimal prediction terms, which induces large errors in the true posterior mean and covariance. To overcome these algorithm limitations, the UKF uses an optimal prediction of the input instead of using a first-order linearization approximation [41]. The technique that eliminates the first-order approximations is called Unscented Transformation (UT).

The UKF estimates in two steps illustrated by the flowchart in Figure 6.8. The first step uses the UT to evaluate and estimate the existing value of the initial guess of the parameters as illustrated in Figure 6.9. That is, the UT converts the vector of estimated parameters into a statistical spread of the parameter values, denoted as sigma points or sigma vectors. In the second step, the parameterized vector is estimated using a Kalman gain (weighting) technique which combines the statistical spread of parameter values (sigma points) to yield the new estimate of the vector of estimated parameters and covariance matrix. This two step process is repeated iterative using a new collected sample each iteration until convergence is reached in the estimation of the vector of parameters. Each collected sample is taken from measured dynamic time response data at an arbitrarily selected instant in time.

Ideally, every set of measured data would yield the true parameter values since we are estimating parameters that theoretically do not change in value (constant → stationary process). However, due to noise and the possibility that a parameter is unidentifiable within the instant of time the data is collected, each iterative step (k) needs to use new data from a different instant in time. Furthermore, the convergence is heavily dependent on the ability to drive the system through sufficiently 'dynamically exciting trajectories' in generating the collected measured data samples. The following section details the theory of the UKF and further highlights the main differences between EKF and the UKF.

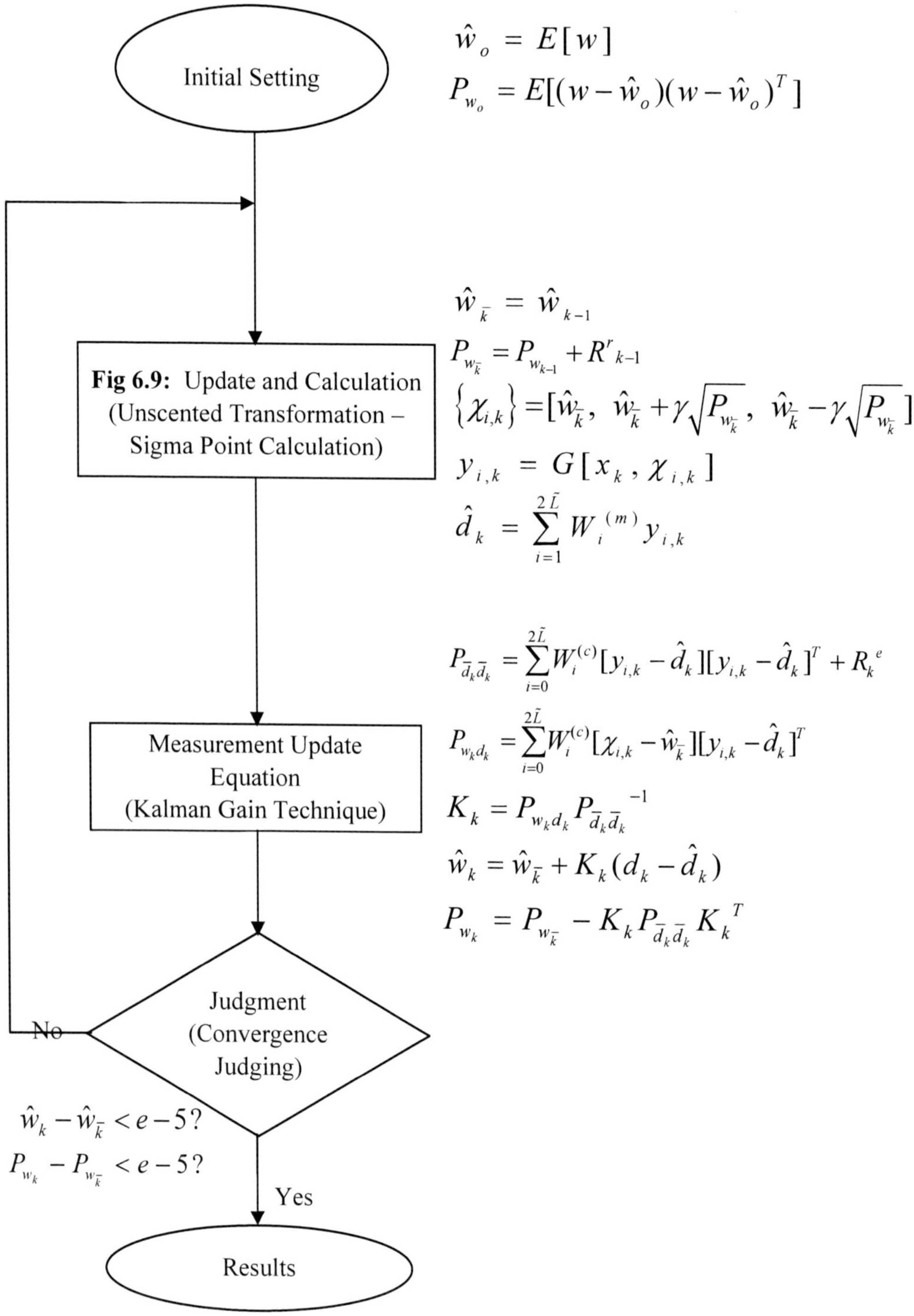

Figure 6.8. Flowchart for the UKF.

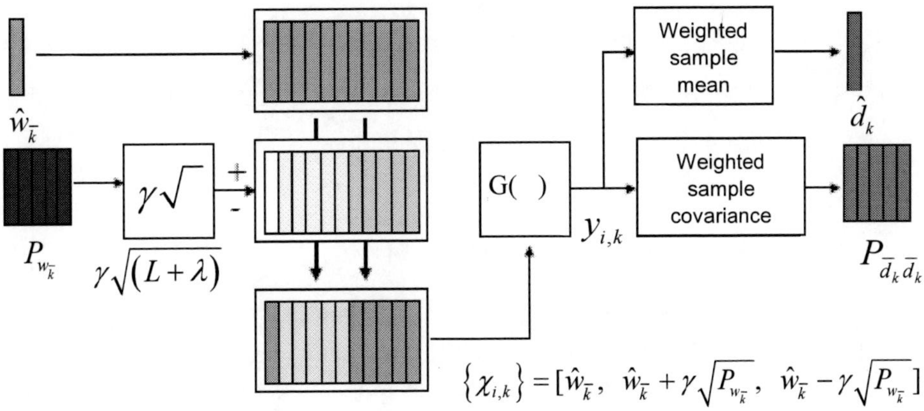

Figure 6.9. Block Diagram of Unscented Transformation [51].

UKF Parameter Estimation

The UKF finds its roots in the 1995 development of the Unscented Transformation (UT) used for filtering nonlinear systems [48, 49]. Julier and Uhlmann [49] later applied the UT to extend Kalman filtering in the development of a generalized method for estimation of nonlinear systems. The classic nonlinear mapping takes the form

$$y_k = G(x_k, w_k) \tag{6.80}$$

where x_k is the vector of measured states and y_k is the updated output vector (calculated from measured states (x_k) and estimated parameters (w_k)) of the kth estimation. In this chapter, the y_k is equated to the nonlinear mapping G expressed as a function of the measured states and estimated parameters governed by the PKM models derived in the prior sections. The dimensions of the function $G(x_k, w_k)$ is dependent on the dimensions of the system and the choice of the nonlinear mapping, i.e., it is a vector defined to be the right-hand side of equation (6.56) for the force-based implementation and a scalar function defined by the left-hand side of equation (6.78) for the energy-based implementation, parameterized by w_k. Thus, all other dimensions of the vectors and matrices used in the UKF approach are determined accordingly. Herein, the estimated parameters of vector (w_k) are the unknown set of PKM inertia parameters being determined from the estimation procedure. The error of the mapping is given by $e_k = d_k - G(x_k, w_k)$ where the goal is to minimize the expected squared error. The d_k is the measured (desired) output. The estimation procedure for the parameter can be defined by the following expressions.

$$\begin{aligned} w_k &= w_{k-1} + u_k \\ d_k &= G(x_k, w_k) + e_k = y_k + e_k \end{aligned} \tag{6.81}$$

where w_k corresponds to a stationary process with an identity state transition matrix, driven by the noise u_k. The output y_k corresponds to the nonlinear observation on w_k. The estimated w_k parameters are updated for each kth iteration.

UKF Parameter Estimation Procedure

The UKF addresses the problems faced in using of the EKF when applied to nonlinear systems. The distinguishing difference of UKF from other estimation methods is that it uses the UT method to generate sigma points for updating the estimated parameters and it is efficient, derivative-free for on-line learning. Using a carefully chosen set of sigma points for generating random sets of the estimated parameters, the UKF captures the true mean and covariance of the Gaussian Random Variables (GRV). Furthermore, the UT is a method for calculating statistics for random variables that undergoes a nonlinear transformation such as the nonlinear mapping G.

In implementing the UKF, one first assumes the vector of the estimated parameters (w_k) as a vector of random variables which has a mean $\hat{w}_k$ and covariance $P_{w_{\bar{k}}}$. To acquire suitable random characteristics of the estimated parameters, sigma points are generated from the current (kth) estimate of the parameters. These sigma points ($\chi_{i,k}$ vectors) constitute a statistical set of $i = 0,1,\ldots,2\tilde{L}$ estimated parameter vectors distributed about w_k. These are determined using the following equations and as illustrated in Figures 6.8 and 6.9.

$$\begin{aligned} \chi_{i,k} &= \hat{w}_{\bar{k}} && \rightarrow i = 0 \\ \chi_{i,k} &= \hat{w}_{\bar{k}} + (\sqrt{(\tilde{L}+\lambda)P_{\hat{w}_{\bar{k}}}})_i = \hat{w}_{\bar{k}} + (\gamma\sqrt{P_{\hat{w}_{\bar{k}}}})_i && \rightarrow i = 1\ldots,\tilde{L} \\ \chi_{i,k} &= \hat{w}_{\bar{k}} - (\sqrt{(\tilde{L}+\lambda)P_{\hat{w}_{\bar{k}}}})_{i-\tilde{L}} = \hat{w}_{\bar{k}} - (\gamma\sqrt{P_{\hat{w}_{\bar{k}}}})_{i-\tilde{L}} && \rightarrow i = \tilde{L}+1, 2\tilde{L} \end{aligned} \tag{6.82}$$

where the $\tilde{L}$ is the dimension of the estimated parameter w_k. Equation (6.82) yields a statistically generated spread of $2\tilde{L}+1$ sigma points ($\chi_{i,k}$ vectors) which are the ith columns of matrix $\{\chi_{i,k}\}$. The corresponding output, y_k results in set of $y_{i,k}$ upon each substitution of the ith sigma point into the nonlinear mapping G. That is for each sigma point i at the kth estimation iteration yields a

$$y_{i,k} = G\left(x_k, \chi_{i,k}\right) \tag{6.84}$$

To generate the new estimate, the UKF uses the following weights $W_i^{(m)}$

$$W_0^{(m)} = \lambda/(\tilde{L}+\lambda)$$
$$W_0^{(c)} = \lambda/(\tilde{L}+\lambda)+(1-\alpha^2+\tilde{\beta}) \quad (6.84)$$
$$W_i^{(m)} = W_i^{(c)} = 1/\{2(\tilde{L}+\lambda)\} \rightarrow i=1,...,2\tilde{L}$$

where $\lambda = \alpha^2(\tilde{L}+\kappa)-\tilde{L}$ is a scaling factor. The α determines the spread of the sigma points defined by the sigma vectors ($\chi_{i,k}$) around the mean $\hat{w}_k$ and is usually set to a small positive value (e.g., 0.001). The κ is an additional scaling factor and is usually set to zero or $(3-\tilde{L})$. The $\tilde{\beta}$ is used to incorporate prior knowledge of the distribution of x (for Gaussian distributions, $\tilde{\beta}$ =2 is optimal). The values of α, κ and $\tilde{\beta}$ selected herein were based on the experience and recommendations of reference [51].

The $(\sqrt{(\tilde{L}+\lambda)P_{\hat{w}_{\bar{k}}}})_i$ is the ith column of the matrix square root (e.g., lower triangular Cholesky factorization of matrix form $(\sigma\sigma^T)$ which yields vectors with consistent units in $\hat{w}_{\bar{k}}$). The sigma points ($\chi_{i,k}$) calculation is the checkpoint of whether the current parameter values are in the convergence range or not. This calculation is shown in Figures 6.8 and 6.9 via block diagram of the UT illustrating the generation of $\hat{d}_k$ and $P_{\bar{d}_k\bar{d}_k}$, and a flowchart of the UKF algorithm. The mean and covariance for output $\hat{d}_k$ are approximated using the weighting sample.

$$\hat{d}_k = \sum_{i=0}^{2\tilde{L}} W_i^{(m)} y_{i,k} \quad (6.85)$$

$$P_{\bar{d}_k\bar{d}_k} = \sum_{i=0}^{2L} W_i^{(c)}[y_{i,k}-\hat{d}_k][y_{i,k}-\hat{d}_k]^T + R_k^e$$
$$P_{w_k d_k} = \sum_{i=0}^{2L} W_i^{(c)}[\chi_{i,k}-\hat{w}_{\bar{k}}][y_{i,k}-\hat{d}_k]^T \quad (6.86)$$

The R_k^e is the measurement noise covariance matrix. In UKF procedure, there are several important points to check to use this method [51]. First, the convergence rate and tracking performance are directly influenced by the choice of the process noise covariance $E\left[u_k u_k^T\right] = R_k^r$. Units for R_k^r are the same as $\left[w_k w_k^T\right]$. Basically, as R_k^r gets larger, the parameters are converging quickly. To choose R_k^r value, there are several options. One option is the randomly choosing a diagonal matrix. Another option is the Robbins-Monro stochastic approximation, which is

$$R_k^{\,r} = (1-\alpha)R_{k-1}^{r} + \alpha K_k^{w}(d_k - G(x_k, \hat{w}))(d_k - G(x_k, \hat{w}))^T (K_k^{w})^T \tag{6.87}$$

where $R_k^{\,r}$ is also restricted to be a diagonal matrix, implying independence assumption on the parameters. Wan *et al.* [51] also found that a Robbins-Monro stochastic approximation method converges better, in general. Thus, in this chapter, equation (6.87) is employed to calculate and update $R_k^{\,r}$. Secondly, complete convergence can be achievable when the covariance P_{w_k} goes to zero. Determining P_{w_k} will produce a good basis for excellent convergence.

Implementation of UKF: PKM Dynamic Parameter Identification

This section presents a simulated implementation of the unscented Kalman filter (UKF) for parameter identification of PKM system parameters. The UF Special 6-6 PKM system was used to illustrate the performance of the identification algorithm in both simulated and experimental implementations. Implementation of the UKF parameter identification algorithm was done for two different forms of the PKM model base: force-based model (equation 6.56) and energy-based model (equation 6.78). The key difference between these two implementations is in the measured data, i.e., d_k measured (desired) output and x_k measured states. The force-based model's measured data includes the actuator forces, and the platform's linear and angular positions, velocities and accelerations. The energy-based model's measured data includes the actuator forces, change in strut lengths, and the platform's linear and angular position and velocities. For the experimental implementation, the states were calculated from the measured strut lengths, time, and numerical derivatives determined from filtered link length measured data. The velocity and acceleration data can be either measured directly or determined from differentiated measured position data. The differentiation process can further exacerbate the problem of noise and/or measurement error by the introduction of an additional source of numerical errors.

Since the data acquired from experiments always contains some form of noise and/or measurement errors, a comparative analysis of the UKF parameter identification algorithm's performance for the two different model-bases was conducted to determine if there is any benefit in using the energy-based model approach other than reduced modeling effort in deriving the theoretical model. The anticipated benefit is reduced noise effects due to lower order derivatives and/or the number of measurements required in the energy-based method. If the results show a significant enhancement in performance, then a greater potential exists for the use of the UKF identification method for online estimation. For an industrial setting, online estimation can be the critical factor that potentially enables the engineer to further improve a machine's performance by reducing the operational time and machining errors.

PKM Simulations (Force-Based)

For the force-based PKM model, the measured data are the actuator forces, and the platform's linear and angular positions, velocities and accelerations. To obtain the emulated measured actuator forces for a given state response, a simulation data-generating program of the model (equation 6.56) was used in which the emulated measured states were entered into the program as given. The emulated measured states are (refer to equations (6.4) through (6.8))

$$x_k = \left(q_k^{T}, \dot{q}_k^{T}, \ddot{q}_k^{T}\right)^T \tag{6.88}$$

The emulated measured (desired) output data is

$$d_k = \left(F_a\right)_{measured} \tag{6.89}$$

and the calculated output is

$$y_k = G\left(x_k, w_k\right) = right\text{-}hand\ side\left(equation\ 7.56\right) \tag{6.90}$$

Again, it is noted that for implementation purposes, the center of mass locations were assumed known, i.e., not included in the set of parameters to be identified. Herein, the CAD-based values that were determined by reference [52] are used. In the analysis and implementation, it was also assumed that all struts are of modular design and have the same inertia parameters, thus reducing the number of parameters in the identification set. Hence, the parameters identified (w_k) in the implementations are

$$w_k = \left[m_e, m_b, m_p, I_{xxb}, I_{yyb}, I_{zzb}, I_{xxp}, I_{yyp}, I_{zzp}, I_{xxe}, I_{yye}, I_{zze}\right]_k \tag{6.91}$$

As stated above, the data-generating program consists of the system model where the 'true' values for the inertia parameters (CAD-based values in the simulated implementation) were used for generating the emulated measured data. Herein, two different tests were conducted: freefalling test and actively controlled strut test. The freefalling test was used to generate emulated measured data. The actively controlled strut test emulates an experimental implementation in which the strut forces are actively controlled. The freefalling test might be considered more useful in cases where the strut force data is very noisy or contains significant errors. After acquiring the emulated measured data from the data-generating program for each of these tests, the data was fed into the UKF algorithm for estimating the mass parameters. To further build confidence in the UKF parameter estimation method for PKMs, friction is introduced into the system and the results are compared to the frictionless case. As a representative model of friction found in mechanical systems, the modified friction model [53] was selected and is defined as follows.

$$F_{frict} = \begin{cases} Fric(i,1)\cdot \dot{L}_i + Fric(i,2) & for \quad \dot{L}_i > 0.002 \\ \left(2\cdot Fric(i,2)/0.004\right)\cdot \dot{L}_i & for \quad -0.002 < \dot{L}_i < 0.002 \\ Fric(i,1)\cdot \dot{L}_i - Fric(i,2) & for \quad \dot{L}_i < -0.002 \end{cases} \tag{6.92}$$

where $Fric(i,1)$ = viscous friction coefficient for ith strut

$Fric(i,2)$ = Coulomb friction coefficient for ith strut

$\dot{L}_i$ = velocity of individual strut (Unit: m/s)

Table 6.1 provides the estimated coefficients for the friction model obtain using single strut tests of reference [31, 54]. Figure 6.10 shows the vertical strut friction test bed and a representative plot of the results obtained for strut 1 [31, 54].

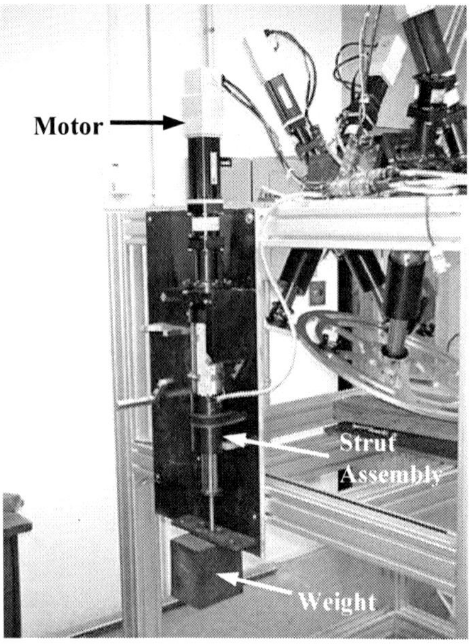

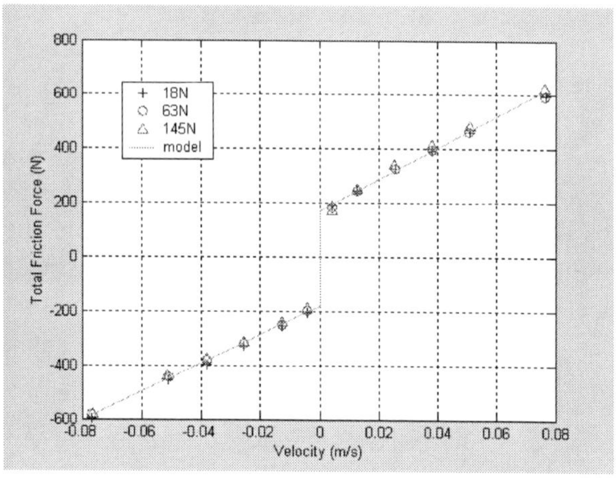

(a) The Vertical Strut Friction Test Stand(b) Strut 1 Friction Data

Figure 6.10. Single Strut Friction Test [54].

Table 6.1. Calculated Friction Value from Individual Strut Testing for 'Without Gearbox' Case

	Strut 1 *	Strut 2**	Strut 3**	Strut 4 *	Strut 5**	Strut 6**
Viscous (N*s/m)	1258 (3333)	1012 (2681)	1099 (2901)	3517 (8930)	1140 (3023)	1168 (3100)
Coulomb (N)	50 (190)	49 (186)	58 (220)	131 (330)	49 (186)	53 (202)

Note, * Estimated friction coefficients obtained from horizontal single strut experiments.

** Estimated friction coefficients obtained from scaled reference [54] estimates

( ) stands for the friction value obtained from Reference [54]

PKM Simulations (Energy-Based Model)

The model-base and the measured data are the differences between the actual implementation of force-based model and energy-based model in UKF parameter identification. All other assumptions hold for both energy-based and force-based implementations. In the implementation cases presented, it is noted that the energy-based model results are for only the actively controlled strut tests. The energy-based model's data-generating program is the same as the one used for force-based model with an additional emulated measured data set of $(\Delta L_i)_k$ for each strut i and time step k. Hence, the emulated measured states for equation (6.78) are

$$\mathrm{x}_k = \left(\mathrm{q_k}^{\mathrm{T}}, \dot{\mathrm{q}}_{\mathrm{k}}^{T}\right)^{T} \tag{6.93}$$

and the emulated measured actuator and change in strut lengths ((desired) measured output) force data yields the measured change in energy (work done by the actuators)

$$\mathrm{d}_k = \left(\sum_{i=1}^{n} \left(F_{a_i}\right)_k \left(\Delta L_i\right)_k\right)_{measured} \tag{6.94}$$

and the calculated output is

$$\mathrm{y}_k = G\left(\mathrm{x_k}, w_k\right) = h_{k+1}\left(\mathrm{x_{k+1}}, w_{k+1}\right) - h_k\left(\mathrm{x_k}, w_k\right) \tag{6.95}$$

where the parameters identified (w_k) are the same set as used for force-based.

$$w_k = \left[m_e, m_b, m_p, I_{xxb}, I_{yyb}, I_{zzb}, I_{xxp}, I_{yyp}, I_{zzp}, I_{xxe}, I_{yye}, I_{zze}\right]_k \tag{6.96}$$

Results of Simulated UKF

The same physical setup and conditions for both the force-based modeling and energy-based model implementations were used for each simulation case. For the freefalling test, the movable platform starts at the 'home' position (initial condition). The 'home' position is defined as the PKM configuration for which the mass center of the movable platform is located at [x = 0.00, y = 0.00, z = - 0.35] meters from the inertial frame {W} as defined in Figures 6.5 and 6.6. The physical dimensions of the PKM are as defined in Figure 6.4.

Table 6.2. Initial Conditions

Position of the mass center of the movable platform	[0 0 -.35] m
Velocity of the mass center of the movable platform	[0 0 0] m/sec
Acceleration of the mass center of the movable platform	[0 0 0] m/sec^2
Angular velocity of the mass center of the movable platform	[0 0 0] rad/sec
Angular acceleration of the mass center of the movable platform	[0 0 0] rad/sec^2

For the actively controlled strut test, the movable platform starts at the 'home' position (initial condition). The mass center of the movable platform is then commanded to move sinusoidally about the 'home' position in the x-direction of the inertial frame with amplitude of one inch (case 1). It should be noted that the maximum width of UF-PKM workspace is approximately 4 inches in the X_w and Y_w plane. For case 2, the movable platform moves sinusoidally with the same amplitude about the 'home' position in the z-direction of the inertial frame. For case 3, the center of mass of the movable platform moves sinusoidally about the 'home' position in the positive/negative x and positive/negative y direction. The frequency of the movement is 4*π Hz (approximately 45 % of the maximum speed capability of the UF PKM). The initial conditions of all motions are contained in Table 6.2. The trajectories (cases) used in the simulations and their parameter values are specified in Table 6.3.

Table 6.3. Trajectories in Simulation (Four cases)

Case	1	2	3	4
Direction	**x-direction**	**z-direction**	**xy-direction**	**Freefalling**
Direction and magnitude of the movement	[0.0254 0 0] m	[0 0 0.0254] m	[0.0254 0.0254 0] m	Freefalling
Movement frequency of the mass center of the movable platform	4*π Hz	4* π Hz	4* π Hz	NA

Again, note it has been assumed that all struts have the same inertia parameters and that the center of mass locations and kinematic parameters are known. The rule of notation used in the simulations are as follows:

- All headers and titles in the tables and figures relating to force-based model use the notation type: FTYPE1_TYPE2_TYPE3
    - For FTYPE1
        - FA denotes 'active controlled strut tests'
        - FF for 'freefalling tests'
        - Note that for FF type, there is no TYPE2.
    - For TYPE2
        - X for x-directional movement, Z for z-directional movement, and XY for xy-directional movement
    - For TYPE3
        - W indicates 'with friction' and WO represents 'without friction'

- All notations relating to energy-based model follow the same rule as force-based model with the exception of using E instead of F for denoting the TYPE1 case.

Simulation Results (Force-Based, Without Friction)

Tables 6.4 and 6.5 show the simulation results for the force-based model, without friction. For mass parameters, compared to the true value, the x-direction movement (FA_X_WO) and the xy direction movement (FA_XY_WO) yield the best estimates. The estimated mass values converged within 0.04% of its true value. The estimated mass parameters for z direction movement (FA_Z_WO) and freefalling case (FF_WO) however converged to within 16.5% and 11.4%, respectively. Table 6.5 shows the value of convergence for inertia parameters. In all cases, the movable platform parameter estimates (I_{xxp}, I_{yyp}, I_{zzp}) had better convergence than strut parameter estimates (I_{xxe}, I_{yye}, I_{zze}, I_{xxb}, I_{yyb}, I_{zzb}). The maximum percentage difference for platform parameters is 0.8%, however, the maximum percentage difference for strut parameters is 26.2%. The largest percentage difference can be seen on $I_{yye.}$ This could be observed in two ways. One, the numerical value assigned in the simulation was very small compared to other parameter values (i.e., the largest parameter in the simulation was 7.7214, which is roughly 142,000 times greater than I_{yye} value (inertia about longitudinal axis of ball spline)). And/or, the selected test trajectories (data) do not yield sufficient dynamic excitation for identification of the I_{yye} parameter. The dynamic excitation effects were explored further by randomizing or spreading the response data inputs (x_k, d_k).

Figures 6.11-6.12 are representative plots of the identification iterative results using the force-based model, with and without friction (see reference Oh [32] for additional details). The results for the without friction case show a parameter convergence tendency towards their true values for the actively controlled strut cases (FA_X, FA_XY). For m_e case, the results of both x directional movement and xy directional movement converged faster and with greater accuracy than the other two (z and freefalling) trajectories. This tendency can also be found in other mass cases (m_b, m_p). For freefalling test, it shows generally poorer convergence accuracy for all cases. This can be attributed to the limitation of the strut and the workspace of the UF Special 6-6 PKM on its range of motion as well as the PKM configuration approaching a geometric condition in which struts are essentially parallel. That is, since the range of strut lengths is limited, the 'freefalling' movement has only a limited amount of data since the motion occurred too fast to capture in detail. This data was not sufficient to run the unscented Kalman filter algorithm for extended period without either reusing data sets or truncating the iterations, hence the iterations were truncated.

Simulation Results (Force-Based, With Friction)

Tables 6.6 and 6.7 show the values of convergence for the force-based model, with friction results. For the mass properties, the estimated values are in error from the true values by less than 3% (x-direction), 2.6% (z-direction) and 16.2% (xy-direction). The convergence error between the estimated value and true value in xy-direction exhibited a large increase compared to the without friction case. For m_e case, z-direction case is the fastest and closest to the true value. The x-direction cases and xy-direction converge close enough to be considered good, but show slower convergence than the z-direction trajectory. Similar trends were found on m_b and m_p case.

Table 6.4. Mass Value (Units – kilogram) – Force-Based, Without Friction

Parameters	True Value	FA_X_WO	FA_X_WO % Diff	FA_Z_WO	FA_Z_WO % Diff	FA_XY_WO	FA_XY_WO % Diff	FF_WO	FF_WO % Diff
m_e	1.79	1.79	-0.00	1.80	-0.11	1.79	-0.00	1.62	8.65
m_b	7.72	7.72	0.04	6.46	16.47	7.72	0.04	6.89	10.49
m_p	4.82	4.82	0.01	4.91	-1.78	4.82	0.00	4.25	11.34

Table 6.5. Inertia Value (all times $1x10^{-6}$, Units – Nm^2) – Force-Based, Without Friction

Parameters	True Value ($x10^{-6}$)	FA_X_WO	FA_X_WO % Diff	FA_Z_WO	FA_Z_WO % Diff	FA_XY_WO	FA_XY_WO % Diff	FF_WO	FF_WO % Diff
Ixxb	563500.00	563709.70	-0.04	510318.60	9.44	63709.70	-0.04	504657.10	10.44
Iyyb	69400.00	60508.83	12.81	60516.44	12.80	60508.83	12.81	60517.37	12.80
Izzb	563500.00	503280.70	10.69	509689.80	9.55	503280.70	10.69	504657.30	10.44
Ixxp	90761.00	90041.92	0.79	90042.65	0.79	90041.92	0.79	90042.02	0.79
Iyyp	181283.60	180022.00	0.70	180023.30	0.70	180022.00	0.70	180022.00	0.70
Izzp	90761.21	90025.98	0.81	90026.65	0.81	90025.98	0.81	90026.00	0.81
Ixxe	18269.40	18040.12	1.26	17771.91	2.72	18040.12	1.26	18039.45	1.26
Iyye	54.20	40.00	26.19	40.00	26.19	40.00	26.19	40.00	26.19
Izze	18269.40	18045.12	1.23	18313.76	-0.24	18045.12	1.23	18045.98	1.22

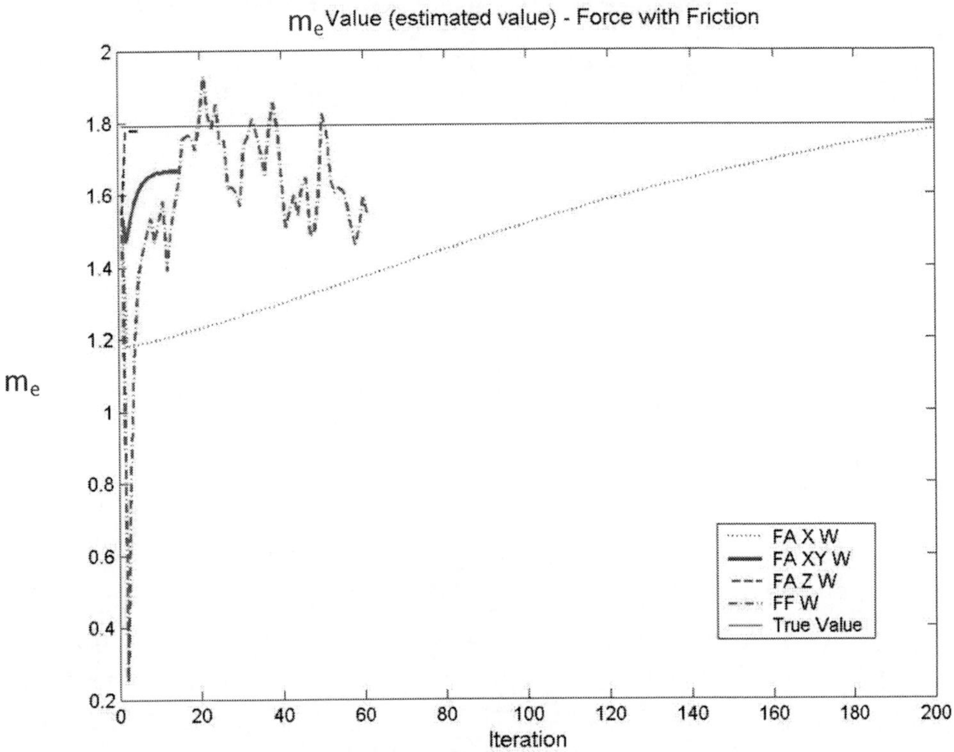

Figure 6.11. m_e Value (Force with Friction).

Simulation Results (Energy-Based, Without Friction)

Comparing only the cases without friction (Tables 6.4 and 6.8), the energy-based model the resulting percentage differences between the true and estimated values all remained under 7.3% for the mass properties. In general, the convergence performance rate and accuracy using energy-based model case was worse than that achieved using force-based model case with exception of the m_b estimate for the FA_Z_WO case. It is also observed in comparing energy-based (without friction) results with the force-based (with friction) results, the convergence accuracies exhibited to a large degree the same order of magnitude in percentage error. One notable difference is that the energy-based model achieved much better convergence accuracy, yielding a lower percentage difference between true and estimated parameter value of I_{yye}. The percentage differences between true and estimated for the other inertia parameters are of either similar or higher magnitudes then that obtained using the force-based model. For m_e case, the convergence tendencies in the x and xy directional movement are found to exhibit the similar trend of convergence. This trend can also be seen in m_b and m_p cases. The z-directional movement shows more of a linear tendency in convergence.

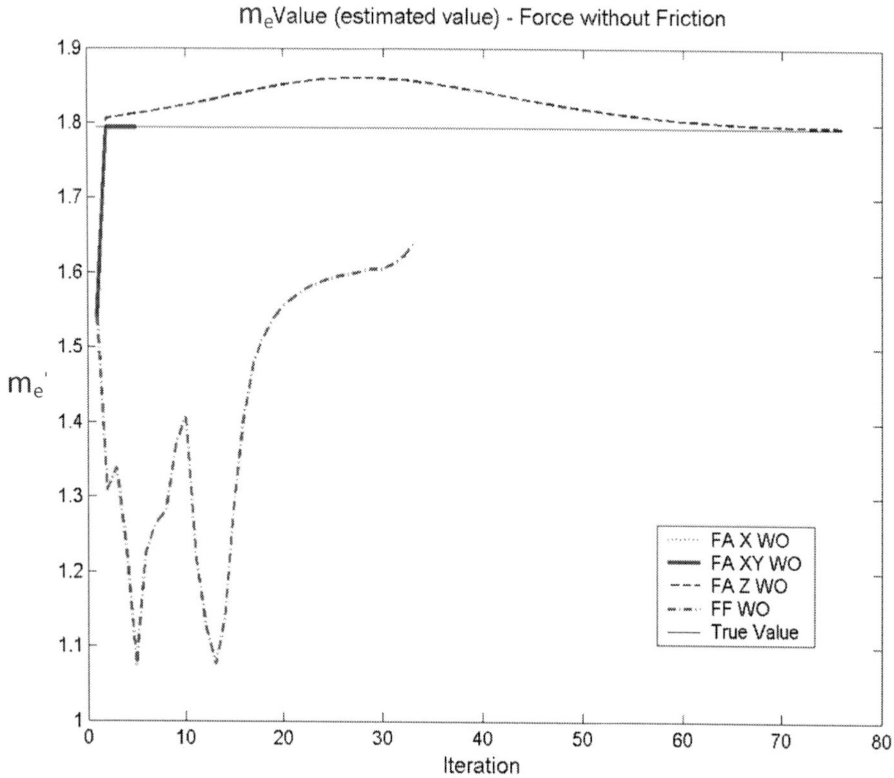

Figure 6.12. m_e Value (Force without Friction).

Simulation Results (Force-Based, Without Friction, Original/Random/Spread)

The kth measured data (x_k, d_k) consists of the system states and actuator forces measured at a given instant in time. In theory, given a perfect model-base and noise-free data, one could use a single set of measured data (x_k, d_k) (i.e., $(x_k, d_k) = (x_{k+1}, d_{k+1})$ for all k) and run the UKF algorithm until w_k converges to the true estimate. Also, since w_k is an estimate of non-time varying parameters of the system, then any kth set of measured data (x_k, d_k) should yield the same w_k estimate. However, depending on the values of the measured states, it is possible that some w_k parameters would not be identifiable because at the given instant in time their contribution to the dynamics response is zero or essentially negligible. Thus, in practice, at each kth iteration a new set of measured data (x_k, d_k) is used so as to ensure the effects of all identifiable parameters are captured in the estimation process. The use of multiple data sets (x_k, d_k) is also necessary when these sets are obtained from experiments, that in general contain noise and contributions from the unmodeled dynamics.

Table 6.6. Mass Value (Units – kilogram)– Force-Based, With Friction

Parameters	True Value	FA_X_W	FA_X_W % Diff	FA_Z_W	FA_Z_W % Diff	FA_XY_W	FA_XY_W % Diff	FF_WO	FF_WO % Diff
m_e	1.79	1.78	0.83	1.78	0.79	1.67	7.07	1.55	13.86
m_b	7.72	7.50	2.92	7.86	-1.85	7.08	8.25	7.18	7.01
m_p	4.82	4.92	-1.97	4.94	-2.54	5.60	-16.11	4.69	2.80

Table 6.7. Inertia Value (all times $1x10^{-6}$, Units – Nm^2) – Force-Based, With Friction

Parameters	True Value ($x10^{-6}$)	FA_X_W	FA_X_W % Diff	FA_Z_W	FA_Z_W % Diff	FA_XY_W	FA_XY_W % Diff	FF_WO	FF_WO % Diff
Ixxb	563500.00	599181.80	-6.33	517260.50	8.21	543644.50	3.52	510242.50	9.45
Iyyb	69400.00	60511.60	12.81	60515.85	12.80	60505.87	12.82	60516.53	12.80
Izzb	563500.00	503588.90	10.63	504016.90	10.56	543107.00	3.62	509748.30	9.54
Ixxp	90761.00	90041.95	0.79	90042.00	0.79	90036.97	0.80	90042.80	0.80
Iyyp	181283.60	180022.00	0.70	180022.00	0.70	180020.50	0.70	180023.60	0.70
Izzp	90761.21	90025.99	0.81	90026.00	0.81	90023.17	0.81	90026.80	0.81
Ixxe	18269.40	21865.73	-19.69	18040.25	1.25	19875.89	-8.79	17795.81	2.59
Iyye	54.20	40.00	26.19	40.00	26.19	40.40	25.44	40.00	26.19
Izze	18269.40	18045.15	1.23	18045.15	1.23	20824.28	-13.98	18289.91	-0.11

Table 6.8. Mass Value (Units – kilogram)– Energy-Based, Without Friction

Parameters	True Value	EA_X_WO	EA_X_WO % Diff	EA_Z_WO	EA_Z_WO % Diff	EA_XY_WO	EA_XY_W % Diff
m_e	1.79	1.84	-2.56	1.74	2.86	1.85	-3.24
m_b	7.72	7.92	-2.55	7.57	1.95	7.97	-3.25
m_p	4.82	4.55	5.71	4.50	6.48	4.47	7.29

There are infinity possible data sets (x_k, d_k) from which one can select. However there is no assurance, that the selected data sets will yield a $w_k \rightarrow$ 'true' values, the question arises as to which collection of data sets (x_k, d_k) should one choose. That is, what motion trajectory should the PKM be driven to follow so as to yield a sufficiently dynamically exciting response from which to extract (x_k, d_k). In search of a guarantee, researchers have developed various methodologies for generating sufficiently exciting trajectories [19, 20]. As in reference [51], van der Merwe *et al.* used a free-fall approach. Per the study conducted by Oh [32], it was determined that sinusoidal trajectories about the 'home' position provide sufficiently dynamically exciting data. However, to further substantiate the capabilities and limitations in using UKF and to explore the dynamic excitation effects, the force-based model UKF method was implemented where the response data time incremental sets were either fed into the algorithm in a random time order or simply in an increased spread in the time increment of the selected data. The prior sections' results used response time ordered (x_k, d_k) data sets as depicted by the original case. To restate, since the time increments of the previous simulations from one time incremental step to the next were very small (0.001 second), the dynamic variation of the response data was limited as well throughout the estimation process. Thus, randomizing or spreading the (x_k, d_k) data sequence was used as a means for validating that the prior results had sufficient dynamic excitation and to gauge how robust the UKF method. The term 'random' means that the sequence of the (x_k, d_k) data sets given to the estimation algorithm is randomly chosen and the term 'spread' represents that the sequence of (x_k, d_k) data sets given to the estimation algorithm is chosen. e.g., from every 10 incremental steps (i.e. instead of using data number 1, 2, 3, … , using 1, 11, 21, 31, …).

For the force-based model (without friction) using randomized data, the resulting percentage difference in x and xy direction shows good convergence. However, similar to the case using sequential data, the z-directional movement and freefalling tests do not yield high convergence. The results for 'spread' case show better performance in the percentage difference compared to the 'random' case. Thus, the x and xy directional movements show superiority in convergence performance where the z direction exhibited the worst performance.

Comparing the 'random' with 'spread' cases, convergence results are at times indistinguishable between the two cases. That is, m_e, m_b and m_p convergence results in x direction and xy direction cases are similar. Data from the original (sequential) set and the spread case are comparable to each other, and the random case exhibits a slower convergence tendency. The z-direction case and freefalling cases do not have good convergence when using the data in the random nor the spread case, again experiencing the same limitations as cited earlier.

Comparison between Force- and Energy-Based with Noisy Data

As an additional comparison between the force-based model and the energy-based model cases without friction, noise was added into the UKF estimation. A 1% random noise was added to the UKF measured data (x_k) of the states. The force-based model case was also ran with 5% noise. As a representative case, the results of the xy-directional trajectory (without friction) simulation show that as the noise increases the difference between the estimates and the CAD-based 'true' values increases as expected. The mass parameters are estimated within 5.4% for force-based model and 37.8% for energy-based model when the 1% noise is added. This result shows that the force-based model outperforms the energy-based approach and hence is the preferred model-base to be used for the UKF. Additional convergence results for the other mass parameters are provided in reference Oh [32].

Summary of PKM Dynamic Parameter Identification Results Using the UKF Method

Summarizing, the above provided comparisons on the selection of the model-base for UKF parameter identification in terms of estimation convergence accuracy and rate trends, for cases without friction, with different motion trajectories, and with and without noise. It was found that the x-directional case exhibited the lowest percentage difference between estimated and true parameter values. The xy-direction cases had similar trends for the x-direction cases; however, when friction was added to the system, the results were not as reliable as those found for the x-direction case. The convergence tendency for the z-directional movement and the 'freefalling' test were both poor, in general. Mostly, 'freefalling' tests suffered from the limited range of motion in which to capture a sufficient amount of data as well as the directional symmetry of the strut axes (also seen for the z-directional movement).

For the z-direction cases, energy-based model is preferred to the force-based model due to a lower percentage difference between the estimated and true values of the relatively small parameters. However, for x-direction and xy-direction case, force-based model is superior to the energy-based model in terms of accuracy of convergence. Hence, a significant improvement using energy-based method was not generally realized. For data set types (original/random/spread), the results indicate that original data or spread data usage is the preferred for obtaining better estimation results, as well as it requires less data processing effort.

Experimental Validation of UKF for PKMs

This section presents the experimental setups and the estimated inertial parameter results for the UF-PKM. Since the energy-base method did not perform as well as expected, only the force-based UKF approach is used in this chapter for the experimental validation study. An ultimate goal for this section is to determine how well the estimation program performs in a real PKM application with machine and instrumentation limitations. Since the strut lengths and individual strut forces are the most readily available data that can be acquired from the machine controller and sensors, the accuracy of this data is a critical for extracting other system states such as strut velocities and accelerations. Shamblin [54] noted that the encoder reading for acquiring strut length data and current reading for acquiring the strut force contained significant noise resulting in measurement inaccuracies. This noise effect was further amplified as numerical approximations were used in determining the strut velocities and accelerations. Therefore, to implement the UKF identification procedure, sources of error and limitations are expected such as those listed below.

- Signal Noise
- Kinematic Errors
- Measurement Errors
- Controller Parameter and Modeling Errors
- Directional Control Problem
- Following Error
- Poor Encoder/ Current Readings
- Limited Sensing and Actuation

Hence, in the experimental implementation of UKF, corresponding uncertainties and reconciliation techniques had to be taken into consideration throughout the analysis and identification of the system parameters. As discussed in reference [55], the effects of limited sensing were considered in regards to assuming the geometric center of the platform could be assumed the same as the center of mass. Theoretical results indicate that errors in this approximation would result in errors in the resulting parameter estimation. However, due to the simplicity and the full knowledge of the PKM's components properties, the implementation of the UKF parameter identification algorithm was done assuming all centers of masses per reference [52]. Assuming ideal kinematics, the limited sensing issues (e.g., non-direct measurement of platform motion) were circumvented using kinematics and numerical derivatives to determine the platform position, velocity, and acceleration.

Experimental Setup

The UF PKM (refer to Figures 6.4) is actuated by Aerotech BM200 3-phase brushless motors with encoder feedback. The motors are powered by Aerotech BA20 amplifiers using pulse-width modulation to provide current control. The motion control software is the open-architecture, PC-based Enhanced Machine Controller, developed by the National Institute of Standard and Technology (NIST), and runs under Real-Time Linux on a Pentium II PC. The interface between the amplifiers, motors, and software is a Servo-to-Go motion controller card. At the present level of controller development, only translational motions of the platform are achievable.

In the experiments, the platform was moved in coordinated motion by executing a single G-code command or a series of commands from a script file. The built-in logging feature was used to record strut lengths and motor current values. From the data, actual platform position for each time step was obtained using an iterative forward kinematics solver [47]. The platform velocity and acceleration were calculated using the PKM kinematics and the strut length, velocities and acceleration. The strut velocities and accelerations were determined numerically applying the 4-point central difference smoothing equations ([54], equations (6.97) and (6.98)) to filtered position data.

$$\left(\dot{L}_k\right)_i = \left(\frac{\left(-L_{k+2} + 8L_{k+1} - 8L_{k-1} + L_{k-1}\right)_i}{12\Delta t}\right) \tag{6.97}$$

$$\left(\ddot{L}_k\right)_i = \left(\frac{\left(-L_{k+2} + 16L_{k+1} - 30L_k + 16L_{k-1} - L_{k-2}\right)_i}{12\left(\Delta t\right)^2}\right) \tag{6.98}$$

where $\left(L_k\right)_i$ is the absolute strut length for the ith strut at kth iteration step (equation (6.11)) and Δt is the incremental time step.

The motor current readings were converted to the measured $\mathbf{F}_a$ using the motor torque constant, and transmission ratio of the ballscrew. The friction forces along the strut axes were estimated using strut velocity data with a Coulomb and viscous friction model. In prior research on the UF-PKM [54], the model's friction coefficients were obtained experimentally through system identification of each individual strut.

Challenges to Experimental Validation and Reconciliation

Numerous sources of error presented difficulties in experimentally validating the identification of the dynamic parameters. Much of the error was indirectly caused by the limitations of the servo controller. The PKM's controller uses a PID algorithm to generate the control effort from the encoder measured errors. In tuning the system's controller, the derivative term was found necessary for minimizing oscillations that would occur if only P (proportional) or PI (proportional – integral) control was used. The derivative term also amplified the high frequency vibrations that arose from encoder noise. The result is that the actual machine motion is characterized by small amplitude, high frequency positional oscillations in addition to the smooth commanded motion.

In addition to the undesired oscillations are the 'false' movements caused by finite encoder resolution. As the motor rotates, the position is read as a series of step moves by the encoder. When the true position is between encoder pulses, the position as read by the encoder is seen to oscillate between the two pulses, introducing oscillations that are not actually occurring. Naturally, differentiating position data to obtain velocity and especially, acceleration, amplifies both the real and false oscillations. This noise propagation is such that acceleration caused solely by encoder noise is of similar magnitude to the acceleration of the commanded motion. For this study, an 8th order Butterworth filter was utilized to eliminate positional noise of frequencies greater than 10Hz, as the commanded machine movements were of much lower frequency. This alleviated much of the problem, but noise at less than 10Hz still existed and became indistinguishable from actual motion.

Errors due to noise are also manifested in empirically determined constants such as friction coefficients. During a friction test, although the commanded velocity is constant, the actual velocity and the corresponding control effort are oscillatory. This contributes to random error when measuring the electrical current required to produce a constant strut velocity. Measurement of the motor torque constant (used for converting current measurements into strut forces) was similarly affected.

Another source of error is unmodeled transmission dynamics. For analysis purposes, the transmission of motor rotation to linear strut movement is assumed constant and kinematically precise. The actual machine uses a flexible rubber coupling between the motor shaft and ballscrew, acting as a spring-damper system. Thus, oscillations occurring in the motor would not be completely transmitted to the platform motion, as is assumed in the simulation. Additionally, the ballscrew itself acts as a rotational spring, though its effect on error is much less than the flexible coupling.

Compounding the previous errors in the friction and torque constants is the relatively small magnitude of measured dynamic forces in relation to friction forces. Thus, to reduce friction within the system, 3-to-1 precision gearboxes were removed from the original UF-PKM setup. Removing the gearbox also had the effect of reducing motor inertia effects ($F_{m/b}$). Furthermore, because of the dominance of the friction with the UF-PKM, these and other unmodeled dynamics effects were assumed negligible in the implementation of the UKF identification method. Due to time limitations, only two sets of single strut friction coefficients were updated with new experimental results (strut 1 and 4). The single strut friction tests were performed using a horizontal single strut configuration. This eliminated the effects from gravity. The resultant estimated coefficients for the friction model using single strut tests are shown on Table 6.1. Using these results compared to reference [54], reference [54]'s estimates were scaled to provide new estimates for the remaining struts, thus accounting for the change in the gear ratios with the new experimental setup.

Friction Extraction and Estimation Stages

Since the friction coefficients (both viscous and Coulomb) also need to be estimated using the UKF algorithm, the total number of parameters increased to 24 (3 masses, 9 principal inertias and 12 friction coefficients). The

$$\begin{aligned} w_k = [w_k|_{simulated}, &Fric(1,1), Fric(1,2), Fric(2,1), Fric(2,2), Fric(3,1), Fric(3,2), \\ &Fric(4,1), Fric(4,2), Fric(5,1), Fric(5,2), Fric(6,1), Fric(6,2)] \end{aligned} \tag{6.99}$$

where it is assumed that all struts have identical mass and inertia properties. The principal axes of inertia for each strut are defined along the $\{\hat{s}_2, \hat{s}_3, \hat{s}_{23}\}_i$ axes, and the center of mass locations are those estimated in reference [52]. The parameter identification was first executed for the entire set of 24 parameters. Using the following UKF initial estimates to start the estimation process (Reference [54] single strut friction estimates plus within 90% of CAD-based mass and inertia), the parameter estimation did not converge. This led to implementing a three-stage identification approach.

- **Stage 1.** Conducting an estimation of friction followed by inertia parameter estimation. Stage 1 has two steps:
    - *Stage 1-A.* The initial guess for the friction estimation is made (e.g., the prior strut estimates of Table 6.1 [54] and/or the author's single strut results). At this stage, the CAD-based inertia parameters were used as the 'true' inertia values, thus reducing the estimation vector to consist of only the viscous and Coulomb friction parameters. All three trajectories (x, z, and xy-direction) were tested for estimating the friction parameters.

$$w_k\big|_{Stage\,1-A} = [Fric(1,1), Fric(1,2), Fric(2,1), Fric(2,2), Fric(3,1), Fric(3,2), Fric(4,1), Fric(4,2), Fric(5,1), Fric(5,2), Fric(6,1), Fric(6,2)]^T \quad (6.100)$$

- *Stage 1-B.* After obtaining the estimated friction values for each trajectory, the inertia parameters are estimated with the estimated friction parameters from Stage 1-A used as the 'true' friction values. At this step, the estimation vector now consists of only the inertia parameters with initial guess within 90% of the CAD-based mass and inertia values. All three trajectories are conducted as well to estimate the inertia parameters.

$$w_k\big|_{Stage\,1-B} = [m_e, m_b, m_p, I_{xxb}, I_{yyb}, I_{zzb}, I_{xxp}, I_{yyp}, I_{zzp}, I_{xxe}, I_{yye}, I_{zze}]_k^T \quad (6.101)$$

Note, that in Stage 1-B the estimated friction parameters from FA_X (Stage 1-A) were used as the 'true' friction values in estimating the inertia parameters in FA_X. Also, the estimated friction parameters from FA_Z (Stage 1-A) were used as the 'true' friction values in estimating the inertia parameters in FA_Z.

- **Stage 2.** Stage 2 is essentially a re-execution of Stage 1 using a 'new' initial settings; conducting an estimation of friction followed by inertia parameter estimation using the estimated friction parameters of Stage 1-A as the initial guess for Stage 2-A. That is, for x-directional movement; the previously estimated friction parameters of Stage 1-A from the x-directional movement case are used as the expected true values for the friction estimation, and similarly done for z and xy-directional movements. Stage 2 has two steps:
  - *Stage 2-A.* The initial guess for the friction estimation is the estimated friction parameter values from Stage 1-A. At this stage, the CAD-based inertia parameters were used for the 'true' inertia values, thus again reducing the estimation vector to consist of only the viscous and Coulomb friction parameters. All three trajectories (x, z, and xy-direction) were tested for estimating the friction parameters.

$$w_k\big|_{Stage\,2-A} = w_k\big|_{Stage\,1-A} \quad (6.102)$$

  - *Stage 2-B.* After obtaining the estimated friction values for each trajectory, the inertia parameters then are estimated with applying the estimated friction parameters from Stage 2-A used as the 'true' friction values. That is, the estimation vector now consists of only the inertia parameters with initial guess within 90% of the CAD-based mass and inertia values). All three trajectories are conducted as well to estimate the inertia parameters.

$$w_k\big|_{Stage\,2-B} = w_k\big|_{Stage\,1-B} \quad (6.103)$$

Note, that the estimated friction parameters from FA_X (Stage 2-A) were used as the friction values to estimate the inertia parameters in FA_X. Also the estimated friction parameters from FA_Z (Stage 2-A) were used as the friction values to estimate the inertia parameters in FA_Z.

- **Stage 3.** Both the friction and inertia parameters were estimated simultaneously. i.e.,

$$w_k\Big|_{Stage\ 3} = \left[\left(w_k\Big|_{Stage\ 1-B}\right)^T, \left(w_k\Big|_{Stage\ 1-A}\right)^T\right]^T \tag{6.104}$$

  The data used in this stage as the initial estimate is the combined values from Stage 2-A and Stage 2-B. For example, the estimated friction values from Stage 2-A and the estimated inertia values from Stage 2-B in FA_X were combined into one set and used as the initial parameter values to estimate both the friction and the inertia parameters in FA_X, FA_Z, and FA_XY runs for Stage 3.

Experimental PKM Dynamic System Identification Results Using UKF Method

In implementing the three-stage identification approach, several observations are made for estimating the parameters. For friction case, the direct measurement from the individual struts using the single strut tests does not capture the spherical and Hooke joint friction values. This can be delineated from the estimation results for the different input trajectories. E.g., the X- and XY- motions move the struts over different joint ranges while the Z-motion yields joint motion closer to the X-motion case. Selecting as initial values the estimates of prior estimations determined along different motion trajectories quickly narrowed the possible results towards finding the true values of the system. The inertia parameters exhibited a much faster convergence to a 'true' value. To conclude, the single strut estimates have not fully captured the strut friction. Thus, the new friction estimates are considered more accurate. The inertia estimates showed very good, accurate convergence especially when considering the fact that friction forces dominated the data. However, the differences between friction estimates per the different trajectories still needs further resolve as to the determination of the true rotational friction differences of each spherical and Hooke joint's value. Hence, the accurate modeling for spherical and Hooke joints are highly recommended for better explanation of these possible phenomena. For greater details, refer to Oh [32].

To conclude, the recommended identification methodology for actual implementation on PKM's is the multistage iterative approach using the force-based UKF model algorithm. This is particularly true for systems with a large amount of friction. The following highlights some of the critical issues.

- First, one should not ignore the importance of the initial guess for the friction parameters (the usual dominance found in most ballscrew driven machines). Thus, the tests with different trajectories (i.e. x, z, and xy-directional movement) must be conducted for finding the possible friction parameter values. (Stages 1 and 2)
- Since the estimation program could fall into the local minimum for the highly nonlinear systems, the estimated friction parameter values obtained from the previous step must be tested again with other trajectories. (e.g., the estimated

friction parameters from x-directional movement test must be performed with xy- and z-directional movement test).

- The final friction parameters, that are found to show estimation convergence in all other trajectories, should be taken into consideration as the closer value to the true value.

References

[1] Stewart, D., A Platform with Six Degrees of Freedom, *Proceedings of the Institute for Mechanical Engineering*, London, Vol. 180, pp. 371-386, 1965.

[2] Fassi, I., and Wiens, G., Multiaxis Machining: PKMs and Traditional Machining Centers, *Journal of Manufacturing Process*, Vol. 1, No. 2, pp 79-92, 2000.

[3] An, C. H., Atkeson, C. G., and Hollerbach, J.M., Estimation of Inertial Parameters of Rigid Body Links of Manipulator, *Proceedings of 24th Conference on Decision and Control*, pp. 990-995, 1985.

[4] Olsen, H. B. and Bekey, G.A., Identification of Parameters in Modeling of Robots with Rotary joints, *IEEE* 1985.

[5] Pradeep, K. K. and Kanade, T., Parameter Identification of Robot Dynamics, *Proceedings of 24th Conference on Decision and Control*, pp. 1754-1760. Dec. 1985.

[6] Olsen, H. and Bekey, G., Identification of Robot Dynamics, *Proceedings of 1986 IEEE International Conference on Robotics and Automation*, Vol. 2, pp 1004-1010, San Francisco, CA, April 1986.

[7] Gautier, M., Numerical Calculation of the Base Inertial Parameters of Robots, *Proceedings of IEEE Conference on Robotics and Automation*, Cincinnati, pp. 1020-1025, 1990.

[8] Canudas de Wit, C. and Aubin, A., Robot Parameter Identification via Sequential Hybrid Estimation Algorithm, *Proceedings of the 1991 IEEE International Conference on Robotics and Automation*, pp. 952-957, Apr. 1991.

[9] Ghodoussi, M. and Nakamura, Y., Principal Base Parameters of Open and Closed Kinematic Chains, *Proceedings of the IEEE International Conference on Robotics and Automation*, pp. 84-89, Apr. 1991.

[10] Bennis, F., Khalil, W., and M. Gautier, Calculation of the Base Inertial Parameters of Closed-Loops Robots, *Proceedings of the 1992 IEEE International Conference on Robotics and Automation*, May 1992.

[11] Lin, S., Minimal linear Combinations of the Inertia Parameters of a Manipulator, *IEEE Transactions on Robotics and Automation*, Vol. 11 No. 3, pp. 360-373, Jun. 1993.

[12] Ji, Z., Study of the Effect of Leg Inertia in Stewart Platforms, *Proceedings of the 1993 IEEE International Conference on Robotics and Automation*, Atlanta, GA, Volume 1, IEEE Computer Society Press, May 1993.

[13] Armstrong-Helouvry, B., Dupont, P., and De Wit, C., A Survey of Models, Analysis Tools and Compensation Models for the Control of Machines with Friction, *Survey Paper*, Elsevier Science Ltd, pp. 1083-1102, 1994.

[14] Lin, S., An Approach to the Identifiable Parameters of a Manipulator, *Journal of Robotics Systems,* Vol. 11, No. 7, pp. 641-656, 1994.

[15] Fregolent, A. and Sestieri, A., Identification of Rigid Body Inertia Properties from Experimental Data, *Mechanical Systems and Signal Processing*, Vol. 10, No. 6, pp. 697-709, Jan. 1996.

[16] Tinone, H. and Aoshima, N., Parameter Identification of Robot Arm with Repetitive Control, *Journal of Control*, Vol. 63, No. 2, pp. 225-238, 1996.

[17] Swevers, J., Ganseman, C., Bilgin, D. Tukel, De Schtter, J., and Van Brussel, H., Optimal Robot Excitation and Identification, *IEEE Transactions on Robotics and Automation*, Vol. 13, No. 5, pp. 730-740, Oct. 1997.

[18] Yang, H. and Masory, O., Vibration Analysis of Stewart-Platform, *Proceedings of the 9th Conference on Recent Advances in Robotics*, Florida Atlantic University, April, 1996.

[19] Armstrong, B., On Finding Exciting Trajectories for Identification Experiments Involving Systems with Nonlinear Dynamics, *The International Journal of Robotics Research*, December 1989, Vol. 8, No. 6, pp. 28-48.

[20] Indri, M., Calafiore, G., Legnani, G., Jatta, F. and Visioli, A., Optimized Dynamic Calibration of Scara Robot, *Proceedings of 15th Triennial World Congress (IFAC)*, Barcelona, Spain, 6 pp., 2002.

[21] Kang, C. and Han, J., Force Analysis of A Stewart Platform-Based Force/Torque Sensor Using A Linearization Technique, *Proceedings of the ASME Dynamics Systems and Control Division*, Vol. 58, pp. 747-752, 1996.

[22] Wang, S. and Ehmann, K. F., Error Model and Accuracy Analysis of A Six-DOF Stewart Platform, *Manufacturing Science and Engineering*, Vol. 2-1, pp. 519-530, 1995.

[23] Sika, Z., Kocandrle, P., Stejskal, V., An Investigation of Properties of the Forward Displacement Analysis of the General Optimization Methods, *Mech., Mach. Theory*, Vol. 33, No. 3, pp. 245-253, 1998.

[24] Takeda, Y. Funabashi, H., Shen, G. and Ichikawa, K. and Hirose, K., Stiffness Analysis of a Spatial Six-Degree-of-Freedom In-Parallel Actuated Mechanism with Rolling Spherical Bearings, *Year 2000 Parallel Kinematic Machines International Conference*, Sep. 2000.

[25] Seong, S., and David, B., Parameter Estimation to Improve Accuracy of Mechanical System Simulations Using Rigid Body Codes, *Computer & Structure*, Dec. 1997.

[26] Siddhartha, S. S., Beale, D.G., and Wang, D., *A General Method for Estimation Dynamic Parameters of Spatial Mechanisms*, 1998.

[27] Christopher, G. A., Chae, An H., and Hollerbach J.M., Estimation of Inertial Parameters of Manipulator Loads and Links, *The International Journal of Robotics Research*, Vol. 5, No. 3, Fall 1986.

[28] Bhattacharya, S., Hatwal, H., and Ghosh, A., An On-Line Parameter Estimation Scheme for Generalized Stewart Platform Type Parallel Manipulators, *Mechanical Machine Theory*, Vol. 32, No. 1, pp. 79-89, 1995.

[29] Register, R.B., Mason, P.A.C., and Wiens, G.J., "Identification of PKM Strut Friction Using An Extended Kalman Filter", *2000 Florida Conference On Recent Advances In Robotics (web-proceedings)*, Boca Raton, FL, May 4-5, 2000, 8 pp.

[30] Wiens, G.J., Hardage, D.S., Register, R.B., Shamblin, S.A., Wells, R.A., Plymale, D.L., and Bieg, L.F., "Dynamic Modeling, Characterization and System Identification of Hexapod Machine Tools", *Proceedings of the 2001 NSF Design, Manufacturing & Industrial Innovation Research Conference*, January 7-10, 2001, Tampa, FL, 5 pp. (CD-ROM)

[31] Wiens, G.J., Shamblin, S.A., and Oh, Y.-H., "Characterization of PKM Dynamics in Terms of System Identification", Journal of Multi-body Dynamics, Special Issue on *Parallel Kinematic Machines*, Vol. 216, Part K, 2002, pp. 59-72.

[32] Oh, Young-Hoon, "Dynamic Parameter Identification of Parallel Kinematic Mechanisms using Unscented Kalman Filter", *PhD dissertation*, University of Florida, Fall 2003.

[33] Wiens, G., and Walker, C., "Hexapod Machine Tool Error Analysis with Inclusion of System Dynamics", *1998 IMECE Proceedings of the ASME: Manufacturing Science and Engineering*, Anaheim, CA, November 15-20, 1998, MED-Vol. 8, pp. 705-713.

[34] Wiens, G., and Hardage, D., "Dynamics and Controls of Hexapod Machine Tools", *Proceedings of the First European-American Forum on Parallel Kinematic Machines: Theoretical Aspects and Industrial Requirements*, Milan, Italy, August 31-September 1, 1998, 7 pgs. (Invited Position Paper). Also published in Parallel Kinematic Machines: Theoretical Aspects and Industrial Requirements, Springer-Verlag, 1999, pp. 217-225, (Editors: C.R. Boër, L. Molinari-Tosatti, K.S. Smith).

[35] Wiens, G.J., and Johnston, P.R., "Dynamic Performance of Parallel Kinematic Machines", *Transactions of the North American Manufacturing Research Institution of SME*, Vol. XXVII, 1999, pp. 311-316.

[36] Hardage, D.S., and Wiens, G.J., "Modal Analysis and Modeling of a Parallel Kinematic Machine", *1999 IMECE Proceedings of the ASME: Manufacturing Science and Engineering*, Nashville, TN, November 14-19, 1999, MED-Vol. 10, pp. 857-862.

[37] Shamblin, S.A., and Wiens, G.J., "Characterization Of Dynamics In Special 6-6 PKMs", *Proceedings of ASME Biennial Mechanisms and Robotics Conference*, DETC2000 /MECH6710, September 10-13, 2000, Baltimore, Maryland, 8 pp. (CD-ROM)

[38] Shamblin, S.A., and Wiens, G.J., "Characterization of Dynamics in PKMs", *Proceedings of Year 2000 Parallel Kinematic Machines International Conference* (2000-PKM-IC), September 14-15, 2000, Ann Arbor, Michigan, pp. 24-33. (Plenary Presentation)

[39] Wiens, G.J., "Parameter Sensitivity In System Identification Of Parallel Mechanisms", *Proceedings of ASME 2002 Design Engineering Technical Conferences and Computers and Information in Engineering Conference*, Montreal, Canada, September 29-October 2, 2002, 10 pp.

[40] Wiens, G.J. and Hardage, D.S., "Structural Dynamics and System Identification of Parallel Kinematic Machines", *Proceedings of International Design Engineering Technical Conferences,* DETC2006-99671, September 10-13, 2006, Philadelphia, PA, 10 pp.

[41] Wan, E. A. and van der Merwe, R., The Unscented Kalman Filter for Nonlinear Estimation, Proceedings of Symposium 2000 on Adaptive Systems for Signal Processing, Communication and Control *(AS-SPCC), IEEE,* Lake Louise, Alberta, Canada, Oct, 2000.

[42] Griffis, M., and Duffy, J., A forward Displacement Analysis of a Class of Stewart Platforms, *Journal of Robotic Systems*, Vol. 6, No. 6 pp 703-720, Apr. 1989.

[43] Baiges-Valentin, I., J., Dynamic Modeling of Parallel Manipulators, *Ph.D. Dissertation*, University of Florida, Gainesville, FL., 1995.

[44] Johnston, P.R. Dynamic Analysis and Performance of Parallel Mechanisms, *Master's Thesis*, University of Florida, Gainesville, FL, 1997.

[45] Walker, C., Error Analysis of Hexapod Machine Tools with Inclusion of System Dynamics, *Master's Thesis*, University of Florida, Gainesville, FL, 1997.

[46] Tsai, L.W., Robot Analysis: *The Mechanics of Serial and Parallel Manipulators*, John-Wiley & Sons, Inc., 1999.

[47] Register, R.B. Characterization and Identification of Joint Friction in Generalized Stewart Platform Mechanisms, Master's Thesis, University of Florida, Gainesville, FL, 2000.

[48] Julier, S.J., and Uhlmann, J.K., and Durrant-Whyte, H., A New Approach for Filtering Nonlinear Systems, *Proceedings of the American Control Conference*, pp 1628-1632, 1995.

[49] Julier, S.J., and Uhlmann, J.K., A General Method for Approximating Nonlinear Transformations of Probability Distributions, *Technical Report*, RRG, Department of Engineering Science, University of Oxford, November 1996.

[50] Julier, S.J., and Uhlmann, J.K., A New Extension of the Kalman Filter to Nonlinear Systems, Proceedings of AeroSense: *The 11th International Symposium on Aerosoace/Defence Sensing, Simulation and Control*, 1997.

[51] Wan, E. A. and van der Merwe, R., *Kalman Filtering and Neural Networks*, Chapter 7 – The Unscented Kalman Filter, John Wiley & Sons, Inc., 2001.

[52] Johns, S., 'SAMM Special 6-6 PKM Inertia Parameters', University of Florida Space, *Automation, and Manufacturing Mechanisms Laboratory Internal Report*, Gainesville, Florida, Technical Report No. Space, Automation and Manufacturing Mechanisms-University of Florida-5, 1999.

[53] Armstrong, H., *Control of Machines with Friction*, Kluwer Academic Publishers, Boston, MA, 1991.

[54] Shamblin, S., System Identification and Dynamics Characterization in Parallel Kinematic Machines, *Master's Thesis*, University of Florida, Gainesville, FL, 2002.

[55] Oh, Y.H. and Wiens, G.J.., "Parameter Sensitivity in Sensor Location of Parallel Kinematic Mechanisms", *Proceedings of 2002 Florida Conference on Recent Advances in Robotics*, Miami, FL., May 23-24, 2002, 7pp.

In: Robotics: State of the Art and Future Trends
Editors: G. Legnani and I. Fassi
ISBN: 978-1-62100-403-5

Chapter 7

FRACTIONAL-ORDER CONTROLLERS FOR ROBOT MANIPULATOR

Hadi Delavari*[*,1] and Danial Mohammadi Senejohnny[†,2]
[1] Department of Electrical Engineering, Hamedan University of Technology, Hamedan, Iran
[2]Young Researchers Club, Islamic Azad University, Tehran-Markaz Branch, Tehran, Iran

Abstract

The increasing use of robot manipulators in delicated fields such as medical surgery, industrial automation, welding and military section as a few to mention, has raised the need for a well tracking precision. In this chapter, a robust Fractional Order Sliding Mode Control (FOSMC) beside an adaptive Fractional Order PID control (FOPID) as two paradigms of Fractional Order Controller's (FOC's) are proposed for robot manipulator trajectory control. The performance and the significance of the controllers are investigated based on simulation results.

1. Introduction

Robotics is a relatively young field of modern technology that crosses traditional engineering boundaries. Understanding the complexity of robots and their applications requires knowledge of electrical engineering, mechanical engineering, systems and industrial engineering, computer science, economics, and mathematics. New disciplines of engineering, such as manufacturing engineering, applications engineering, and knowledge engineering have emerged to deal with the complexity of the field of robotics and factory automation.

An official definition of such a robot comes from the Robot Institute of America (RIA): A robot is a reprogrammable multifunctional manipulator designed to move material, parts, tools, or specialized devices through variable programmed motions for the performance of a variety of tasks.

[*] Corresponding Author's E-mail address: hdelavary@gmail.com
[†] E-mail address: d.senejohnny@gmail.com

According to recent literatures, an equation contains fractional-order derivative/integration is referred to as Fractional-Order Ordinary Differential/integral Equation (FO-ODE). FO-ODE is defined as an equation containing fractional-order derivatives, this type of differential equation could be replaced by their counterparts in system modelling [1] or controller design [2]. In this chapter the matter of Fractional Order Controller (FOC) for robot manipulators will be discussed.

In recent years, as the need for achieving greater accuracy in tracking problems there has been a surge of interest in using robust control techniques to tackle this problem. Many approaches have been proposed in the literature such as computer torque method [3], optimal control [4, 5], sliding mode control [6, 7], adaptive control [8] and intelligent approaches [9-11] to mention few, but the methods provided for control action in this manuscript are designed with the sense of FO-ODE this is the reason that distinguishes FOC with comparison to the classical methods and put control engineering into a new era [12-17].

This chapter is organized as follows: In the following section, explication of fractional derivative operands and geometrical interpretation of fractional operator is presented. In section 3 manipulator model used in simulation results is introduced. In 4th and 5th section, two Fractional Order Control (FOC) methods will be proposed and proven to be mathematically stable. Section 6 investigates problem of optimization for the purpose of comparison. Numerical simulations related to the proposed control schemes carried out by MATLAB showed in section 7. Finally this chapter concluded by drawing a conclusion in section 8.

2. About Fractional Calculus

Fractional integration and fractional differentiation are generalizations of notions of integer-order integration and differentiation, and include *n*-th derivatives and *n*-fold integrals (*n* denotes an integer number) as particular cases. Because of this, it would be ideal to have such geometric interpretations of fractional-order operators [19], which will provide also a link to known classical interpretations of integer-order differentiation and integration.

2.1. Geometric Interpretation of Fractional Integration: Shadows on the Wall

Let us consider the left-sided Riemann-Liouville fractional integral [18] of order α

$$ {}_0I_t^{\alpha} f(t) = \frac{1}{\Gamma(\alpha)} \int_0^t f(\tau)(t-\tau)^{\alpha-1} d\tau, \qquad t \geq 0 \tag{2.1}$$

and write it in the form

$$ {}_0I_t^{\alpha} f(t) = \int_0^t f(\tau) dg_t(\tau) \tag{2.2}$$

where $g_t(\tau)$ is given by

$$g_t(\tau) = \frac{1}{\Gamma(\alpha+1)}\left\{t^\alpha - (t-\tau)^\alpha\right\} \tag{2.3}$$

The function $g_t(\tau)$ has an interesting scaling property. Indeed, if we take $t_1 = kt$ and $\tau_1 = k\,\tau\,(k>0)$, then

$$g_{t_1}(\tau_1) = g_{kt}(k\,\tau) = k^\alpha g_t(\tau) \tag{2.4}$$

Now let us consider the integral (2.2) for a fixed *t*. Then it becomes simply a Stieltjes integral, and we can utilize G. L. Bullock's idea [20].

Let us take the axes τ, g and *f*. In the plane (τ, g) we plot the function $g_t(\tau)$ for $0 \le \tau \le t$. Along the obtained curve we "build a fence" of the varying height $f(\tau)$, so the top edge of the "fence" is a three-dimensional line $(\tau, g_t(\tau), f(\tau))$, $0 \le \tau \le t$.

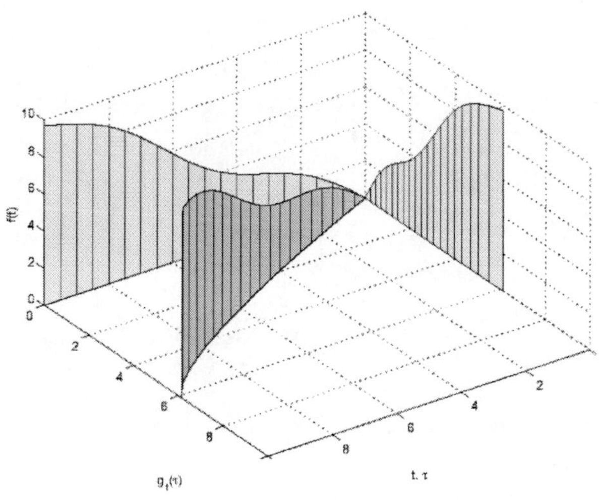

Figure. 1. The "fence" and its shadows: ${}_0I_t^1 f(t)$ and ${}_0I_t^\alpha f(t)$, for $\alpha = 0.75$ and $f(t) = t + 0.5\sin(t)$, $0 \le t \le 10$.

This "fence" can be projected onto two surfaces (see Fig. 1):

- the area of the projection of this "fence" onto the plane (τ, f) corresponds to the value of the integral

$$ {}_0I_t^1(t) = \int_0^t f(\tau)d\tau;\, t \ge 0 \tag{2.5}$$

- the area of the projection of the same "fence" onto the plane (g, f) corresponds to the value of the integral (2.2), or, what is the same, to the value of the fractional integral (2.1).

In other words, our "fence" throws two shadows on two walls. The first of them, that on the wall (τ, f), is the well-known "area under the curve $f(\tau)$ ", which is a standard geometric interpretation of the integral (2.5). The "shadow" on the wall (g, f) is a geometric interpretation of the fractional integral (2.1) for a fixed t.

Obviously, if $\alpha = 1$, then $g_t(\tau) = \tau$, and both "shadows" are equal. This shows that classical definite integration is a particular case of the left-sided Riemann-Liouville fractional integration even from the geometric point of view [19].

2.2. Fractional Derivative Operands

In this chapter, we use Caputo fractional operators ${}_a^cD_t^\alpha$ as our main tool, so its fractional derivatives is defined as

$$ {}_a^cD_t^\alpha f(t) = \frac{1}{\Gamma(\alpha - n)} \int_0^t \frac{f^{(n)}(\tau)}{(t-\tau)^{\alpha-n+1}} d\tau, \qquad n-1 < \alpha < n \tag{2.6} $$

where a and t are the limits of operation, and $\Gamma(.)$ is the well known Euler's Gamma function. The other prevalent and well known definitions are Grünwald–Letnikov (GL) and the Riemann–Liouville (RL) [18, 21, 22].

The GL definition is

$$ {}_aD_t^\alpha f(t) = \lim_{N \to \infty} \left[\frac{t-a}{N}\right]^{-\alpha} \sum_{j=0}^{N-1} (-1)^j \binom{\alpha}{j} f(t - j\left[\frac{t-a}{N}\right]) \tag{2.7} $$

where $[.]$ is a flooring-operator, while the RL definition is given by

$$ {}_aD_t^\alpha f(t) = \frac{1}{\Gamma(n-\alpha)} \frac{d^n}{dt^n} \int_a^t \frac{f(\tau)}{(t-\tau)^{\alpha-n+1}} d\tau, \qquad n-1 < \alpha < n \tag{2.8} $$

3. Robot Manipulator Dynamic Modeling

A manipulator is defined as an open kinematics, chain of rigid links. Each degree of freedom of manipulator is powered by independent torques. In the absence of friction or other disturbances, dynamic of an n-link rigid robotic manipulator system can be described by the following second order nonlinear vector differential equation

$$ M(q)\ddot{q} + C(q, \dot{q})\dot{q} + G(q) = U \tag{3.1} $$

where $q,\dot{q},\ddot{q} \in R^n$, q joint variable n-vector and U is n-vector of generalized forces. $M(q) \in R^{n\times n}$ is symmetric and positive definite inertia matrix, $C(q,\dot{q})$ is centripetal and coriolis vector, and $G(q)$ is the gravitational torque vector.

As shown in Fig. 2 a two-degree of freedom polar robot manipulator has one rotational and sliding joint in the (x, y) plane. Neglecting the gravity force and normalizing the mass and length of the arm, a nominal model of two-degree of freedom polar robot can be expressed as follows:

$$\begin{aligned} \dot{x}_1(t) &= x_2(t) \\ \dot{x}_2(t) &= \left[\left(\eta x_1(t) + M(x_1(t)+a)x_4^2(t)\right) + u_1(t)\right] / (\eta + m) \\ \dot{x}_3(t) &= x_4(t) \\ \dot{x}_4(t) &= -2\left[M(x_1(t)+a)x_2(t)x_4(t) + \eta x_1(t) + u_2(t)\right] / \left[J_1 + J_2 + \eta x_1^2(t) + M(x_1(t)+a)^2\right] \end{aligned} \tag{3.2}$$

where η is the mass of motional link, M is the payload, J_1 and J_2 are moments of inertia of the motional link with respect to the vertical axis through c and o, respectively.

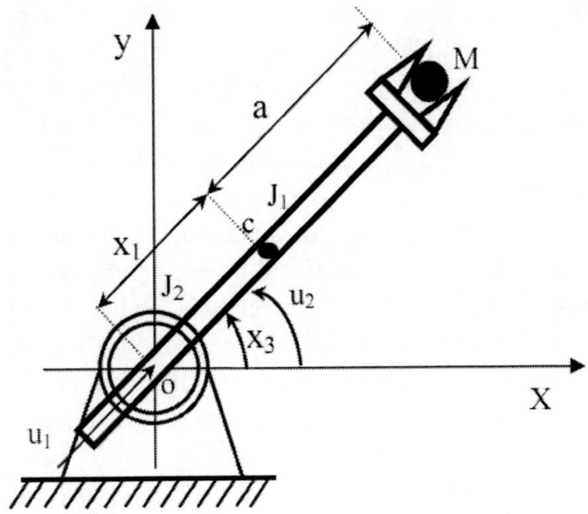

Figure 2. A two-degree of freedom polar manipulator.

Suppose the robot manipulator uncertain dynamic model can be described by a coupled second-order nonlinear system of the form

$$\begin{aligned} \dot{x}_{2k-1}(t) &= x_{2k}(t) \\ \dot{x}_{2k}(t) &= f_k(x,t) + b_k(x)u_k(t) + \delta_k(t), \qquad k = 1,2 \end{aligned} \tag{3.3}$$

where $X_k = \left[x_{2k-1}, x_{2k}\right]^T$ are measurable states vector of the system, $u_k(t) \in R$ is control action vector, $\delta_k(t)$ stands for uncertainty of the dynamic, including frictions and disturbance, $b_k(x)$ and $f_k(x,t)$, are the control gains and the nonlinear dynamics of the system, respectively, here and in sequel k runs as $k = 1, 2$.

Let us define

$$e_{2k-1}(t) = x^{d}_{2k-1}(t) - x_{2k-1}(t) \tag{3.4}$$

as an error signal between the desired and actual outputs.

Already in [42] and [2] adaptive FOPID and fractional PD$^{\alpha}$ sliding mode control schemes has proven and showed to outperform the integer types for robot manipulators control, respectively. In this chapter utilizing the robustness properties of fractional order controller we will propose a two new fractional order controller, PDD$^{\alpha}$ sliding surface and adaptive FOPID. In the next section the proposed Adaptive Fractional-Order PID Controller (FOPID) will be studied.

4. Adaptive Fractional-Order PID Controller

Fractional-order PID (FOPID) controllers which relies on the extension of derivate and integral actions in a PID controllers to their fractional-order counterparts is receiving wide attention in the recent years both from academic and industrial point of view. In fact, in principle, they provide more flexibility in the controller design, with respect to the standard PID controllers, because they have five parameters to select (instead of three) [12]. However, this also implies that the tuning of the controller can be much more complex.

The concept of FOPID controllers was proposed by Podlubny in 1997 [23]. He also demonstrated the better response of this type of controller, in comparison with the classical PID controller, when used for the control of fractional order systems [12]. In [24] Ziegler-Nichols tuning rules for FOPID are reported. In [25] the authors applied an optimal FOPID tuned by Particle Swarm Optimization (PSO) algorithm to control the Automatic Voltage Regulator (AVR) system. Application of these controllers on heat diffusion and irrigation canal is studied in [26, 27]. In [28] two fractional order proportional integral controllers (FOPI and FO[PI]) are proposed and designed for a class of fractional order systems. Very recently, a new set of tuning rules based on a first-order-plus-dead-time model of the process is presented in [29]. The tuning rules have been devised in order to minimize the integrated absolute error with a constraint on the maximum sensitivity. Also it is shown that, if the controller is restricted to a PI structure, then the use of a fractional integral action is not useful in improving the performance and therefore a standard PI controller is the best option. Conversely, the use of a fractional derivative action allows improving the control performance.

In the robotics context, it is known that PID controllers are sensitive to uncertainties which arise from imprecise knowledge of the kinematics and dynamics, from joint and link flexibility, actuator dynamics, friction, sensor noise, and unknown loads, in this section we will carry on with developing an adaptive Fractional-order PID (FOPID) controller scheme.

4.1. Definitions

According to the aforementioned dynamic model

$$\begin{aligned} \dot{x}_{2k-1}(t) &= x_{2k}(t) \\ \dot{x}_{2k}(t) &= f_k(x,t) + b_k(x)u_k(t) + \delta_k(t) \end{aligned} \tag{4.1}$$

where $X_k = [x_{2k-1}, x_{2k}]^T$ are states vector of the system. In addition, it is assumed that there exist two positive upper bounds $f_k^u(.)$ and δ_k^u satisfying $|f_k(x,t)| \le f_k^u(.)$ and $|\delta_k(t)| \le \delta_k^u$. Define $X_k^d = \left[x_{2k-1}^d, \dot{x}_{2k-1}^d\right]^T$ and assume that both x_{2k-1}^d and $\dot{x}_{2k-1}^d$ are bounded, i.e. $\|X_k^d\|_\infty = \sup_{t\ge0}\|X_k^d(t)\| < \infty$ (it means that the desired states are bounded). Then the error vector of system becomes $E_k = X_k^d - X_k = [e_{2k-1}, \dot{e}_{2k-1}]^T$ where $e_{2k-1}(t) = x_{2k-1}^d(t) - x_{2k-1}(t)$ and $\dot{e}_{2k-1}(t) = \dot{x}_{2k-1}^d(t) - \dot{x}_{2k-1}(t)$. Suppose that we can choose a gain vector $K_k = [k_0, k_1]^T$ such that roots of $s^2 + k_1 s + k_0 = 0$ are in the open left-half complex plane. Now let a feedback linearization controller be given by

$$u_k^* = b_k^{-1}(x)\left(-f_k(x,t) - \delta_k(t) + K_k^T E_k + \ddot{x}_{2k-1}^d\right) \tag{4.2}$$

Substituting (4.2) into (4.1), we have

$$\ddot{e}_{2k-1} + k_1\dot{e}_{2k-1} + k_0 e_{2k-1} = 0 \tag{4.3}$$

Consequently, from (4.3), we have $e_{2k-1}(t) \to 0$ as $t \to \infty$, i.e., $x_{2k-1} \to x_{2k-1}^d$ asymptotically. However, it is noted that the relative models in (4.1) are assumed to be unknown in this study. Throughout the chapter, the following assumption is made.

Let the constraint set Ω_k^x for the state X_k be assumed as

$$\Omega_k^x = \{X_k \in R^2 : \|X_k\| \le M_k^x\} \tag{4.4}$$

where M_k^x is a pre-specified parameter.

It is desired that the state trajectory of system X_k never reach the boundary of Ω_k^x during the control procedure. For simplicity of analysis, we may choose $M_k^x \ge \|X_k^d\|_\infty$, also according to inequality $\|X_k\| \le M_k^x$, the choice for M_k^x must be in a way that both the inequalities hold.

The fractional form of a PID controller, with input $e_{2k-1}(t)$ and output u_k^{FOPID}, is generally given as

$$u_k^{FOPID} = k_k^P e_{2k-1}(t) + k_k^I D^{\lambda_k} e_{2k-1}(t) + k_k^D D^{\mu_k} e_{2k-1}(t) \quad (4.5)$$

where $k_k^P, k_k^I, k_k^D \in R$ and $\lambda_k \in (-1,0]$, $\mu_k \in [0\,1)$ denote order of fractional integral and derivative action, respectively.

For tuning purpose three PID control gains in (4.5) k_k^P, k_k^I, and k_k^D, as an adjustable parameters will be updated online by an adequate adaptation mechanism derived via classical Lyapunov theorem based on sliding mode and gradient method, section 4.3, also the order of derivative and integral actions μ_k and λ_k will be determined offline based on genetic algorithm, section 6.

First of all by merit of a supervisory controller, the stability of the closed-loop FOPID control system under external disturbance $\delta_k(t)$ is guaranteed and in addition for tuning purpose three FOPID adjustable parameters in (4.5) will be updated online by an adequate adaptation mechanism motivated from the sliding mode control [30], where further investigated in sequel.

4.2. Supervisory Controller Design

Now let the control input in (4.1) be given by

$$u_k = u_k^{FOPID} + u_k^s \quad (4.6)$$

where u_k^{FOPID} is the FOPID controller as shown in (4.5) and u_k^s is the extra supervisory controller that will be fired only when the states of the system exceeds some bound (4.4) and guarantees the stability of the system.

To design the supervisory controller u_k^s, substituting (4.6) into (4.1), we have, in view of (4.2)

$$\begin{aligned} \dot{x}_{2k}(t) &= f_k(x,t) + b_k(x)\left(u_k^{FOPID} + u_k^s\right) + \delta_k(t) \\ &= f_k(x,t) + b_k(x)\left(u_k^{FOPID} + u_k^s + u_k^* - u_k^*\right) + \delta_k(t) \\ &= \ddot{x}_{2k-1}^d + K_k^T E_k - b_k(x)\left(u_k^* - u_k^{FOPID} - u_k^s\right) \end{aligned} \quad (4.7)$$

This implies that

$$\ddot{e}_{2k-1} = -K_k^T E_k + b_k(x)\left[u_k^* - u_k^{FOPID} - u_k^s\right] \quad (4.8)$$

Let

$$A_k = \begin{bmatrix} 0 & 1 \\ -k_0 & -k_1 \end{bmatrix},\ B_k = \begin{bmatrix} 0 \\ 1 \end{bmatrix} \quad (4.9)$$

be a companion form. From (4.8), we have

$$\dot{E}_k = A_k E_k + B_k b_k(x)\left[u_k^* - u_k^{FOPID} - u_k^s\right] \tag{4.10}$$

Now consider a Lyapunov function candidate [7]

$$V_k^e = \frac{1}{2} E_k^T P_k E_k \tag{4.11}$$

where P_k is a positive definite symmetric matrix satisfying the Lyapunov equation [7]

$$A_k^T P_k + P_k A_k = -Q_k \tag{4.12}$$

and Q_k is a positive definite symmetric matrix selected by designer [7]. Define

$$V_k^M = (1/2)\lambda_{\min}(P_k)\left(M_k^x - \left\|X_k^d\right\|_\infty\right)^2 \tag{4.13}$$

where $\lambda_{\min}(P_k)$ denotes the minimum eigenvalue of P_k. Note that if $\left\|X_k\right\| \geq M_k^x$, then, from (4.11), we have

$$\begin{aligned} V_k^e &\geq \frac{1}{2}\lambda_{\min}(P_k)\left\|E_k\right\|^2 \geq \frac{1}{2}\lambda_{\min}(P_k)\left(\left\|X_k\right\| - \left\|X_k^d\right\|\right)^2 \\ &\geq \frac{1}{2}\lambda_{\min}(P_k)\left(M_k^x - \left\|X_k^d\right\|_\infty\right)^2 = V_k^M \end{aligned} \tag{4.14}$$

Hence if $V_k^e < V_k^M$, then $\left\|X_k\right\| < M_k^x$, Moreover, the time derivative of V_k^e along the trajectories of the closed-loop system (4.10) satisfies

$$\begin{aligned} \dot{V}_k^e &= \frac{1}{2}E_k^T\left(A_k^T P_k + P_k A_k\right)E_k + E_k^T P_k B_k b_k(x)\left(u_k^* - u_k^{FOPID} - u_k^s\right) \\ &= -\frac{1}{2}E_k^T Q_k E_k + E_k^T P_k B_k b_k(x)\left(u_k^* - u_k^{FOPID} - u_k^s\right) \\ &\leq -\frac{1}{2}E_k^T Q_k E_k + \left|E_k^T P_k B_k b_k(x)\right|\left(\left|u_k^*\right| + \left|u_k^{FOPID}\right|\right) - E_k^T P_k B_k b_k u_k^s \end{aligned} \tag{4.15}$$

From (4.2) and presumes of $\left|b_k^{-1}(.)f_k(.)\right| \leq f_k^u(.)$ and $\left|b_k^{-1}(.)\delta(.)\right| \leq \delta_k^u$ we could say

$$u_k^* \leq f_k^u(.) + \delta_k^u + \left|b_k^{-1}(x)K_k^T E_k\right| + \left|b_k^{-1}(x)\left(\ddot{x}_{2k-1}^d\right)\right| \tag{4.16}$$

Hence, if the supervisory controller is chosen as

$$u_k^s = \begin{cases} 0 & if\ V_k^e < V_k^M \\ \text{sgn}\left(E_k^T P_k B_k b_k\right)\left[f_k^u(.) + \delta_k^u + \left|b_k^{-1}(x) K_k^T E_k\right| + \left|b_k^{-1}(x)\left(\ddot{x}_{2k-1}^d\right)\right| + \left|u_k^{FOPID}\right|\right] & if\ V_k^e \geq V_k^M \end{cases} \quad (4.17)$$

then, from (4.17) and (4.16), we can guarantee that $\dot{V}_k^e < 0$ in (4.15) if $V_k^e \geq V_k^M$ [30-32].

In the next subsection we will carry on with developing an adaptation law for the proposed FOPID controller scheme.

4.3. Adaptation Law Synthesis

Based on concepts of sliding mode control, a proper adaptation law based on the use of gradient method is presented in this section to minimize designed sliding condition for updating FOPID control gains [30].

Let us define a designed signal x_k^r as:

$$\dot{x}_k^r = \ddot{x}_{2k-1}^d + K_k^T E_k \quad (4.18)$$

This may be achieved using a sliding mode controller. A sliding surface can be defined as:

$$S_k = x_{2k} - x_k^r \quad (4.19)$$

When the system settles at the sliding mode, i.e. $S_k = 0$ therefore:

$$x_{2k} = x_k^r \quad (4.20)$$

Replacement of relation (4.20) into (4.18) and in view of (4.1) approaches:

$$\ddot{x}_{2k-1} = \ddot{x}_{2k-1}^d + K_k^T E_k \quad (4.21)$$

Or in a scalar format:

$$\ddot{e}_{2k-1} + k_1 \dot{e}_{2k-1} + k_0 e_{2k-1} = 0 \quad (4.22)$$

The errors approaches zero when k_0 and k_1 are assigned appropriately. To investigate the stability a Lyapunov function will be candidate as follows:

$$V_k = \frac{1}{2} S_k^2 \quad (4.23)$$

In order to stay in the sliding mode the differentiation of the Lyapunov function must be negative definite. Satisfying the necessity in (4.24) guarantees states to keep staying in the surface $S_k = 0$.

$$V_k = S_k \dot{S}_k < 0 \tag{4.24}$$

From (4.24) it is guaranteed that $S_k(t) \to 0$ as $t \to \infty$.

To derive an adequate adaptation mechanism for tuning three FOPID control gains, a gradient search method is used to minimize the sliding condition $S\dot{S}$ in (4.24) as possibly in this study. The gradient search algorithm is calculated in the direction opposite to the energy flow.

The derivative of the sliding surface is:

$$\dot{S}_k = \dot{x}_{2k} - \dot{x}_k^r = \left[f_k(x,t) + b_k(x) u_k(t) + \delta_k(t) - \dot{x}_k^r \right] \tag{4.25}$$

Substituting (4.6) into (4.25) and multiplication of S_k to both sides of equation (4.25) yields:

$$S_k \dot{S}_k = S_k \left[f_k(x,t) + b_k(x)\left(u_k^{FOPID} + u_k^s\right) + \delta_k(t) - \dot{x}_k^r \right] \tag{4.26}$$

Based on the gradient method and the chain rule, the FOPID coefficients will be updated as

$$\begin{aligned}
\dot{k}_k^P &= -\gamma_k \frac{\partial S_k \dot{S}_k}{\partial k_k^P} = -\gamma_k \frac{\partial S_k \dot{S}_k}{\partial u_k^{FOPID}} \frac{\partial u_k^{FOPID}}{\partial k_k^P} = -\gamma_k S_k e_{2k-1}(t) \\
\dot{k}_k^I &= -\gamma_k \frac{\partial S_k \dot{S}_k}{\partial k_k^I} = -\gamma_k \frac{\partial S_k \dot{S}_k}{\partial u_k^{FOPID}} \frac{\partial u_k^{FOPID}}{\partial k_k^I} = -\gamma_k S_k \left[\int_0^t e_{2k-1}(\tau) d\tau \right] \\
\dot{k}_k^D &= -\gamma_k \frac{\partial S_k \dot{S}_k}{\partial k_k^D} = -\gamma_k \frac{\partial S_k \dot{S}_k}{\partial u_k^{FOPID}} \frac{\partial u_k^{FOPID}}{\partial k_k^D} = -\gamma_k S_k \left[\frac{d}{dt} e_{2k-1}(t) \right]
\end{aligned} \tag{4.27}$$

where $\gamma_k > 0$ is the learning rate. Notice that if the learning rate γ_k or the initial values of FOPID control gains are not selected adequately, the resulted FOPID controller will probably make the states of system divergent i.e $\|X_k\| > M_k^x$. Fortunately, the supervisory controller as in (4.17) will play an important role that provides an extra input to pull the states back to the pre-specified constraint set Ω_k^x i.e $\|X_k\| < M_k^x$ and also guarantees the stability of the system. GA can be used (section 6) to search the parameter space to find appropriate values of the adaptive FOPID fractional differentiation and integration action i.e. μ_k , λ_k in (4.5).

The block diagram of the robust adaptive FOPID control with a supervisory controller for uncertain robot manipulators is depicted in Figure 3.

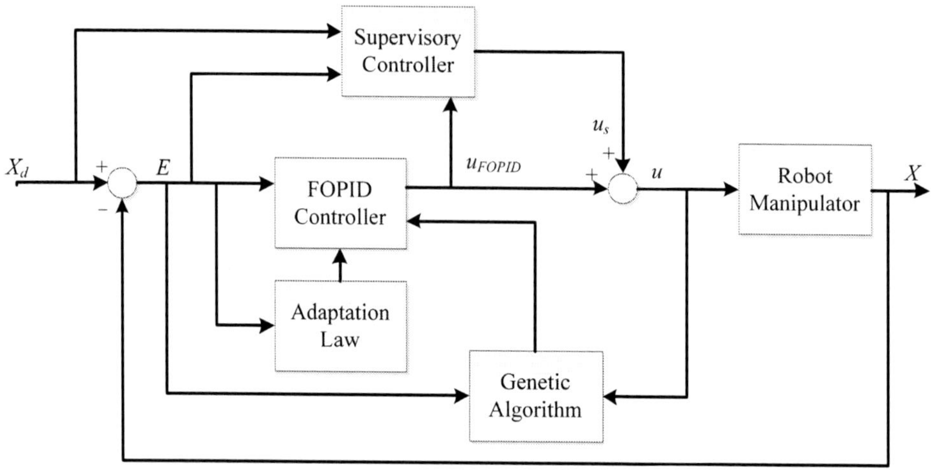

Figure 3. Robust fractional adaptive FOPID control.

5. Fractional – Order Sliding Mode Controller

Fractional-order sliding mode control will be studied in this section, since it offers merits of sliding mode and fractional controllers simultaneously. Mainly, there are two different approaches to design a fractional-order sliding mode controller. In [2], authors have employed fractional-order sliding surfaces PD^{α} instead of traditional surfaces PD and obtained classical reaching law via classical Lyapunov theory. Such a fractional-order sliding mode control is investigated on different case studies and it is contended that outperforms conventional sliding mode controllers [2] [33-35]. In the other hand, in [36] fractional-order sliding mode control is implemented based on a fractional reaching law i.e. $D^{q}s = -k\ \mathrm{sgn}(s)$ instead of $\dot{s} = -k\ \mathrm{sgn}(s)$, and PID surface sliding modes has been already used in [40, 41], where sliding surface has been chosen as $S_k(e_{2k-1}(t)) = k_p e_{2k-1}(t) + k_i \int e_{2k-1}(t)dt + k_d \dot{e}_{2k-1}(t)$, this is a special case of general form $S_k(e_{2k-1}(t)) = k_p e_{2k-1}(t) + k_i D^{\lambda} e_{2k-1}(t) + k_d D^{\mu} e_{2k-1}(t)$, when $-1 < \lambda < 0,\ 0 < \mu < 1$. It can be seen that PID surface sliding mode is a special case of $PI^{\lambda}D^{\mu}$ surface sliding mode, when $\lambda = -1$ and $\mu = +1$. Here in order to utilize the robustness properties of fractional order controller, we used the term ${}^{c}D^{\alpha} e_{2k-1}(t)$ in sliding surface formulation which means, α order derivative of error dynamic in (3.4) with the sense of Caputo fractional derivative definition in section 2.1.

5.1. FOSMC Design

The sliding mode design approach consists of two steps. The first involves the design of a switching function, $S = 0$, such that the sliding motion satisfies the design specifications. Second one is concerned with selection of a control law which will enforce the sliding mode.

The following Fractional PDD^{α} sliding surface is proposed

$$S_k(e_{2k-1}(t)) = \lambda_k e_{2k-1}(t) + \dot{e}_{2k-1}(t) + {}^cD^{\alpha} e_{2k-1}(t) \tag{5.1}$$

where $\lambda_k > 0$.

Taking the time derivative from both sides of (5.1), results:

$$\dot{S}_k(e_{2k-1}(t)) = \lambda_k \dot{e}_{2k-1}(t) + \ddot{e}_{2k-1}(t) + {}^cD^{\alpha} \dot{e}_{2k-1}(t) \tag{5.2}$$

With the sense of $e_{2k-1}(t) = x^d_{2k-1}(t) - x_{2k-1}(t)$ we have

$$\dot{S}_k(e_{2k-1}(t)) = \lambda_k \dot{e}_{2k-1}(t) + (\ddot{x}^d_{2k-1}(t) - \dot{x}_{2k}(t)) + {}^cD^{\alpha} \dot{e}_{2k-1}(t) \tag{5.3}$$

Substituting (4.1) into (5.3) results

$$\dot{S}_k(e_{2k-1}(t)) = \lambda_k \dot{e}_{2k-1}(t) + \left[\ddot{x}^d_{2k-1}(t) - f_k(x,t) - b_k(x)u_k(t) - \delta_k(t)\right] + {}^cD^{\alpha} \dot{e}_{2k-1}(t) \tag{5.4}$$

The control input can be given as $u_k = u_k^{eq} + u_k^{sw}$, where u_k^{eq} and u_k^{sw} are the equivalent control and the switching control, respectively. The equivalent control, u_k^{eq}, proposed by Utkin [37] is based on the nominal (estimated) plant parameters, while the switching control, u_k^{sw}, ensures the discontinuity of the control law across sliding surface.

By forcing $\dot{S}_k(t)$ to zero substituting in (5.4), the overall control signal results:

$$u_k(t) = b_k^{-1}(x)\left[\lambda_k \dot{e}_{2k-1}(t) + {}^cD^{\alpha} \dot{e}_{2k-1}(t) - f_k(x,t) + \ddot{x}^d_{2k-1}(t)\right] + u_k^{sw} \tag{5.5}$$

where u_k^{sw} is the switching function and is introduced to be directly as

$$u_k^{sw} = -b_k^{-1}(x)\left[k_k^s \operatorname{sgn}(S_k(t))\right] \tag{5.6}$$

where $k_k^s \in R$ is a switching feedback control gain and

$$\operatorname{sgn}(S_k(t)) = \begin{cases} +1 & if\ S_k(t) > 0 \\ 0 & if\ S_k(t) = 0 \\ -1 & if\ S_k(t) < 0 \end{cases} \tag{5.7}$$

Substitution of (5.5) into (5.4), results:

$$\dot{S}_k(e(t)) = -k_k^s \operatorname{sgn}(s) - \delta_k(t) \tag{5.8}$$

For the purpose of stability analysis assume the following Lyapunov function

$$V_k = \frac{1}{2} S_k^2 \tag{5.9}$$

Taking time derivative from (5.9) we have

$$\begin{aligned} \dot{V}_k &= S_k \dot{S}_k = S_k \left(-k_k^s \operatorname{sgn}(S_k) - \delta_k(t) \right) \\ &= S_k \left(-k_k^s - |\delta_k(t)| \right) \operatorname{sgn}(S_k) \\ &\leq -|S_k| \left(k_k^s + |\delta_k(t)| \right) \end{aligned} \tag{5.10}$$

Since $|S_k| > 0$ if $k_k^s > -|\delta_k(t)|$ i.e. $k_k^s > -\delta_k^u$ we have $\dot{V}_k < 0$; hence $\dot{V}_k$ is negative definite. From the above analysis sliding condition and also the global asymptotic stability is guaranteed.

When the control law is chosen as (5.5), chattering phenomena will occur as soon as the state hits the sliding surface because of discontinuity in *signum* function. To reduce the chattering a *saturation* function (5.12) is used instead of the *signum* function [7]. Hence, the alternative control signal become as

$$u_k(t) = b_k^{-1}(x) \left[\lambda \dot{e}_{2k-1}(t) + {}^{c}D^{\alpha+1} e_{2k-1}(t) - f_k(x,t) + \ddot{x}_{2k-1}^{d}(t) - k_k^s sat(S_k(t)) \right] \tag{5.11}$$

where

$$sat(S_k(t)) = \begin{cases} \operatorname{sgn}(S_k(t)) & if \ |S_k(t)| \geq 1 \\ S_k(t) & if \ |S_k(t)| < 1 \end{cases} \tag{5.12}$$

Block diagram of proposed robust FOSMC is depicted in Figure 4.

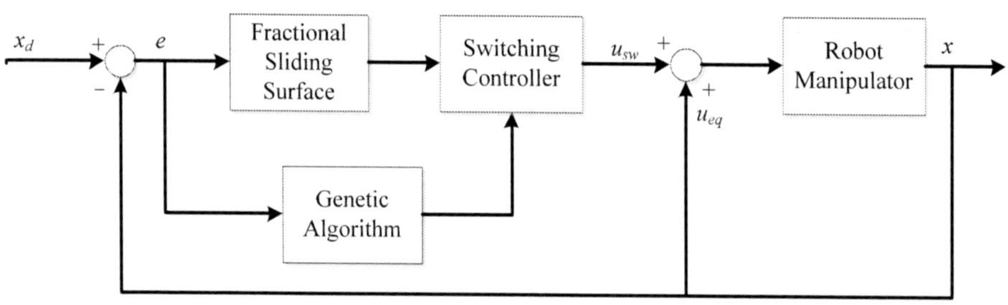

Figure 4. Robust fractional PDD$^{\alpha}$ sliding mode control.

6. Genetic Algorithm

An objective function is used to evaluate the fitness of the string to the environment [38]. Accordingly GA is used to search the parameter space to find appropriate values of the adaptive FOPID i.e. μ_k, λ_k in (4.5) and FOSMC, switching controller parameter i.e. k_k^s in (5.11).

The cost function is defined in such a way that the selected parameters to minimize the error to provide a good tracking performance. The cost function (the criteria for optimization) is defined as follows:

$$\int \left[w_1 \left(e_{2k-1}^2 \right) + w_2 \left(u_{2k-1}^2 \right) \right] dt \tag{6.1}$$

$$\int w_3 \left(e_{2k-1}^2 \right) dt, \qquad k = 1, 2 \tag{6.2}$$

where e_{2k-1} is defined in (3.4), and w_1, w_2, w_3 are the weighting factors. This solution is optimal only for the control task for which the optimization has been performed, beside the GA fitness function is chosen by means of repeated simulation, further details given in section 7.

7. Simulation Results

A comparison based upon numerical simulation generated by Computer Aided Design (CAD) is a straightforward way to demonstrating which one of the Fractional Order Controllers (FOC) outperforms the other one. In control engineering definition of outperformance is robustness to manipulator uncertain dynamic from one hand, and tracking the desired trajectory on the other.

Robot Manipulator: During simulation studies, parameters of polar manipulator are chosen as:

$$M = 1.5\,kg\ ,\ \eta = 1\,kg\ ,\ J_1 = J_2 = 1\,kg\ m^2,\ a = 1\,m$$

Initial conditions are

$$\left[x_1(0), x_2(0), x_3(0), x_4(0) \right]^T = \left[-0.15, -0.2,\ 0.35, 0.5 \right]^T$$

Disturbances Considering as, [2, 39]

$$\begin{aligned} \delta_1(t) &= 0.4\ \cos(5\pi t)\, Nm \\ \delta_2(t) &= 0.6 \cos(5\pi t)\ Nm \end{aligned} \tag{7.1}$$

It must be noted we can consider other type of disturbance instead of (7.1), but we just need to have the upper bound of that disturbance, then according to (4.16) and (5.10) the overall system will be stable.

The robot manipulator joints are driven according to the following desired trajectory:

$$\begin{aligned} x_1^d &= 0.5 \sin\left(\frac{\pi}{10} t\right) m \\ x_3^d &= 2\pi \sin\left(\frac{\pi}{10} t\right)\ rad \end{aligned} \tag{7.2}$$

FOPID: The corresponding upper bounds in (4.16) related to robot manipulator dynamic and disturbance can be formulated as follows:

$$\begin{aligned}
&\left|(\eta+m)\ \left(0.4\ \cos(5\pi t)Nm\right)\right| \le 0.4(\eta+m) = \delta_1^u \\
&\left|\left(J_1+J_2+\eta x_1^2(t)+M\,(x_1(t)+a)^2\right)\left(0.6\cos(5\pi t)\ Nm\right)\right| \le 0.6\left(J_1+J_2+\eta x_1^2(t)+M\left(|x_1(t)|+a)^2\right)\right) = \delta_2^u \\
&\left|\eta x_1(t)+M\,(x_1(t)+a)x_4^2(t)\right| \le \eta|x_1(t)|+M\,|x_1(t)|x_4^2(t)+aMx_4^2(t) = f_1^u(.) \\
&\left|M\,(x_1(t)+a)x_2(t)x_4(t)+\eta x_1(t)\right| \le M\,|x_1(t)||x_2(t)||x_4(t)|+aM\,|x_2(t)||x_4(t)|+\eta|x_1(t)| = f_2^u(.)
\end{aligned} \tag{7.3}$$

These choices just sense theoretically, and mean that there is upper bound for the terms $f_k^u(.)$ and δ_k^u in (4.16).

We choose $k_0 = 6$ and $k_1 = 9$ so that the roots of $s^2 + k_1 s + k_0 = 0$ are in the left-half complex plane. From (4.9), we have

$$A_1 = A_2 = \begin{bmatrix} 0 & 1 \\ -6 & -9 \end{bmatrix} \tag{7.4}$$

In comparison with equation (4.12) by choosing $Q_1, Q_2 = diag\,[3,3]$ (i.e. it must be symmetric and positive definite) a matrix gain P_1, P_2 can be found as:

$$P_1 = P_2 = \begin{bmatrix} 3.417 & 0.25 \\ 0.25 & 0.195 \end{bmatrix} \tag{7.5}$$

The value of $M_k^x, k = 1,2$ is deliberately selected to be large enough such that the system state X_k never reach the boundary of $\Omega_k^x, k = 1,2$.

Primary setting of FOPID coefficients are chosen as:

$$k_1^P(0) = 15,\ k_2^P(0) = 10,\ k_1^I(0) = 35,\ k_2^I(0) = 25,\ k_1^D(0) = 20,\ k_2^D(0) = 15, \tag{7.6}$$

Whilst the learning rate has been selected as $\gamma_1 = \gamma_2 = 5$.

Here GA will be used to minimize the cost function in (6.1) and find an optimum value of the adaptive FOPID controller. The parameters of the GA with the author experience are chosen as:

Population size = 150, Crossover probability = 0.7,
Generations = 90, Mutation probability =0.02,
Search space of μ_k and λ_k is a real space of $[0,1)$ and $(-1,0]$, respectively.

The outcome of Genetic Algorithm verifies that FOPID achieves less cost function. Let $w_1 = 2, w_2 = 1$, an optimal parameters are averaged over some different run and found as: $\lambda_1 = -0.901$, $\mu_1 = 0.921$, $\lambda_2 = -0.931$ and $\mu_2 = 0.932$.

The FOPID adaptive parameters k_1^P, k_2^P, k_1^I, k_2^I, k_1^D, k_2^D updated based on the gradient method investigated in section 4.3, Fig. 5.

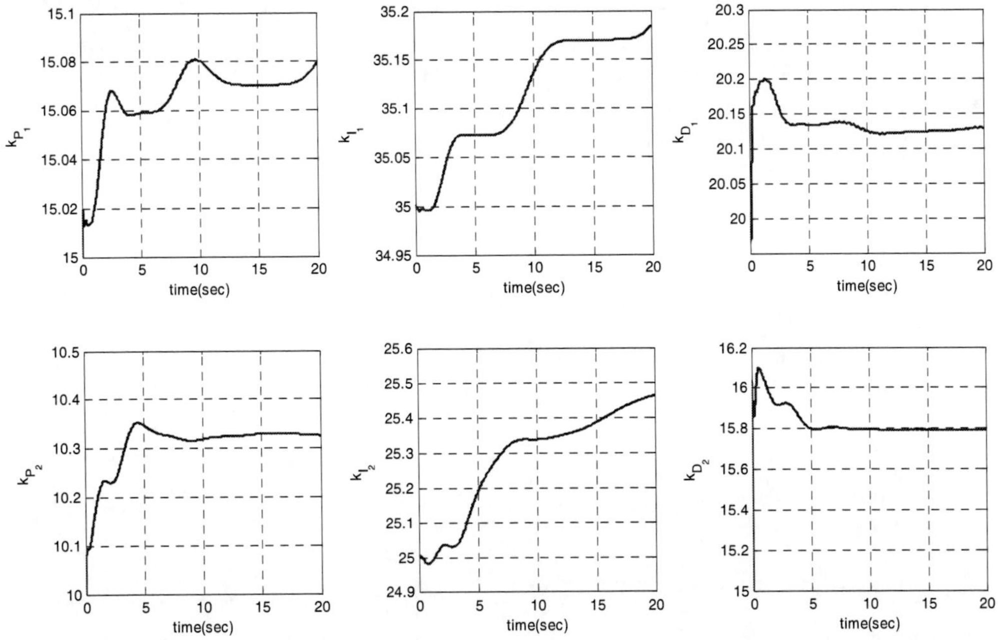

Figure 5. Adaptive FOPID control parameters.

Tracking action of $x_1^d = 0.5\sin(\frac{\pi}{10}t)\,m$ and $x_3^d = 2\pi\sin(\frac{\pi}{10}t)\ rad$ by using adaptive FOPID controller has been shown in Fig. 6, where control signal and tracking error related Figs are also attached.

FOSMC: Here GA will be used to minimize the cost function in (6.2) and find an optimum value of the FOSMC controller. The GA parameters specified as follow

Population size = 150,	Crossover probability = 0.7,
Generations = 90,	Mutation probability =0.02,

k_1^s and k_2^s belongs to $[0, 10]$.

These are chosen from the author experience without losing the generality. Let $w_3 = 1$ and optimal parameters of the FOSMC with fractional PDD$^{\alpha}$ are obtained with GA, $k_1^s = 6.981$ and $k_2^s = 3.875$, and the others parameters are chosen as $\lambda_1 = 10$, $\lambda_2 = 10$, $\alpha = 0.78$.

Simulation result pertinent to FOSMC scheme has shown in Fig. 6.

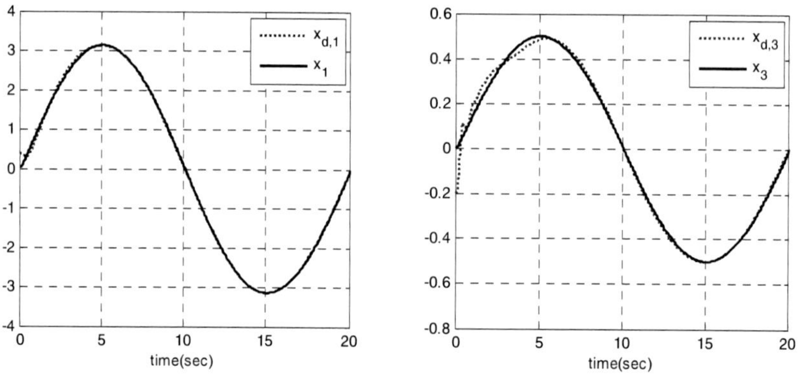

(a) Tracking action of desired state ($x_1(t), m.$, $x_3(t), rad.$)

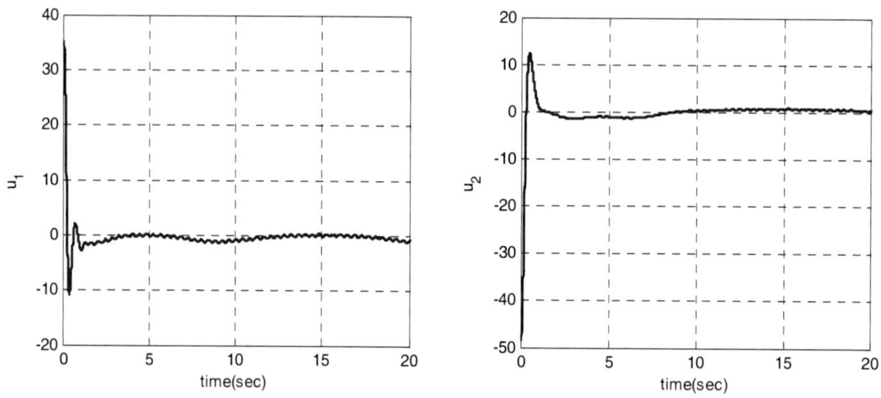

(b) Control effort (u_1(t),N. and u_2(t),N.m)

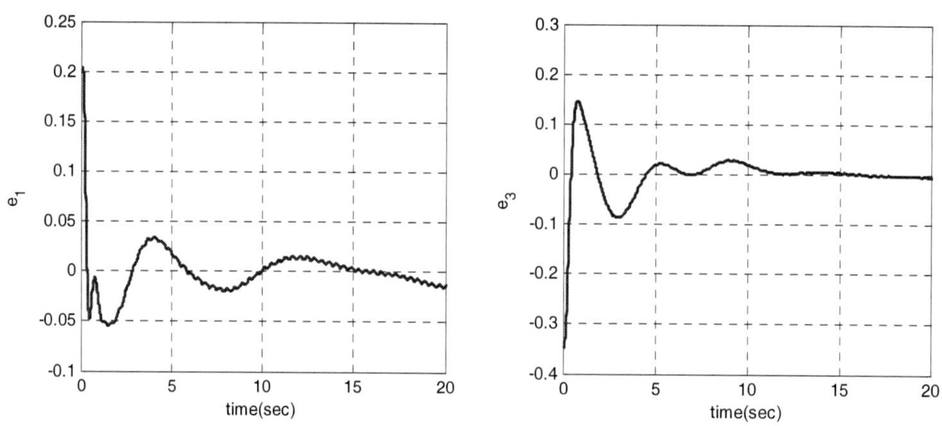

(c) Tracking Errors ($e_1(t), m.$, $e_3(t), rad.$)

Figure 6. Adaptive FOPID.

Fast tracking response is observed by employing the proposed PDD$^{\alpha}$ FOSMC, Fig 7 (a). Also the chattering of u_1 and u_2, by using saturation function instead of signum function [7], are shown minimized in Fig. 7 (b) . Accordingly the Zeno phenomenon (chattering with infinite frequency around the sliding surface) is suppressed by using saturation function instead of signum function in (5.11), Fig. 8.

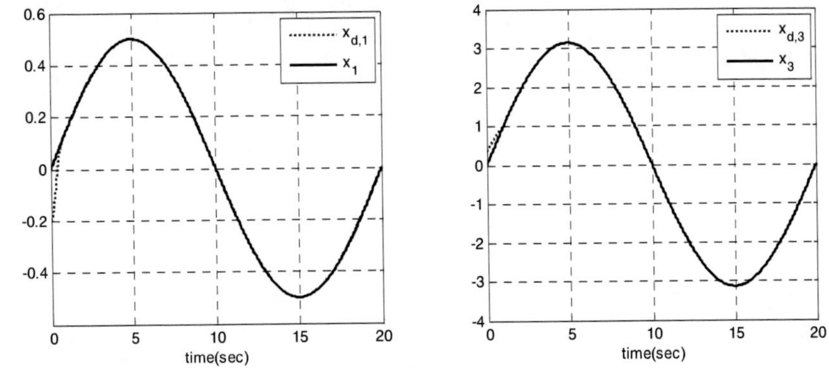

(a) Tracking action of desired state ($x_1(t), m.$, $x_3(t), rad.$)

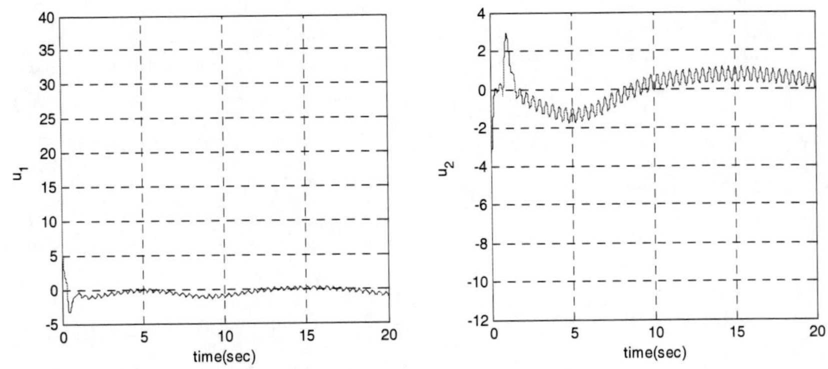

(b) Control effort (u_1(t),N. and u_2(t),N.m)

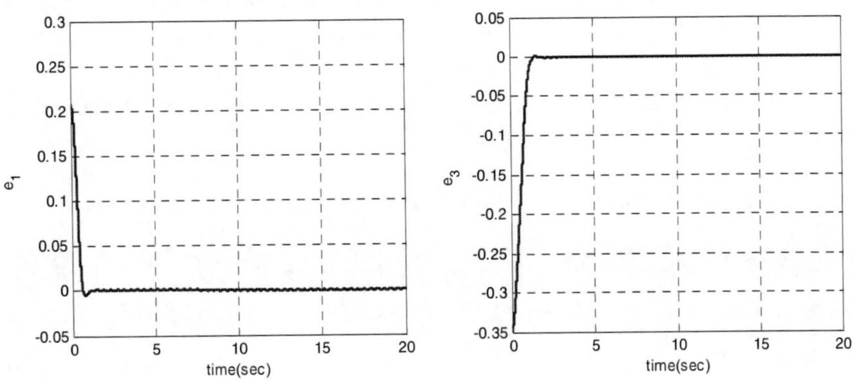

(c) Tracking Errors ($e_1(t), m.$, $e_3(t), rad.$)

Figure 7. FOSMC

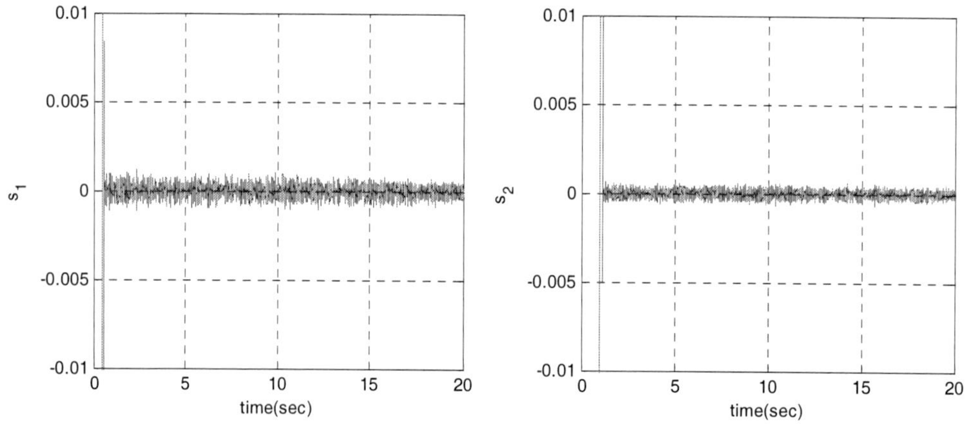

(b) Signum function

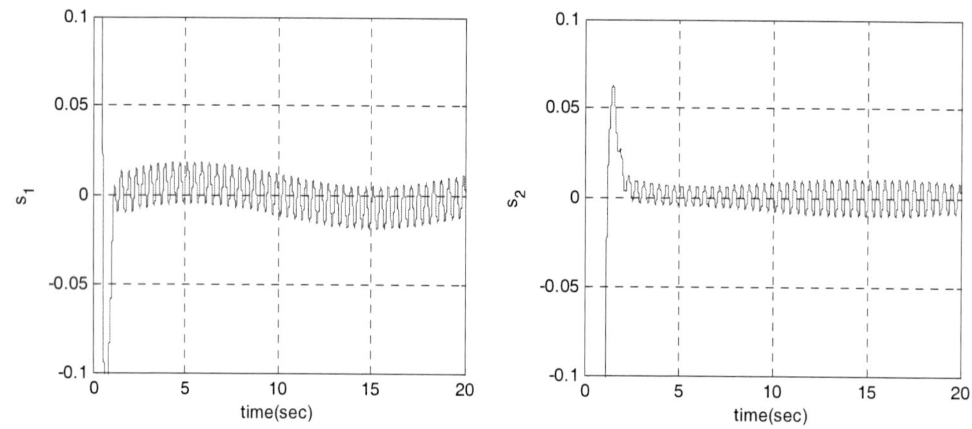

(b) Saturation function

Figure 8. Chattering effect.

Comparison: From Table 1, one can readily see that the absolute errors ($E_{1,abs}$, $E_{3,abs}$) and absolute control efforts ($U_{1,abs}$, $U_{2,abs}$) of employed PDD$^{\alpha}$ FOSMC are less than adaptive FOPID. Also, simulation results verify results, by demonstrating a better tracking performance of a nonlinear system and robustness against model parameter uncertainty by PDD$^{\alpha}$ FOSMC.

Table 1. Results of fractional order controller's performances

Controller	$E_{1,abs}$	$E_{3,abs}$	$U_{1,abs}$	$U_{2,abs}$
Adaptive FOPID	1.35×10^{-2}	2.14×10^{-2}	0.8173	1.0956
PDD$^{\alpha}$ FOSMC	4.1×10^{-3}	1.11×10^{-2}	0.6055	0.7163

8. Conclusion

A trajectory control of a robot manipulator using adaptive FOPID controller and sliding mode control with fractional PDD$^{\alpha}$ surface has been studied in this chapter. Three adaptive FOPID control gains in (4.5), k_k^P, k_k^I, and k_k^D, are adjustable parameters and updated online with an adequate adaptation mechanism and the λ_k and μ_k determined offline, using genetic algorithm (GA). The stability of the proposed adaptive FOPID control system has guaranteed under external disturbance using supervisory controller and Lyapunov theory. Beside this, fractional sliding mode control is the other issue studied in this chapter. In this method a fractional PDD$^{\alpha}$ sliding surface has been chosen and the pertinent control action derived and shown that starting from any initial conditions of the state dynamic, system outputs will asymptotically track the desired trajectories with a fast and robust response. Simulation results verify that the proposed fractional PDD$^{\alpha}$ sliding mode controller provides better tracking and robustness capabilities than Adaptive FOPID.

References

[1] W. Deng, C. Li, Chaos synchronization of the fractional Lü system, *Physica A* **353** (2005), 61–72.

[2] H. Delavari, R. Ghaderi, A. Ranjbar, S. Momani, Fuzzy fractional order sliding mode controller for nonlinear systems, *Communication in Nonlinear Science and Numerical Simulation* **15** (2010) 963–978.

[3] M.W. Spong, M. Vidyasagar, Robot Dynamics and Control, Wiley, New York, 1989.

[4] Green, A., and Sasiadek, J.Z., Dynamics and trajectory tracking control of a two-link robot manipulator, *J. Vibr. Control* **10** (10) 1415–1440, 2004.

[5] Green, A., Sasiadek, J.Z., Fuzzy and optimal control of a two-link flexible manipulator, in: *Proceedings of the IEEE/ASME International Conference on Advanced Intelligent Mechatronics*, pp. 1169–1174, 2001.

[6] S.H. Huh, Z. Bien, Robust sliding mode control of a robot manipulator based on variable structure-model reference adaptive control approach, *IET Control Theory Applications*, vol. 1, 1355–1363, 2007.

[7] J.J.E Slotine, W. Li, *Applied Nonlinear Control*, Prentice Hall, United States of America, 1991.

[8] M. Zeinali, L. Notash, Adaptive sliding mode control with uncertainty estimator for robot manipulators, *Mechanism and Machine Theory*, vol.45, 80–90, 2010.

[9] O. Barambones, V. Etxebarria, Robust neural control for robotic manipulators, *Automatica* **38** 235–242, 2002.

[10] B.K. Yoo, W.C. Ham, Adaptive control of robot manipulator using fuzzy compensator, *IEEE Trans. Fuzzy Syst.* **8** 186–198, 2000.

[11] H. D. Patiño, R. Carelli, B. R. Kuchen, Neural Networks for Advanced Control of Robot Manipulators, *IEEE Transactions on Neural Networks*, vol.13, 345-354, 2002.

[12] Podlubny, Fractional-order systems and PI$^{\lambda}$D$^{\mu}$ controllers, *IEEE Transactions on Automatic Control* **44** (1999) 208–214.

[13] D. Valerio, Fractional robust system control, Ph.D. Thesis, University of Lisbon (P), 2005.

[14] B.M. Vinagre, C.A. Monje, A.J. Calderon, J.I. Suarez, Fractional PID controllers for industry application. A brief introduction, *Journal of Vibration and Control* **13** (2007) 1419–1429.

[15] A.J. Calderon, B.M. Vinagre, V. Feliu, Fractional order control strategies for power electronic buck converters, (2006) *Signal Processing*, **86** (10), pp. 2803-2819.

[16] B. M. Vinagre, I. Petráš, I. Podlubny, Y. Q. Chen, Using Fractional Order Adjustment Rules and Fractional Order Reference Models in Model-Reference Adaptive Control, *Nonlinear Dynamics*, 2002-07-01, Springer Netherlands.

[17] C.A. Monje, B.M. Vinagre, V. Feliu, Y. Q. Chen, Tuning and auto-tuning of fractional order controllers for industry applications, (2008) *Control Engineering Practice*, **16** (7), pp. 798-812.

[18] Podlubny, Fractional Differential Equations. Academic Press, San Diego (1999).

[19] Podlubny, Geometric and Physical Interpretation of Fractional Integration and Fractional Differentiation, fractional calculus and applied analysis, Vol. 5, pp 367-386, 2002.

[20] G.L. Bullock, A geometric interpretation of the Riemann-Stieltjes integral. *American Mathematical Monthly* **95**, No 5 (May 1988), 448-455.

[21] K. Miller, B. Ross, *An Introduction to the Fractional Calculus and Fractional Differential Equations*, Wiley, New York, 1993.

[22] K. Oldham, J. Spanier, *The Fractional Calculus*, Academic Press, New York.

[23] I.Podlubny, L. Dorcak, I. Kostial, On Fractional Derivatives, Fractional-Order Dynamic Systems and $PI^{\lambda}D^{\mu}$ controllers, *Proceedings of the 36th Conference on Decision & Control,* San Diego, California, USA, December 1997.

[24] D. Vale´rio, J.S. Costa, Tuning of fractional PID controllers with Ziegler–Nichols-type rules, Signal Processing, Vol. 86,(2006) pp. 2771–2784.

[25] M. Zamani, M. Karimi-Ghartemani, N. Sadati, M. Parniani, Designof fractional order PID controller for an AVR using particle swarm optimization, *Control Engineering Practice* **17**(2009)1380–1387.

[26] S. Jesus, J.A. Tenreiro Machado, Fractional control of heat diffusion systems, *Nonlinear Dyn* (2008) 54: 263–282.

[27] J. Domingues, D. Valerio, J. S. Costa, Rule-based fractional control of an irrigation canal, *ASME Journal of Computational and Nonlinear Dynamics*, 2010. Accepted.

[28] Y. Luo, Y. Q. Chen, C. Y. Wang, Y. G. Pi, Tuning fractional order proportional integral controllers for fractional order systems, *Journal of Process Control* **20** (2010) 823–831.

[29] F. Padula, A. Visioli, Tuning rules for optimal PID and fractional-order PID controllers, *J. Process Control* (2010), doi:10.1016/j.jprocont.2010.10.006.

[30] W.D. Chang, J. J. Yan, Adaptive robust PID controller design based on a sliding mode for uncertain chaotic systems, *Chaos Solitons and Fractals* **26**(1), 167–175 (2005).

[31] Wang LX, *Adaptive fuzzy systems and control: design and stability analysis*, New Jersey: Prentice-Hall; 1994.

[32] Wang LX, A course in fuzzy systems and control, New Jersey: Prentice-Hall; 1997.

[33] H. Delavari, R. Ghaderi, A. Ranjbar, S. Momani, Fractional order control of a coupled tank, *Nonlinear Dynamics* Springer, 2010, Volume 61, Number 3, 383-397.

[34] H. Delavari, M.R. Faieghi, A. Ranjbar, Fractional Order Sliding Mode Controller for Inverted-Pendulum, *4th IFAC workshop on fractional derivative and its application*, Badajoz, Spain, 2010.

[35] H. Delavari, A. Ranjbar N., R. Ghaderi, S. Momani, Fractional Sliding Mode Control for *Synchronization of Chaotic Systems, Fractional Differentiation and its Applications*, Ankara, Turkey, 05 - 07 November, 2008.
[36] M.O. Efe, Fractional order sliding mode control with reaching law approach, *Turk J Elec Eng & Comp Sci*, Vol.18, No.5, 2010.
[37] V.I. Utkin, Sliding modes in optimization and control problems. New York: Springer-Verlag; 1992.
[38] Goldberg DE. Genetic algorithms in search, optimization and machine learning, San Mateo (CA): Addison-Wesley; 1989.
[39] P.R. Ouyang, W.J. Zhang, M. M. Gupta, An adaptive switching learning control method for trajectory tracking of robot manipulators, *Mechatronics* **16** (2006) 51–61.
[40] Y. Stepanenko, Y. Cao, Ch.Y. Su, Variable Structure Control of Robotic Manipulator with PID Sliding Surfaces, *International Journal of Robust and Nonlinear Control*, vol. 8, 79*Đ*90 (1998).
[41] lyas Eker, Second-order sliding mode control with experimental application, *ISA Transactions*, **49** (2010) 394-405.
[42] H. Delavari, R. Ghaderi, A. Ranjbar N., S.H. HosseinNia, S. Momani, *Adaptive Fractional PID Controller for Robot Manipulator*, accepted for FDA2010, University of Extremadura, Badajoz, Spain.

In: Robotics: State of the Art and Future Trends
Editors: G. Legnani and I. Fassi
ISBN: 978-1-62100-403-5

Chapter 8

Visual Control of Robotic Manipulators: Fundamentals

Paulo J. S. Gonçalves [*] ***and Pedro M. B. Torres*** [†]
[*,†] Polytechnic Institute of Castelo Branco, School of Technology,
Av. Empresário, 6000-767 Castelo Branco, Portugal
[*] Technical University of Lisbon, IDMEC / IST,
Av. Rovisco Pais, 1049-001 Lisboa, Portugal

Abstract

Robotic Manipulators are highly complex dynamic systems that can use a large amount of sensors to recognize its work environment. Laser, infrared or ultrasound sensors are used to measure distances. Force sensors are used to recognize contact between the robot and other objects. Cameras can be used to visualize the robot's global work environment.

Nowadays, robotic manipulators are used in factories to perform tasks in a specially designed work environment. Force and distance sensors are mainly used in these setups. Introducing cameras in this setup could, in a hopefully near future, make robot manipulators completely autonomous when performing tasks in unknown, or unable, to be precisely, modelled environments.

In the past decades, the use of computer vision and image processing methods has been constrained due to its high demanding computational costs. In the late 70's dozens of seconds were needed to extract simple object characteristics, e.g., edges. At that time, vision was used to detect the object position in the 3D space. This position was then used by the robot controller to drive the robot to it. Since the early 90's, due to the increasing capacity of computers, the vision sensor data can be used as sensor feedback to close the control loop. Nowadays the camera manufacturers can supply fast cameras that can achieve sample times up to one millisecond, i.e. comparable to distance or force sensors. This achievement, followed by the recent developments in Visual Control, opens a wide area for robotic applications. Visual Control, also called Visual Servoing, has as its main objective the increase of flexibility and precision of the robot manipulator, when performing tasks. The vision sensor can give the robot

[*] E-mail address: paulo.goncalves@ipcb.pt
[†] E-mail address: pedrotorres@ipcb.pt

a general overview of its working environment, which can be used as control input. Visual Control aims to cancel the robot's manipulator position error, using as feedback, in the control loop, the data provided by the vision sensor.

The vision data can be directly or indirectly used in the control loop. The large majority of Visual Servoing approaches use an inner robot controller, with joint variables feedback, to stabilize the robot around the control action arriving from the outer visual control loop. The vision data, i.e., visual features, that characterize the object to be manipulated can be used directly in the control loop, e.g., the 2D pixels of the object or the object image moments. Visual features can also be used indirectly, e.g., the 3D data points of the object that describes its pose. 2D and 3D visual features can be combined to achieve the called hybrid features. The application of the above presented visual features derives different control laws, which are called image-based Visual Servoing, position-based Visual Servoing and hybrid Visual Servoing. In these approaches, the model of the visual motor interaction is already known. More recently the visual motor interaction has been estimated analytically or by learning, opening a new area for Visual Control of uncalibrated robotics systems.

In this chapter, Visual Servoing approaches that can be applied to robot manipulators, are presented and compared, while the corresponding control laws are derived. The first visual control laws presented, describe the model of the interaction between the object (seen by the camera) and robot motions, i.e., visual motor interaction. Approaches where the visual motor interaction is estimated are also presented.

The first section of the chapter presents Visual Servoing approaches. These approaches are supported in screw theory, and the mathematical formulation of visual servoing will start from this background.

In the next section, Visual Servoing approaches are presented. Within the Visual Servoing framework there are two main categories to obtain the Visual Motor Interaction: based on a model already known; based on an estimated model. The first category can be divided into how the image features are obtained: Image Based; Position Based; Hybrid. The second category can be divided into how the model is estimated: analytically, using classical approaches like Broyden's update; by learning using fuzzy modeling or neural networks.

The last section of the chapter presents a brief review of the applications developed worldwide, using Visual Servoing approaches to control robotic manipulators, both in industry and academia. Experimental results of Learning approaches implementation on real robotic manipulators are presented, allowing the reader to identify real implementation issues regarding the visual frame rate, the error introduced by the visual data, and its implications to robot control.

Keywords: Visual Servoing, Robotic Manipulators, Fuzzy Control, Fuzzy Modeling.
AMS Subject Classification: 65D, 68T, 93C.

PART I: FUNDAMENTALS

1. Introduction

Visual Servo Control, *visual servoing* [27,29], defines closed loop robot control using cameras to extract visual information from the robot's work environment. From the desired robot position and the visual information obtained from the camera, the visual servo control law gives the desired velocities that move the robot. The use of vision systems in robotics aims to increase flexibility and precision of robots. One of the first works that used vision to control a robot was presented by Shirai and Inoue [45], almost 40 years ago. The role of computer vision is to provide the robot with information about its work environment, to be used in the control loop.

Information from the robot's work environment can be used in control by two ways. The first approach called *Open-Loop-Robot-Control*, separates the vision part from the control part. The start and final positions of the robot are obtained using computer vision (pose estimation algorithms) and the robot's movement is controlled only on the joint or cartesian space. This approach needs camera calibration and the model of the object to manipulate. The second approach called *visual servoing*, uses directly or indirectly visual information to close the robot control loop. In this approach it is often used an inner controller, with joint feedback, to stabilize the robot around the control action that arrives from the outer vision control loop. The large majority of visual servoing applications are based on this approach.

Direct visual information uses pixel information, e.g., 2D points describing features of interest on the object to manipulate. Indirect visual information implies extraction of relevant object characteristics from 2D coordinates in the image space, e.g., the 3D pose of the object related to the camera frame.

Typical visual servoing applications can be classified as i) *positioning* the robot end-effector to an object; ii) *tracking* an object, e.g., keeping a constant distance from the object to the robot.

The object to manipulate is an important part of the visual control loop and its characterization is of special importance to formulate the control problem. This visual information can be obtained from one or more cameras looking at the robot and the object *eye-to-hand*, or being placed in the end-effector *eye-in-hand.* Object characteristics can be obtained: i) using 2D data expressed directly in pixel coordinates from the image space, ii) using 3D data to obtain the object pose, iii) fusing 2D and 3D data. To use these object characteristics, visual features, in the control loop, it is mandatory to obtain a relation between the feature space and the world, or camera frames. This relation (interaction model) has been over the years, an exciting field of research, and it is the base of visual servoing approaches. The first and second approaches are called *image-based visual servoing* and *position-based visual servoing*, respectively [29]. The third approach is called *hybrid visual servoing* [24], being the first reported work presented in [38].

Interaction models between the visual features of the object and camera coordinates frames can be grouped in two. The models known a priori, using analytical relations are in the first group. The models estimated numerically are in the second group. These models are used to generate movement commands to the robot and are called, amongst other terms [24], by image Jacobian. The other important relation is between the visual features and the robot joint coordinates. To build this model, the robot Jacobian and the relation between the camera and end-effector frames must be known. These relations will be the focus of the next sections.

2. Screw Theory in Visual Servo Control

Let us consider the case in which a camera is attached to the robot end-effector watching an object. The object is represented in an image, by a set of 2D points, and from these a set of image features can be obtained. If the robot moves, each image features moves across the image. A feature may be image coordinates (2D features) of one point of the object or a set of six coordinates (3D features) representing the pose (position and orientation) of the object with respect to the camera.

Visual servo control is centered in the mathematical relation of the variation of the object features, obtained from the image, and the robot velocities, joint or end-effector. This relation is often called *Jacobian* or *Interaction Matrix*. The Jacobian can be decomposed in the Image Jacobian, and the Robot Jacobian. The Image Jacobian relates the variation of object features in the image and the camera velocities. The Robot Jacobian relates the end-effector velocities to the joint velocities. Note that when the camera position is different from the end-effector position, a relation between these two coordinate frames should be obtained. These relations are obtained using the screw theory, for 2D features and 3D features, suited respectively for image-base and position-based visual servo control.

2.1. Image-based Visual Servo Control

For eye-in-hand image-based visual servoing, it is necessary to derive the relation between the velocity of the robot and of one image point. The corresponding Jacobian will be constructed as follows. Considering a generic point in space, $\mathbf{x}_{3D}$ depicted in figure 1, related to the coordinate frame $\mathbf{F}_c$, that is a feature, $\mathbf{s}$, of an object:

$$\mathbf{x}_{3D} = \begin{bmatrix} X & Y & Z \end{bmatrix}^T \tag{1}$$

Considering that the frame $\mathbf{F}_c$ is moving with velocity $\mathbf{v} = \begin{bmatrix} \upsilon_x\ \upsilon_y\ \upsilon_z\ \omega_x\ \omega_y\ \omega_z \end{bmatrix}^T$, i.e., the first three for translational velocities and the last three for angular velocities. The point velocity $\dot{\mathbf{x}}_{3D}$, is given by:

$$\dot{\mathbf{x}}_{3D} = \begin{bmatrix} -\mathbf{I}_3 & S(\mathbf{x}_{3D}) \end{bmatrix} \cdot \mathbf{v} \tag{2}$$

where $S(\mathbf{x}_{3D})$ is the skew-symmetric matrix associated to vector $\mathbf{x}_{3D}$ defined by:

$$S(\mathbf{x}_{3D}) = \begin{bmatrix} 0 & -Z & Y \\ Z & 0 & -X \\ -Y & X & 0 \end{bmatrix} \tag{3}$$

Without loss of generality the distance from the camera frame $\mathbf{F}_c$ and the image plane π, can be considered unitary, and from figure 1, it can be observed that $tan(\alpha) = \frac{Y}{Z} = \frac{y}{1}$ and that $tan(\beta) = \frac{X}{Z} = \frac{x}{1}$. The image point $\mathbf{m}$ can be obtained by the following:

$$\mathbf{m} = \begin{bmatrix} x & y \end{bmatrix}^T = \begin{bmatrix} \frac{X}{Z} & \frac{Y}{Z} \end{bmatrix}^T \tag{4}$$

The image point $\mathbf{m}$, in metric coordinates, can also be computed from the coordinates of the point measured in pixels and the camera intrinsic parameters [20].

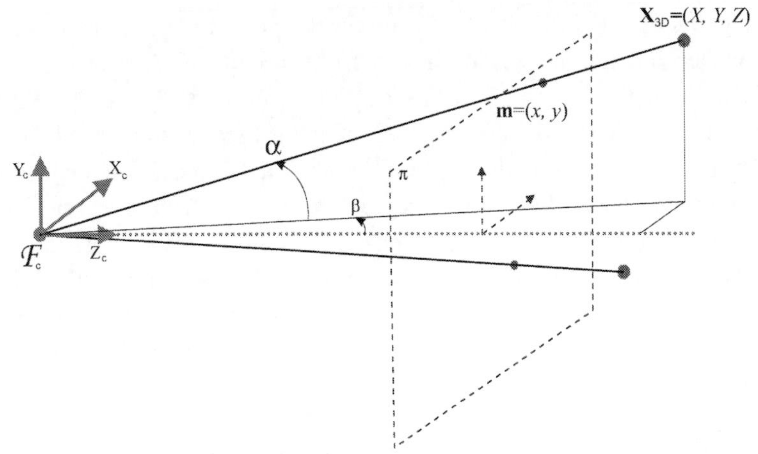

Figure 1. Image formation from the Pin Hole Camera Model.

Differentiating **m** in time, its velocity is obtained,

$$\dot{\mathbf{m}} = \begin{bmatrix} \dot{x} & \dot{y} \end{bmatrix}^T = \begin{bmatrix} \frac{\dot{X}\cdot Z - X\cdot\dot{Z}}{Z^2} \\ \frac{\dot{Y}\cdot Z - Y\cdot\dot{Z}}{Z^2} \end{bmatrix} = \frac{1}{Z}\cdot\begin{bmatrix} 1 & 0 & -\frac{X}{Z} \\ 0 & 1 & -\frac{Y}{Z} \end{bmatrix}\cdot\begin{bmatrix} \dot{X} \\ \dot{Y} \\ \dot{Z} \end{bmatrix} \tag{5}$$

$$\dot{\mathbf{m}} = -\mathbf{J}_\upsilon(\mathbf{m})\cdot\dot{\mathbf{x}}_{3D} \tag{6}$$

$$\mathbf{J}_\upsilon(\mathbf{m}) = \frac{1}{Z}\begin{bmatrix} -1 & 0 & \frac{X}{Z} \\ 0 & -1 & \frac{Y}{Z} \end{bmatrix} \tag{7}$$

Knowing that $\mathbf{x}_{3D} = Z\cdot\begin{bmatrix} x & y & 1 \end{bmatrix}^T$ and using equations (2) and (6), the following relation between the velocities of the moving frame **v** and the image point in the image, can be obtained:

$$\dot{\mathbf{m}} = -\mathbf{J}_\upsilon(\mathbf{m})\cdot\begin{bmatrix} -\mathbf{I}_3 & S(\mathbf{x}) \end{bmatrix}\cdot\mathbf{v} \tag{8}$$

$$\dot{\mathbf{m}} = \begin{bmatrix} -\frac{1}{Z} & 0 & \frac{X}{Z^2} & x\cdot y & -\left(1+x^2\right) & y \\ 0 & -\frac{1}{Z} & \frac{Y}{Z^2} & 1+y^2 & -x\cdot y & -x \end{bmatrix}\cdot\mathbf{v} \tag{9}$$

It can be concluded, that:

$$\dot{\mathbf{m}} = \begin{bmatrix} \mathbf{J}_\upsilon(\mathbf{m}) & \mathbf{J}_{(\upsilon,\omega)}(\mathbf{m}) \end{bmatrix}\cdot\mathbf{v} \tag{10}$$

Considering that the object image feature, **s**, is the 2D point, **m**, of the object, i.e., $\mathbf{s} = \mathbf{m}$, and that the camera frame $\mathbf{F}_c$ is moving, $\mathbf{v} = {}^c\dot{\mathbf{r}}_c$, equation (10) can be rewritten. The following equation defines the image Jacobian, $\mathbf{J}_i$, relating the features velocities, $\dot{\mathbf{s}}$, to the camera velocities, ${}^c\dot{\mathbf{r}}_c$:

$$\dot{\mathbf{s}} = \mathbf{J}_i\cdot{}^c\dot{\mathbf{r}}_c = \begin{bmatrix} -\frac{1}{Z} & 0 & \frac{x}{Z} & x\cdot y & -\left(1+x^2\right) & y \\ 0 & -\frac{1}{Z} & \frac{y}{Z} & 1+y^2 & -x\cdot y & -x \end{bmatrix}\cdot{}^c\dot{\mathbf{r}}_c \tag{11}$$

In (11) the image Jacobian, $\mathbf{J}_i$, is obtained from the 2D point $\mathbf{m}$ and its depth Z, related to the camera frame. Depth is not measured directly from the image. Amongst other techniques it can be calculated using structure from motion techniques [13] (where 3D information can be obtained from two or more consecutive images). It can also be used a more practical scheme, i.e., use a constant value, the depth Z^*, obtained at the desired position of the robot. To move a robot towards the desired position using visual servo control, the robot usually receives the end-effector velocities in the end-effector frame, ${}^e\dot{\mathbf{r}}_e$, or its joint velocities $\dot{\mathbf{q}}$. Since the feedback for the visual control law is in the image space, the camera velocities in the camera frame, ${}^c\dot{\mathbf{r}}_c$, must be used. Assuming that the camera is attached to the end-effector, and using the transformation matrix ${}^c\mathbf{W}_e$, depicted in (12) the end-effector to camera transformation is obtained.

$$ {}^c\dot{\mathbf{r}}_c = {}^c\mathbf{W}_e \cdot {}^e\dot{\mathbf{r}}_e \tag{12} $$

The transformation matrix, ${}^c\mathbf{W}_e$, contains the translation , ${}^c\mathbf{t}_e$, and the rotation, ${}^c\mathbf{R}_e$, between the camera and end-effector frames:

$$ {}^c\mathbf{W}_e = \begin{bmatrix} \mathbf{I}_{3\times3} & S({}^c\mathbf{t}_e) \\ \mathbf{0}_{3\times3} & \mathbf{I}_{3\times3} \end{bmatrix} \begin{bmatrix} {}^c\mathbf{R}_e & \mathbf{0}_{3\times3} \\ \mathbf{0}_{3\times3} & {}^c\mathbf{R}_e \end{bmatrix} = \begin{bmatrix} {}^c\mathbf{R}_e & S({}^c\mathbf{t}_e)\cdot{}^c\mathbf{R}_e \\ \mathbf{0}_{3\times3} & {}^c\mathbf{R}_e \end{bmatrix} \tag{13} $$

In, ${}^c\mathbf{W}_e$, the rotation, ${}^c\mathbf{R}_e$, and the translation ${}^c\mathbf{t}_e$, are obtained when the camera is rigidly attached to the end-effector.

During visual servo control, the robot receives from the visual control law, as a control action, the desired joint velocities, $\dot{\mathbf{q}}$. The relation between the end-effector velocities and the joint velocities must be known. It is described in (14) where the Robot Jacobian, $\mathbf{J_R}$, is calculated in the end-effector frame:

$$ {}^e\dot{\mathbf{r}}_e = {}^e\mathbf{J_R}(\mathbf{q}) \cdot \dot{\mathbf{q}} \tag{14} $$

The relation between the object's features in the image and, the end-effector coordinates and joint coordinates, is described in (15) and in (16), respectively. The Jacobian matrix for image-based visual servo control is defined by comparison between (16) and (17). In the next sections, these equations are used to obtain the visual control law.

$$ \dot{\mathbf{s}} = \mathbf{J}_i \cdot {}^c\mathbf{W}_e \cdot {}^e\dot{\mathbf{r}}_e \tag{15} $$

$$ \dot{\mathbf{s}} = \mathbf{J}_i \cdot {}^c\mathbf{W}_e \cdot {}^e\mathbf{J_R} \cdot \dot{\mathbf{q}} \tag{16} $$

$$ \dot{\mathbf{s}} = \mathbf{J}_{2D} \cdot \dot{\mathbf{q}} \tag{17} $$

2.2. Position-based Visual Servo Control

The feature considered by the position-based visual servo algorithm is a vector of six parameters representing the pose of one object with respect to the camera attached to the robotic manipulator end-effector. It is necessary to calculate the jacobian which describes the relation between the robot motion and the image feature.
The pose of the object, ${}^c\mathbf{T}_o$, as seen by the camera(s) is related to the current camera(s) frame $\mathbf{F}_c$, as depicted in figure 2. In figure 2 are also present other coordinate frames, $\mathbf{F}_0$, $\mathbf{F}_o$, $\mathbf{F}_{c^*}$, respectively the world, object and desired camera(s) frames. This feature can

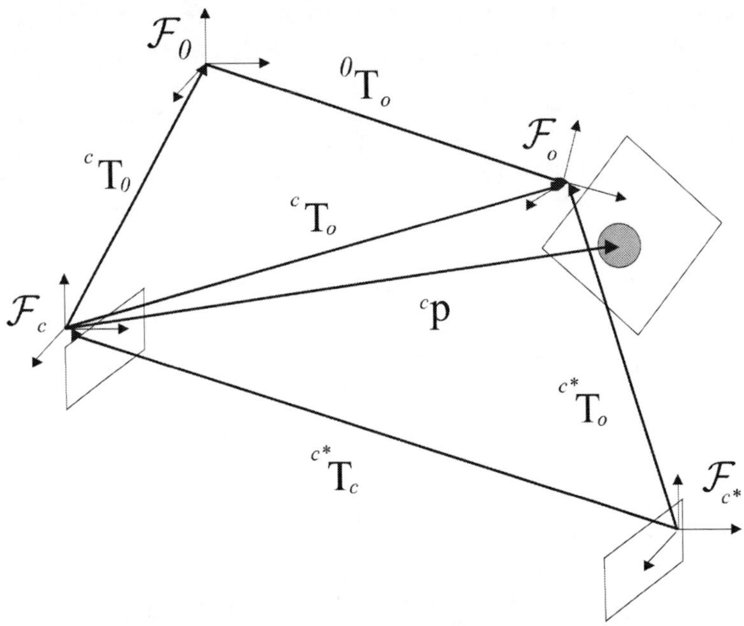

Figure 2. Possible object features for position-based visual servoing.

be used in case the camera(s) is(are) located in 3D space looking the robot and object, or when it is located in the end-effector looking the object. In figure 2 it is also presented other possibility to choose the object feature based in its pose, i.e., the current pose of the camera related to the camera pose in the desired position, ${}^{c^*}\mathbf{T}_c$, that can be used when the camera is in the end-effector looking the object. Other feature is also possible to choose from, ${}^c\mathbf{T}_0$. In this text the first feature is used, i.e., ${}^c\mathbf{T}_o$.

$$ {}^c\mathbf{T}_o = \begin{bmatrix} {}^c\mathbf{R}_o & {}^c\mathbf{t}_o \\ \mathbf{0}_{1\times 3} & 1 \end{bmatrix} \Rightarrow \mathbf{s} = \begin{pmatrix} {}^c\mathbf{t}_o^T & {}^c\mathbf{u}_o^T\theta \end{pmatrix}^T \tag{18} $$

where rotation ${}^c\mathbf{R}_o$ is expressed by the vector ${}^c\mathbf{u}_o\theta$, where ${}^c\mathbf{u}_o$ is the rotation axis and θ the amplitude of the rotation which superimpose the object frame to the camera frame. (19) and (20) are both defined in [20].

$$ \theta = \arccos\left(\frac{1}{2}\,Trace({}^c\mathbf{R}_o)\right) \tag{19} $$

$$ {}^c\mathbf{u}_o\theta = \frac{1}{2\cdot\sin(\theta)}\begin{pmatrix} r_{32}-r_{23} \\ r_{13}-r_{31} \\ r_{21}-r_{12} \end{pmatrix} \text{ ,if } \begin{cases} 0 \le \theta < \pi \\ \pi < \theta < 2\pi \end{cases} \tag{20} $$

where r_{ij} are the elements of the rotation matrix ${}^c\mathbf{R}_o$.
In the following, the camera is looking at the object. The Jacobian between the velocity of the object features and the joint velocities is obtained in two parts. The first part aims to

obtain the three translational parameters and the second one the three rotational parameters. The image Jacobian, relatively to the translation ${}^{c}\mathbf{t}_{o}$, is defined by:

$$ {}^{c}\dot{\mathbf{t}}_{o} = \left[\begin{array}{cc} -\mathbf{I}_{3\times 3} & S({}^{c}\mathbf{t}_{o}) \end{array}\right] \cdot \mathbf{v} \tag{21} $$

Considering the rotation, defined by the vector ${}^{c}\mathbf{u}_{o}\,\theta$:

$$ (\dot{{}^{c}\mathbf{u}_{o}\,\theta}) = \left[\begin{array}{cc} \mathbf{0}_{3\times 3} & \mathbf{L}_{\omega} \end{array}\right] \cdot \mathbf{v} \tag{22} $$

following [37] the rotational Jacobian, $\mathbf{L}_{\omega}$, can be obtained,

$$ \mathbf{L}_{\omega}\left({}^{c}\mathbf{u}_{o},\theta\right) = \mathbf{I}_{3\times 3} - \frac{\theta}{2}\cdot S({}^{c}\mathbf{u}_{o}) + \left(1 - \frac{\mathrm{sinc}\,(\theta)}{\mathrm{sinc}^{2}\left(\frac{\theta}{2}\right)}\right)\cdot\left(S({}^{c}\mathbf{u}_{o})\right)^{2} \tag{23} $$

where,

$$ \mathrm{sinc}(\theta) = \begin{cases} 1 & \text{if } \theta = 0 \\ \frac{\sin(\theta)}{\theta} & \text{if } \theta \neq 0 \end{cases} \tag{24} $$

The global Jacobian used in position-based visual control, $\mathbf{J}_{3D}$, for the feature ${}^{c}\mathbf{T}_{o}$, is defined by:

$$ \dot{\mathbf{s}} = \left(\begin{array}{c} {}^{c}\dot{\mathbf{t}}_{o} \\ (\dot{{}^{c}\mathbf{u}_{o}\,\theta}) \end{array}\right) = \left[\begin{array}{cc} -\mathbf{I}_{3\times 3} & S({}^{c}\mathbf{t}_{o}) \\ \mathbf{0}_{3\times 3} & \mathbf{L}_{\omega} \end{array}\right]\cdot \mathbf{v} \tag{25} $$

$$ \dot{\mathbf{s}} = \mathbf{J}_{3D}\cdot \mathbf{v} \tag{26} $$

2.3. Hybrid Visual Servo Control

The feature of the object used in this approach fuses the image and the position based approaches. The object pose used is ${}^{c^*}\mathbf{R}_{c}$, defined in the previous section. The Jacobian is calculated in the remainder of this section.

To the 2D information of the point seen in the image $\mathbf{m}$, defined in equation (4), it is added a third coordinate. Defining the variable $z = \ln(Z)$, where $\ln(Z)$ is the natural logarithm of the depth Z, the extended metric coordinates of a image point, $\mathbf{m}_{e}$, are obtained:

$$ \mathbf{m}_{e} = \left[\begin{array}{ccc} x & y & z \end{array}\right]^{T} = \left[\begin{array}{ccc} \frac{X}{Z} & \frac{Y}{Z} & ln(Z) \end{array}\right]^{T} \tag{27} $$

Note that was used $\ln(Z)$ instead of Z as the new feature for the extended metric coordinates. With this choice it was aimed to obtain a third coordinate for translation to add to the previous x and z, and also to obtain a simple Jacobian, as seen by the added third line of (28), $\left[\begin{array}{ccc} 0 & 0 & 1 \end{array}\right]$, when compared to (7).

Differentiating the point $\mathbf{m}_{e}$ over time, its velocity is obtained:

$$ \dot{\mathbf{m}}_{e} = \left[\begin{array}{ccc} \dot{x} & \dot{y} & \dot{z} \end{array}\right]^{T} = \left[\begin{array}{c} \frac{\dot{X}\cdot Z - X\cdot\dot{Z}}{Z^{2}} \\ \frac{\dot{Y}\cdot Z - Y\cdot\dot{Z}}{Z^{2}} \\ \frac{\dot{Z}}{Z} \end{array}\right] = \frac{1}{Z}\cdot\left[\begin{array}{ccc} 1 & 0 & -\frac{X}{Z} \\ 0 & 1 & -\frac{Y}{Z} \\ 0 & 0 & 1 \end{array}\right]\cdot\left[\begin{array}{c} \dot{X} \\ \dot{Y} \\ \dot{Z} \end{array}\right] \tag{28} $$

$$ \dot{\mathbf{m}}_{e} = -\mathbf{L}_{\upsilon}(m_{e})\cdot\dot{\mathbf{x}}_{3D} \tag{29} $$

Using equations (2) and (29), is obtained the following relation between the velocities of the moving frame **v**, where the camera is located, and the velocity of the point in the image:

$$\dot{\mathbf{m}}_e = \begin{bmatrix} -\frac{1}{Z} & 0 & \frac{x}{Z} & x \cdot y & -\left(1+x^2\right) & y \\ 0 & -\frac{1}{Z} & \frac{y}{Z} & 1+y^2 & -x \cdot y & -x \\ 0 & 0 & -\frac{1}{Z} & y & x & 0 \end{bmatrix} \cdot \mathbf{v} \tag{30}$$

where Z is obtained using computer vision approaches [20].
Following, it can be concluded, that:

$$\dot{\mathbf{m}}_e = \begin{bmatrix} \mathbf{L}_{\upsilon}(\mathbf{m}_e) & \mathbf{L}_{(\upsilon,\omega)}(\mathbf{m}_e) \end{bmatrix} \cdot \mathbf{v} \tag{31}$$

It is now needed to include the vector ${}^{c^*}\mathbf{u}_c\theta$ that represents the rotation, as used in equation (22). The result is the image Jacobian for hybrid visual servoing, $\mathbf{J}_{2\frac{1}{2}D}$, that includes six features to control a six-degree freedom robot. The first three for translation and the other three for rotation.

$$\dot{\mathbf{s}} = \begin{pmatrix} \dot{\mathbf{m}}_e \\ (\dot{{}^{c^*}\mathbf{u}_c\theta}) \end{pmatrix} = \begin{bmatrix} \mathbf{L}_{\upsilon}(\mathbf{m}_e) & \mathbf{L}_{(\upsilon,\omega)}(\mathbf{m}_e) \\ \mathbf{0}_{3\times3} & \mathbf{L}_{\omega}\left({}^{c^*}\mathbf{u}_c,\theta\right) \end{bmatrix} \cdot \mathbf{v} \tag{32}$$

$$\dot{\mathbf{s}} = \mathbf{J}_{2\frac{1}{2}D} \cdot \mathbf{v} \tag{33}$$

3. Visual Servoing Approaches

Visual information is nowadays extensively used in robotic applications. The visual information from the work environment can be used to control mobile robots and/or fixed manipulators, performing a given task. In the following sub-sections, both classical and learning based visual servo control are presented, along with their main goals.

3.1. Image Based Visual Servoing

In image based visual servoing, visual features of an object are obtained directly from its image, e.g., 2D points of interest of the object. Only simple algorithms are used to extract object characteristics, e.g., when using simple primitives as points, lines or ellipses [12] More examples of object characteristics, based on image moments are presented in [7].

The global control architecture is shown in figure 3, where the block *Robot inner loop* law is a PD control law [2]. The dashed lines represent the variables needed to compute the total Jacobian, which are x, y, Z and $\mathbf{q}$ in (16).

The robot joint velocities $\dot{\mathbf{q}}$ to move the robot to a predefined point in the image, $\mathbf{s}^*$ are derived from the use of the *Visual control law*. To achieve this goal, the error between the desired and the actual image features position in the image, $\mathbf{e} = \mathbf{s} - \mathbf{s}^*$, should be zero. Further, the image features velocity at the desired position should also be zero, to guarantee that the robot manipulator stops. Thus, the image features error velocity is related to the image features velocity as follows:

$$\dot{\mathbf{e}} = \dot{\mathbf{s}} - \dot{\mathbf{s}}^* = \dot{\mathbf{s}} \tag{34}$$

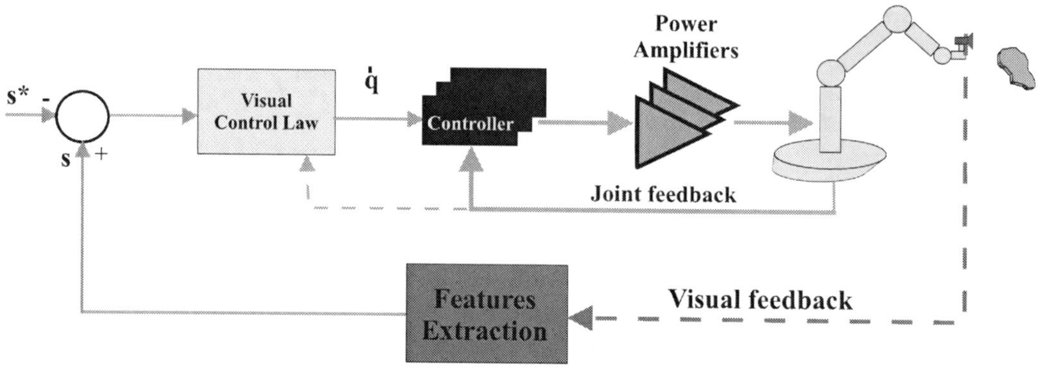

Figure 3. Image Based Visual Servoing.

where it was assumed that the object is static in the desired position, $\dot{\mathbf{s}}^* = 0$.
In this chapter, an exponential decay of the image features error during the servoing is chosen to be applied, as suggested in [12]:

$$\dot{\mathbf{e}} = -\mathbf{K_p} \cdot \mathbf{e} \tag{35}$$

where $\mathbf{K_p}$ is a positive gain, that is used to increase or decrease the decay of the velocity error. From (16), (34) and (35), the proportional visual control law is defined as [19]:

$$\dot{\mathbf{q}} = -\mathbf{K_p} \cdot \hat{\mathbf{J}}_{2D}^{-1}(x, y, Z, \mathbf{q}) \cdot (\mathbf{s} - \mathbf{s}^*) \,. \tag{36}$$

where $\hat{\mathbf{J}}_{2D}$ is the estimated Jacobian matrix of $\mathbf{J}_{2D} = \mathbf{J}_i \cdot {}^c\mathbf{W}_e \cdot {}^e\mathbf{J}_\mathbf{R}$. During the control loop the image Jacobian (11) needs to be estimated at each time step, i.e., an estimate of the depth Z must be obtained. Please note that in (36), when the number of image features and robot joint coordinates are equal, i.e. $\dim(\mathbf{s}) = \dim(\mathbf{q})$, the Jacobian $\hat{\mathbf{J}}_{2D}$ is a square matrix and its inverse $\hat{\mathbf{J}}_{2D}^{-1}$ can be computed at each iteration of the video rate. When redundant features are used, $\dim(\mathbf{s}) > \dim(\mathbf{q})$, a pseudo-inverse Jacobian $\hat{\mathbf{J}}_{2D}^{+}$ must be employed.

In closed loop, the image features error decreases if the exponential decay is specified and if the Jacobian is non-singular. When $\dim(\mathbf{s}) \neq \dim(\mathbf{q})$, the following condition must be satisfied to assure asymptotic global stability of the image-based control law [12]:

$$\mathbf{J}_{2D} \cdot \hat{\mathbf{J}}_{2D}^{+} > 0 \tag{37}$$

Stating that the matrix must be positive-definite. Note that a matrix $\mathbf{A}(n \times n)$ is positive-definite if, $\mathbf{x}^T\mathbf{A}\mathbf{x} > 0, \forall \mathbf{x} \neq 0 \in \Re^n$. As $\hat{\mathbf{J}}_{2D}$ must be estimated at each visual sample time and depends on Z, the global convergence is not assured in practice because Z is not constant during the servoing. Local convergence can only be assured in the neighbourhood of the desired position, only if its value, Z^*, is used as a constant value during the servoing. Z^* can be coarsely measured along the camera optical axis as the distance between the camera frame and the desired position of the feature, by moving the robot to the desired position.

3.2. Position Based Visual Servoing

The global control architecture is shown in figure 4, where the block *Robot inner loop* law is a PD control law [2]. Using the same methodology from the previous section to obtain

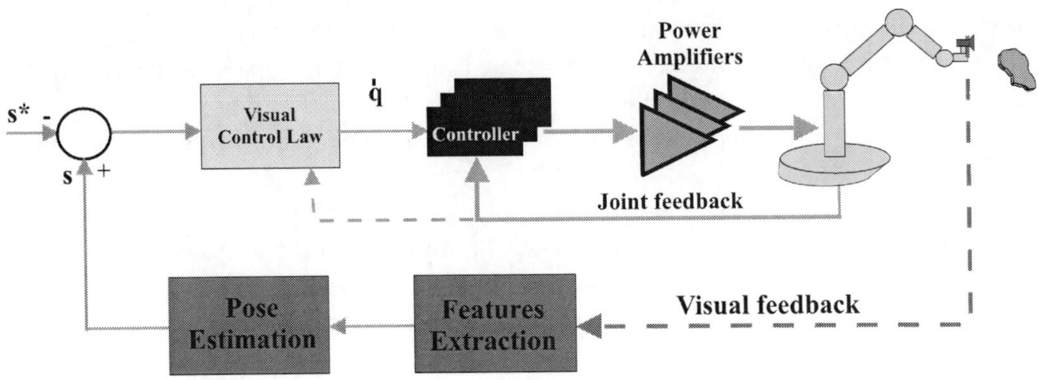

Figure 4. Position Based Visual Servoing.

the visual control, the following is derived:

$$\dot{\mathbf{q}} = -\mathbf{K_p} \cdot (\hat{\mathbf{J}}_{3D} \cdot {}^{c}\mathbf{W}_e \cdot {}^{e}\mathbf{J_R})^{-1} \cdot (\mathbf{s} - \mathbf{s}^*) . \tag{38}$$

where $\hat{\mathbf{J}}_{3D}$ is the estimated Jacobian matrix, i.e., it depends on the feature ${}^{c}\mathbf{T}_o$ estimation.

3.3. Hybrid Visual Servoing

Using the same methodology from the previous section to obtain the visual control law, the following is derived:

$$\dot{\mathbf{q}} = -\mathbf{K_p} \cdot (\hat{\mathbf{J}}_{2\frac{1}{2}D} \cdot {}^{c}\mathbf{W}_e \cdot {}^{e}\mathbf{J_R})^{-1} \cdot (\mathbf{s} - \mathbf{s}^*) \tag{39}$$

where $\hat{\mathbf{J}}_{2\frac{1}{2}D}$ is the estimated Jacobian matrix, i.e., it depends on both, depth Z and rotation ${}^{c^*}\mathbf{u}_c\theta$, estimations.

This approach, presented by Malis [38], decouples the rotation and translation movements of the robot, as shown in equation (32). It is characterized by a global convergence to the object features seen in the image. After that, several approaches have been proposed. In [11] it was proposed a method that also decouples the translation and rotation, where special focus was given to the camera 3D trajectory. Another approach separates rotation and translation of the z axis from the other axes of the camera frame [9, 10]. Hybrid visual servoing has been evolving and hybrid controllers have been developed, *Hybrid Switched Systems*. These systems can commute between different controllers during the task execution, to obtain the best result to control the robot for given tasks. Such systems have been proposed in [8, 15, 16, 25, 39].

3.4. Estimation of Visual-Motor Interaction

Apart from the previous approaches, where the robot-camera model is already known, in other cases it must be estimated. This type of visual servoing systems can deal with unknown robot kinematics and/or unknown camera parameters. They are based on an on-line or off-line estimation, prior to the execution of the task, and can be estimated analytically or by learning. Examples of obtaining the robot-camera model analytically are presented in [28, 30], while by learning they are presented in [47]. This type of systems are also defined in the previous referenced literature by *uncalibrated*. By using this type of robot-camera models, the system becomes independent of robot type, camera type or even camera location.

In this chapter, the robot-camera model estimation by learning is addressed, using fuzzy techniques, to obtain a controller capable of controlling the robotic system. An inverse fuzzy model is used to derive the inverse robot-camera model, in order to compute the joints and end-effector velocities in a straightforward manner. The inverse fuzzy model is applied directly as a real-time controller.

From the modeling techniques based on soft computing, fuzzy modeling is one of the most appealing. In visual servoing, fuzzy logic has been used to learn the robot-camera model [47, 48]. As the robotic manipulator together with the visual system is a nonlinear system, which is only partly known, it is advantageous to use fuzzy modeling as a way to derive a model (or an inverse model, as in this case) based only on measurements. Various techniques can be applied to derive such models, fuzzy clustering, neural learning methods or orthogonal least squares (see e.g. [22] for an overview). Fuzzy clustering is most often used to derive fuzzy models based on the data obtained from measurements of dynamic systems. This approach is used in this chapter and it is presented in detail in [21]. The next subsections present the fuzzy modeling approach and how the uncalibrated fuzzy model was obtained.

3.5. Fuzzy Modeling

Fuzzy modeling often follows the approach of encoding expert knowledge expressed in a verbal form in a collection of if–then rules, creating the model structure. Parameters in this structure can be adapted using input-output data. When no prior knowledge about the system is available, a fuzzy model can be constructed entirely on the basis of systems' measurements. In the following, we consider data-driven modeling based on fuzzy clustering, which is described in [46]. This approach proved to be better than other well-known methods, such as ANFIS (Adaptive Neural Fuzzy Inference Systems), which was presented in [32].

Assume that data from an unknown MISO (multiple-input and single-output) system $y = F(\mathbf{x})$ is observed. The aim is to use this data to construct a deterministic function $y = f(\mathbf{x})$ that can approximate $F(\mathbf{x})$. The function f is represented as a collection of fuzzy if–then rules. Depending on the form of the propositions and on the structure of the rule base, different types of rule-based fuzzy models can be distinguished.

For the robot under control, the direct robot-camera model states that the image features will move between consecutive time steps k and $k+1$, in the image space. This variation of the image features is defined as $\delta\mathbf{s}(k+1) = \mathbf{s}(k+1) - \mathbf{s}(k)$. Given a predefined robot velocity $\dot{\mathbf{q}}(k)$, and knowing its actual position $\mathbf{P}(k)$ in the world coordinate frame at the current time step k, the image features will move from $\mathbf{s}(k)$ to $\mathbf{s}(k+1)$.

In this chapter, rule-based MISO models of first order Takagi-Sugeno [49] type are considered. This type of models is described next, followed by a brief description of the identification procedure using fuzzy clustering.

Takagi–Sugeno fuzzy model

First order, MISO, Takagi-Sugeno rule-based models, are considered in this chapter. The models have n inputs and one single output. For systems with more outputs, several MISO models are needed to model the overall system, e.g., two MISO models are needed to model systems of two outputs.

The fuzzy model is divided in rules. Each rule is applied to a sub-domain of the input variables. This sub-domain is called a cluster and is obtained using clustering approaches [23]. In each cluster there is a rule that describes the input-output relation. In conclusion, the rules of a Takagi-Sugeno fuzzy model describe local input-output relations, typically in a linear form, where the output y_i of each rule is a linear combination of the inputs $\mathbf{x}$:

$$R_i : \textbf{If } x_1 \text{ is } A_{i1} \textbf{ and } \ldots \textbf{and } x_n \text{ is } A_{in}$$
$$\textbf{then } y_i = \mathbf{a}_i\mathbf{x} + b_i\,, \quad i = 1,2,\ldots,K. \tag{40}$$

Here R_i is the ith rule, $\mathbf{x} = [x_1,\ldots,x_n]^T$ is the input (antecedents) variables. $A_{i1},\ldots,A_{in}$ are fuzzy sets defined in the antecedent space, that must be estimated. y_i is the rule output variable. $\mathbf{a}_i$ and b_i are the consequent parameters of the linear combination for each rule, that must be estimated. K denotes the number of rules in the rule base. For each cluster i, the correspondent rule R_i gives an output y_i, that must be combined to obtain the only output of the MISO system. The aggregated output of the model, $\hat{y}$, is calculated by taking the weighted average of the rule consequents of each rule:

$$\hat{y} = \frac{\sum_{i=1}^{K} \beta_i y_i}{\sum_{i=1}^{K} \beta_i}, \tag{41}$$

where β_i is the degree of activation of the ith rule, stating for a given input what rules are the most active.

$$\beta_i = \Pi_{j=1}^{n} \mu_{A_{ij}}(x_j), \quad i = 1,2,\ldots,K, \tag{42}$$

and $\mu_{A_{ij}}(x_j) : \mathbb{R} \rightarrow [0,1]$ is the membership function of the fuzzy set A_{ij} in the antecedent of R_i.

Fuzzy Model Parameters Identification by fuzzy clustering

To identify the first order Takagi-Sugeno model defined in (40), i.e., the antecedent and consequent parameters, the regression matrix $\mathbf{X}$ and an output vector $\mathbf{y}$ are constructed from the available input-output data:

$$\mathbf{X}^T = [\mathbf{x}_1,\ldots,\mathbf{x}_N], \quad \mathbf{y}^T = [y_1,\ldots,y_N]. \tag{43}$$

Here $N \gg n$, is the number of samples used for identification, taken at each time step. In this chapter, N is the number of time samples needed by the robot to perform a predefined trajectory. The objective of identification is to construct the unknown nonlinear function $\mathbf{y} = f(\mathbf{X})$ from the data, where f is the first order Takagi-Sugeno model in (40).

The number of rules K and the antecedent fuzzy sets A_{ij} are determined by means of fuzzy clustering in the product space of the inputs and the output, $[x_1 \times \ldots \times x_n \times y]$. Hence, the data set Φ to be clustered is composed from $\mathbf{X}$ and $\mathbf{y}$:

$$\Phi^T = [\mathbf{X}, \mathbf{y}]. \tag{44}$$

Given Φ, where the N input-output data samples are stored, and an estimated number of clusters K, the Gustafson-Kessel clustering algorithm [23] is applied to compute the fuzzy partition matrix $\mathbf{U} = [\mu_{ik}]_{K\times N}$, stating the degree of membership of each sample data, $k = 1...N$, to each cluster, $i = 1...K$.
Clustering provides a description of the system in terms of its local characteristic behavior in regions of the input-output data, Φ, where each cluster defines a rule. Unlike the popular fuzzy c-means algorithm [6], the Gustafson-Kessel algorithm applies an adaptive distance measure suited for highly non-linear systems like robotic manipulators. As such, it can find hyper-ellipsoid regions in the input-output data, Φ, that can be efficiently approximated by hyperplanes described by the first order Takagi-Sugeno fuzzy rules, i.e. with dimension $n+1$ (n inputs and one output).

The fuzzy sets in the antecedent of the rules are obtained from the partition matrix $\mathbf{U}$, whose ik^{th} element $\mu_{ik} \in [0,1]$ is the membership degree of the data object $\mathbf{z}_k = [\mathbf{x}_k{}^T y_k]^T$ to the cluster i. To be used in the first order Takagi-Sugeno model defined in (40), one-dimensional fuzzy sets A_{ij} must be obtained from the multidimensional fuzzy sets defined point-wise in the ith row of the partition matrix by projections onto the space of each input variable $x_j, j = 1 \ldots n$:

$$\mu_{A_{ij}}(x_{jk}) = \text{proj}_j(\mu_{ik}), \tag{45}$$

where proj is the point-wise projection operator [35]. The point-wise defined fuzzy sets A_{ij} are approximated by suitable parametric functions, like gaussian membership functions [46], in order to compute $\mu_{A_{ij}}(x_j)$ for any value of x_j. At this point the antecedent parameters for each rule, A_{ij}, of the first order Takagi-Sugeno model have been estimated.

The consequent parameters for each rule are obtained as a weighted ordinary least-square estimate. In the following, are presented the steps to formulate the problem, following [46].
The consequents for each rule are placed in the following vectorial form, $\theta_i^T = \left[\mathbf{a}_i^T; b_i\right]$. The input-output data is extended and defined by $\mathbf{X}_e = [\mathbf{X}; \mathbf{1}]$. A diagonal matrix of weights in $\mathbb{R}^{N\times N}$, must be defined $\mathbf{W}_i$, having the degree of activation $\beta_i(\mathbf{x}_k)$, as its kth diagonal element.
Assuming that the columns of $\mathbf{X}_e$ are linearly independent and $\beta_i(\mathbf{x}_k) > 0$ for $1 \leq k \leq N$, the weighted least-squares solution of $\mathbf{y} = \mathbf{X}_e\theta + \varepsilon$, where ε is a residue, becomes:

$$\theta_i = \left[\mathbf{X}_e^T \mathbf{W}_i \mathbf{X}_e\right]^{-1} \mathbf{X}_e^T \mathbf{W}_i \mathbf{y}. \tag{46}$$

At this point the consequent parameters for each rule, $\mathbf{a}_i$ and b_i, of the first order Takagi-Sugeno model have been estimated.

3.5.1. Uncalibrated Fuzzy Visual Servoing

To obtain an accurate Jacobian, $\mathbf{J}$, a perfect modelling of the camera, the chosen image features, the position of the camera related to the world, and the depth of the target related to the camera frame must be accurately determined. Even when a perfect model of the Jacobian is available, it can contain singularities, which hampers the application of a control law. Remind that the Jacobian must be inverted to send the joint velocities to the robot. When the Jacobian is singular, the control cannot be correctly performed.

To overcome the difficulties regarding the Jacobian, a new type of differential relationship between the features and camera velocities was proposed in [48]. This approach states that the current joint variation, $\delta\mathbf{q}(k)$, depends on the image features variation that the robot must perform at the next time step, $\delta\mathbf{s}(k+1)$, and the current position of the robot manipulator, $\mathbf{q}(k)$:

$$\delta\mathbf{q}(k) = \mathbf{F}_k^{-1}(\delta\mathbf{s}(k+1), \mathbf{q}(k)). \tag{47}$$

In visual servoing, the goal is to obtain a joint velocity, i.e., the joint variation $\delta\mathbf{q}(k)$, capable of driving the robot according to a desired feature position, $\mathbf{s}(k+1)$, with an also desired feature variation, $\delta\mathbf{s}(k+1)$, from any position in the joint spaces. This goal can be accomplished by modelling the inverse function $\mathbf{F}_k^{-1}$, using inverse fuzzy modelling as proposed in this chapter and presented in Section 3.5.. This new approach to visual servoing allows to overcome the problems stated previously regarding the Jacobian and the calibration of the robot-camera model.

PART II: APPLICATIONS

4. Applications and Implementation

4.1. Review of worldwide applications

Early ages in Visual Servo Control of Robots have stayed in research laboratories, and often the object features in the image were 2D or 3D points. The robot work environment was highly structured and often the object to manipulate had large dots in a regular background, [29]. To use simple image processing algorithms and to lower the image control loop computational time, the illumination was also kept nearly constant. In such scenarios and with the available hardware a sample time of 50/60Hz was often obtained, and in special cases, not for common manipulators, 300Hz was also obtained [31].

Up to 2002 [31], the applications of visual servoing and visual guidance included: Aircraft landing guidance; Underwater robot control; Catching ball; Part mating; Tracking and grasping a toy train; Picking parts from fast moving conveyor; Fruit picking; Weld seam tracking; Telerobotics, etc.

With the increasing capability of hardware, i.e., video cameras, frame grabbers and computers, and along with the development of computer vision and image processing methods, the scenarios where visual servoing is being applied have largely evolved. Recently, manipulators have reached 500 Hz in the visual control loop [43] and in special applications 1000Hz have been achieved [26]. This is a major breakthrough and allows the visual loop to reach the 1KHz sample frequency of common position or force control. Force and Visual Servo Control [4, 17] have been applied together with success in several applications to enhance the visual control in the contact zone of the object.

Visual servoing has been applied not only to robotic manipulators, but also to parallel robots [1] autonomous ground robots [5] and aerial robots [42], just to name some examples. This area of control is expanding to even more challenging fields, like humanoids' control [14,50] in unknown environments. These unknown environments made the control problem more difficult to tackle due to uncertainties that can appear in the environment. New approaches have been proposed to estimate the interaction model between the robot and the environment seen through the image(s). These approaches [21,36,44,47] are based on estimation or learning of the well known Jacobian matrix in Visual Servo Control.

Visual Servoing is also evolving to new types of images, e.g., medical images. In this applications, surgical robotics [3,18,33,34,40], the visual loop is used to position ultrasound probes, to track organs, and to assist surgeons performing surgical tasks.

4.2. Implementation on Robotic Manipulators of Learning Approaches

In this section of the chapter are shown the results of the presented fuzzy modeling approach, *uncalibrated visual servoing* applied to position-based visual servo control.

4.2.1. Experimental Setup

To validate the proposed approach, a five-degree of freedom EurobTec IR52C Robotic Manipulator was used. The visual features were obtained from the images using a stereo vision system, with two *U-Eye* cameras, in eye-to-hand configuration, i.e., looking the robotic manipulator end-effector.

The experimental setup is presented in figure 5. At the IR52C robot end-effector it is attached an object with colored LEDs, depicted in figure 6, that is seen by the stereo vision system. For each of the three LEDs, its 3D coordinates, ${}^c\mathbf{t}_o$, are obtained. Note that, during the servoing, the LEDs are all in the same plane and therefore its rotation relative to the stereo vision frame (located in the left camera), ${}^c\mathbf{R}_o$, is the same. The visual processing is performed in the PC2, that acquires and process [41] the images from both cameras and sends the 3D visual features to the network, using UDP protocol, at each time step.

In PC1 is implemented the visual servoing control law, that receives the UDP packets coming from the PC2. Computer PC2, as in [41], acquires images from cameras, performs color segmentation, and extracts the LEDs 3D coordinates, ${}^c\mathbf{t}_o$, in real-time, corresponding to the 3D position of the LED. Since three LEDs are used in this setup, nine features are obtained and therefore ${}^c\mathbf{t}_o$ is a (9×1) vector. Note that, as stated before, the rotation, ${}^c\mathbf{R}_o$, is the same for all three LEDs. In the experiments presented in the following sections, the rotation, ${}^c\mathbf{R}_o$, was kept constant and therefore was the same for all the robot trajectories performed. For these reasons the (9×1), ${}^c\mathbf{t}_o$, is the feature vector, $\mathbf{s}$, sent to the network.

4.2.2. Fuzzy Modeling Results

To obtain the first order Takagi-Sugeno model defined in (40), the regression matrix (43) must be obtained from the input-output data of the robot. In section 3.5.1. was defined that the outputs of the model are the robot joint variations $\delta\mathbf{q}(k)$. These joint variations should be capable of driving the robot according to a desired feature position, $\mathbf{s}(k+1)$. Since the robot has five joints, five MISO models must be estimated, according to section 3.5..

To obtain the data for model identification, the robot was moved in its 3D workspace, within the field of view of the cameras, making a 3D spiral with a center point, depicted in figure 7. The variables needed for identification, $\delta\mathbf{q}(k)$ and $\delta\mathbf{s}(k+1)$, were stored during

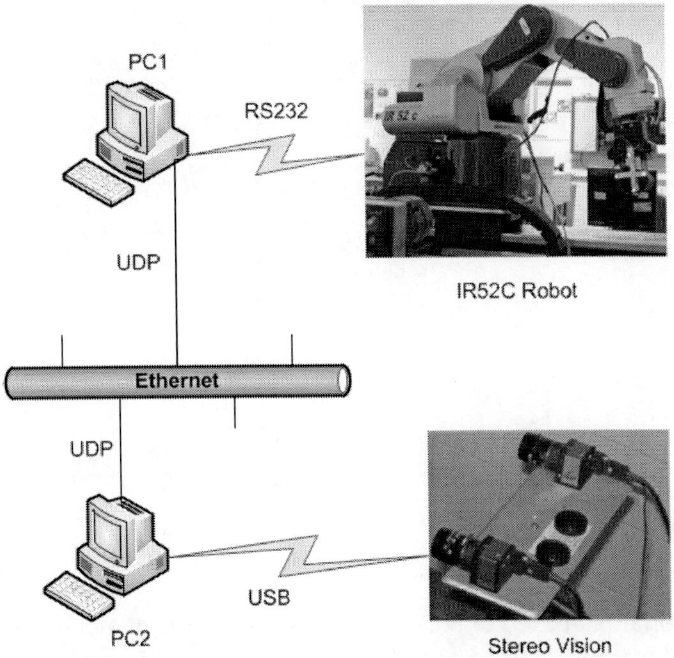

Figure 5. The eye-to-hand experimental setup.

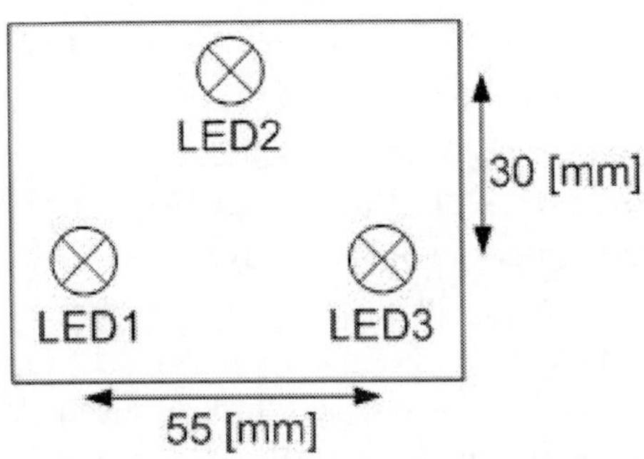

Figure 6. Configuration of the markers placed on the end-effector.

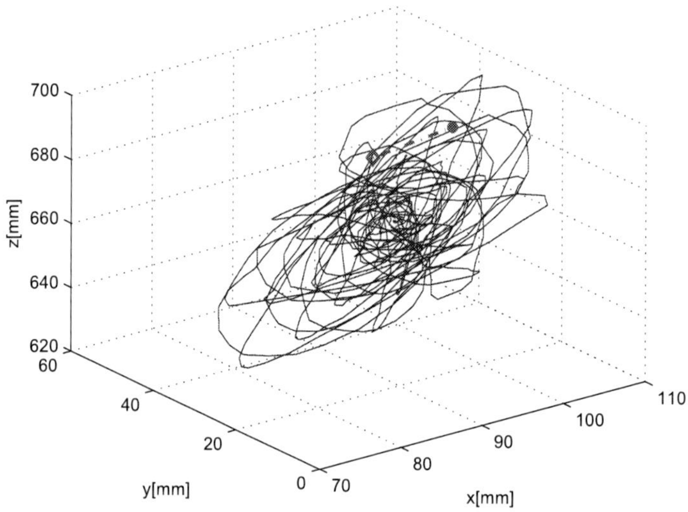

Figure 7. 3D Spiral path used to collect the data for the fuzzy model identification, solid thin line. Task linear trajectory used to test the visual servo control law, dash-dotted thick line.

the 3D spiral path, i.e., the off-line training task. This procedure allows to cover a wide range of values for $\delta\mathbf{q}(k)$ and $\delta\mathbf{s}(k+1)$:

$$\delta\mathbf{s}(k+1) = \mathbf{s}^* - \mathbf{s}(k+1) \tag{48}$$

$$\delta\mathbf{q}(k) = \frac{\mathbf{q}^* - \mathbf{q}(k)}{\Delta t} \tag{49}$$

where Δt is the visual sample time (0.1 seconds) and in the training phase:
$\mathbf{s}^* = [92.0, 25.0, 667.5, 66.5, 55.0, 669.0, 115.0, 57.5, 689.0](mm)$ are the 3D coordinates of the LEDs, at the center of the 3D spiral;
$\mathbf{q}^* = [-18.3545, -72.2546, -51.2007, -53.5447, -169.4038](^\circ)$ are the robot manipulator joint values, at the center of the 3D spiral.

In figures 8, 9 and 10 are depicted, over time, the model inputs, $\delta\mathbf{s}(k+1)$, and the model outputs $\delta\mathbf{q}(k)$, respectively.
To estimate the model accuracy two descriptors are used: VAF (Variance Accounted For), defined in (50), and MSE (Mean Squared Error), defined in (51). A perfect match occurs, when VAF is 100% and MSE is 0.

$$VAF = 1 - \frac{var(\delta\mathbf{q} - \widehat{\delta\mathbf{q}})}{var(\delta\mathbf{q})} \times 100\% \tag{50}$$

where $\widehat{\delta\mathbf{q}}$ is the output of the estimated fuzzy model, and $var(\ldots)$ is the variance of a vector.

$$MSE = \frac{1}{n}\sum(\widehat{\delta\mathbf{q}} - \delta\mathbf{q})^2 \tag{51}$$

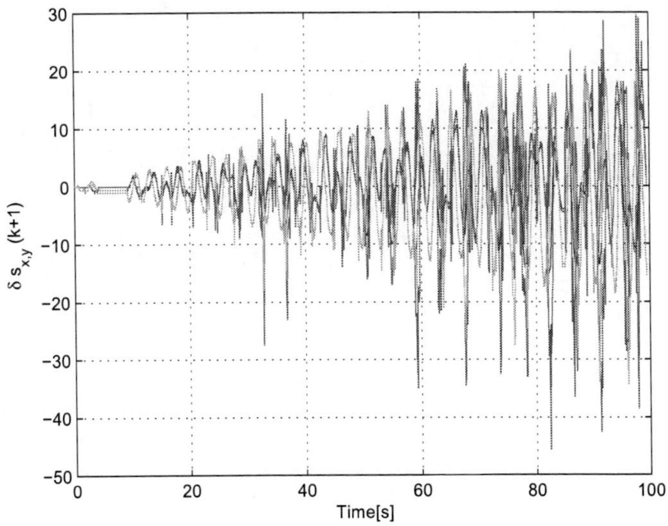

Figure 8. The input data for model identification, $\delta \mathbf{s}_{xy}(k+1)$, i.e., in the x and y directions.

	Rules	VAF	MSE $[(°/s)^2]$
Joint 1	3	98,2%	0,23
Joint 2	3	97,3%	0,93
Joint 3	3	94,4%	2,81
Joint 4	3	93,2%	1,21
Joint 5	3	98,2%	0,23

Table 1. Results of the off-line fuzzy model, obtained for each joint.

where n is the size of the vectors. The values for VAF and MSE must be obtained for each robot joint using all the n samples of data collected for each joint, since five MISO fuzzy models are estimated as stated previously.

In Table 1 are presented the values of VAF and MSE for the off-line fuzzy modeling for each one of the five robot joints, i.e., one MISO model for each joint, using real data from the robot and vision systems. With only three rules, excellent values of VAF and MSE were obtained, meaning that is possible to obtain a good model for estimating the joint velocities, suitable to be used as a model based inverse controller. Figure 10, shows how close the output of the model is to the training data. In more detail, figure 11 depicts the result for joint 4, the worst VAF presented in table 1.

4.2.3. Control Results

This section presents the obtained control results, using the Off-Line fuzzy model based control, defined in figure 12. These results, that validate the proposed approach, were obtained from the real robot, when performing the Uncalibrated Visual Servoing using the fuzzy model previously presented.

To validate the estimated fuzzy models for controlling the robot manipulator, several

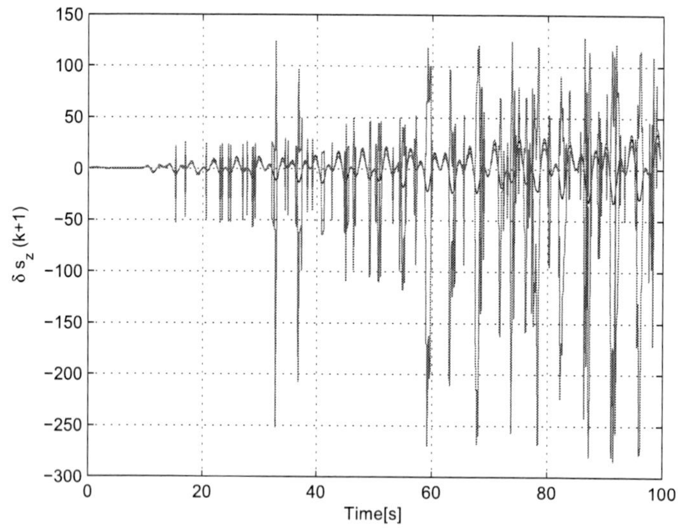

Figure 9. The input data for model identification, $\delta \mathbf{s}_z(k+1)$, i.e., in the z direction.

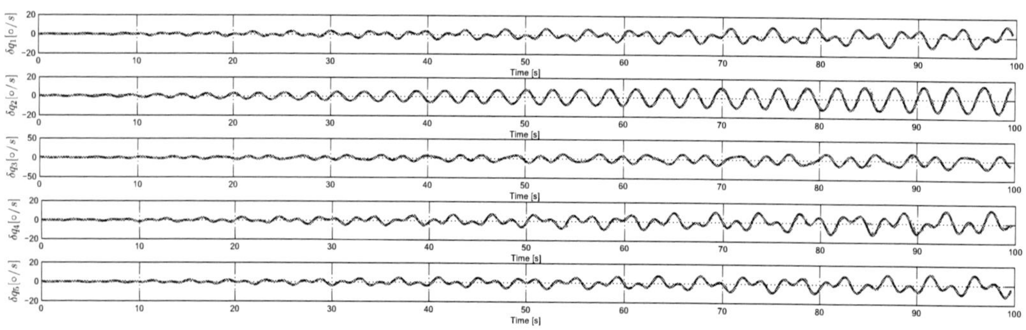

Figure 10. The output data, $\delta \mathbf{q}(k)$. For model identification, solid-line, and the estimated fuzzy model output, dash-dotted line.

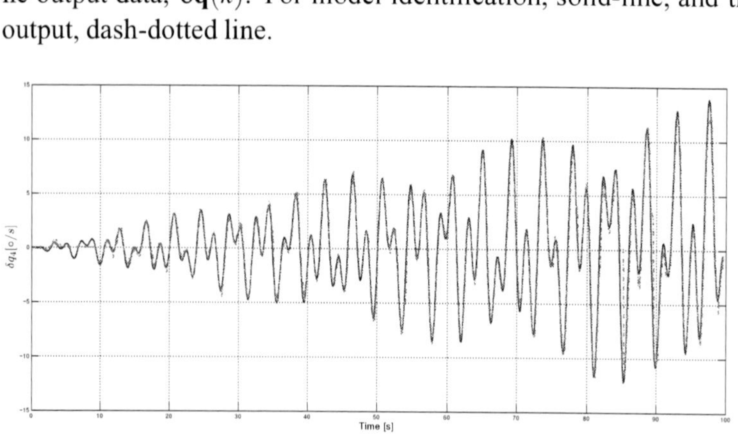

Figure 11. The output data, $\delta q_4(k)$. For model identification, solid-line, and the estimated fuzzy model output, dash-dotted line.

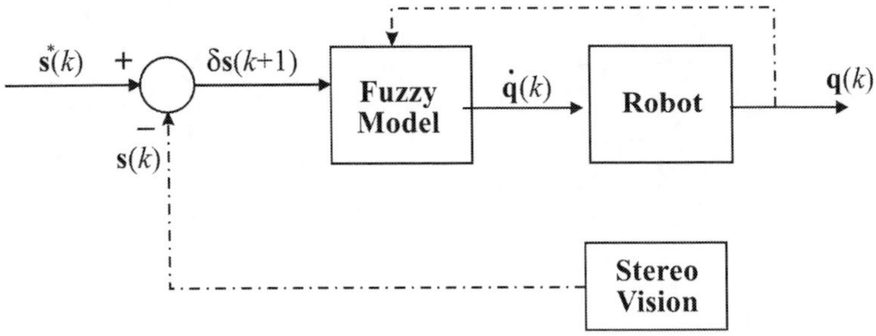

Figure 12. Uncalibrated Visual Servo Control Loop.

positions were set within the robot workspace seen by the stereo vision system. Positioning tasks were given to the robot manipulator to fulfil. The operator gives, to the robot, the desired 3D coordinates of the object, either by a-priori knowledge or by moving the robot using a classical Cartesian control law. In this work the second approach was used.

The experiment was conducted following the general procedure:

1. Move the robot to the desired position, using a classical Cartesian control law, where the stereo vision system can see the object, i.e., the three LED system.
2. Extract the desired feature vector $\mathbf{s}^*$ of the object, i.e., the 3D coordinates of the three LEDs.
3. Move the robot to the initial position, using a classical Cartesian control law, where the stereo vision can extract the initial feature vector of the object, $\mathbf{s}_i$.
4. Start the Fuzzy Uncalibrated Visual Servoing to reach $\mathbf{s}^*$.

Although in general the target position, $\mathbf{s}^*$, to be tracked may varies over the time, in this experiment it was kept fixed.

Amongst all the positioning tasks given to the robot, in figure 7 is depicted one of the tested, the dash-dotted line. This trajectory has the following taught desired position, $\mathbf{s}^* = [101.0\,, 26.0\,, 659.0\,, 76.5\,, 54.0\,, 667.0\,, 123.5\,, 60.0\,, 677.5](mm)$, and the following initial position, $\mathbf{s}_i = [107.5\,, 38.0\,, 670.0\,, 81.0\,, 65.5\,, 675.0\,, 129.5\,, 72.5\,, 690.0](mm)$.

Results are quite satisfactory, despite some initial oscillations of the 3D position error. The robot stabilizes and stops at the desired position with a small error, i.e., within 3 [mm]. This error was due to the precision of the robot (1 [mm] maximum) and the stereo vision system (4 [mm] maximum, in the depth direction). Figure 13, figure 14 and figure 15, show the error for each LED, with respect to the 3D coordinates X, Y, Z, respectively, obtained in one of several trajectories performed with the robot. The error is measured between the current robot position (obtained from the stereo vision system) and the taught desired position, for each one of the 3D directions. The visual servo control approach can stabilize the robot, as depicted in figures 13 to 15, which shows the evolution of the visual features error, during the trajectory.

This approach to visual servo control, based on fuzzy modeling to obtain an uncalibrated visual servo control system, is suitable to control the robot manipulator. The controller obtained presented excellent results, with errors within 3[mm] of the desired position, using the presented experimental setup.

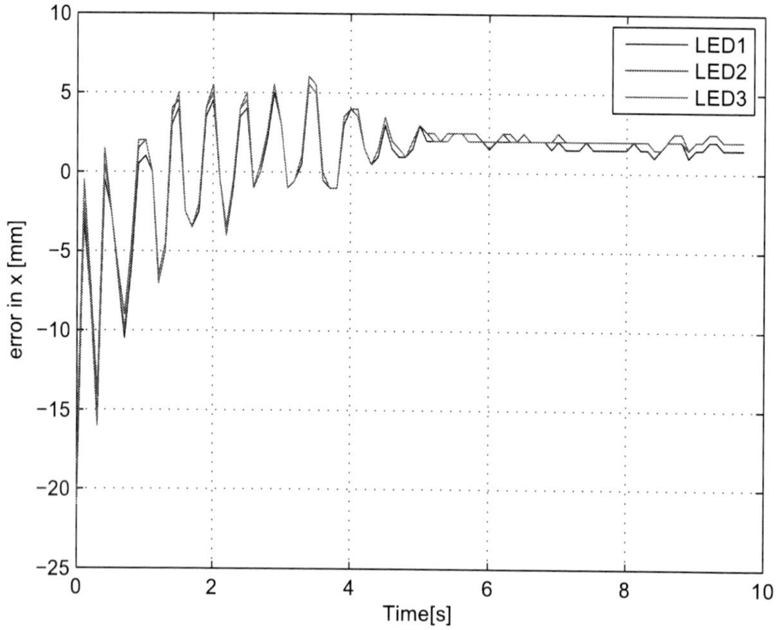

Figure 13. Evolution of the error of the position on the X coordinate.

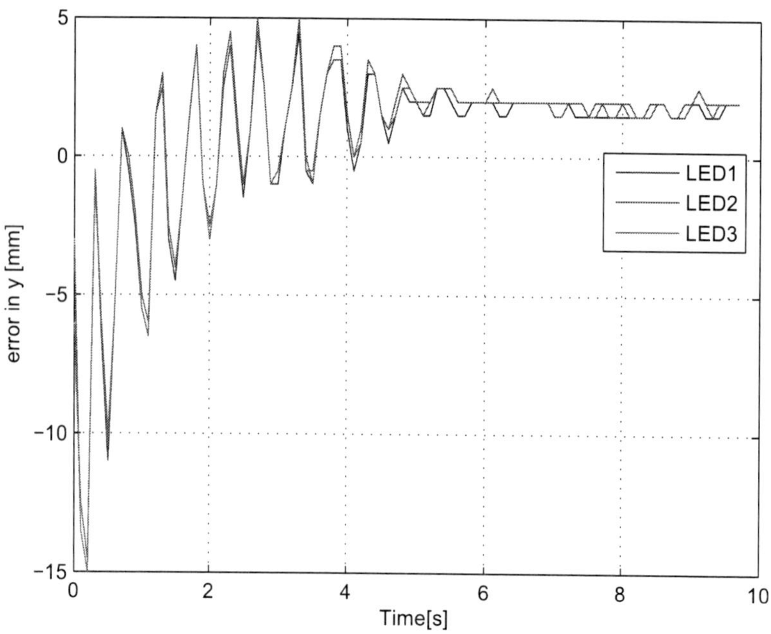

Figure 14. Evolution of the error of the position on the Y coordinate.

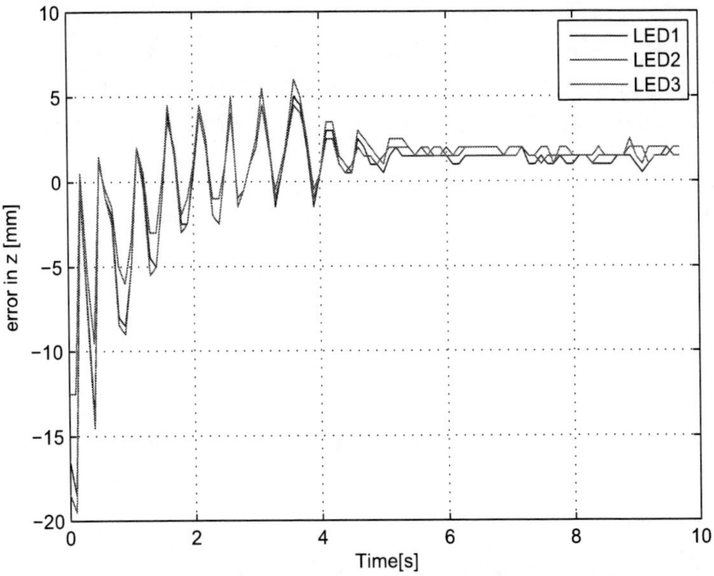

Figure 15. Evolution of the error of the position on the Z coordinate.

5. Conclusions

This chapter presented the fundamentals of Visual Control of Robotic Manipulators. The chapter focuses on Visual Servoing approaches that can be applied to robot manipulators. A detailed mathematical formulation was given for each of the approaches presented in the chapter: Image Based, Position Based, Hybrid, and Fuzzy Uncalibrated Visual Servoing.

Visual control laws that describe the mathematical model of the interaction between the object (seen by the camera) and robot motions, i.e., visual motor interaction, were presented. An approach where the visual motor interaction is estimated, using fuzzy models, was also presented. The Fuzzy based approach was presented and validated using an experimental setup developed for visual servoing experiments, under real working conditions. These experimental results allows the reader to identify real implementation issues regarding the visual frame rate, the error introduced by the visual data, and its implications to robot control.

A section of the chapter was dedicated to a brief review of the applications developed worldwide, using Visual Servoing approaches to control robotic manipulators, both in industry and academia.

Acknowledgments

To the Portuguese Science Foundation, FCT, for the funding to IDMEC through the Associated Laboratory in Energy, Transports, Aeronautics and Space (LAETA). To the FCT project: PTDC/EME-CRO/099333/2008.

References

[1] L. Angel, A. Traslosheros, J. M. Sebastian, L. Pari, R. Carelli, and F. Roberti, *Vision-based control of the robotenis system*, Recent Progress in Robotics: Viable Robotic Service to Human, Lecture Notes in Control and Information Sciences, vol. 370, 2008, pp. 229–240.

[2] K.J. Aström and T. Hägglund, *Pid controllers: Theory, design, and tuning*, Instrument Society of America, Research Triangle Park, USA, 1995.

[3] A. Ayadi, B. Bayle, P. Graebling, and J. Gangloff, *An image-guided robot for needle insertion in small animal. accurate needle positioning using visual servoing* , IROS, 2008, pp. 1453–1458.

[4] J. Baeten, H. Bruyninckx, and J. De Schutter, *Integrated vision/force robotic servoing in the task frame formalism*, International Journal of Robotics Research **22** (2003), no. 10, 941–954.

[5] H.M. Becerra, G. Lopez-Nicolas, and C. Sagues, *A sliding mode control law for mobile robots based on epipolar visual servoing from three views* , IEEE Transactions on Robotics **27** (2011), no. 1, 175–183.

[6] J. Bezdek, *Pattern recognition with fuzzy objective functions*, Plenum Press, New York, 1981.

[7] F. Chaumette, *Image moments: a general and useful set of features for visual servoing* , IEEE Transactions on Robotics and Automation **20** (2004), no. 4, 713–723.

[8] G. Chesi, K. Hashimoto, D. Prattichizzo, and A. Vicino, *Keeping features in the camerać field of view: a visual servoing strategy*, Proceedings of the 15th International Symposium on Mathematical Theory of Networks and Systems (South Bend, USA), 2002.

[9] P. Corke and S. Hutchinson, *A new hybrid image-based visual servo control scheme*, Proceedings of the 39th IEEE Conference on Decision and Control (Sidney, Australia), 2000, pp. 2521–2526.

[10] P. Corke and Seth Hutchinson, *A new partitioned approach to image-based visual servo control*, IEEE Transactions on Robotics and Automation **17** (2001), no. 4, 507–515.

[11] K. Deguchi, *Optimal motion control for image-based visual servoing by decoupling translation and rotation*, Proceedings of the IEEE International Conference on Intelligent Robots and Systems, 1998, pp. 705–711.

[12] B. Espiau, F. Chaumette, and P. Rives, *A new approach to visual servoing in robotics* , IEEE Transactions on Robotics and Automation **8** (1992), no. 3, 313–326.

[13] O. Faugeras, *Three dimensional computer vision, a geometric viewpoint*, MIT Press, Cambridge, Massachusetts, USA, 1993.

[14] F. Gamarra, K. Pinpin, C. Laschi, and P. Dario, *Forward model applied in visual servoing for a reaching task in the icub humanoid robot*, Applied Bionics and Biomechanics Journal, Taylor and Francis **6** (2009), no. 3, 345–354.

[15] N. Gans and S. Hutchinson, *An asymptotically stable switched system visual controller for eye in hand robots*, Proceedings of the IEEE/RSJ International Conference on Intelligent Robots and Systems (Las Vegas, USA), 2003, pp. 735–742.

[16] N. Gans and Seth Hutchinson, *A switching approach to visual servo control*, Proceedings of the 2002 IEEE International Symposium on Intelligent Control (Vancouver, Canada), 2002, pp. 770–776.

[17] G.J. Garcia, J.A. Corrales, J. Pomares, and F. Torres, *Survey of visual and force/tactile control of robots for physical interaction in spain*, Sensors - open access journal (2009), 9689–9733.

[18] R. Ginhoux, J. Gangloff, M. de Mathelin, L. S., M. Sanchez, and J. Marescaux, *Active filtering of physiological motion in robotized surgery using predictive control*, IEEE Transactions on Robotics **21** (2005), no. 1, 67–79.

[19] P.J. Sequeira Gonçalves and J.R. Caldas Pinto, *Camera configurations of a visual servoing setup, for a 2 dof planar robot*, Proceedings of the 7th International IFAC Symposium on Robot Control (Wroclaw, Poland), September 2003, pp. 181–187.

[20] P.J.S. Gonçalves, *Controlo visual de robots manipuladores*, Tese de Doutoramento, Instituto Superior Tecnico, Lisboa, Portugal, 2005.

[21] P.J.S. Gonçalves, L.F. Mendonça, J.M.C. Sousa, and J.R. Caldas Pinto, *Uncalibrated eye-to-hand visual servoing using inverse fuzzy models*, IEEE Transactions on Fuzzy Systems **16** (2008), no. 2, 341–353.

[22] Serge Guillaume, *Designing fuzzy inference systems from data: an interpretability-oriented review*, IEEE Transactions on Fuzzy Systems **9** (2001), no. 3, 426–443.

[23] D. E. Gustafson and W. C. Kessel, *Fuzzy clustering with a fuzzy covariance matrix*, Proceedings of the IEEE Conference on Decision and Control (San Diego, USA), 1979, pp. 761–766.

[24] K. Hashimoto, *A review on vision-based control of robot manipulators*, Advanced Robotics **17** (2003), no. 10, 969–991.

[25] K. Hashimoto and T. Noritsugu, *Potential switching control in visual servo*, Proceedings of the IEEE International Conference on Robotics and Automation (San Francisco, USA), 2000, pp. 2765–2770.

[26] Y. He, Z. Ye, D. She, R. Pieters, B. Mesman, and H. Corporaal, *1000 fps visual servoing on the reconfigurable wide simd processor*, Proc. ASCI Conference 2010, Nov 2010.

[27] J. Hill and W. Park, *Real time control of a robot with a mobile camera*, 9th International Symposium on Industrial Robots, 1979, pp. 233–246.

[28] K. Hosoda and M. Asada, *Versatile visual servoing without knowledge of true jacobian*, Proceedings of the IEEE/RSJ International Conference on Intelligent Robots and Systems, 1994, pp. 186–193.

[29] S. Hutchinson, G. Hager, and P. Corke, *A tutorial on visual servo control*, IEEE Transactions on Robotics and Automation **12** (1996), no. 5, 651–670.

[30] M. Jägersand, O. Fuentes, and R. Nelson, *Experimental evaluation of uncalibrated visual servoing for precision manipulation*, Proceedings of the IEEE International Conference on Robotics and Automation, 1994, pp. 186–193.

[31] F. Janabi-Sharifi, *Visual servoing: Theory and applications*, Opto-Mechatronic Systems Handbook, Ed. H. Cho, CRC Press (2002), 15.1–15.25.

[32] J.S.R. Jang, *Anfis: Adaptive-network-based fuzzy inference system*, IEEE Transactions on Systems, Man, and Cybernetics **3** (1993), no. 23, 665–685.

[33] A. Krupa, J. Gangloff, C Doignon, M. de Mathelin, G. Morel, J. Leroy, L. Soler, and J. Marescaux, *Autonomous 3-d positioning of surgical instruments in robotized laparoscopic surgery using visual servoing*, IEEE Trans. on Robotics and Automation **19** (2003), no. 5, 842–853.

[34] A. Krupa, G. Morel, and M. de Mathelin, *Achieving high precision laparoscopic manipulation through adaptive force control*, Advanced Robotics **18** (2004), no. 9, 905–926.

[35] R. Kruse, J. Gebhardt, and F. Klawonn, *Foundations of fuzzy systems*, John Wiley and Sons, Chichester, 1994.

[36] J.-T. Lapreste, F. Jurie, M. Dhome, and F. Chaumette, *An efficient method to compute the inverse jacobian matrix in visual servoing*, IEEE Int. Conf. on Robotics and Automation, ICRA'04 (New Orleans, Louisiana), vol. 1, April 2004, pp. 727–732.

[37] E. Malis, *Contributions à la modélisation et à la commande en asservissement visuel*, These de l'Université de Rennes I, 1998.

[38] E. Malis, F. Chaumette, and S. Boudet, *2 $\frac{1}{2}$ d visual servoing*, IEEE Transactions on Robotics and Automation **15** (1999), no. 2, 238–250.

[39] N. Mansard and F. Chaumette, *Tasks sequencing for visual servoing*, Proceedings of the IEEE International Conference on Inteligent Robots and Systems (Sendai, Japan), 2004, pp. 992–997.

[40] R. Mebarki, A. Krupa, and F. Chaumette, *2d ultrasound probe complete guidance by visual servoing using image moments*, IEEE Transactions on Robotics **26** (2010), no. 2, 296–306.

[41] P.M.R. Morgado, J.R. Caldas Pinto, J. M. M. Martins, and P.J.S. Gonçalves, *Cooperative eye-in-hand/stereo eye-to-hand visual servoing*, Proceedings of RecPad 2009 - 15th Portuguese COnference on Pattern Recognition (Aveiro, Portugal), 2009.

[42] Alexandra Moutinho, Luiz Mirisola, Jose Raul Carreira Azinheira, and Jorge Manuel Miranda Dias, *Project diva: guidance and vision surveillance techniques for an autonomous airship*, pp. 77–120, Nova Science Publishers, Inc., 2008.

[43] A. Namiki, T. Senoo, S. Mizusawa, and M. Ishikawa, *High speed visual feedback control for grasping and manipulation*, Visual Servoing Via Advanced Numerical Methods, LNCIS 412, G. Chesi and K. Hashimoto Eds, Springer (2010), 39–53.

[44] J.M. Sebastian, L. Pari, L. Angel, and A. Traslosheros, *Uncalibrated visual servoing using the fundamental matrix*, Robotics and Autonomous Systems **57** (2009), no. 1, 1–10.

[45] Y. Shirai and H. Inoue, *Guiding a robot by visual feedback in assembly tasks*, Pattern Recognition **5** (1973), 99–108.

[46] J.M. Sousa and U. Kaymak, *Fuzzy decision making in modeling and control*, World Scientific Publishing Co., Singapore, 2002.

[47] I.H. Suh and T W. Kim, *A visual servoing algorithm using fuzzy logics and fuzzy-neural networks*, Mechatronics **10** (2000), 1–18.

[48] I.H. Suh and T.W. Kim, *Fuzzy membership function based neural networks with applications to the visual servoing of robot manipulators*, IEEE Transactions on Fuzzy Systems **2** (1994), no. 3, 203–220.

[49] T. Takagi and M. Sugeno, *Fuzzy identification of systems and its applications to modelling and control*, IEEE Transactions on Systems, Man, and Cybernetics **15** (1985), 116–132.

[50] N. Vahrenkamp, C. Boge, K. Welke, T. Asfour, J. Walter, and R. Dillmann, *Visual servoing for dual arm motions on a humanoid robot*, IEEE/RAS International Conference on Humanoid Robots (Humanoids), 2009.

In: Robotics: State of the Art and Future Trends
Editors: G. Legnani and I. Fassi

ISBN: 978-1-62100-403-5

Chapter 9

FLEXIBLE APPLICATIONS OF ROBOTIC MANIPULATORS BASED ON VISUAL-FORCE CONTROL AND COOPERATION WITH HUMANS

J. A. Corrales*[*], *G. J. García, F. Torres, F. A. Candelas, J. Pomares and P. Gil
University of Alicante, Spain

Abstract

Most of the tasks developed by robotic manipulators nowadays in industries are limited by their lack of context-awareness. In fact, robotic manipulators are usually applied on repetitive tasks where no sensors are required and no changes of the robot behavior are permitted. This chapter proposes a new robotic system based on multiple sensors and human cooperation in order to overcome this limitation. In particular, the proposed robotic system uses a camera and a force sensor installed at the end-effector in order to interact with the environment. The images from the camera are processed by a visual servoing technique to control the trajectory of the robot while the measurements of the force sensor are used to control the physical interaction of the robot with the manipulated objects. In addition, the proposed system includes a human tracking system based on a motion capture system, an Ultra-wideband (UWB) localization system and a hierarchy of bounding volumes in order to guarantee the safety of the human operators who collaborate with the robots during the development of the task. All these components have been combined in order to develop a flexible robotic system which has been used in different assembly/disassembly applications which are also described in this chapter.

Introduction

Robotic manipulators are more and more widespread in current industries due to their positional precision, repeatability and durability. They are usually used in repetitive assembly tasks where all the components are always in the same position and only a position control of

[*] E-mail address: jcorrales@ua.es

the robot is needed. Nevertheless, the application of robotic manipulators which can react to changes in their environment may improve the productivity of the current industrial processes. In particular, context-aware robotic manipulators do not need to be completely isolated from the rest of the production line and thus the cost of the manufacturing process can be reduced not only in terms of space (by removing the current fenced robotic cells) but also in terms of time (by redefining the production processes so that much more subtasks can be performed simultaneously). In addition, this context-awareness of robotic manipulators enables the participation of new agents during the production such as human operators who can develop those subtasks which cannot be completed by robots. Thereby, more flexible and complex tasks can be developed in industrial robotic workspaces.

However, the development of more flexible tasks requires that robots have a better knowledge of their environment so that they can change their behavior according to the movements of the dynamic components of the product and the movements of the human operators who collaborate during the task. This context-awareness can be achieved by adding sensors in the robotic workspace. In particular, visual and force sensors are the most commonly used sensor technologies since they allow robots to follow surrounding objects and detect contact with their environment, respectively. Therefore, the controller of the production system should take into consideration the visual and force information obtained from these sensors in order to drive the trajectory of the robot. Two different techniques can be applied in order to do this: visual servoing and force control.

Visual servoing is a robot control technique which determines the movements which have to be performed by a robot according to the changes of the images registered by a camera installed in the workspace (over the robot or in a fixed position). This technique can be used to follow a pre-established trajectory. However, previous trajectory tracking systems based on visual servoing present a time-dependent behavior which cannot complete the desired trajectory if the robot's motion is stopped by an unexpected event. This situation is common in human-robot interaction tasks since the human can approach the robot too much, making impossible the completion of the initially planned trajectory due to collision risks. For this reason, the authors of this chapter have developed a time-independent tracking system based on visual-servoing which is able to correct this deficiency by completing the initially planned trajectory after the collision risk disappears.

Force control is another important element for the development of context-aware robot controllers. Force sensors obtain the force and torque which are applied by the robot in the environment and thus a more precise contact task can be performed. Force control allows the robot to detect whether assembly tasks have been correctly performed or whether the robot has collided with some unexpected obstacle. The authors propose to use an impedance control scheme for forces in combination with a visual-servoing path tracker in order to develop more flexible assembly tasks.

In addition, the collaboration of human operators can be very important in order to gain more flexibility in the development of industrial tasks. In particular, humans are excellent agents for those subtasks which require dexterity and/or intelligence. Human operators have usually developed these subtasks in special places of the production line which are isolated from the robots due to risk of collisions between them. On the contrary, the development of context-aware manipulators makes possible their direct collaboration in the same workspace since the behavior of the robots could be adapted to the movements of the humans. In this case, the robot needs very precise information about the position of the human operators and thus the authors propose the use of a human tracking system composed by an inertial motion capture system and an Ultra-WideBand (UWB) localization system. This tracking system not only registers the global position of the human operators who wear it but also the relative position of the limbs of their bodies.

These three components (visual-servoing path tracker, force control and human-tracking system) have been combined by the authors for the development of several flexible assembly/disassembly tasks. This chapter does not only contain the theoretical foundations of these techniques previously developed by the authors but also a review of the main real applications where they have been applied. In particular, this chapter is organized in the following sections: section 2 describes a visual servoing technique for controlling the path followed by the robot, section 3 describes a force control technique for controlling the force which is applied by the robot to the manipulated objects, section 4 describes the human tracking system which verifies that there is no collision danger during human-robot cooperation, section 5 enumerates some applications where these techniques have been combined and finally, section 6 draws the main conclusions of the presented research.

Visual Servoing for Path Tracking

In a robotic task, the robot has to be frequently positioned at a fixed location with respect to the objects in the scene. However, the positions of these objects are not always controlled. Therefore, it is not possible to establish beforehand the precise location of the robot end-effector which will accomplish the task correctly. Visual servoing is a technique that allows positioning a robot with respect to an object using visual information [1]. Typically, visual servoing systems are position-based and image-based classified [2]. Position-based visual servoing requires the computation of a 3D Cartesian error for which a perfect CAD-model of the object and a calibrated camera are necessary. These types of systems are very sensitive to modeling errors and noise perturbations. In image-based visual servoing the error is directly measured in the image. This approach ensures the robustness with respect to modeling errors, but generally an inadequate camera motion in the 3D Cartesian space is obtained [3]. In addition, image-based visual servoing systems present singularities and/or local minima problems in large displacements tasks [1]. To overcome these drawbacks maintaining the good properties of image-based visual servoing robustness with regard to modeling and camera calibration errors, the tracking of a sufficiently sampled path between the two distant poses can be performed.

Basically, the image-based visual servoing approach consists of extracting visual data from an image acquired from a camera and comparing it with the visual data obtained at the desired position of the robot. By minimizing the error between the two images, it is possible to control the robot to the desired position (see the visual servoing task represented in Figure 2). A classical image-based visual servoing task [1] can be described by an image function, $\mathbf{e}_t$, which must be regulated to 0:

$$\mathbf{e}_t = \mathbf{s} - \mathbf{s}^* \quad (1)$$

where **s** is a M x 1 vector containing M visual features corresponding to the current state, while $\mathbf{s}^*$ denotes the visual features values in the desired state.

$\mathbf{L}_s$ represents the interaction matrix which relates the variations in the image with the variations in the velocity of the camera:

$$\dot{\mathbf{s}} = \mathbf{L}_s \cdot \dot{\mathbf{r}} \tag{2}$$

where $\dot{\mathbf{r}}$ indicates the velocity of the camera.

By imposing an exponential decrease of $\mathbf{e}_t$ ($\dot{\mathbf{e}}_t = -\lambda \mathbf{e}_t$) it is possible to obtain the following control action for a classical image-based visual servoing [1]:

$$\mathbf{v}_c = -\lambda \hat{\mathbf{L}}_s^+ \left(\mathbf{s} - \mathbf{s}^*\right) \tag{3}$$

where $\hat{\mathbf{L}}_s^+$ is the pseudoinverse of an approximation of the interaction matrix.

As mentioned above, using this control law it is only possible to achieve a desired location from an initial one. However, in order to perform the positioning task in a complex scene (like the scenes described in this chapter), the desired location must be achieved by tracking a given desired trajectory. This trajectory guarantees that the robot does not collide with the objects in the workspace. The notation of the trajectory to be tracked is described next.

From an off-line stage, the discrete trajectory of the features in the image to be tracked by the robot is obtained $\boldsymbol{T} = \left\{ {}^k\mathbf{s} / \mathrm{k} \in 1..\mathrm{N} \right\}$, with ${}^k\mathbf{s}$ being the set of M points or features observed by the camera at instant k, ${}^k\mathbf{s} = \left\{ {}^k\boldsymbol{f}_i / \mathrm{i} \in 1..\mathrm{M} \right\}$.

Figure 1 shows the robot 3D trajectory and the image trajectory that the robot must track when the robot performs one of the disassembly tasks described next in the Applications Section of the chapter. In

Figure 1a, the 3D trajectory which the robot end-effector has to perform during the task is represented. In this figure, the four laser points which will be employed by the visual servoing system are shown. ${}^k\mathbf{P}_i$ / i=1...M are the 3D coordinates (with respect to the camera coordinate frame) of the laser points extracted by the camera in instant k. The projective coordinates in the image of a given point ${}^k\mathbf{P}_i$ can be obtained as ${}^k\boldsymbol{f}_i = \mathrm{pr}_\xi({}^c\mathbf{M}_o\ {}^k\mathbf{P}_i)$, where pr_ξ denotes the perspective projection model according to the intrinsic parameters ξ, and ${}^c\mathbf{M}_o$ represents the pose of the object frame with respect to the camera frame. In

Figure 1b, the set of M image features observed by the camera at each instant k, ${}^k\mathbf{s}$, during the trajectory is represented.

The systems proposed up to now to track trajectories in the image using visual servoing [3] [4] [5], employ a temporal reference, $\mathbf{s}^*(\mathrm{t})$:

$$\mathbf{v}_c = -\lambda \hat{\mathbf{L}}_s^+ \left(\mathbf{s} - \mathbf{s}^*(\mathrm{t})\right) \tag{4}$$

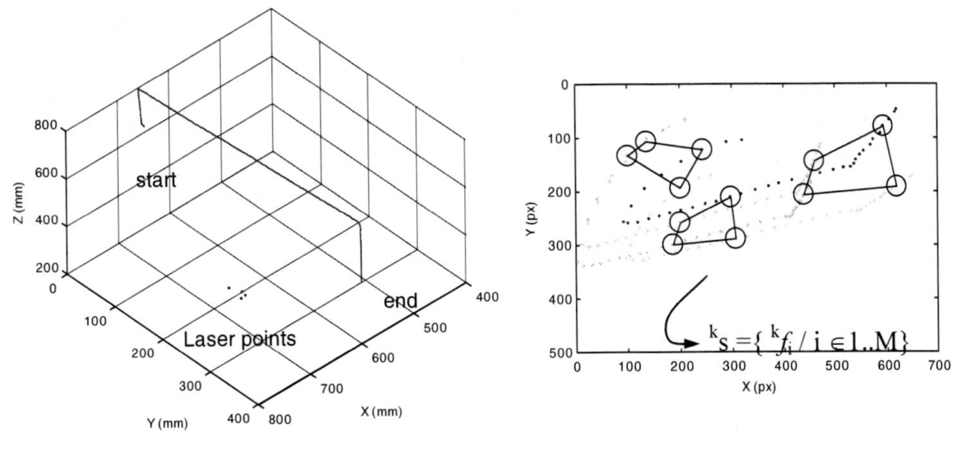

(a) (b)

Figure 1. Trajectory notation: (a) 3D trajectory, (b) Image trajectory.

This control action is similar to the one employed by an image-based visual servoing (3). However, in (4) the desired features are a time-dependent function with provides image features of the desired trajectory during the tracking. These systems do not guarantee the correct tracking of the path when the robot interacts with its environment. If the robot interacts with an object placed in its workspace, the system continues sending visual references to it even though the robot cannot move. Once the obstruction ends, the references that have been sent up to the moment are lost, not allowing the correct tracking of the path. Furthermore, it is necessary to specify the required tracking velocity to apply the visual servoing system to real applications. However, using the previous visual servoing systems it is not possible to specify the desired velocity that the robot will employ to track the trajectory [6]. Thus, the method developed to track trajectories in the image must resolve the problems found in the methods proposed up to now. To do so, the proposed visual servoing system must be able to track the desired trajectory using a time-independent method and considering a given desired tracking velocity $|\mathbf{v}_d|$.

The set of visual features observed at the initial camera position are represented by ${}^0\mathbf{s}$. From this initial set of image features it is necessary to find an image configuration which provides the robot with the desired velocity, $|\mathbf{v}_d|$. To do this, the system iterates over the desired trajectory $\boldsymbol{T}$. For each image configuration ${}^k\mathbf{s}$ stored in $\boldsymbol{T}$, the corresponding camera velocity is determined considering an image-based visual servoing system (at this first stage $\mathbf{s} = {}^0\mathbf{s}$):

$$ {}^k\mathbf{v} = -\lambda \hat{\mathbf{L}}_s^+ \left(\mathbf{s} - {}^k\mathbf{s}\right) \tag{5} $$

This process continues until the velocity $|{}^k\mathbf{v}|$ which is the nearest to $|\mathbf{v}_d|$ is obtained. At this moment, the set of features ${}^k\mathbf{s}$ will be the desired features to be used by an image-based visual servoing system. However, the visual features, ${}^j\mathbf{s}$, which provide the desired velocity are between ${}^{k}\mathbf{s}$ and ${}^{k-1}\mathbf{s}$. To obtain the correct image features the method described in [7] is employed.

Therefore, once the control law represented in Equation (5) is executed, the system searches again for a new image configuration which provides the desired velocity. This process continues until the complete trajectory is tracked. During motion, a new set of visual features **s** is obtained at each iteration. This set of features **s** is employed to compute the new desired image features for the visual servoing controller.

Force Control for Contact Interaction

So far, most of the applications developed to combine visual and force information employ hybrid control [8] [9]. In such a case, the workspace must be precisely known, so that a division of the directions controlled by force and those controlled by position (i.e., using a visual servoing system) is first carried out. With a view to increasing the versatility of the system and improving its response to the uncertainties that arise during a manipulation task, a special visual impedance control system has been developed. It uses the information obtained from the time-independent image path tracker described in previous Section. This system has an external visual feedback, in contrast to others approaches [10] which have the vision and force systems at the same level of the control hierarchy.

The objective of the impedance control is to carry out the combined control of the movement of the robot and its interaction force. It can be stated that the case described in this chapter is active control, in which force feedback is used to control the movements of the joints. The following impedance equation enforces an equivalent mass-damper-spring behaviour for the pose displacement when the end-effector exerts a force $\boldsymbol{F}$ on the environment:

$$\boldsymbol{F} = \mathbf{I}\Delta\ddot{\boldsymbol{x}}_{vc} + \mathbf{D}\Delta\dot{\boldsymbol{x}}_{vc} + \mathbf{K}\Delta\boldsymbol{x}_{vc} \quad (6)$$

where $\boldsymbol{x}_c$ is the current end-effector pose, $\boldsymbol{x}_v$ is the reference trajectory, $\Delta\boldsymbol{x}_{vc} = \boldsymbol{x}_v - \boldsymbol{x}_c$, $\mathbf{I}\in\Re^{nxn}$ is the inertial matrix, $\mathbf{D}\in\Re^{nxn}$ is damping matrix and $\mathbf{K}\in\Re^{nxn}$ is the stiffness. They are diagonal matrices and characterize the desired impedance function. These matrices are empirically computed.

To implement this controller, an external visual feedback has been introduced in the impedance control scheme with an internal movement feedback. Thus, a position-based impedance control system called accommodation control [11] in which the desired impedance is limited to pure damping **D** (in [12] the main aspects of the stability of this type of control is shown) has been chosen for this task. In this case:

$$\boldsymbol{F} = \mathbf{D}\Delta\dot{\boldsymbol{x}}_{vc} \quad (7)$$

Therefore, the control law obtained will be:

$$\dot{\boldsymbol{x}}_c = \dot{\boldsymbol{x}}_v - \mathbf{D}^{-1}\boldsymbol{F} \quad (8)$$

where the term $\dot{\boldsymbol{x}}_{\mathrm{v}}$ is obtained by using the Equation (5):

$$\dot{\boldsymbol{x}}_{\mathrm{v}} = -\lambda \cdot \mathbf{C}_{\mathrm{v}} \cdot \hat{\mathbf{L}}_{\mathrm{s}}^{+} \cdot \left(\mathbf{s} - {}^{j}\mathbf{s}\right) \tag{9}$$

where the velocity is mapped from the camera coordinate frame to the robot end-effector frame by using the matrix $\mathbf{C}_{\mathrm{V}}$. This matrix relating the camera coordinate frame and the robot end-effector coordinate frame is fixed in an eye-in-hand visual servoing configuration. $\mathbf{C}_{\mathrm{V}}$ is empirically computed in the experiments shown in this paper.

Figure 2 shows the proposed control impedance scheme. In this figure the inverse kinematics component implements a conventional pseudo inverse algorithm, $\dot{\boldsymbol{q}}_{\mathrm{c}} = \mathbf{J}_{\mathrm{q}}^{+}\dot{\boldsymbol{x}}_{\mathrm{c}}$, where $\dot{\boldsymbol{q}}_{\mathrm{c}}$ are the corresponding joint velocities and $\mathbf{J}_{\mathrm{q}}^{+}$ is the pseudo inverse of the Jacobian matrix.

Therefore, by defining the desired damping **D**, the data from both sensors are combined as shown in Equation (8). This way the trajectory generated by the movement flow-based visual servoing system is corrected according to the robot's interaction.

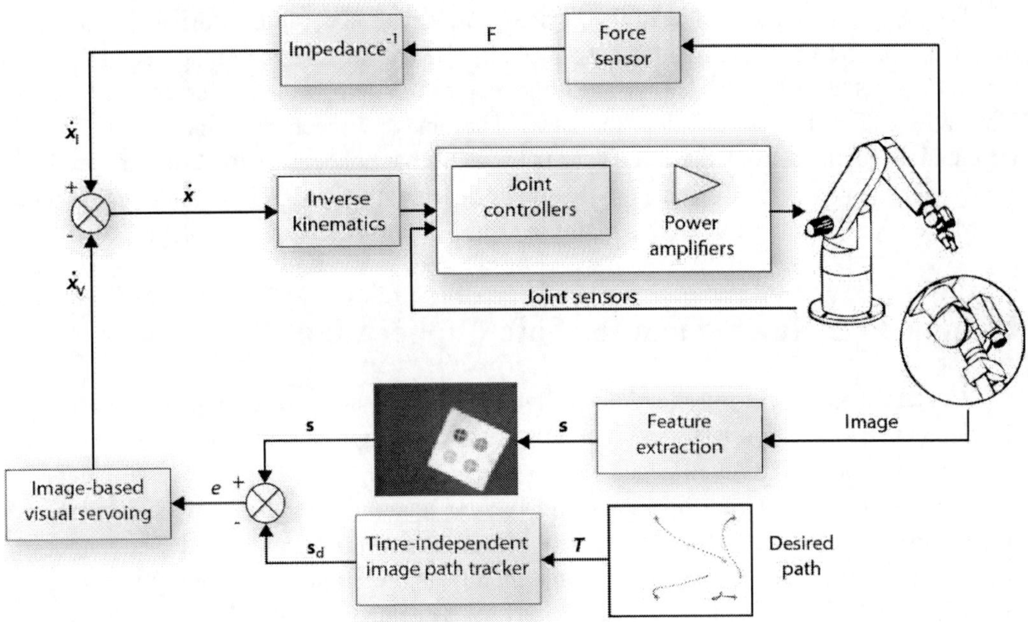

Figure 2. Impedance control scheme.

So far, approaches to fuse visual and force information do not consider the possibility of both sensors providing contradictory information at a given moment of the task (i.e., the visual servoing system establishes a movement direction that is impossible according to the interaction information obtained from the force sensor).

To assure that a given task in which it is required an interaction with the setting is correctly developed, the system must carry out a variation of the trajectory in the image, depending on the spatial restrictions imposed by the interaction forces. To solve this problem, the systems described in previous works have been applied [13] [14].

In the first method, given a collision with the workspace and having recognized the normal vector *n* of the contact surface [13], the transformation that the camera must undergo to fulfil the spatial restrictions is determined. This transformation, given by a rotation matrix **R** and a translation *t*, is calculated so that it represents the nearest direction to the one obtained from the time-independent image path tracker, and which is contained in the plane of the surface. Therefore, to guarantee that visual information will be coherent with that obtained from the force sensor, the image trajectory must be modified, so that the new trajectory is the one obtained from the previous mentioned transformation. To do so, the concept of projective homography matrix **G** is introduced. This matrix relates the points viewed from a camera configuration with the same points observed from another position of the camera. Computing matrix **G** at each iteration (i.e, each time an image is captured) according with the transformation obtained after a collision is detected (through the rotation matrix **R** and the translation *t*), the new trajectory in the image space is obtained. This homography matrix **G** is computed from two images, a fixed one (the image acquired at the moment of the collision), and a variable one (the image acquired at each iteration until the task is completed).

The second method to change the desired image path according to the forces measured from the force sensor of the robot is described in [14]. The advantage of this second method is that represents an uncalibrated visual-force control system which does not require any kinematic calibration to develop the task. The image trajectory is modified from the information obtained from the force sensor by using the concept of force-image interaction matrix. This matrix relates changes in the image space with changes in the interaction forces. In order to estimate the value of this matrix a Gauss-Newton method is employed.

Human Tracking System for Safe Cooperation

As described in the introduction section, the participation of human operators in the development of industrial tasks can increase their flexibility by performing those parts of the task which require the intelligence and/or the dexterity of a human. Nevertheless, the cooperation between human operators and robotic manipulators can be dangerous for the human operators because of the possibility of collisions. In order to avoid this danger, this chapter proposes a human tracking system which guarantees that the robotic system is aware of the human's location so that the robot trajectory is adapted accordingly.

This human tracking system has to determine not only the global position of the human operator in the environment but also the location of all her/his limbs since cooperative tasks may require a too close approximation of the human. Due to this fact, the developed human tracking system combines the measurements of an inertial motion capture system and an Ultra-wideband (UWB) localization system in order to compute the precise position and orientation of the complete body of the human operator who cooperates with the robot.

A motion capture system (*GypsyGyro-18* [15] from *Animazoo*) based on inertial sensors has been used because of its advantages over other motion capture technologies [16]: it does not suffer from magnetic interferences (unlike magnetic systems), optical occlusions do not affect its precision (unlike vision systems) and it is comfortable to wear (unlike mechanical systems). This inertial motion capture system is composed of a suit with 18 IMUs (Inertial Measurement Units with 3 gyroscopes, 3 accelerometers and 3 magnetometers) which are used to compute on real-time the relative orientations between the limbs of the human operators. Thereby, this system returns a skeleton which represents hierarchically with high precision the relative configuration (position and orientation) of the main parts of the human body: head, neck, arms, forearms, hands, thorax, thighs, legs and feet. Although these measurements are very precise for representing the position of every body part with respect to the root node of the skeleton (situated in the middle of the hips), the global position of this root node accumulates an important drift due to inaccuracies on the steps detection.

In order to correct the drift of the global position of the skeleton obtained from the inertial motion capture system, a global localization system based on UWB signals (*Ubisense RTLS* [17] from *Ubisense*) has been used. This localization system is based on two devices: four sensors which are installed at fixed positions of the industrial environment and a small tag (similar to a credit card) which is worn by the human operator. This tag sends UWB signals which are processed by the four sensors in order to compute an estimation of the global position of the tag in the environment by implementing a combination of AoA (Angle of Arrival) and TDoA (Time-Difference of Arrival) techniques.

Since the global position measurements obtained from both tracking systems have complementary features, a fusion algorithm is proposed to combine their benefits. On the one hand, the inertial motion capture system has a high sampling rate (30-120Hz) but their measurements can accumulate drift errors due to incorrect steps detection. On the other hand, the UWB localization system obtains more precise global position measurements but its sampling rate is considerably lower (5-10Hz). Because of this complementariness, the authors have decided to develop a fusion algorithm based on a Kalman filter [18][19] in order to combine the global position measurements of both systems. Figure 3 depicts a scheme of the main steps of this fusion algorithm.

Firstly, the fusion algorithm computes the transformation matrix ${}^{U}T_{I}$ (4x4 matrix with rotation and position) which relates the coordinate systems of the inertial motion capture system (frame *I*) and the UWB system (frame *U*). This transformation matrix is used to represent the measurements of both systems in the same coordinate system (frame *U*). Then, the algorithm begins to register global position measurements from both systems. When a measurement $p_{inertial}^{I}$ from the inertial system is obtained, the algorithm transform it to the *U* frame by applying the matrix ${}^{U}T_{I}$. Next, the measurement $p_{inertial}^{U}$ is applied in the prediction step of the Kalman filter in order to obtain a-priori estimation $\hat{p}^{-}$ of the global human position. When a measurement p_{uwb}^{U} from the UWB system is obtained, the algorithm applies it in the correction step of the Kalman filter, it obtains an estimation $\hat{p}$ of the current position of the human and it recalculates the transformation matrix ${}^{U}T_{I}$. The matrix ${}^{U}T_{I}$ is recalculated with the new UWB measurement in order to correct the drift accumulated by the previous measurements of the inertial system. In addition, not only the position estimations computed in the correction step are used by the tracking system but also the a-priori estimations of the prediction step. Thereby, the tracking system combines the precision of the UWB system (by correcting the accumulated drift in the correction step of the filter) with the high sampling rate of the inertial system (by updating the human position in the prediction step each time a new inertial measurement is registered).

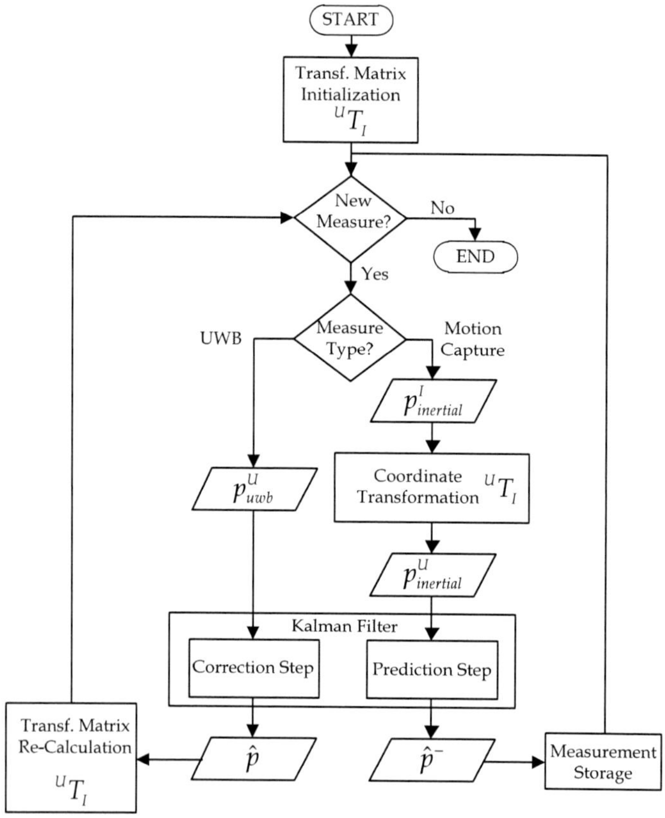

Figure 3. Scheme of the fusion algorithm for the human tracking system.

This fusion algorithm obtains a precise estimation of the global location of the human operator in the industrial environment. This global location is applied to the root node of the skeleton of the inertial motion capture system in order to compute precisely the global position of all the body parts of the human operator. Nevertheless, this skeleton only represents the lengths of the bones of the human but it does not take into consideration their other dimensions. In order to overcome this drawback, the authors have defined a group of bounding volumes which cover every bone of the human skeleton and whose dimensions are big enough to contain inside the corresponding surfaces of the human body.

The authors propose the use of bounding volumes [20] to represent the surface of the human body instead of applying a detailed triangle mesh representation in order to reduce the computational cost needed to compute the distance between the human and the robot, and thus guarantee a real-time processing of the measurements. Nevertheless, in order to reduce even more this computational cost, the authors have developed a hierarchy of bounding volumes [21] [22], not only for the human operator's body but also for the robotic manipulator (Mitsubishi PA-10). This hierarchy is divided into three different levels, which represent different levels of precision and computational efficiency (see Figure 4 for a graphical and tabular representation of this hierarchy for the human and see Figure 5 for the representation of this hierarchy for the robot). The first level is composed of one AABB (Axis-Aligned Bounding Box) which covers all the body of the human operator and another AABB which covers all the structure of the robotic manipulator. The second level is also composed of AABBs, but each AABB represents a functional part of their bodies (arms/torso/legs for the human and base/arm/forearm for the robot). Finally, the third level of the hierarchy presents the maximum level of detail and it is obtained by covering each bone of the human's and robot skeletons by a SSL (Sphere-Swept Lines) volume.

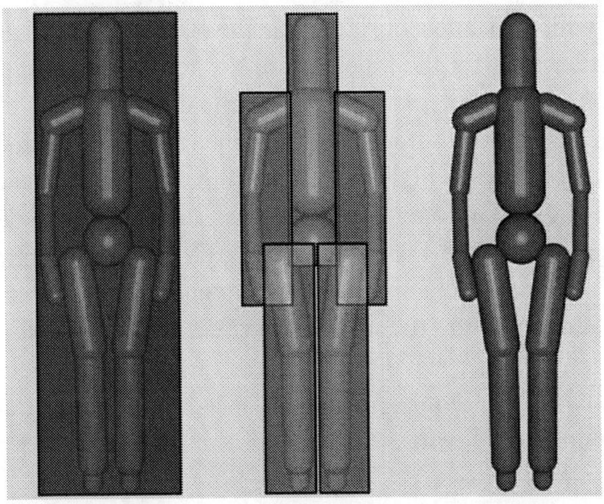

Level 1	Level 2	Level 3
Global AABB	Left Lower Limb AABB	Left Thigh SSL Left Leg SSL Left Foot SSL
	Right Lower Limb AABB	Right Thigh SSL Right Leg SSL Right Foot SSL
	Torso-Head AABB	Abdomen SSL Thorax SSL Neck SSL Head SSL
	Left Upper Limb AABB	Left Shoulder SSL Left Arm SSL Left Forearm SSL Left Hand SSL
	Right Upper Limb AABB	Right Shoulder SSL Right Arm SSL Right Forearm SSL Right Hand SSL

Figure 4. Hierarchy of bounding volumes of the human operator.

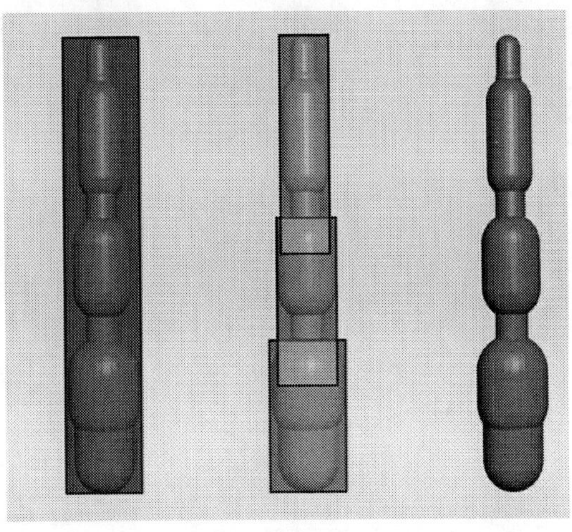

Level 1	Level 2	Level 3
Global AABB	Base AABB	Base SSL S1 SSL
	Arm AABB	S2 SSL S3 SSL
	Forearm AABB	E1 SSL E2 SSL W1 SSL W2 SSL

Figure 5. Hierarchy of bounding volumes of the robotic manipulator.

The bounding volumes of each level of this hierarchy will be used to compute an approximation of the human-robot distance depending on the required precision at each moment of the human-robot interaction task. First of all, the distance computation algorithm will use the first level of the hierarchy (global AABBs) in order to obtain an upper threshold for the human-robot distance. Next, if this distance is bigger than a pre-established threshold, this means that the human and the robot are far away enough to discard the other levels of the bounding volume hierarchy. If this distance is smaller than this threshold, the system will use the second level of the hierarchy (local AABBs) in order to obtain a more precise approximation of the human-robot distance. Similarly to the first step, if this distance is bigger than a second pre-established threshold, the distance computation will have finished since no more accuracy is required. If the second-level distance approximation is not sufficient, the third level of the hierarchy will be used in order to obtain the most precise approximation of the human-robot distance. This hierarchical organization will reduce the number of pairwise distance tests between the bounding volumes and the third level of the hierarchy will only be used when the human and the robot are really too close.

The control system of the robot will take into consideration the value of the distance between the human and the robot in order to adapt the trajectory of the robot accordingly. In particular, in this research two different safety strategies have been implemented. In the first strategy, when the human-robot distance is below a safety threshold, the robot will be stopped and will not be moved until the distance is again bigger than the threshold. In the second strategy, the robot will go away from the human operator in a linear trajectory in order to keep the safety distance. In both cases, the initial visual servoing path tracking controller (with the force controller in case of physical interaction) will be stopped when the safety threshold is exceeded and thus the behaviour of the robot is context-aware and flexible with regard to the human operator's movements. This behaviour is depicted in Figure 6, where the main elements of the robot controller are shown. Some applications where this safety strategy has been applied are described in detail in the following section.

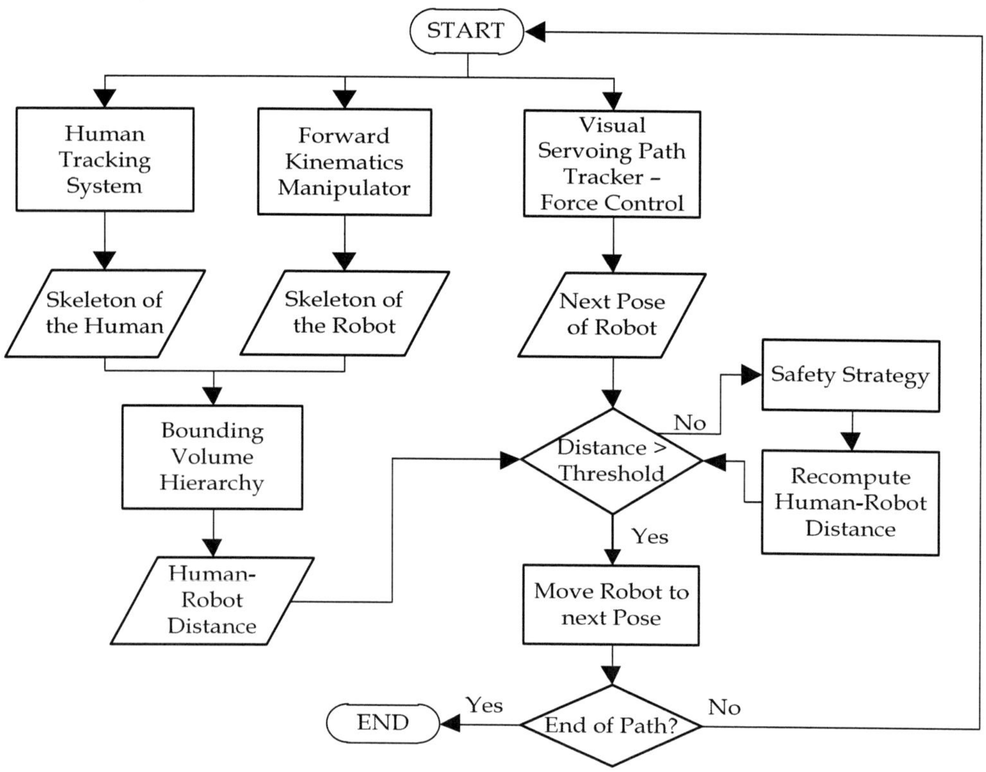

Figure 6. Main diagram of the robot controller.

Applications

The robotic system described in the previous sections has been applied in several assembly and disassembly tasks in order to verify the efficacy and efficiency of the proposed techniques. In the following subsections, some of these applications are described in detail.

Application 1: Disassembly of a streetlamp for bulb replacement

The first application consists on the disassembly of a streetlamp in order to change its blown bulb by a new one. The experimental setup of this application is shown in Figure 7. It is composed by a Mitsubishi PA10 robotic manipulator which disassembles the spherical diffuser of the streetlamp in order to access to the blown lamp, a human operator who leaves a new bulb over a turning table and a storage box where the blown bulb will be thrown away.

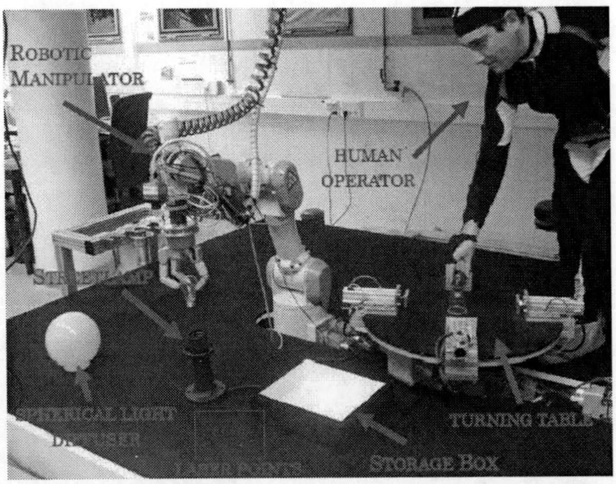

Figure 7. Experimental setup of the disassembly task of a streetlamp.

This task can be divided in the following phases which are shown in Figure 8. Phases of the disassembly task of a streetlamp for bulb replacement.

1. *Disassembly of the streetlamp (Figure 8.* Phases of the disassembly task of a streetlamp for bulb replacement.
2. *a):* First of all, the robotic manipulator grasps the spherical light diffuser of the streetlamp with a three-fingered Barrett hand and turns it until it is released from the base. This moment is detected by a change in the force/torque pattern of force sensor installed at the end-effector of the robot. This is an example of force control in a contact task. Then, the robot takes off the diffuser and leaves it over the floor.
3. *Transportation of the blown bulb (Figure 8.* Phases of the disassembly task of a streetlamp for bulb replacement.
4. *b):* The robot unscrews the blown bulb by analyzing again the force/torque pattern of the force sensor. Next, it uses the visual servoing path tracking technique in order to follow the trajectory between the streetlamp and the storage box. The visual servoing algorithm employs four laser points as image features to be followed by an eye-in-hand

camera (installed at the end-effector). Finally, the robot leaves the blown light inside the storage box.

5. *Collaboration of the human operator (Figure 8.* Phases of the disassembly task of a streetlamp for bulb replacement.
6. *c):* While the robot manipulator is performing phase 2, the human operator enters the workspace in order to leave a new bulb over the turning table. In this case, the human operator collaboration is needed since the storage box of new bulbs is out of the robot's workspace. During this phase, the human tracking system computes constantly the distance between the human and the robot by applying the hierarchy of bounding volumes described above. When this distance is smaller than a safety threshold, the robot stops tracking the trajectory of phase 2 and activates a safety behavior which takes the robot away from the human in a linear path in order to keep the human-robot distance below the threshold.
7. *Installation of the new bulb and assembly of the streetlamp (8d):* When the human operator goes away from the workspace, the safety behavior is deactivated and phase 2 is completed by leaving the bulb inside the storage box. Next, the robot takes the new bulb which is over the table and screws it in the streetlamp's base. Finally, the robot installs again the light diffuser. All these subtasks are performed by executing a visual servoing path tracker for controlling the robots movements and a force control for determining the end of screwing activities.

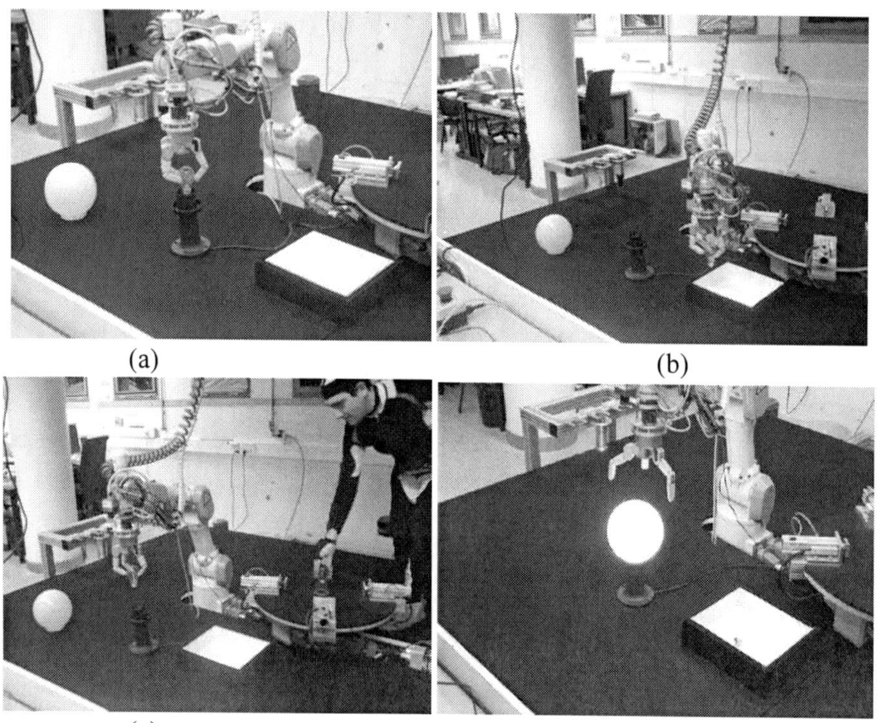

(a) (b)

(c) (d)

Figure 8. Phases of the disassembly task of a streetlamp for bulb replacement.

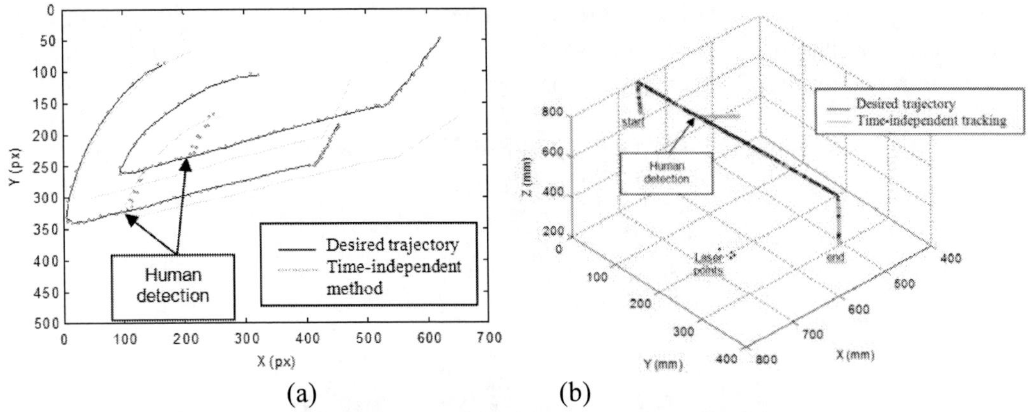

Figure 9. Trajectories of the robot end-effector in (a) image space and (b) 3D space.

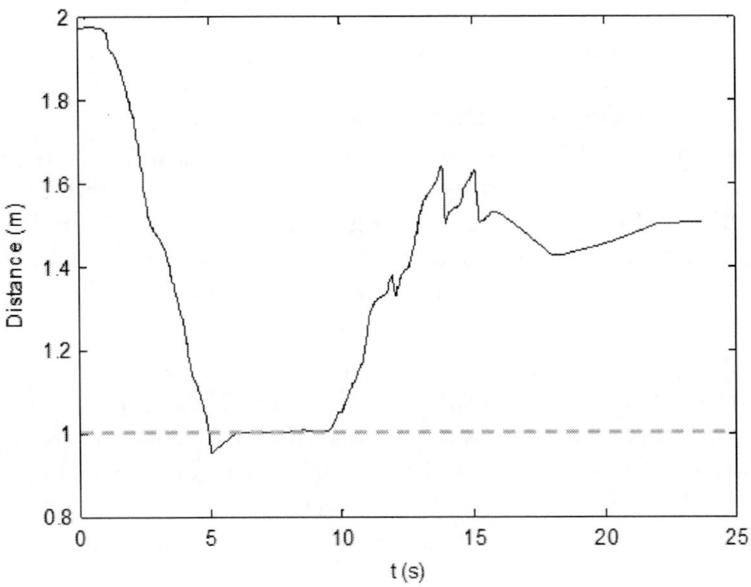

Figure 10. Minimum distance between the human operator and the robotic manipulator.

The time-independent path tracker based on visual servoing is able to track the trajectory of the robot even after the execution of the safety behavior during phases 2 and 3 of the task. The end-effector of the robot follows all the intermediate steps of the initially established trajectory not only in the image space of the visual features but also in the tridimensional space (Figure 9b). Thereby, the robot is able to leave the blown bulb in the storage box although the path tracker is blocked by the safety behavior for several seconds.

On the other hand, the safety behavior is activated during the execution of phase 3 and guarantees that the minimum distance between the human operator and the robotic manipulator is higher than a safety threshold (1m). As it is shown in Figure 10, when the human enters the workspace, the human-robot distance decreases until instant t=5s when it goes below the safety threshold. In this moment, the visual servoing path tracker is stopped and the safety strategy is activated. This strategy moves the robot away from the human and keeps the human-robot distance over the safety threshold. The safety strategy is deactivated approximately at instant t=10s when the human operator goes away from the workspace and the human-robot distance is again greater than the safety threshold.

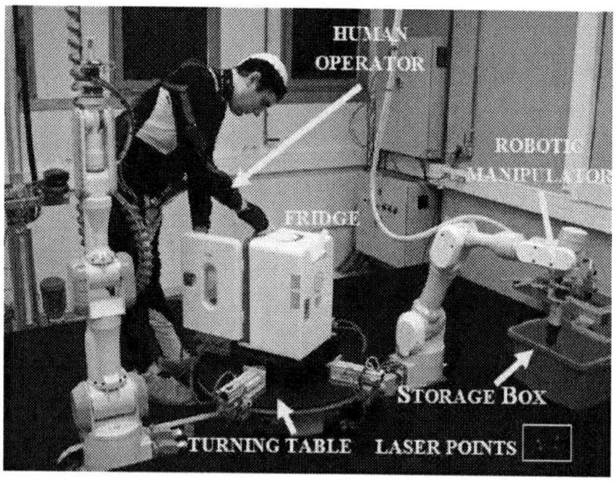

Figure 11. Experimental setup of a disassembly task of an electrical appliance (fridge).

Application 2: Disassembly of an electrical appliance

The second developed application is also based on the disassembly of an element composed by several parts. In this case, the robot has to disassembly a small electrical appliance (a fridge). The collaboration of the human is needed to open the fridge door and remove a tray which is inside since the robot has not enough dexterity to complete the task itself. Figure 11 depicts the experimental setup of this task.

This task can be divided into the following three phases:

1. *Positioning of the robotic manipulator over the rear lid (Figure a):* First of all, the robotic manipulator has to position its end-effector over the rear lid of the fridge in order to unscrew its screws. In order to do this, the robotic manipulator follows a pre-established path by executing the time-independent visual servoing path tracker. This technique uses the eye-in-hand camera of the robot's end-effector and follows four laser points on the floor.
2. *Collaboration of the human operator for tray removal (Figure b and Figure c):* While the robotic manipulator is performing phase 1, the human operator enters the workspace to open the door of the fridge and extract the interior tray. When the human tracking system determines that the human-robot distance is smaller than a safety threshold, the system activates the safety behavior in order to keep this safety distance. The path tracking process is reactivated when the human goes away and the human-robot distance is again greater than the threshold.
3. *Unscrewing of the rear lid (Figure d):* When the robotic manipulator finishes the path tracking of phase 1, the robot is situated just in front of the rear lid of the fridge. Then, the eye-in-hand camera segments the holes of the screws and begins to unscrew each of the four screws which hold the rear lid. During the screwing process, the robot controller analyses the force/torque pattern of the force sensor installed in the end-effector to determine the moment when each screw is completely unscrewed.

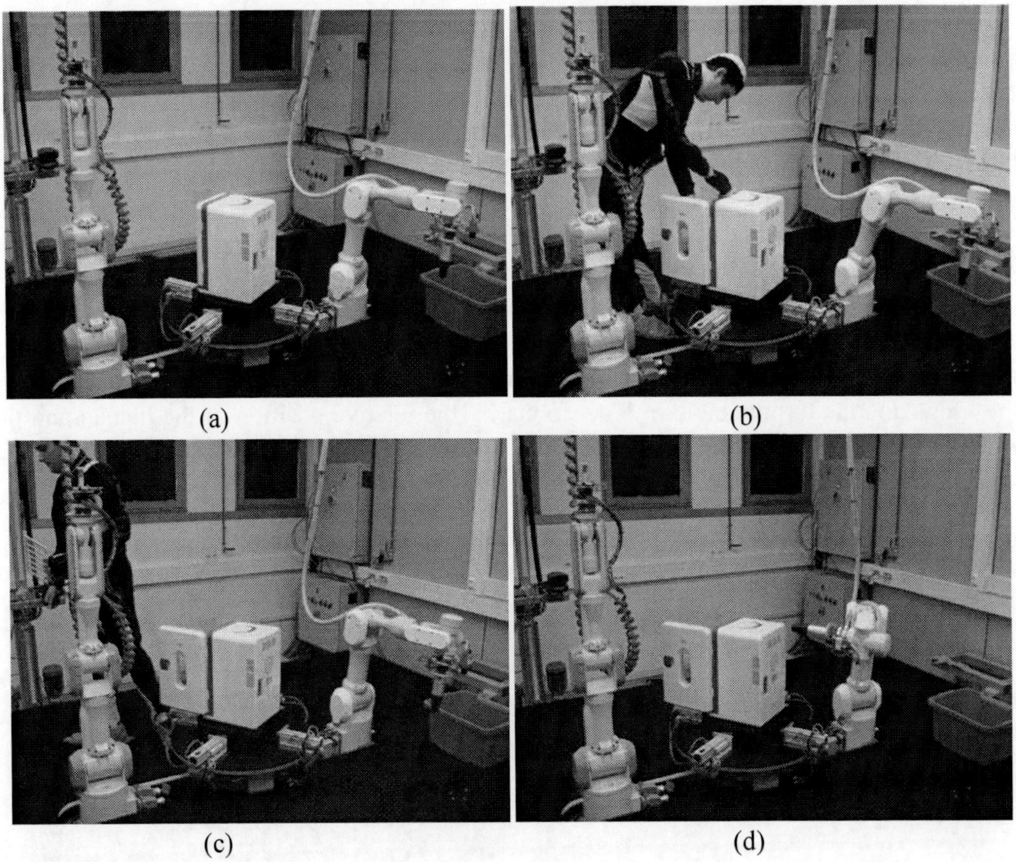

Figure 12. Phases of a disassembly task of an electrical appliance (fridge).

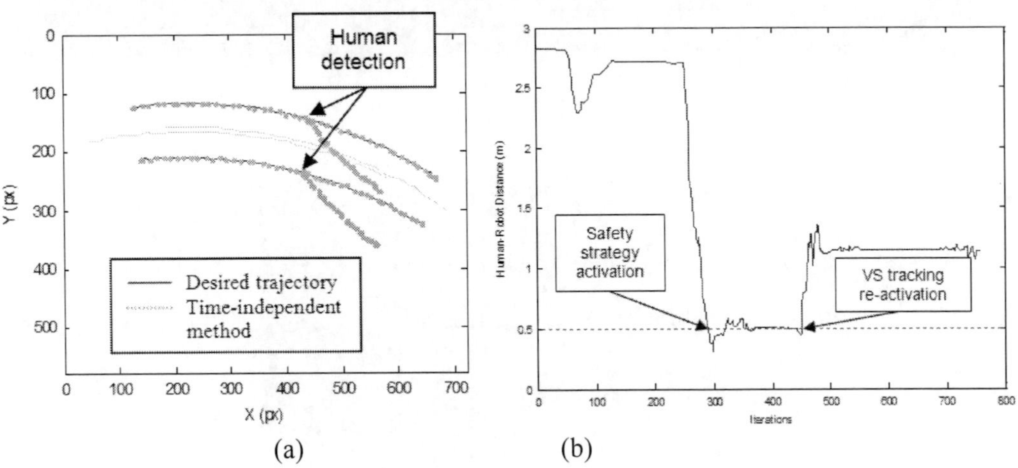

Figure 13. Evolution of the robot's movement: (a) tracking of the visual features in the image space and (b) minimum human-robot distance.

As in the previous application, the time-independent path tracker is able to follow all the intermediate steps of the trajectory in phase 1 even after having been stopped during the execution of the safety strategy (see Figurea). In fact, the safety strategy is activated when the human operator approaches the fridge and the human-robot distance is below the safety threshold (0.5 m in this application). When the human-robot distance is again higher than the safety threshold, the visual servoing path tracker is reactivated (see Figureb).

The developed robotic system uses the human tracking system in order to update precisely the human's skeleton location and the hierarchy of bounding volumes in order to compute the minimum human-robot distance. For instance, Figure 14 shows a sequence of same pictures where the bounding volumes of the human and the robot (SSLs) are shown during the human collaboration (phase 2 of the task). It has also been added the 3D model of the fridge and the turning table in order to clarify the relative position of the human and the robot. By using these bounding volumes, the safety strategy can be implemented since the system is able to compute on real-time the human-robot distance.

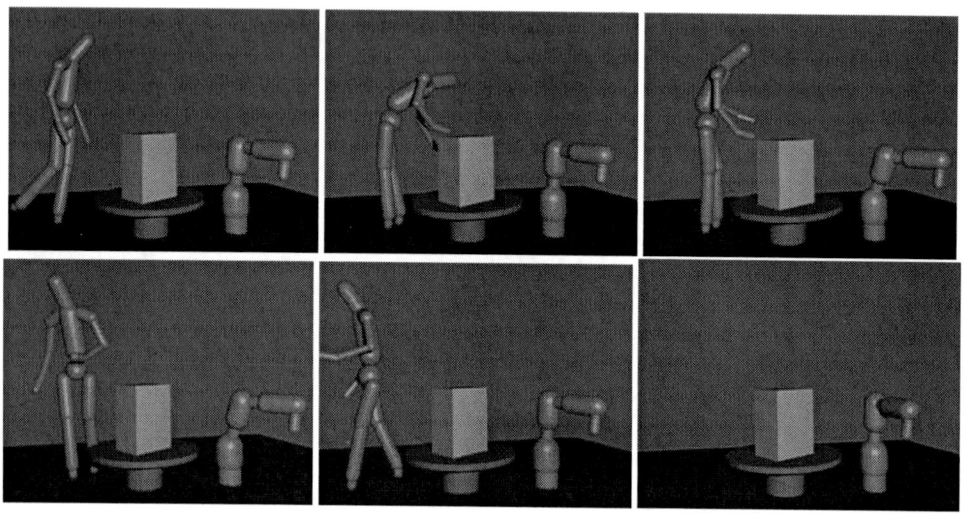

Figure 14. Sequence of bounding volumes (level 3) generated during the human-robot interaction phase of the task.

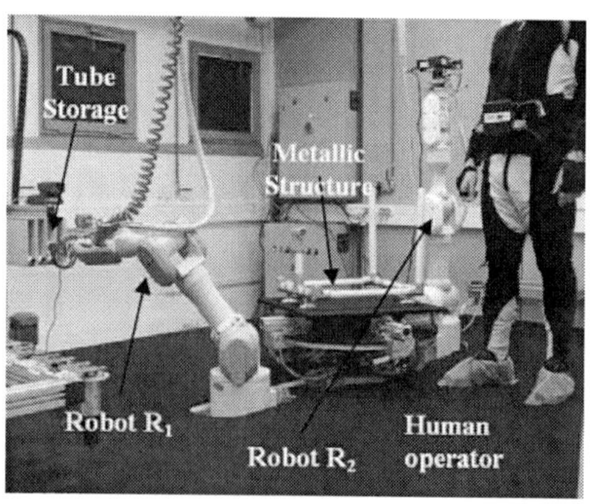

Figure 15. Experimental setup of an assembly task of a metallic structure.

Application 3: Assembly of a metallic structure

The last application is an assembly process of a metallic structure constructed by connecting several tubes. In this case, the human collaboration is required to install the connectors between the tubes because they are composed of two parts and they are very difficult to be handled by a robot. In addition, a second manipulator is also needed in order to perform several actions of the assembly task in parallel. Figure shows the experimental setup of this task: the two Mitsubishi PA-10 robots which cooperate during the task, the storage box where the tubes of the structure are stored, the metallic structure which is situated over the turning table and the human operator who collaborates in the task.

This task can be divided into the following three phases:

1. *Robot cooperation for horizontal tube installation (Figure 16a):* The main goal of robot 2 in this task is to hold the tube whose ends have to be connected by the human operator. Nevertheless, this tube is inside the storage box which is out of its workspace. Due to this fact, robot 1 has to take a tube out from the storage box and handle it to robot 2. Robot 1 has a parallel gripper for extracting the tubes from the box while robot 2 has a three-fingered Barrett hand for holding firmly the tube. When robot 2 grasps the tube, it moves it between two vertical tubes of the metallic structure where the human operator should add the connectors.
2. *Insertion of a vertical tube in the structure (Figure 16b, Figure c and Figure d):* While robot 2 is holding the horizontal tube, robot 1 extracts a new tube from the storage box in order to insert it vertically in an empty connected of the structure. The movement of robot 1 is controlled by an impedance control with a position reference of a time-independent path tracker, as described in section 3. As in the previous tasks, robot 1 uses the eye-in-hand camera to track laser points which are projected on the floor, near the robot. However, in the case, in these insertion tasks is required an impedance control since the robot has to modify the predefined image trajectory to be compliant with the contacts between the tube and the connector.
3. *Collaboration of the human for connector installation (Figure b and Figure c):* While phase 2 is being executed, the human operator enters the workspace for the installation of two T-connectors at the ends of the horizontal tube which is hold by robot 2. When the human-robot distance is below a safety threshold (1m), the safety behavior is activated and the visual servoing path tracker of phase 2 is stopped. When the human abandons the workspace after having screwed the two connectors of the horizontal tube, robot 1 continues the impedance control of phase 2 and inserts the vertical tube in the structure (see Figure d).

Figure 17a shows the evolution in the 3D Cartesian space of the end-effector of the robot during the development of phase 2. This plot compares the original trajectory planned for the visual tracker with the real one obtained with the method of Section 3, which modifies the image path accordingly to the measured forces. Figure 17b shows the forces measured by the force sensor located at the end-effector of the robot during the insertion of the tube.

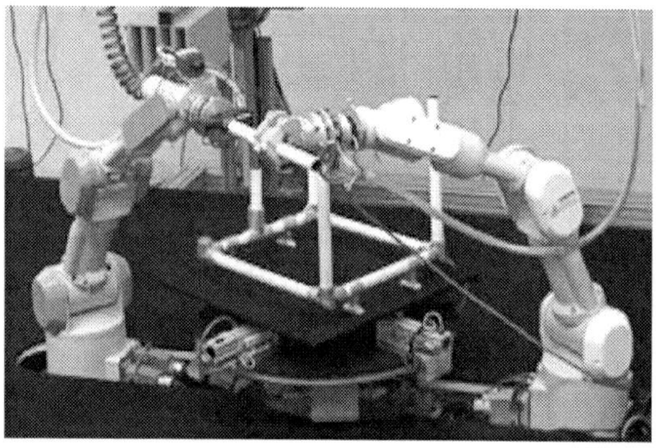

(a)

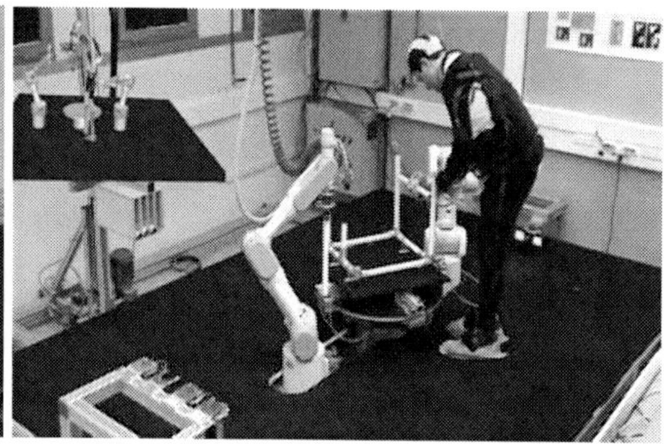

(b)

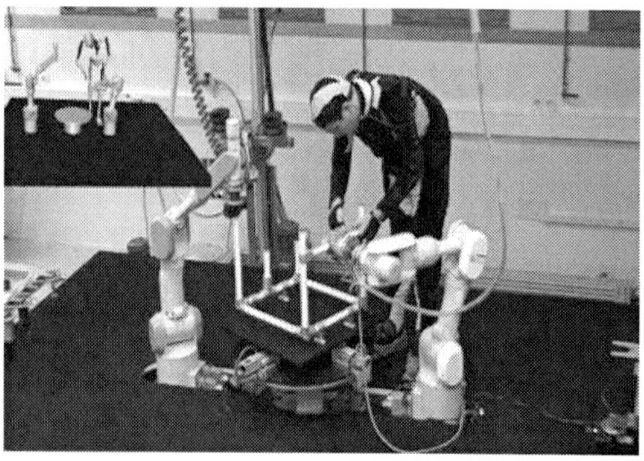

(c)

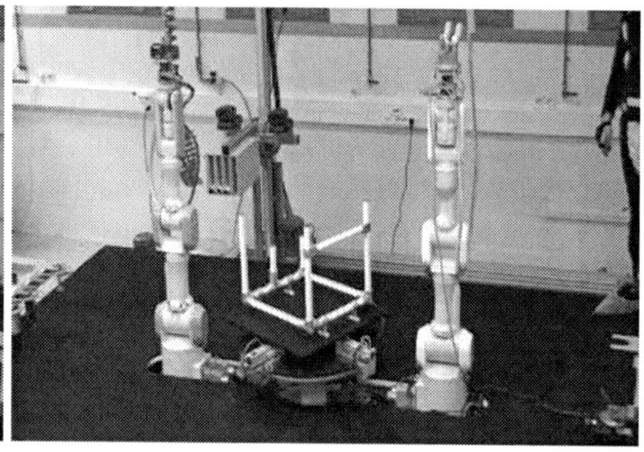

(d)

Figure 16. Phases of an assembly task of a metallic structure.

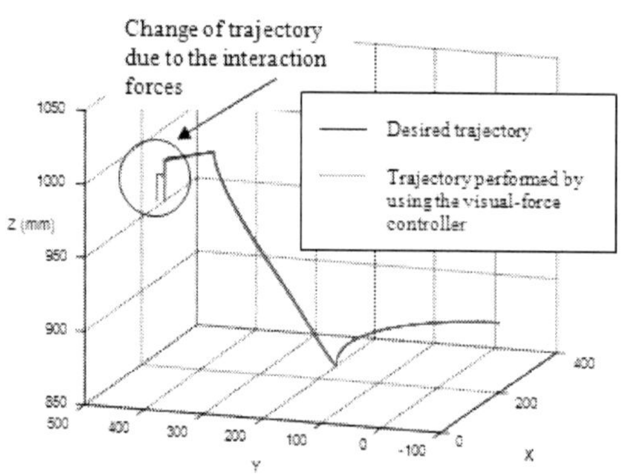

(a)

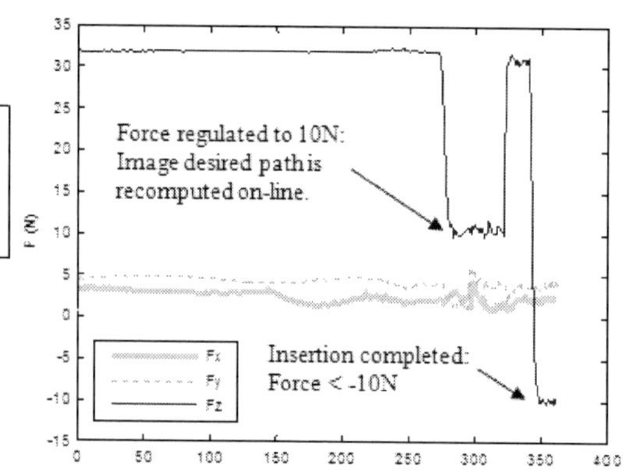

(b)

Figure 17. Evolution of the impedance control in a tube insertion task: (a) 3D evolution of the end-effector and (b) Forces measured at the robot end-effector.

Conclusions

This chapter presents the integration of several sensor-based techniques which are used to develop flexible assembly/disassembly tasks with robotic manipulators. In particular, the developed system is composed of three main components: visual servoing path tracker, force control system and human tracking system.

The visual servoing path tracker deals with the control of the robot's movements so that the end-effector of the robot follows a pre-established path. The main advantage of this method over traditional position-based robotic controllers is that the trajectory of the robot can be adapted to changes of the object's positions because the trajectory depends on visual features and not on absolute positions. Thereby, the manipulated object can be positioned in any point of the robot's workspace and a completely structured environment is not required. In addition, the time-independency feature of the developed algorithm assures that the desired path will always be completed even though there are unexpected situations where the normal robot's behavior has to be stopped.

The force control system complements the visual servoing path tracker in situations where contact with the environment is required. The use of the force control permits to modify the initial planned path of the object in case that the robot contacts an object. In addition, the analysis of the force/torque patterns applied by the end-effector of the robot helps to determine when contact tasks (such as a screwing activity or an insertion task) are finished.

Finally, the human tracking system obtains precise information about the location of the human operator who collaborates with the robots in the development of the tasks. This tracking system is composed of two elements: the sensor fusion system and the hierarchy of bounding volumes. The sensor fusion system combines the measurements of an inertial motion capture system and a UWB system so that a skeleton with the location of all the human operator's body parts is computed. This skeleton is covered by the hierarchy of bounding volumes and thus an efficient approximation of the human-robot distance can be computed. This distance is analyzed by the robot controller and if it is smaller than a safety threshold, it activates a safety strategy of the robot which avoids collisions with the human operator.

Although this robotic system increases the flexibility of the tasks where robotic manipulators work, the developed safety strategies are very simple since they involve the complete stop of the visual servoing path tracker while there is some danger of collision. In future research, the authors will try to include the safety strategy inside the visual servoing path tracker so that the safety behavior will reconfigure the original path so that the delay in the development of the task is minimum while the human operator's safety is guaranteed.

Acknowledgments

This work is supported by the Spanish Ministry of Science and Innovation through the research project DPI2008-02647 ('Intelligent Manipulation through Haptic Perception and Visual Servoing by Using an Articulated Structure situated over a Robotic Manipulator').

References

[1] S. Hutchinson, G. G. Hager and P. I. Corke, "A tutorial on visual servo control", *IEEE Transactions on Robotics and Automation*, **12** (5), 651-670 (1996).

[2] F. Chaumette and S. Hutchinson, "Visual Servo Control, Part I: Basic Approaches", *IEEE Robotics and Automation Magazine*, **13** (4), 82-90 (2006).

[3] Y. Mezouar and F. Chaumette, "Path Planning For Robust Image-based Control", *IEEE Transactions on Robotics and Automation*, **18** (4), 534-549 (2002).

[4] G. Chesi and Y. S. Hung, "Global path-planning for constrained and optimal visual servoing", *IEEE Transactions on Robotics*, **23**, 1050-1060 (2007).

[5] E. Malis, "Visual servoing invariant to changes in camera-intrinsic parameters", *IEEE Transactions on Robotics and Automation*, **20**, 72-81 (2004).

[6] J. Pomares and F. Torres, "Movement-flow based visual servoing and force control fusion for manipulation tasks in unstructured environments". *IEEE Transactions on Systems, Man, and Cybernetics—Part C*, **35** (1), 4-15 (2005).

[7] G. J. Garcia, J. Pomares and F. Torres, "Automatic robotic tasks in unstructured environments using an image path tracker", *Control Engineering Practice* , **17**, 597-608 (2009).

[8] J. Baeten, and J. De Schutter, "Hybrid Vision/Force Control at Corners in Planar Robotic-Contour Following", *IEEE/ASME Transactions on Mechatronics*, **7** (2), 143-151 (2002).

[9] B. J. Nelson and P. K. Khosla, "Force and vision resolvability for assimilating disparate sensory feedback", *IEEE Transaction on Robotics and Automation*, **12** (5), 714-731 (1996).

[10] Y. Zhou, B. J. Nelson and B. Vidramaditya, "Fusing force and vision feedback for micromanipulation", *Proceedings of the IEEE International Conference on Robotics and Automation*, 1998 (Leuven, Belgium), pp. 1220-1225.

[11] D. E. Whitney, "Force feedback control of manipulator fine motions", *Journal of Dynamic Systems, Measurement and Control*, 97-97 (1977).

[12] E. Malis, G. Morel, and F. Chaumette, "Robot control from disparate multiple sensors", *International Journal of Robotics Research*, **20** (5), 364-378 (2001).

[13] J. Pomares, G. J. Garcia and F. Torres, "A robust approach to control robot manipulators by fusing visual and force information", *Journal of Intelligent Robotic Systems*, **48**, 437-456 (2007).

[14] G. J. Garcia, J. Pomares and F. Torres, "Robot guidance by estimating the force-image interaction matrix", *IFAC International Workshop on Intelligent Manufacturing Systems*, May 2007 (Alicante, Spain).

[15]Animazoo, "GypsyGyro-18 motion capture system", 2011 at http://www.animazoo.com

[16] G. Welch and E. Foxlin, "Motion tracking: no silver bullet but a respectable arsenal", *IEEE Computer Graphics and Applications*, **22** (6), 24-38 (2002).

[17]Ubisense, "Ubisense RTLS", 2011 at http://www.ubisense.net

[18] J. A. Corrales, F. A. Candelas and F. Torres, "Hybrid tracking of human operators using IMU/UWB data fusión by a Kalman filter", *3rd ACM/IEEE International Conference on Human-Robot Interaction*, Mar. 2008 (Amsterdam, Holland) pp. 193-200.

[19] J. A. Corrales, F. A. Candelas and F. Torres, "Sensor data integration for indoor human tracking", *Robotics and Autonomous Systems*, **58** (8), 931-939 (2010).

[20] J. A. Corrales, F. A. Candelas and F. Torres, "Safe human-robot interaction based on dynamic sphere-swept line bounding volumes", *Robotics and Computer-Integrated Manufacturing*, **27** (1), 177-185 (2011).

[21] J. A. Corrales, F. Torres and F. A. Candelas, "Safe Human-Robot Interaction based on a Hierarchy of Bounding Volumes", *IEEE ICRA 2010 Workshop on Multimodal Human-Robot Interfaces*, May 2010 (Anchorage, Alaska) pp. 74-79.

In: Robotics: State of the Art and Future Trends
Editors: G. Legnani and I. Fassi
ISBN: 978-1-62100-403-5

Chapter 10

ROBOTICS IN REHABILITATION - PART I: REQUIREMENTS AND CONTROL ISSUES

Nicola Pedrocchi[1,*], *Matteo Malosio*[1], *Federico Vicentini*[1], *Lorenzo Molinari Tosatti*[1], *Marco Caimmi*[1,2] *and Franco Molteni*[2]
[1]Institute of Industrial Technologies and Automation, National Research Council, Italy
[2]Villa Beretta, Fondazione Valduce

Abstract

Robotics in rehabilitation present many aspects in the domain of physical human-robot interaction. The design and the application of robots in neuro-motor recovery of impaired people pose, in fact, demanding requirements in terms of robot control, safety, dependability and openness to sensor interfaces for the provided solutions. Conditions and modalities of the use of robots in rehabilitation is still debated because of the relatively recent availability of suitable machines and, mostly, because of the manifest difficultie in setting clinical trials for experimentations and validations. Nevertheless, the rehabilitation community comes to a general agreement about the effica y of natural movements during training sessions. As a resulting requirement, robotic systems in rehabilitation have to display properties beyond the state-of-the-art in terms of kinematic topology, available power and, more importantly, in control architectures. Such architectures, in particular, should be suitable for allowing the implementation of the most advanced interaction control strategies and the integration of very helpful subsystems, like sensors, motion tracking and analysis systems, virtual and augmented reality, multiple interfaces. Focusing on the rehabilitation of the upper limb, the use of robotic manipulators as end-effector based assisting machines enables a wide range of investigations in the fiel of interaction control. Users (in this case patients) are therefore considered entirely part of the system from both an application (safety and dependability) and a modeling perspective. In this work, reviews of human-robot interaction requirements, control strategies and modeling are presented. Such background introduces a case study about the use of a manipulator coupled with several subsystems for the rehabilitation of the upper limb.

*E-mail address: nicola.pedrocchi@itia.cnr.it

PACS 05.45-a, 52.35.Mw, 96.50.Fm.
Keywords: Rehabilitation Robotics, Soft Robotics, Human Robot Interaction, Exoskeletons.

1. Introduction

Service robotics has been widely reckoned as one of the leading trends for the advances in both design and application of robotic devices. In particular, medical and, more specifically, rehabilitation robotics pose significan challenges in terms of requirements. Key aspects to be considered at both design and usage phases include, in fact: the interaction with humans that experience a particular temporary condition of impairment; the safety of applications and machines that are used in clinical contexts; the wide range of interaction modes that may occur during different phases of the rehabilitation process; the need of adaptability of robot applications to the peculiar conditions of treated patients and their physical features; the essential accordance and effica y of any robot application to the medical and therapeutic purposes. Unlike other medical fields rehabilitation robotics presents relatively limited experience from a clinical evaluation view point. In addition to the short period since the introduction of rehabilitation robots, the difficult in setting up large trials prevents a comprehensive validation of methodologies, guidelines and standards. Having all these medical aspects poorly contributed to build sound and consolidated clinical scenarios where robots are to be introduced, technological requirements in design and development of robotics applications are, as a consequence, displaying a wide range of approaches, whose detailed description is beyond the scope of this work.

Nevertheless, a common framework of requirements can be traced for both the design of robots and the control strategies that are implemented. Regardless the type of machines that are used in rehabilitation robotics, the main aspect here introduced includes the physical robot-patient interaction, which in turn can be declined according to a perspective of safe interaction and to the control strategies that are actually implemented in order to achieve effective behavior for the rehabilitation purposes. Therefore, the methodological framework where robotics in rehabilitation is placed is that of physical Human-Robot Interaction (pHRI). Significantl , rehabilitation happens to be an application fiel for pHRI robotics where all its major features are exploited, namely the mechanics and control of robots in close functional contact with users (patients), the safety and dependability perspective in robot design and usage, the emotional/relational involvement of users with the training personnel, with the environment (including, for instance, Virtual Reality experience) and, definitel , with the robotic machines. The pHRI aspects in the design process, which predominantly takes into account the clinical purposes and requirements previously drafted, is presented in a companion paper [1]. The pHRI aspects related instead to the safety and dependability are introduced in this paper from a general view point, along with the detailed discussion about the control issues that must be taken into account in designing applications for rehabilitation. In particular, both control techniques and dynamical modeling are key points in the development of pHRI applications, especially for those involving complex interactions. Complexity in interactions, within the scope of this work, is mainly due to the conditions of human users (impaired) and the focus of dynamics/control discussion onto the upper limb, which is fairly more articulated than the lower limb. Therefore, all the technology involved is considered according to the medical application requirements that, in turn, are strongly influence by the modality and purposes of the rehabilitation process.

An introduction to the rehabilitation context and open issues is given in the following Section 2. in order to outline the background conditions for the design of robots and, mostly, the development of control strategies. Due to the close interaction with users, the concepts of safety and dependability are reviewed in Section 3. while the low-level robot control issues are reviewed in Section 4. together with some modeling consideration about the human limb. The human side of the interacting system, in fact, is all the time involved not only as a user, yet generating a number of remarkable requirements in terms of safety and control performances, but also as an active part of the whole application. A case study is therefore presented in Section 5., where the implementation of a system responding to the functional requirements outlined along this work and relying on the control techniques most suitable for the interaction with users is presented.

2. The Context of Robot-assisted Rehabilitation

Rehabilitation may be seen as a set of interventions aiming at improving and maintaining the subject's motor control during a process of restoring motor functions in people with neurological impairments. Motor control would be more precisely define as *acting ability* because any movement is always due to an *ideomotor* neural representation of a specifi task, with a specifi goal and in a specifi context. ''Moving'' means to perform an action, although two theoretically identical movements, when performed in various environments and contexts, may become distinct actions characterized by specifi sensations, proprioceptive information and motor aspects. A hypothetical subject performing the two above mentioned movements, define as identical, is actually having two different cognitive and sensory experiences and the actions related to motor control mechanisms are therefore different as well [2, 3, 4, 5]. In 1991, Mulder, introducing his model of the human motor behavior, stated:

> "*motor behavior is the result of an integrated and adaptive biological system. Cognitive, perceptual, and motor mechanisms are not independent elements, but are viewed as inseparable parts of this functional system. The human body is a wonderful biological machine designed for adaptive and flexible functioning in an ever-changing context.*" [6]

These are key-points that should be taken into account during the designing process of rehabilitation robots. Rehabilitation robots, in fact, should not be considered as machines just acting on the patient for movement assistance but rather as learning machines that continuously interact with the patient and the environment during the performance of an action. An ideal robot should therefore perceive the patient's intention to move, be aware of the chosen movement strategy and monitor the limbs motion during the execution of a task. During motion, the robot should also *decide* whether and when to assist the action or, contrarily, to even resist the movement.

State-of-the-art machines, however, are far from such advanced adaptive behaviors that are meant to encompass algorithms based on known principles of the human motor behavior. Instead, for any required action (*e.g.* reaching for an object), a movement strategy is given in a pre-programming phase while feedback available during motion are used for online motor control. Within this framework a robot is adaptive only in the sense of providing an interactive behavior with both the patients and the environment through sensors and controls. Biological signals acquired are, in particular, processed through algorithms specificall developed for the interpretation of pathological behaviors. For example, a hypo-activation (observed through the analysis of the electromyographic signal) in an agonist muscle during a given movement may either trigger an NFES stimulator or may be interpreted as the need for extra movement assistance. Or, contrarily, could even trigger a stop of the robot until a large enough signal is read. These are all valid possibilities and a combination of these three interventions may also be a valid solution. Therefore, the right solution should be chosen on the basis of a sound theory regarding the neurological re-learning process but always taking into account the patient's condition and residual abilities - both physical and neurological - for the customization of the treatment. Such a right solution, in turn, should derive from a consistent medical methodological approach that, however, at the state of the art presents multiple branches.

There are in fact two main philosophies regarding robot-aided rehabilitation. The firs states that robots should assist the patient in the performance of functional movements and they should therefore allow movements to be as natural as possible. The rationale relies upon the evidence that patients mostly improve in the movements they practice. The other states that robots should be compliant with patient interaction and provide a minimum degree of movement assistance. This approach relies on the evidence that a preserved relationship between patient effort and resulting movement together with the preserved possibility to make errors improve the re-learning process [7, 8, 9, 10]. One approach focuses on motion features, while the second focuses more on motor schema learning: the main differences lie in the way the neural feedback is carried through the movements performed by patients and assisted by robots. As a result, the right choice in design and control of robots becomes predominant in the achievement of functional objectives in both approaches. In addition, any approach should nevertheless always be chosen by taking into account the patient's general condition and his residual motor abilities.

Summarizing, depending on the philosophy followed, robots may be more or less compliant and may either supply minimum assistance or totally assist the patient during movement. In all cases, however, the monitoring of the patient status is always advisable. Patients in fact will hardly respond uniformly to a determined robotic treatment and the neuro-muscular response should therefore be instrumentally assessed. Abnormal neuro-muscular patterns underlying the movement could be, in fact, present but not observable without a proper instrumental evaluation [11]. It is becoming more and more evident that a properly designed and controlled robot for rehabilitation should also display the ability to sense the patient's cognitive and neurological condition during exercises. As a complex integrated system capable of continuously adjusting for the patient's changing response and environmental circumstances, the robot has to fulfil a number of functional requirements. In particular, robots:

(i) should allow smooth movements on curved trajectories with nonlinear velocity profiles resembling the human motor behavior;

(ii) should be equipped with force sensors for measuring and controlling the patient/robot interaction;

(iii) should be equipped with an integrated surface electro-myographic (EMG) acquisition system;

(iv) should be equipped with a biomimetic controller programmed according to an integrated and adaptive biological model;

(v) should provide the patient with proper visual and haptic feedback.

Considering the firs functional requirement, the motion paths and profile are key features of the trajectories to be performed from a clinical viewpoint, remarkably posing a corresponding foundational set of requirements for the development of suitable control strategies. On one side, under the assumption that robots should assist patients in performing smooth movements, preferably functional and activities-of-daily-living (ADL) movements [12], it still remains unclear *which* particular movements should be considered. While treatments for lower limbs usually display standard motion laws because walking is a basic and slightly varying motion task, in the case of the upper limb the definitio of suitable movements is not trivial. Upper limbs are incredibly adaptive organs that allow the performance of high-level interactive movements; are capable of performing numerous functional tasks in an infinit number of kinematic solutions; are sophisticated multi-joint systems suitable for heavy jobs as well as for fin movements, like the exploration and the manipulation of objects. Certainly not all movements may be trained and this variability makes the selection of a reduced number of movements necessary. A selection in fact should be preferably done on the basis of key determinants of the movement control [13] having a clear shared framework methodology not being posed yet. In addition, once the target movements have been selected , the definitio of the rightly tuned motion profil is still a hard task from a control view point because of the consequences of the control output on the fina functional movements and, ultimately, on the therapeutic outcomes. By combining the motion profiling *i.e.* basically the domain of position/velocity control, with the interaction with the patient, the number of degrees of freedom and features to be combined into a control strategy become larger and more cumbersome. By introducing interaction control, the set of requirements and issues broadens towards the dependability of the system, *i.e.* the way the system induces confidenc in the user and appears to be safe enough.

3. Physical Human-Robot Interaction

Recent industrial robotics trends have very much given consideration to the issues of pervasive presence of robots within human populated environments. Both research and industry-driven actions (like recent PHRIDOM [14], PHRIENDS [15] and SMErobot [16] projects) have been mainly devoted to the enhancement of pHRI concepts, mainly under an industrial perspective. In fact, the key pHRI figure include the cooperative execution of tasks by humans and robots and the sharing of the same workspace. The notion of *interaction* is therefore declined in terms of *contact*, mirroring the task cooperation and the workspace sharing scenarios according to the notion of contact *by planning* or *by accident*. The former is primarily involved in most of the low-power tasks, like manipulation, handling, assembly, etc or, actually, the robot-assisted rehabilitation and other medical tasks.

The latter has been considered as the major issue to be discussed and formalized in relationship with the requirements of safety. Most of industrial scenarios, as a matter of fact, involve accidental contacts and related countermeasures (device design, control strategies, application algorithms). This arises from the need of removing limiting factors for production and/or fl xibility of robotic cells, like fences and blunt stops as solely response to rising alarm on safety (*e.g.* getting closer to the robot). Being the robot technology able to provide more fl xible solutions, *i.e.* modulating the response of the system according to actual conditions, for instance just limiting speed or re-planning the execution trajectories, the further step is represented by the standardization of guidelines and conditions which are used in sharing the same workspace. Revision of ISO 10218-2 standard, ''Application and Use of Robots in the Work Places'', is intended to address the work place safety requirements from an end-user standpoint. Nevertheless, the discussion and the formulation of the standard cumulated a significan delay (expected 2006) because of the sensitive matter related to safety and reliability of mechanical and control architectures. Such scenario does not even further display easiness when involving medical devices and related standards.

3.1. Safety Issues

Safety issues are mainly devoted - also from a standardization viewpoint - to the prevention of the risks of collision occurring between robots and users. Yet not completely removable, energy transfer by the robot, resulting in serious human injuries, has to be limited. Many severity indexes of injuries and damages criteria in pHRI are proposed and evaluated in robot safety assessment ([17] and citations therein). As a result, last decade was devoted to the exploration of safety countermeasures in all aspects of manipulators design, including mechanics, electronics, and software. Safety can be therefore achieved by design, *i.e. intrinsic safety* or *by control* (see Fig. 1). Pursuing a safe behavior by control is therefore accounted as "tactical" rather than "strategical", *i.e.* a set of means operating at runtime rather than casted by design. It extendedly involves control laws as well as sensor-drive collision prevention algorithms, mainly based on reactive collision avoidance and prediction of human behavior. Path planning jointly plays also a role in terms of safety and dependability because the robot trajectories can be purposefully made non-threatening w.r.t. the user. Indeed, control methods and collision reaction algorithms cannot fully cover for poor mechanical features in case of impacts, but they are fundamental for performance and reliability improvements. In addition, safe architectures introduce also additional requirements and child assessments against system faults and human unpredictable behaviors. On the other hand, intrinsic safety is barely detached from the control layer because most of compliant actuation principles are actually driven by suitable controllers rather than being compliant *per se*. Despite any effort in classification safety still remains a closely integrated concept in design and in implementation of suitable controllers and of monitoring applications during task execution.

3.1.1. Intrinsic Safety

Among design principles related to *intrinsic safety*, a major role is played by the reduction of robot effective inertia, suitable both for limited energy transfer in accidental contacts

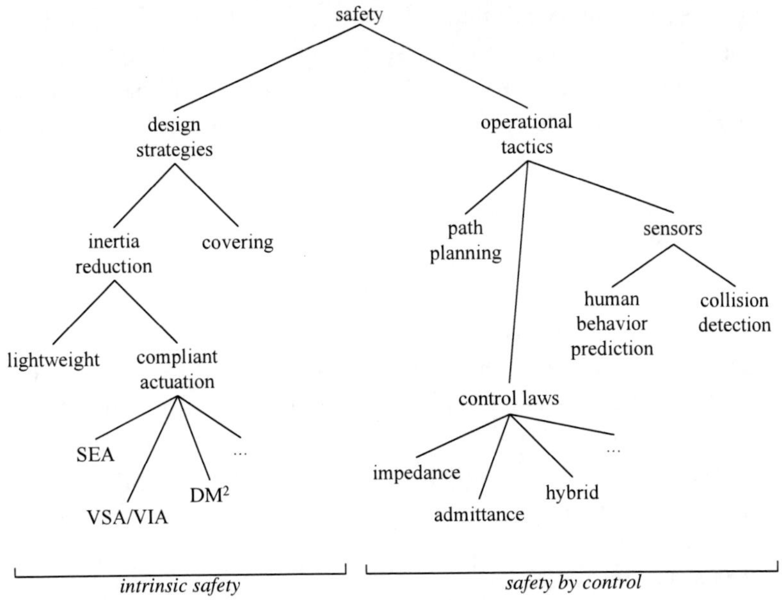

Figure 1. Safety chart.

and for comfortable perception of handled device during cooperation tasks. Effective inertia originates from the mechanical structure and, more importantly, from the rotor dynamics. Therefore **lightweight** manipulators happen to be particularly fi for safe interaction. A typical example is given by DLR/KUKA Roboter GmbH Lightweight Robot [18, 19], which operates a payload equal to its own weight (13,5-16 kg, depending on links cover materials). Light but stiff materials were used for the links, as well as harmonic drives were used for elastic joint transmission together with joint torque control. Alternatively, tendon-driven solutions allow the relocation of all relevant weights (mostly, the motors) at the robot base, like for the Barrett Inc. Whole Arm Manipulator (WAM) [20] originally developed at MIT [21, 22]. Coupled with lightweight, possibly elastic, links, an entire class of **compliant transmissions** is used for mechanically separating the inertias of the motors from those of the links. Within this class, the prominent actuation principles include, among other, the Variable Impedance Actuation (VIA) as an extension of the Variable Stiffness Actuation (VSA) [23, 24], where dynamical features such as stiffness, damping, and gear-ratio of the transmission rapidly and continuously vary during task execution; the Distributed Macro-Mini Actuation (DM^2) methodology [25], where each joint hosts a parallel-pair of actuators, one for heavy low-frequency torque generation, the other for light high-frequency torque actuation, combining all impedances of the manipulator-actuator system; the Series Elastic Actuation (SEA) [26, 27], where a spring of (measured) stiffness is located between the motor and the load.

A common design issue encompasses the reduction of the effective inertia of the whole manipulator by the elastic decoupling of the rotor inertias, see also [28]. In addition, most of the solutions based on elastic joints provide a torque feedback in order to reject high frequency disturbances. From modeling and control viewpoints, however, link deformation must be considered as well as the presence of compliant transmissions, for which deformations can be assumed to be concentrated at the joints. The direct effect is often a significan reduction of control bandwidth and motion speed in response to torque inputs. Indeed, the shift towards the human-centered robotics paradigm poses significan challenges for both design and control whenever the design for intrinsic safety blends with speed and accuracy requirements.

3.1.2. Safety by Control

Among operational tactics for safety by control, *interaction control* laws play a predominant role in planned contact scenarios of pHRI, *collision avoidance/prediction algorithms* add a safety layer in the accidental contact scenarios in addition to energy reduction policies introduced by intrinsic safety, while *dependable path planning* involves the generation of suitable robot behaviors according to human perception and expectations. This latter component stays in between the planned and accidental contact modes since it involves a class of emotional and physical reactions in case of proximity interaction.

A) Control techniques for pHRI.

Many studies for the cooperative task between the human operator and the robot have used the impedance control for the force control of the robot [29, 30, 31, 32]. The control design should in fact ensure stable interaction control [33, 34, 35] and should prevent the introduction of an exceeding amount of energy into the system w.r.t. the safety requirements and/or injury scales for impacts. Passivity-based [36] control laws, for instance, provide a class of model-based formulated controllers that shape the kinetic and potential energy of the robot [37, 38] in order to keep the overall energy transfer limited and prove the global asymptotically stability of the close-loop control.
An example of combined lightweight design and torque-feedback passive-based control is given in [39, 40, 19, 41, 42], where also the monitoring of torques distributed along the robot are applied [43] in order to fast react to unexpected collisions. However, interaction control alone may not be sufficien without further consideration of human behavior. The cooperative system made by human and robot, which consists of an impedance [1] control (Fig. 2) based on the positioning control of the robot, becomes unstable when the human operator increases the stiffness of his/her arm or body.
The instability of the system is caused by the time delay of the movement of the robot or the human operator. In order to stabilize the system, it is necessary to increase the damping or the stiffness coefficients *i.e.* by adjusting the impedance characteristic of the robot. Moreover, when the robot performs cooperative tasks, like carrying a heavy object, or positioning tasks using conventional position control methods, high values of damping or stiffness coefficient should be settled in the impedance characteristic of the robot in order to prevent the control system to become unstable. As a result, the human operator has to carry out the task with an extra load for the large impedance characteristic of the robot. These preliminary considerations outline the need to include the human impedance model directly into the control strategy. Detailed discussion about impedance-based modeling of robot control and user behavior is given in Section 4..

---

[1]A physical system that accepts motion inputs and yields force outputs is define as an *impedance*, system that accepts force inputs and yields motion outputs is define as an *admittance* (Fig. 3).

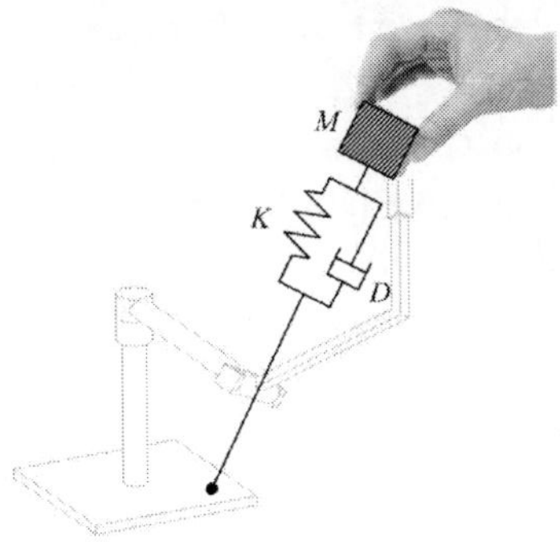

Figure 2. Human-robot system generic impedance model.

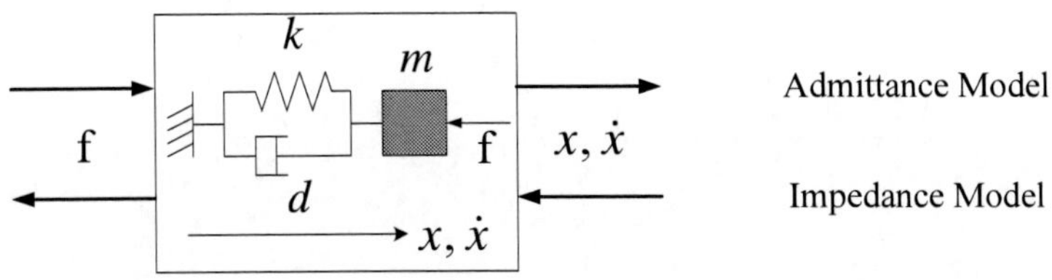

Figure 3. Impedance/admittance second-order models.

B) Path planning.

Path planning in anthropic domain adds the common features of offlin motion planning to the additional features related to intelligibility and acceptability of robot motion by the user. Major objective of path planning becomes in fact to endow the motion trajectory with ''natural'' kinematics [44], i.e. reproducing human-like patterns and transients. In particular, abrupt pattern changes and acceleration transients should avoid threatening/unexpected movements. Recently, motion-related safety has been considered under the *mental* perspective due to the perception and psycho-physical reaction against robot behavior [45, 46]. This class of desired features robot behavior blends with the more general concept of dependability, helping in providing comfortable and trustful interaction by users.

3.2. Dependability

Dependability (Fig. 4) can be considered as the property of displaying easiness, confi dence and reliance to the user when performing interaction tasks. Under such conventionally abstract notion, dependability wraps a set of clearly define child properties [47, 48], which altogether have a great deal of impact over the robot control and resulting behavior. User confidence as a matter of fact, stands on *usability*, which can measure the feeling, comfort and easiness of learning and/or setting up the robot motion, *reliability*, which mainly addresses the continuity of service rarely throwing unexpected exceptions or rarely presenting failures, *availability*, which gives account of readiness of such service, *integrity* and *maintainability*, which are the absence of improper system alterations and the ability to recover from alteration, respectively, *safety*, which has been extendedly discussed before. Most of dependability child properties can be fulfille as a trade-off at whole system level. Safety may hinder very much the availability or usability of a robot. Integrity is necessary for safety, reliability, and availability. Availability, then, cannot be achieved without maintainability and full integrity excludes maintainability.

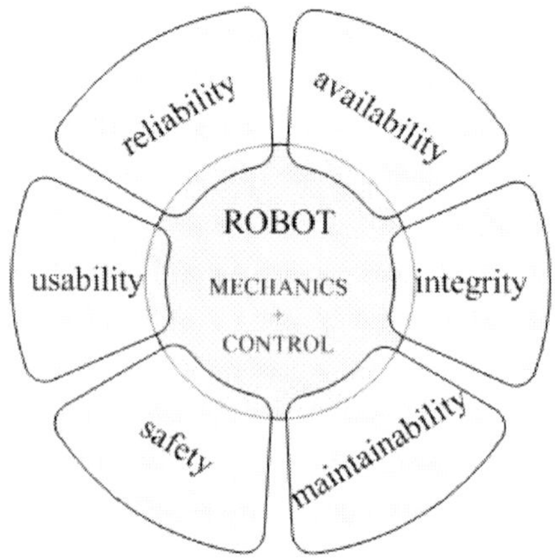

Figure 4. (simplified dependability chart.

Dependable robots for pHRI are therefore required to keep into consideration a number of design and implementation issues related to the access to the robot resources by the user, as well as the control strategies performing a given task address the *way* these resources are used w.r.t. the user expectation, the task requirements and the safety aspects. In anthropic domain, safety should therefore be the underlying [49, 50], measurable [51, 52, 53] system feature, while dependability can comprehensively defin the quality of the interaction, eventually the suitability for pHRI.

4. Robot Control Issues in pHRI

In robot-assisted rehabilitation, especially for upper limb applications, pHRI usually involves a manipulator, regardless the complexity of its kinematical structure, topology and actuation, that exchanges force with a patient. The interaction therefore implicitly requires the control of forces along at least one of the degrees of freedom in the operational space of the manipulator while the other degrees of freedom remain position controlled. Mason formalized this idea in the notion of *Hybrid Control* [54]. Without specifying the particular type of position or force control to be actually used, this notion only makes partitions of the domain of definitio of the total degrees of freedom into two subspaces: in firs space a position control is employed while in the complementary one a force control is employed. In the position control subspace, simple strategies demonstrated to be adequate (*e.g.* . PID) as well as sophisticated enhancements have improved performance (*e.g.* . computed torque control, adaptive control) too. In the force control subspace, two major different strategies - *pure force control* and *impedance control* - are usually adopted. Ideally, a *pure force controller* attempts to make the manipulator act as pure force source, independent of position, while *impedance control* provides a dynamic relationship between the robot's position and the force it exerts. In this latter way, the robot should be controlled in order to equivalently display a mechanical impedance to positional constraints imposed by the environment.

Within the framework of general hybrid control, the pHRI is discussed according to modeling and control issues. In pHRI, both systems, namely the human and the robot, actively cooperate in general in order to achieve a common goal. In the foregoing discussion, a review of interaction control strategies is firstl presented. Then, how a human understands/perceives a robot under coupling conditions is investigated.

4.1. Review of Control Strategies Suitable for pHRI

Preliminarily, the dynamic model of a generic rigid link rigid joint robot manipulator (according to [38]) is described in the joint space by the form:

$$\mathbf{B}_{jnt}(\mathbf{q})\ddot{\mathbf{q}} + \mathbf{C}_{jnt}(\mathbf{q},\dot{\mathbf{q}})\dot{\mathbf{q}} + \mathbf{f}_{v,jnt}(\mathbf{q},\dot{\mathbf{q}}) + \mathbf{g}_{jnt}(\mathbf{q}) = \mathbf{u} - \mathbf{J}^T(\mathbf{q})\mathbf{f}_h, \tag{1}$$

where $\mathbf{q}$ is the vector of joint variables, $\mathbf{B}_{jnt}$ is the inertia matrix, $\mathbf{C}_{jnt}\,\dot{\mathbf{q}}$ is the vector of Coriolis and centrifugal torques, $\mathbf{f}_{v,jnt}$ is the vector of friction torques, $\mathbf{g}$ is the vector of gravitational torques, $\mathbf{u}$ is the vector of driving torques, $\mathbf{f}_h$ is the vector of contact forces exerted by the human on the end effector, and $\mathbf{J}$ is the Jacobian matrix relating joint velocities to the vector of end-effector velocities.

The control problem consists on the definitio of the driving force $\mathbf{u}$. Early strategies were proposed from late 1970s onward (see [55, 56, 57]), and in the last three decades an impressive amount of different strategies has been proposed. Among them, two wide families of schemas can be recognized: *explicit control strategies* and *position based control strategies* (Fig. 5). In the former explicit paradigma, the driving force $\mathbf{u}$ is directly calculated, while in the position-based strategies the new reference position (or velocity) is calculated at each time step and then imposed as the set-point for the inner position control loop. Historically, firs investigations were in the domain of explicit interaction control strategies. However, this approach is often hindered by the real transmissions in industrial robots that introduce several undesired dynamic effects, like friction, backlashes and compliance, which usually hamper an accurate torque generation [58]. Implicit force control schemes are therefore more commonly implemented: rather than directly commanding the joint torques, the force controller exerts its action on the set-points of the inner position control loops, closing force control loops around position control loops. Despite these differences, similar interaction control strategies can be implemented in both strategies.

Further details are given only for the two major strategies, *pure force control* and *impedance control*.

4.1.1. Pure Force Control

This sensor-based interaction control relies on the accurate measure of exchanged generalized forces/torques in joint [59] or operational [55] domain. With the remarkable exception of direct or nearly-direct drive actuation devices, which can provide a sensorless estimate of motor torque from the motor current, accurate force samples are supplied by force/torque sensors, either mounted at robot wrist or distributed along the

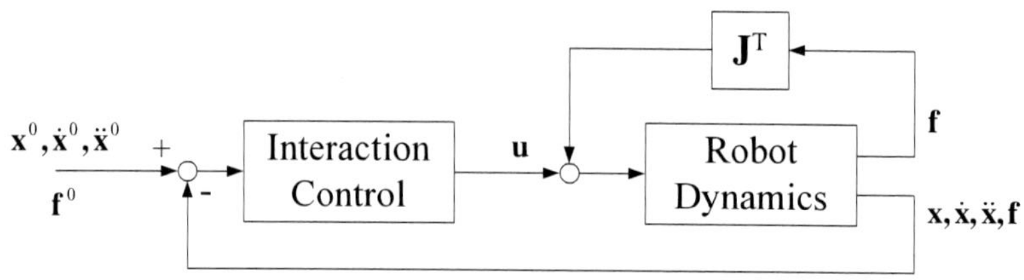

(a) Explicit Interaction Control

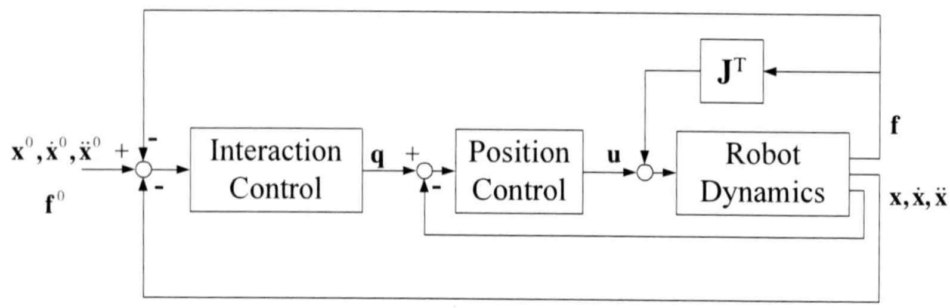

(b) Position Based Interaction Control

Figure 5. The driving force $\mathbf{u}$ can be imposed or directly (a) or indirectly (b) through the integration of an inner control loop.

joints and the robot chain. In literature pure force controls have been exhaustively studied [60, 61, 62, 63, 57], however more often have implemented as the hybrid velocity/force or position/force control [59, 64, 65].

Pure force control strategies requires controlling forces along a single degree of freedom of the operational[2] space of the manipulator, usually considered the direction of force/torque application. Hence, according to (1), the driving force $\mathbf{u}$ is calculated as sum of two terms $\mathbf{u}_f$ and $\mathbf{u}_x$ where the former has not null values is in the force-control directions and the latter has not null values in the the free-movement directions (Fig. 6).

Furthermore, in order to guarantee good system performance at steady state, model-based compensations are required including the static terms, i.e., the manipulator Jacobian and the gravity torques.

$$\begin{aligned} \mathbf{u} &= \mathbf{u}_f + \mathbf{u}_x + \left[\mathbf{J}^T \mathbf{f}_h + \mathbf{g}_{jnt} \right] \\ \mathbf{u}_f &= \mathbf{K}_i \mathbf{J}^T \int (\mathbf{f}^0 - \mathbf{f}_h) dt \\ \mathbf{u}_x &= \mathbf{K}_p \mathbf{J}^{-1} \left(\mathbf{x}^0 - \mathbf{x} \right) + \mathbf{K}_d \mathbf{J}^{-1} \left(\dot{\mathbf{x}}^0 - \dot{\mathbf{x}} \right) \end{aligned} \tag{2}$$

[2]the operational space is define according to the task, e.g. local coordinate frames set on the point of application and along direction of application of forces/torques. Wrenches resultants can therefore result in projected components along the Cartesian space, preserving the notion of single degree of freedom along the exertion of the wrenches themselves where the direction axis define additional constraints.

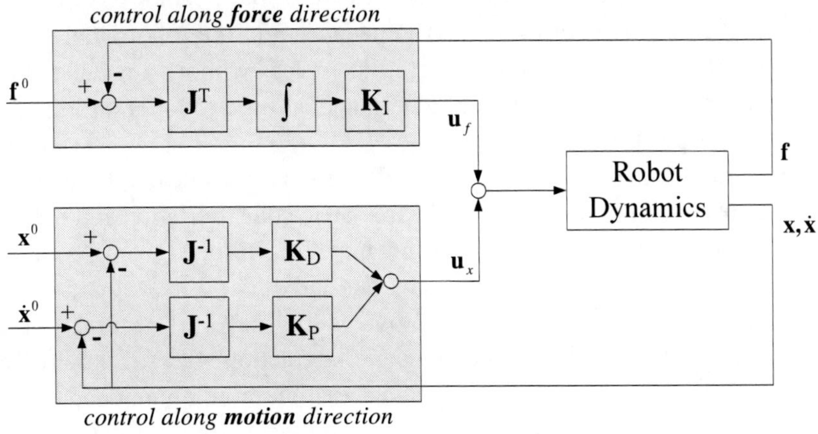

Figure 6. Hybrid force/velocity control scheme.

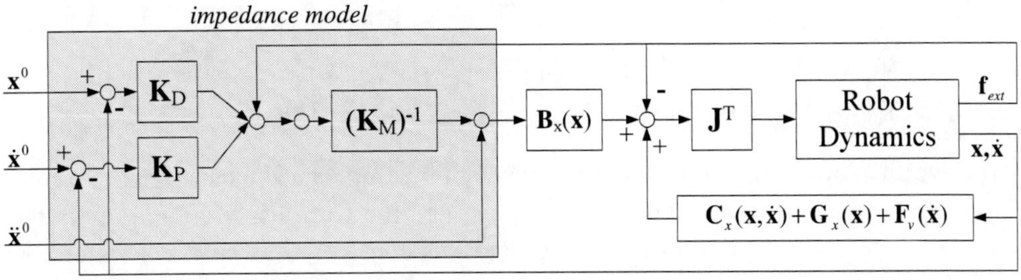

Figure 7. Impedance control scheme. The inverse dynamic control terms (Coriolis, friction and gravity) and the inertia matrix are displayed according to operational space notation.

Regarding $\mathbf{u}_f$, it is worthy to note that the regulation action of the controller is usually based on the integral error, because this allows the system to be more robust. This is also consistent with the industrial practice, where the hard specification on the repeatability of the robot in the presence of model uncertainties strongly suggest the adoption of an integral action.

4.1.2. Impedance Control

Preliminarily, recall that a physical system that accepts motion inputs and yields force outputs is define as an *impedance*, while a system that accepts force inputs and yields motion outputs is define as an *admittance* (Fig. 3).

A robot compliant behavior is reached by imposing a dynamic relationship between the external forces on the robot and the motion itself. The relationship is usually a linear second-order differential equation, *i.e.* . the control try to make the robot react to external forces like a mass-spring-damper equivalent system in the operational space. Therefore, in *Impedance Control* the controller is an impedance and the robot is an admittance (*i.e.* the robot is controlled by imposing directly the forces, without any inner loop), and in *Admittance Control* the controller is an admittance and the robot is an impedance (*i.e.* a position control loop is present).

In general, robotic systems in Impedance Control have stable dynamic interaction with stiff environments but have poor accuracy in free-space due to friction and other unmodeled dynamics. This problem can be mitigated using inner loop torque sensing/control or through hardware modification such as low-friction joints and direct-drive actuators. In contrast to Impedance Control, Admittance Control provides high level of accuracy in non-contact tasks but can result in instability during dynamic interaction with stiff environments. This problem can be eliminated using series elastic actuation or compliant end-effectors, at the cost of reducing bandwidth and performances. In contrast to direct-drives which are used in conjunction with Impedance Control, Admittance Control requires high transmission ratios, like in harmonic drives, for precise motion control. Industrial robotic systems provide in fact good examples of this practice.

Limiting the mathematical analysis to Impedance Control (see Fig. 7), and according to (1), recall that $\mathbf{u}$ is the driving force at joint space:

$$\begin{aligned} \mathbf{u} &= \mathbf{B}_{jnt}\mathbf{J}^{-1}\left(\mathbf{a} - \dot{\mathbf{J}}\,\dot{\mathbf{q}}\right) + \left[\,\mathbf{C}_{jnt}\,\dot{\mathbf{q}} + \mathbf{f}_{v,jnt} + \mathbf{g}_{jnt} + \mathbf{J}^T\left(\mathbf{q}\right)\mathbf{f}_h\right], \\ \mathbf{a} &= \ddot{\mathbf{x}}^0 + \mathbf{K}_M^{-1}\left[\mathbf{K}_D\left(\dot{\mathbf{x}}^0 - \dot{\mathbf{x}}\right) + \mathbf{K}_P\left(\mathbf{x}^0 - \mathbf{x}\right) + f_h\,\right] \end{aligned} \tag{3}$$

It is worth to note that in the classical scheme of the Impedance Control [55] the control is based on a inverse dynamic control strategies, i.e. the Coriolis, friction, gravity terms are compensated and the output of the regulators is considered as an acceleration. This solution guarantees the system decoupling, but a correct estimation of the robot dynamic model is required (meaning adaptive algorithms, robust control solutions, off-line parameter identification) To overcome this problem, Admittance Control strategies are often used.

As conclusion two remarks can be outlined:

(i) the generic term *impedance* sometimes extends to the lower order mechanical models, originally termed as stiffness and compliance controls in place of impedance and admittance, respectively [54, 66, 56]. The pHRI results then in an energy exchange based on the balance of dynamical properties (mostly, the stiffness) of the two coupled systems.

(ii) the coupled compliance can be modifie on line in order to achieve different spatial or time varying behavior, namely depending on the robot configuratio and on the actual values of the dynamical features in the mechanical model over time during the interaction tasks.

4.2. Human Arm Impedance

Although the reviewed hybrid control techniques are generically suitable for any interaction control, the upper limb rehabilitation naturally brings together the manipulator modeling (dynamics and control) with the corresponding modeling of the coupled-by-definitio manipulator that is the human arm. Including the human user, e.g. the patient, into the control loop, the exchange of forces and power cannot merely be reduced to external resultants

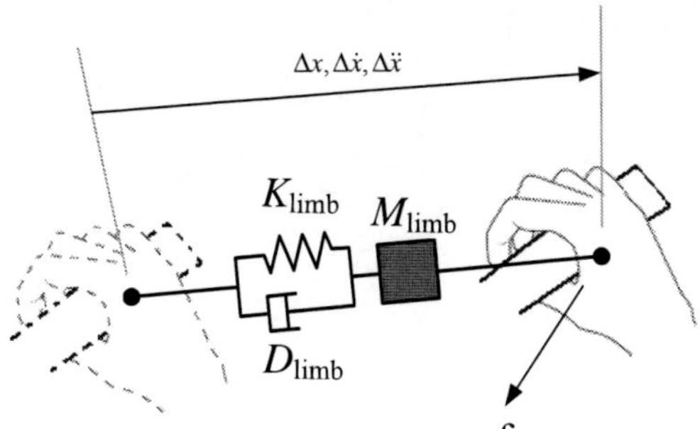

Figure 8. Local impedance model for the human arm.

or disturbances for both the systems considered separate. On the contrary, the dynamical behavior of both systems has to be investigated and implemented into the application in order to achieve an adaptive behavior for the robot. This is particularly important in all the cases, like rehabilitation, that cannot assume constant over time nature of the human-side action, *e.g.* degree of impairment, stage of recovery, ability to drive muscles, modes and performances of neuro-plasticity, wearing and, not secondarily, mood.

In the human arm, a simplifie assumption allows to consider that joint torques of the musculo-skeletal system are generated by an imbalance between tensions in agonist and antagonist muscles. Accordingly, muscles have inherently spring-like properties that can be varied changing the activation levels of muscles. As a result, the equivalent stiffness during movement is an important factor in the study of the control mechanism of biological motion [67]. In addition, several works on human motor control demonstrated that humans are able to recover their original kinematic patterns [68] when subjected to a force fiel that systematically disturbs the arm motion. This ability is accomplished by the adaptation of torques at arm joints in order to compensate the perturbing forces [69, 70]. However, when the perturbation force is suddenly removed, disturbed arms exhibit motion errors due to the previous adaptation action already in place. Therefore, it can be assumed that a basic compensation and learning mechanism is able to exploit the viscoelastic properties of the neuromuscular system [71]. Moreover, the dynamics of neuromuscular system can be modeled as a feedback mechanism that overcomes the action of internal and external perturbations like gravity, the intrinsic dynamics of limb masses or intersegmental torques [72].

Several studies have, in fact, modeled the perception-reaction human system as a mechanical impedance [55, 73, 74], *i.e.* the dynamic behavior of the human body segments is modeled through a second-order dynamic relationship between small forces and position variations (Fig. 8). In details, the control of the impedance of the neuromuscular system is a form of adaptive behavior that the central nervous system (CNS) uses to accommodate perturbations from the environment [75]. In fact, the viscoelastic properties of the neuromuscular system, of the muscles and the refl x loops reduce the "computational" effort at CNS level dedicated to motor control[3]. Such higher level modulation of human impedance

[3]The muscles viscoelasticity is able to generate passive forces against external perturbations and may be considered as a control gain in a peripheral feedback. Humans can tune viscoelasticity by regulating the level of muscle co-contraction and the refl x gain.

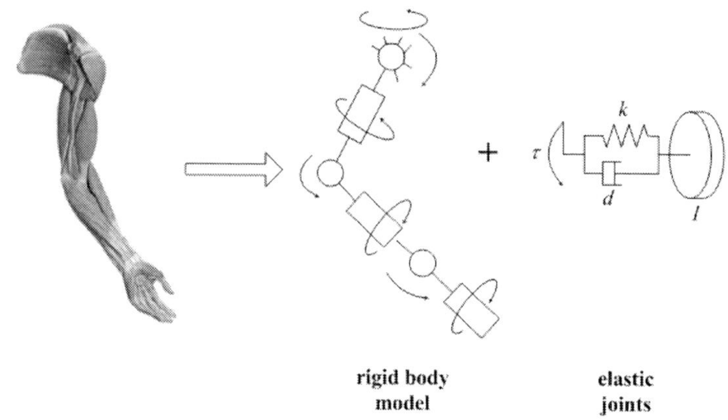

Figure 9. Upper limb model: 7 degree of freedom, rigid-links fl xible-joints.

provides the basis for several theories in human motor control such as the α-model and λ-model equilibrium point theories [76], the virtual trajectory theory [77] and dynamic interaction in manipulation [55].

In conclusion, the mechanical impedance of the neuromuscular system during manipulation tasks determines the reaction forces in response to perturbations from the grasped object. Choosing the mechanical impedance may be one of the ways whereby the central nervous system controls the behavior of the joined human-robot system [75].

4.2.1. Human Upper Limb Dynamic Model

The human mechanical impedance can be voluntarily changed over a wide range of behaviors [78] through, for instance, the level of muscle contractions [4].

A starting point for identifying a model for the human upper limb dynamics able to describe the impedance-based control is the set of equations for classical multi-body dynamics (Fig. 9). In this model, dynamics can be expressed by the second-order nonlinear differential equation

$$\boldsymbol{\Psi}(\ddot{\mathbf{q}}, \dot{\mathbf{q}}, \mathbf{q}) = \boldsymbol{\tau}_{mscl}(\dot{\mathbf{q}}, \mathbf{q}, \mathbf{u}) + \boldsymbol{\tau}_{ext} \tag{4}$$

where, $\mathbf{q}$, $\dot{\mathbf{q}}$, $\ddot{\mathbf{q}}$ are limb joint angular position, velocity, and acceleration vectors, respectively; $\boldsymbol{\tau}_{ext}$ denotes the external force; $\boldsymbol{\tau}_{mscl}$ is the generated torque by the muscles that can be represented as a function of angular position, velocity, and motor command $\mathbf{u}$ generated at supraspinal level in CNS. It is worthy to underline that the number of muscles involved is

[4]The relationship between the neural input to a muscle and its subsequent mechanical behavior is largely complex. For a given neural input the contractile force of a muscle depends on the length of the muscle, the velocity of shortening, the type of muscle, the state of fatigue and the background of training (or of electrical stimulation), among others. However, one fundamental observation is that the neural input to a muscle simultaneously determines the force and the stiffness of the muscle, *i.e.* its resistance to stretch [75]. The physical interaction leads to a learning process during which the human motor control is adapted. In the learning process there is a reduction of the average neural input signal. This reduction is to a large extent due to a decrease in the co-activation of antagonistic muscles. Therefore, the learning process implies adaptability of the mechanical impedance [72].

larger than the number of degrees of freedom of the human upper limbs, i.e. the functions $\boldsymbol{\tau}_{mscl}$ solve also the muscles actuation redundancy.

The set of dynamics in (4) is fairly general and the stiffness, viscosity and inertia parameters of the human upper limbs are implicitly considered. Explicit formulation of dynamical parameters can be therefore displayed taking into account a perturbation imposed to the limbs: Following variational equations, (4) can be reformulated in:

$$\frac{\partial \boldsymbol{\Psi}}{\partial \ddot{\mathbf{q}}}\,\delta\ddot{\mathbf{q}} + \frac{\partial \boldsymbol{\Psi}}{\partial \dot{\mathbf{q}}}\,\delta\dot{\mathbf{q}} + \frac{\partial \boldsymbol{\Psi}}{\partial \mathbf{q}}\,\delta\mathbf{q} = \frac{\partial \boldsymbol{\tau}_{mscl}}{\partial \dot{\mathbf{q}}}\,\delta\dot{\mathbf{q}} + \frac{\partial \boldsymbol{\tau}_{mscl}}{\partial \mathbf{q}}\,\delta\mathbf{q} + \delta\boldsymbol{\tau}_{ext}. \tag{5}$$

Considering the right hand side of the relation, $\partial\boldsymbol{\tau}_{mscl}/\partial\dot{\mathbf{q}}$ and $\partial\boldsymbol{\tau}_{mscl}/\partial\mathbf{q}$ happen to be equivalent damping and stiffness terms for the upper limb dynamics, therefore:

$$\mathbf{D}_{limb}(\mathbf{q}) \equiv -\,\partial\boldsymbol{\tau}_{mscl}/\partial\dot{\mathbf{q}}, \qquad \mathbf{K}_{limb}(\mathbf{q}) \equiv -\partial\boldsymbol{\tau}_{mscl}/\partial\mathbf{q} \tag{6}$$

where $\mathbf{D}_{limb}$ and $\mathbf{K}_{limb}$ are the damping matrix and the upper-limb stiffness matrices for the considered upper limb, respectively.

Considering then the left hand side of (4), the dynamics of a two-link human arm should be modeled as a rigid-links fl xible-joints model (RLFJ) [67]. As a consequence, such dynamics should be described by the following second-order nonlinear differential equation:

$$\boldsymbol{\Psi}(\ddot{\mathbf{q}}, \dot{\mathbf{q}}, \mathbf{q}) = \mathbf{B}_{limb}(\mathbf{q})\,\ddot{\mathbf{q}} + \mathbf{C}_{limb}(\mathbf{q}, \dot{\mathbf{q}})\,\dot{\mathbf{q}} + \mathbf{g}_{limb}(\mathbf{q}) \tag{7}$$

where $\mathbf{B}_{limb}$, $\mathbf{C}_{limb}$ and $\mathbf{g}_{limb}$ are the mass matrix of the multi-body system, the matrix of the Coriolis terms and the gravity forces, respectively.

Then, substituting (7) and (6) into (5), variational model of upper limb results in:

$$\mathbf{B}_{limb}\,\delta\ddot{\mathbf{q}} + \left(\frac{\partial \mathbf{C}_{limb}}{\partial \dot{\mathbf{q}}} + \mathbf{D}_{limb}\right)\delta\dot{\mathbf{q}} + \left(\frac{\partial \mathbf{C}_{limb}}{\partial \mathbf{q}} + \frac{\partial \mathbf{g}_{limb}}{\partial \mathbf{q}} + \mathbf{K}_{limb}\right)\delta\mathbf{q} = \delta\boldsymbol{\tau}_{ext}. \tag{8}$$

One very interesting aspect of this formulation is that equation (8) can be linearized with respect to the unknown parameters like in the classical problem identificatio of robot dynamical parameters [79]. Therefore, in case of availability of measured external forces $\delta\boldsymbol{\tau}_{ext}$, the upper limb dynamics model is explicitly define by solving the linear equation:

$$\mathbf{Y}_{limb}(\ddot{\mathbf{q}}, \dot{\mathbf{q}}, \mathbf{q})\,\boldsymbol{\pi} = \delta\boldsymbol{\tau}_{ext} \tag{9}$$

where $\mathbf{Y}_{limb}$ is the regressor of the dynamic system, $\boldsymbol{\pi}$ is the vector of the set of minimal parameters that describes the upper limb.

4.3. Remarks on Defining and Estimating the Dynamical Model

The estimation of the human arm stiffness requires, in general, a mean, *i.e.* a robot, that injects small perturbations onto the human limb resulting on limb displacements and velocity [73, 67, 80]. External (with respect to the human limb) forces are therefore obtained from a force sensor on the robot. Position data of the limb provide the reference coordinate frames (robot and limb) for kinematic transforms and the free coordinates for the equation set in (9). During the process of estimation, some background conditions have great influence

1. limb joint positions should be accurately measured;
2. upper limb anthropometrical lengths have to be estimated;
3. upper limb adjustable motion ability and range of the stiffness are wider than the ranges of mass and viscosity in the impedance characteristic of the model [73];
4. in the range of low velocity, the stiffness heavily influence cooperative tasks, because the terms of mass and viscosity decrease and the stiffness coefficien is predominant in human-robot cooperative tasks;
5. better stiffness estimation is reached in a range of low velocity;
6. the stiffness of the human user has great influenc on the stability of the control system of the robot.

The correct measurement of the human posture is a fundamental problem for the identificatio of human upper-limb dynamic parameters. An alternative approach is suggested by [80] with the idea of mapping the upper-limb reachable space in the Cartesian domain; the dynamics of the distal part of the human arm is then described on the basis of the position $\mathbf{x}$ of the hand, yielding [73]:

$$\mathbf{M}_{limb}\ (\mathbf{x})\ \ddot{\mathbf{x}} + \mathbf{D}_{limb}\ (\mathbf{x}, \dot{\mathbf{x}})\ \dot{\mathbf{x}} + \mathbf{g}_{limb}\ (\mathbf{x}) + \mathbf{K}_{limb}\,\Delta\,\mathbf{x} = \mathbf{f}_{ext} \tag{10}$$

where $\mathbf{M}_{limb}$ [kg], $\mathbf{D}_{limb}$ [Ns/m], $\mathbf{K}_{limb}$ [N/m] are mass, viscosity and stiffness matrices in Cartesian space; $\mathbf{f}_{ext}$ [N] is the external force and $\Delta\mathbf{x}$ [m] is the difference between the position of hand and the equilibrium point. This simplificatio holds only when the displacement $\Delta\mathbf{x}$ is limited. Following this approach, the whole workspace has to be mapped with several experiments and the resulting values has to be interpolated by fittin functions.

In the closed human-robot loop the key issue is the capability of the two systems in mutual adaptation. As previously noted, human characteristics may be approximated by a variable impedance model. In addition, robot controllers typically run impedance control strategies to manage the pHRI. On one hand, the robot controller should be able to adjust the impedance of the robot according to a particular impedance of the human, depending on the goal of the robot. On the other hand, humans adjust their control properties to the robot impedance in order to maintain the dynamic properties of the overall system [73].

A fina consideration is quoted for the fundamental problem in the closed-loop human-robot control that is *stability*, having a direct impact on the safety of the application. Two main sources of instability in coupled human-robot systems are due to the control of physical interaction and to time delays in the human-in-the-loop control.

5. Case Study: Multi-Sensory Cell for Upper Limb Rehabilitation

As above described, an application for the upper limb treatment beyond the state-of-the-art would likely include the possibility of training complex 6D functional movements - namely those with which the patients are familiar - together with the ability of exchanging

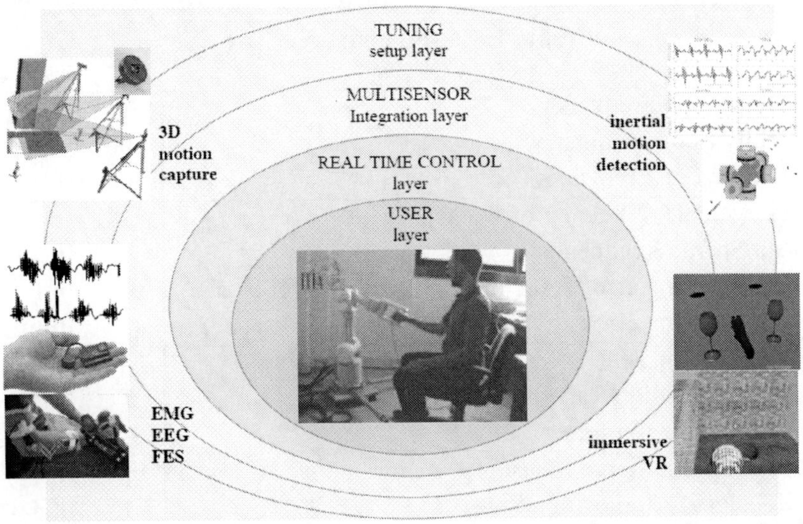

Figure 10. User-centered multisensory-cell with current modular subsystems: functional neuromuscular EEG, EMG diagnostics and FES stimulation (bottom left), markers capture cameras (top left), INS with IMU (top right) and Virtual Environment (bottom right) are integrated into the architectural MULTISENSOR layer and take part of the robot control strategies running into the REAL TIME CONTROL layer. The whole cell behavior is supervised by settings arranged in TUNING layer.

significan power during passive/collaborative sessions. In addition, such an application would also allow multi-modal HRI, an accurate tuning of the setups for adaptable rehabilitation therapies, a modular integration of several hw/sw functional systems.

As a case study for illustrating some of the major requirements and the control techniques previously overviewed, the design of a multi-sensory cell for upper limb rehabilitation based on a serial manipulator is presented. The treatment follows therefore the end-effector approach (as opposite to the exoskeleton companion approach, see [1]) making use of an industrial class redundant manipulator (Mistubishi PA10) because of the possibility to exert enough power along 6D trajectories and to access its open controller for the low level implementation of pHRI control strategies. The design of the multi-sensory cell has been focused on the integration of many different sensors and devices (see Fig. 10) in order to allow the medical staff to setup custom training control parameters and to measure the related recovery level during and after robot-assisted rehabilitation sessions. Such integration requires a fl xible and modular architecture, namely a pool of communication protocols among all the devices. The actual implementation is based on an RPC-based middleware (ZeroC^{TM} Ice^{TM}) where all the modular subsystems presented in the foregoing are networked.

5.1. Set-up Description

A) The robot and the open control architecture.

The robot arm, a Mitsubishi PA10-7 equipped with a force/torque sensor, interacts directly with the patient's arm via a custom handle that can be also arranged for disabling some of

Table 1. Set-up characteristic

Current Loop Frequency	1	[kHz]
Velocity Loop Frequency	500	[Hz]
Position Loop Frequency	500	[Hz]
Max. Velocity (at center of workspace)	1.0	[m/s]
Mean Stiffness (at center of workspace)	25	[kN/m]
Mean Tracking error at 0.1 m/s (at center of workspace)	0.1	[mm]
Mean Tracking error at 1 m/s (at center of workspace)	5	[mm]
Max. Force at TCP	150	[N]
Force Acquisition Rate	1	[kHz]
Low bass filte on force	30	[Hz]
Mean Noise in Force Measure	±2	[N]
EMG signals Acquisition Rate	1	[kHz]
Communication between Control and Safe PLC	400	[Hz]
Communication between Control and Virtual Reality	70	[Hz]
Communication between Control and Camera System	70	[Hz]
Communication between Camera System and Safe PLC	70	[Hz]
Overlap resolution in Virtual Reality	0.5	[mm]

the human wrist degrees of freedom (usually prono-supination). A Linux-based open control system, based on the Xenomai real-time [81] framework allows easy control customization [82, 83]. The hardware and software expandability of desktop PCs allows a simple integration of external devices through dedicated expansion cards and the RPC-middleware-based network infrastructure. Furthermore, open control architectures [84, 85, 86] allow the implementation of advanced control algorithms for customized rehabilitation therapies. Implemented control algorithms are force-based strategies, impedance algorithms [87, 88], artificia potential field [89, 90] on pre-define trajectories and virtual tunnels in Cartesian space.

B) Sensors for biomechanics measures and devices.

The measure of the human activities during the rehabilitation exercises is a key-concept in the multi-sensory room. The setup provides the integration of different sensors:
- an infrared cameras system for 3D tracking passive markers attached to the human arm, hence detecting trajectories and patient's posture,
- an ElectroMyoGrapic (EMG) signals acquisition system,
- a Functional Electro Stimulation (FES) device for the control of the patient activities (integrated but not still tested),
- an Inertial Navigation System (INS).

The sensors for biomechanics measure (camera tracking and EMG) are integrated both for motor coordination analysis and therapeutic purpose. The inertial system provides data used as feedbacks signals for the real-time control and for safety.

C) Virtual Reality engine.

The setup integrates a 3D Virtual Reality (VR) environment engine developed in the last ten years by VR research group of ITIA-CNR (host institution of some of the authors). VR displays to the patient a visual interactive environment, eventually engaging games or set of situations, with the aim of increasing awareness long-term rehabilitation targets, intra-session motivation and concentration in exercises. The use of VR as an advanced and stimulating medium for rehabilitation is not limited to mere visual representation of exercises but also includes the computation of several indicators of patients' performance in order to adapt the complexity and the difficult level of the exercises. Finally, yet utterly significan for a pHRI application, the 3D virtual environment provides also information about collision and physical interaction of virtual objects, enabling in this way a haptic feedback for the patient mediated through the robot (see Section 5.2.).

D) Safe-PLC integration.
Rehabilitation always implies safety requirements, not only in relationship with the performances of machinery used into the application but also, and mainly, with respect to the clinic track of a patient. Safety requirements and applied solutions have in fact to be properly arranged during each diagnostic and therapeutic phase. From an implementation point of view, all the different layers of provided safety are managed by two standard PLC and a safe PLC that monitor the overall system reliability through the continuous verificatio of the robot presence inside a custom-define allowed workspace, *e.g.* a subspace of the complete robot workspace where also the patient is present, properly intervening in case of trespassing (see Section 5.4.).

5.2. Coupled Human-robot-VR Interaction

As an example of the pHRI control implementation, a training modality is presented: the user moves the robot directly holding the end-effector handle and VR replicates the motion in a 3D game scenario (see Fig. 11). The interaction among the different virtual objects (*i.e.* geometric surfaces overlap) is transferred to the patient as a resistive force feedback. The control strategy has been split in three different steps: transformation of the geometrical information given by the VR into an equivalent force, $\mathbf{F}_{env}$; sum of $\mathbf{F}_{env}$ with the force exerted by the human on the tip robot, $\mathbf{F}_h$; then transformation of the resultant force, $\mathbf{F}_{tot}$, in a new reference for the robot driver. In detail, the total force is calculated as the the weighted sum of the virtual force and the force exerted by the human and the weights are function of the overlap depth:

$$\mathbf{F}_{tot} = \lambda_1(d)\mathbf{F}_h + \lambda_2(d)\mathbf{F}_{env}. \tag{11}$$

with $\lambda_1(d) + \lambda_2(d) = 1 \, \forall \, d$.

The weights have been introduced in order to overcome to the low rate frequency in the communication between the control and the VR. Furthermore, the weights allow the limitation of the maximum virtual objects intersection that is a fundamental parameter in order to give a realistic feedback to the user.

A) Haptic Feedback from Virtual Environment.
Let defin k, c and m as the stiffness, damp and mass of the virtual objects in the environment. The $\mathbf{F}_{env}$ is calculated imposing a simple second-order dynamic model relationship

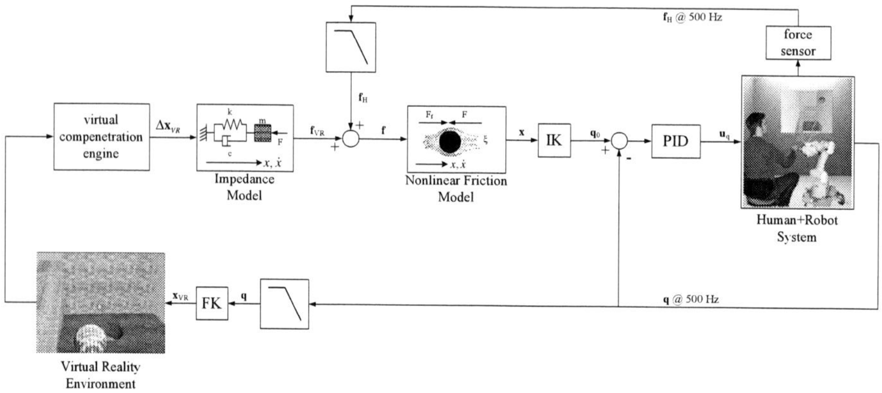

Figure 11. Main loop of interaction control with the user and the Virtual Reality (VR) environment. On the right side, the fast robot control in joint domain whose reference is given by the output of a nonlinear friction model that transforms an input force into Cartesian position/speed - then transformed back into joint space by the inverse kinematics (IK) block. The input force can be just the low-pass filtere signal from the end-effector force sensor (i.e. the user force) or the vectorial resultant of user and virtual interaction with VR world. On the left side, the slower impedance control part of the loop which takes the forward kinematics (FK) transform of robot joints into a visco-elastic model in order to output the corresponding force.

where the input is the overlapping d between virtual objects supposed rigid. The relation is:

$$\mathbf{F}_{env} = \left(k\,d + c\,\|\mathbf{v}\| + m\,\frac{d}{dt}\|\mathbf{v}\| \right) \mathbf{n}; \tag{12}$$

where $\mathbf{n}$ is the contact direction and $\mathbf{v}$ is the velocity of the robot end-effector that corresponds to the velocity of the virtual objects [5].

B) Control of Human-Robot Interaction.
The last step consists in transforming $\mathbf{F}_{tot}$ into a new reference to the robot driver (admittance-like control).

The PA10 robot driver allows two different control modalities: force control and velocity control mode. Since the experimental set-up allows 500 Hz as the maximum communication frequency rate between the external PC-controller and the robot driver, the velocity control modality has been preferred because of control stability. On the other hand, this design choice is rather limiting because it does not allow the development of impedance control strategies integrating the dynamic model of the robot. In order to overcome this limitation, a non-linear friction model in the interaction model has been introduced (see Fig. 11):

$$\ddot{\mathbf{x}} = \frac{\mathbf{F}_{tot} - \left(k + c_1\,\dot{\mathbf{x}} + c_2\dot{\mathbf{x}}^2\right)}{m}. \tag{13}$$

[5] In (12) has been used the magnitude of the robot tool center point velocity $\mathbf{v}$ instead of the firs derivative of the overlap depth d due to the limited communication rate between the robot controller and the VR.

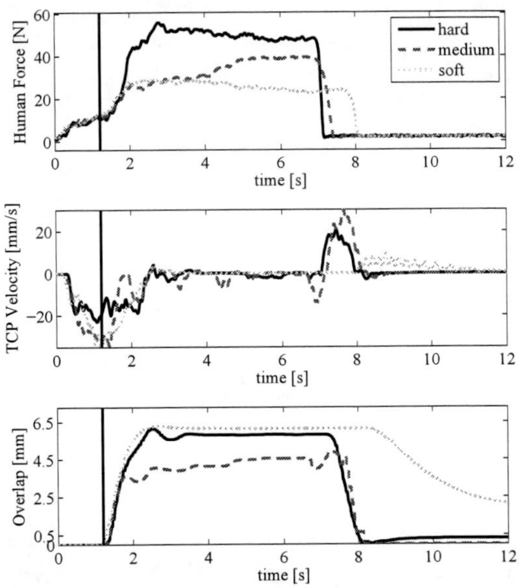

Figure 12. Manual guidance exercise in VR. User move the robot and an haptic feedback is provided on the basis of the contact condition among objects in the VR. Three different levels for the environment stiffness are reported. Vertical bold line points the contact event. Similar overlap depth is reached with different forces on the basis of different environment characteristics. When user let go off the robot (at time 7 [s]), the contact is released progressively with respect to the environment property.

where x is the tool center position of the robot, k a constant offset force, and c_1, c_2 properly define coefficients This model allows to reach an extremely comfortable behavior for the user. In fact, the non-linear friction model guarantees a very smooth modificatio of the motion direction preserving a quite large bandwidth. It is worthy to note that control stability is also preserved by imposing fairly low mass values to the virtual object (less than 2 [kg] with respect to the fact that the robot mass is more than 50 [kg]) despite the dynamic model of the robot is not integrated in the control. However, due to the fact that the resistance is very low at limited velocity, the user experiences an easy interfacing contact with the robot. Then, along with the rapid increase of resistance with velocity, the haptic feedback gains in rendering performances.

An example of the interaction among the robot, a human user and VR is displayed in Fig. 12 (three different levels for the environment stiffness are listed). In this case, the user moves the robot by exerting forces on the end-effector, and as a consequence a simple object is moved in the VR. At time 1.7 [s], there is the contact between the virtual object and another surface in VR, and consequently the velocity decreases to zero and the forces the human exerts increase. On the basis of the environment characteristics the maximum overlap depth and the maximum reachable force can be tuned, although stability remains unaffected.

The environment properties can be modifie during the exercise and a control of transition stability has been implemented by the integration of a simple model based simulation

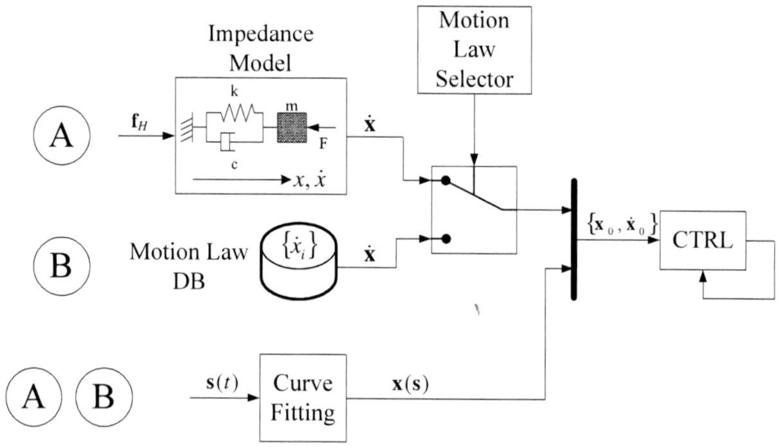

Figure 13. The developed open control architecture allows different training modes that can be switched with robot running. (A) means that the input for the robot movement is imposed by external forces, and (B) means that the motion law is loaded by a database. In the third case, (A) and (B) are active at the same time, the robot is constrained to an imposed path but the movement along it is controlled by the force exerted by the human (see Fig. 14).

engine that forecasts the system evolution on the basis of the new parameters and of the robot state.

5.3. Control of the Robot along a Constrained Trajectory

This kind of control allows to record and execute complex paths and motion laws (*i.e.* trajectories). Paths are described as 3-dimensional analytical splines define by a set of nodes, each of them composed by a spatial position and an orientation (through quaternions) of the robot end-effector. Motion laws can be freely assigned to each path, through a spline-interpolated set of velocities associated to curvilinear coordinate values. The spline-based description of paths and motion laws allows the execution of complex functional rehabilitative trajectories (*e.g.* functional daily-life movements), without being constrained to simple paths (*e.g.* linear or circular movements) or predefine motion-law profile (*e.g.* constant velocity, trapezoidal velocity, etc.). Trajectories and motion laws can be define both processing data acquired by the infra-red camera tracking system and recorded directly moving the robot handle (lead-through programming) like in Fig. 13.

The most interesting aspect of this training mode consists on the fact that the robot can be constrained to the path but the motion law depends on the forces exerted by the patient (see Fig. 14). Similarly to the manual guidance training mode, the control of the force along the trajectories is based on a second order friction model. The main difference is that the motion is a weighted linear combination of the force exerted and a motion law stored. Let defin f_h the magnitude of the $\mathbf{F}_h$, the control equation is define as below (similarly to (13)):

$$\ddot{s} = \lambda_1(f_h)\,\frac{f_h - \left(k + c_1\,\dot{s} + c_2\dot{s}^2\right)}{m} + \lambda_2(f_h)\,\dot{v}_d. \tag{14}$$

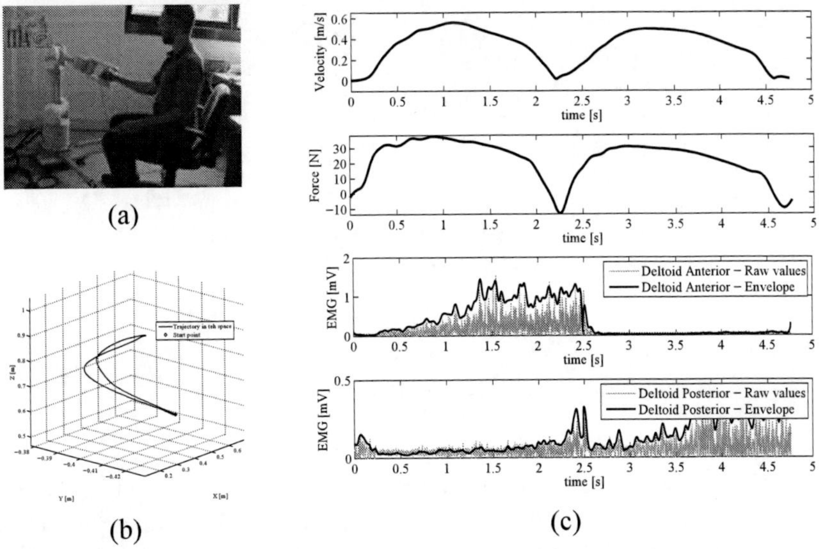

Figure 14. Control of the motion along trajectory. The trajectory (b) is be define processing data acquired by the tracking system and corresponds to the reaching movement performed by the user. The robot movement is constrained to the path, but the motion along the trajectory is imposed by the user (a). In (c) the force imposed by the user is reported. Furthermore, the EMG signals of the Anterior Deltoid and of the Posterior Deltoid are displayed.

where s is the space coordinate along the fi ed 3D path, v_d is the motion law stored, and $\lambda_1(f_h) + \lambda_2(f_h) = 1 \, \forall \, f_h$.

As shown in Fig. 14, the control of the force along the trajectories allow the user to perform smooth and stable movements.

5.4. Safety

Safety issues within the multi-sensory cell are almost entirely related to the risk assessment and the prevention of potential wrong robot motions, including positions, velocity and accelerations, that could harm the interacting user. Intrinsically safe devices, *e.g.* light weight compliant machines [91], may not require dedicated solutions because of the limited torques generated by motor joints and the low residual inertia in case of axes runaway. In the case of fairly more massive robots, like in this work, which are able to provide higher speeds and forces, additional methods and devices are instead vital for automatically preventing that any source of failure hinders a safe interaction with any user. The implemented solution therefore includes three layers of robot monitoring at runtime, all of which end up into a controlled stop and a power cutting (represented with the double switch on the double power lines in Fig. 15).

The firs layer performs a number of watchdog checks over the joints rates and end-effector position errors. The second and third layers require the set plain and Safe PLCs, interconnected over a real time signal bus, and a motion detection sensor able to capture the robot trajectory. Actual implementation involves either the set of cameras for markers

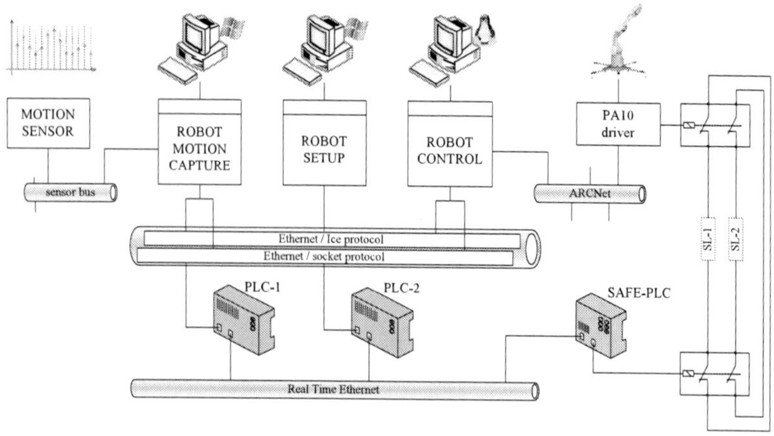

Figure 15. Architectural infrastructure of multi-sensory cell: the motion sensor with related PC (top left) and the robot with related setting and controller PCs (top right) are networked by both a middle-ware layer and low level TCP sockets. Internal dedicated buses are displayed for both modules. The safety PLCs access the data from the sensor and robot over the network via TCP socket and perform the safety algorithms together with the Safe PLC within the real time Ethernet subnet. SL-1 and -2 represent the robot power lines.

tracking, which is used also for the measurement of kinematics of patient's arm, or an INS based upon a Inertial Measurement Unit (IMU) with 3D acceleration and 3D rate sampling. In the case of tracking with cameras, a cluster of markers is placed on a relevant robot link (usually close to the wrist). In the case of the INS, the device is quite compact and is attached to the robot wrist as well. Both systems allow the full kinematics computation, either for derivation or integration of the sampled signals (no details of computational issues are given here). These devices are used as an additional source of sampling for the same trajectory provided by the robot joint sensors. The second layer of safety check is in fact based on the double redundancy of kinematic measurement: the same trajectory is sampled from both the robot and the motion sensor and, additionally, the sampling synchronously redoubled by the two PLC. Double redundant signals are then checked for matching against a given accuracy threshold ϵ_S that depends on the errors due to the calibration inaccuracy ϵ_c, which yield a misalignment between the sensor and robot coordinate frames, the undetected compliance of robot mechanics $\epsilon_r(\mathbf{q})$, which causes the real position of the links carrying the sensor to be different form the nominal one provided by the joints sensors, the latency of transmission of signals eventually carried by asynchronous buses ϵ_t, which causes a time shift of actual values sampled and compared and the intrinsic accuracy of each sensor $\epsilon_{s,i}\ \forall i$. Given the computed trajectories $\{\ddot{\mathbf{x}}_r, \dot{\mathbf{x}}_r, \mathbf{x}_r\}$ from the robot and $\{\ddot{\mathbf{x}}_{mt}, \dot{\mathbf{x}}_{mt}, \mathbf{x}_{mt}\}$ from the motion tracking, the safety stop results in the condition

$$e_{rm} \equiv \|\mathbf{x}_r - \mathbf{x}_{mt}\|, \quad \epsilon \equiv \mathcal{F}\left(\ddot{e}_{rm}, \dot{e}_{rm}, e_{rm},\right)$$

$$\Rightarrow \quad \begin{cases} \epsilon \leq \epsilon_S & \text{ok} \\ \epsilon > \epsilon_S & \text{stop} \end{cases}$$

where

$$\epsilon_S = \epsilon_S \left(\epsilon_c, \epsilon_r(\mathbf{q}), \epsilon_t, \sum_i \epsilon_{s,i} \right)$$

is the mismatching threshold. This safety layer helps in ensuring that the robot is where it was supposed to be.

The third layer provides the verificatio of robot positioning w.r.t the patient, *i.e.* it uses the elaboration of the robot and sensor motion signals for the computation of positional virtual walls around the patient figure The virtual safe volume is of course updated at runtime and displays the minimum thickness around the patient hand holding the robot end-effector.

The full actual implementation is displayed in Fig. 15 where the robot, the sensor and their related supervising PCs, the safety PLCs subsystems are all networked within the communication infrastructure based upon synchronous/asynchronous protocols.

6. Conclusions

The functional requirements in a robot-assisted rehabilitation process, especially for the treatment of the upper limb, are very demanding because of the quite inhomogeneous impairment conditions of patients, the different response, *i.e.* effica y of therapy, by each patient to similar treatments and a very uneven framework of standards, methodologies and types of exercises performed with robots. In addition, robot applications show somehow limited integration of biosignals (like EMG, EEG, etc) to be used as a source of information for adapting the robot behavior during the training sessions. Nevertheless, robot-assisted therapy is widely reckoned to be globally beneficia for patients, despite the actual implementation and a lack of standards in the evaluation of obtained results. Among the many possible rehabilitation strategies, those devoted to the recovery of natural ADL movements for the upper limb are described in this work as an example of the complexity in generating, evaluating and implementing complex 6D motions with an end-effector based machine for rehabilitation. Moreover, due to the nature of the interaction between the robot and the user in rehabilitation, many additional, yet equally important, requirements are posed in the specifi domain of robotics for the design and implementation of robot controls. These requirements have been reviewed in relationship with the notions of safety and dependability. To that extent, the control performances and methodologies are a key issue for obtaining a suited behavior for the very varying conditions faced in rehabilitation application. Therefore, the most relevant models of dynamics and control of both robotic arms and human limbs have been reviewed in order to set robust and compliant controllers. Such set of methodologies, closely derived from the functional and technological requirements, are then presented in an actual implementation of an integrated robotic cell used fro upper limb rehabilitation. Regarding the control issues, the system implements a subset of the control models introduced, discussing moreover the functional integration within the real time loop of modular systems like a virtual reality environment. In this way a sort of closed human-robot loop is set in place. The core machine is actually an industrial-derived manipulator conveniently suited for supporting large libraries of controllers and applications. Such features are mainly achieved through an open control architecture and a modular, Operating

System free and coding language free network architecture.

References

[1] M. Malosio, N. Pedrocchi, F. Vicentini, L. M. Tosatti, M. Caimmi, and F. Molteni, "Robotics in Rehabilitation - Part II: Design of Devices and Mechanisms," in *Manipulators: Theory, types and applications (this issue)* (G. Legnani and I. Fassi, eds.), Nova Science Publishers, 2011.

[2] C. von Hofsten, "An action perspective on motor development," *Trends in Cognitive Sciences*, vol. 8, no. 6, pp. 266–272, 2004.

[3] J. M. Plumert, J. K. Kearney, and J. F. Cremer, "Children's road crossing a window into perceptual–motor development," *Curr Dir Psychol Sci*, vol. 16, no. 5, pp. 255–258, 2007.

[4] E. W. Bushnell and J. P. Boudreau, "Motor development and the mind: The potential role of motor abilities as a determinant of aspects of perceptual development," *Child Development*, vol. 64, no. 4, pp. 1005–1021, 1993.

[5] A. Reichenbach, J.-P. Bresciani, A. Peer, H. H. Bülthoff, and A. Thielscher, "Contributions of the PPC to Online Control of Visually Guided Reaching Movements Assessed with fMRI-Guided TMS," *Cerebral Cortex*, 2010.

[6] T. Mulder, "A Process-Oriented model of human motor behavior: Toward a Theory-Based rehabilitation approach," *Physical Therapy*, vol. 71, no. 2, pp. 157–164, 1991.

[7] S. Masiero, E. Carraro, C. Ferraro, P. Gallina, A. Rossi, and G. Rosati, "Upper limb rehabilitation robotics after stroke: a perspective from the University of Padua, Italy," *J Rehabil Med*, vol. 41, pp. 981–985, Nov. 2009.

[8] P. Morasso, M. Casadio, P. Giannoni, L. Masia, V. Sanguineti, V. Squeri, and E. Vergaro, "Desirable features of a "humanoid" robot-therapist," in *Engineering in Medicine and Biology Society, 2009. EMBC 2009. Annual International Conference of the IEEE*, pp. 2418–2421, 2009.

[9] R. Schmidt and T. Lee, *Motor Control And Learning: A Behavioral Emphasis, Fourth Edition*. Human Kinetics, 4 ed., Jan. 2005.

[10] H. Krebs, N. Hogan, M. Aisen, and B. Volpe, "Robot-aided neurorehabilitation," *Rehabilitation Engineering, IEEE Transactions on* , vol. 6, pp. 75–87, mar 1998.

[11] F. Molteni, M. Caimmi, A. Cazzaniga, G. Gasperini, E. Giandomenico, and C. Giovanzana, "Using surface dynamic electromyography during upper-extremity robotic training," *EUR MED PHYS 2008*, vol. 44, no. Suppl. 1 to No.3, pp. 1–3, 2008.

[12] R. Riener, T. Nef, and G. Colombo, "Robot-aided neurorehabilitation of the upper extremities," *Medical and Biological Engineering and Computing* , vol. 43, pp. 2–10, 2005.

[13] G. Rau, C. Disselhorst-Klug, and R. Schmidt, "Movement biomechanics goes upwards: from the leg to the arm," *Journal of Biomechanics*, vol. 33, no. 10, pp. 1207–1216, 2000.

[14] A. Albu-Schäffer, A. Bicchi, G. Boccadamo, R. Chatila, A. D. Luca, A. D. Santis, G. Giralt, G. Hirzinger, V. Lippiello, R. Mattone, R. Schiavi, B. Siciliano, G. Tonietti, and L. Villani, "Physical Human-Robot Interaction in Anthropic Domains: Safety and Dependability," in *Proc. 4th IARP/IEEE-EURON Workshop on Technical Challenges for Dependable Robots in Human Environments*, 2005.

[15] "PHRIENDS project website: *www.phriends.eu*."

[16] "SMErobot^{TM} project website: *www.smerobot.org*."

[17] S. Haddadin, A. Albu-Schäffer, and G. Hirzinger, "Requirements for safe robots: Measurements, analysis and new insights," *The International Journal of Robotics Research*, vol. 28, pp. 1507–1527, November 2009.

[18] G. Hirzinger, A. Albu-Schäffer, M. Hahnle, I. Schaefer, and N. Sporer, "On a new generation of torque controlled light-weight robots," in *Robotics and Automation, 2001. Proceedings 2001 ICRA. IEEE International Conference on*, vol. 4, pp. 3356–3363, 2001.

[19] A. Albu-Schäffer, S. Haddadin, C. Ott, A. Stemmer, Wimböck, and G. Hirzinger, "The dlr lightweight robot: design and control concepts for robots in human environments," *Industrial Robot: An International Journal*, vol. 34, pp. 376–385, 2007.

[20] B. Rooks, "The harmonious robot," *Industrial Robot: An International Journal*, vol. 33, pp. 125–130, 2006.

[21] K. Salisbury, W. Townsend, B. Ebrman, and D. DiPietro, "Preliminary design of a whole-arm manipulation system (WAMS)," in *Robotics and Automation, 1988. Proceedings., 1988 IEEE International Conference on*, vol. 1, pp. 254–260, Apr. 1988.

[22] W. T. Townsend and J. K. Salisbury, "Mechanical design for whole-arm manipulation," in *Robots and Biological Systems: Towards a New Bionics?* (P. Dario, G. Sandini, and P. Aebischer, eds.), pp. 153–164, Springer-Verlag New York, Inc., 1993.

[23] A. Bicchi and G. Tonietti, "Fast and "soft-arm" tactics [robot arm design]," *Robotics Automation Magazine, IEEE*, vol. 11, pp. 22–33, june 2004.

[24] A. Bicchi, M. Bavaro, G. Boccadamo, D. De Carli, R. Filippini, G. Grioli, M. Piccigallo, A. Rosi, R. Schiavi, S. Sen, and G. Tonietti, "Physical Human-Robot Interaction: dependability, safety, and performance," in *Advanced Motion Control, 2008. AMC '08. 10th IEEE International Workshop on*, pp. 9–14, 2008.

[25] M. Zinn, O. Khatib, B. Roth, and J. Salisbury, "Playing it safe [human-friendly robots]," *Robotics Automation Magazine, IEEE*, vol. 11, no. 2, pp. 12–21, 2004.

[26] M. Williamson, "Series Elastic Actuators," Tech. Rep. AITR-1524, Massachusetts Institute of Technology, 1995.

[27] G. Pratt and M. Williamson, "Series Elastic Actuators," in *Intelligent Robots and Systems 95. 'Human Robot Interaction and Cooperative Robots', Proceedings. 1995 IEEE/RSJ International Conference on*, vol. 1, pp. 399–406, Aug. 1995.

[28] A. Albu-Schäffer, O. Eiberger, M. Grebenstein, S. Haddadin, C. Ott, T. Wimbock, S. Wolf, and G. Hirzinger, "Soft robotics," *Robotics Automation Magazine, IEEE*, vol. 15, pp. 20–30, September 2008.

[29] R. Ikeura and H. Inooka, "Variable impedance control of a robot for cooperation with a human," in *Robotics and Automation, 1995. Proceedings., 1995 IEEE International Conference on*, vol. 3, pp. 3097–3102 vol.3, May 1995.

[30] K. Kosuge and N. Kazamura, "Control of a robot handling an object in cooperation with a human," in *Robot and Human Communication, 1997. RO-MAN '97. Proceedings., 6th IEEE International Workshop on*, pp. 142–147, Oct. 1997.

[31] O. Al-Jarrah and Y. Zheng, "Arm-manipulator coordination for load sharing using compliant control," in *Robotics and Automation, 1996. Proceedings., 1996 IEEE International Conference on*, vol. 2, pp. 1000–1005, Apr. 1996.

[32] H. Kazerooni, "Human power extender: an example of human-machine interaction via the transfer of power and information signals," in *Advanced Motion Control, 1998. AMC '98-Coimbra., 1998 5th International Workshop on*, pp. 565–572, July 1998.

[33] C. H. An and J. M. Hollerbach, "Dynamic stability issues in force control of manipulators," in *American Control Conference, 1987*, pp. 821–827, 1987.

[34] S. Eppinger and W. Seering, "Understanding bandwidth limitations in robot force control," in *Robotics and Automation. Proceedings. 1987 IEEE International Conference on*, vol. 4, pp. 904–909, Mar. 1987.

[35] P. B. Goldsmith, B. A. Francis, and A. Goldenberg, "Stability of hybrid position/force control applied to manipulators with fl xible joints," *Int. Journal of Robotics ans Automation*, vol. 14, no. 4, pp. 89–99, 1999.

[36] S. Stramigioli, *Modeling and IPC Control of Interactive Mechanical Systems: A Coordinate-Free Approach*. Springer-Verlag New York, Inc., 2001.

[37] R. Ortega and M. W. Spong, "Adaptive motion control of rigid robots: A tutorial," *Automatica*, vol. 25, no. 6, pp. 877–888, 1989.

[38] M. W. Spong, S. Hutchinson, and M. Vidyasagar, *Robot Modeling and Control*. John Wiley & Sons, Inc., 2005.

[39] A. Albu-Schäffer, C. Ott, and G. Hirzinger, "A passivity based cartesian impedance controller for fl xible joint robots - part ii: full state feedback, impedance design and experiments," in *Robotics and Automation, 2004. Proceedings. ICRA '04. 2004 IEEE International Conference on*, vol. 3, pp. 2666–2672, 2004.

[40] C. Ott, A. Albu-Schaffer, A. Kugi, S. Stamigioli, and G. Hirzinger, “A passivity based cartesian impedance controller for fl xible joint robots - part i: torque feedback and gravity compensation,” in *Robotics and Automation, 2004. Proceedings. ICRA '04. 2004 IEEE International Conference on*, vol. 3, pp. 2659–2665, 2004.

[41] A. Albu-Schäffer, C. Ott, and G. Hirzinger, “A unifie passivity-based control framework for position, torque and impedance control of fl xible joint robots,” *The International Journal of Robotics Research*, vol. 26, no. 1, pp. 23–39, 2007.

[42] A. Kugi, C. Ott, A. Albu-Schäffer, and G. Hirzinger, “On the passivity-based impedance control of fl xible joint robots,” *Robotics, IEEE Transactions on*, vol. 24, no. 2, pp. 416–429, 2008.

[43] S. Haddadin, A. Albu-Schäffer, A. De Luca, and G. Hirzinger, “Collision detection and reaction: A contribution to safe physical human-robot interaction,” in *Intelligent Robots and Systems, 2008. IROS 2008. IEEE/RSJ International Conference on*, pp. 3356–3363, 2008.

[44] A. D. Santis, B. Siciliano, A. D. Luca, and A. Bicchi, “An atlas of physical human-robot interaction,” *Mechanism and Machine Theory*, vol. 43, no. 3, pp. 253–270, 2008.

[45] P. Rani, N. Sarkar, and C. Smith, “Affect-sensitive human-robot cooperation - theory and experiments,” in *Robotics and Automation, 2003. Proceedings. ICRA '03. IEEE International Conference on*, vol. 2, pp. 2382–2387, 2003.

[46] D. Kulič and E. Croft, “Affective state estimation for human-robot interaction,” *Robotics, IEEE Transactions on*, vol. 23, no. 5, pp. 991–1000, 2007.

[47] J.-C. Laprie, “Dependability of computer systems: concepts, limits, improvements,” in *Software Reliability Engineering, 1995. Proceedings., Sixth International Symposium on*, pp. 2–11, Oct. 1995.

[48] A. Avižienis, J.-C. Laprie, B. Randell, and C. Landwehr, “Basic concepts and taxonomy of dependable and secure computing,” *Dependable and Secure Computing, IEEE Transactions on*, vol. 1, pp. 11–33, jan.-march 2004.

[49] J. Heinzmann and A. Zelinsky, “Quantitative safety guarantees for physical human-robot interaction,” *The International Journal of Robotics Research*, vol. 22, no. 7-8, pp. 479–504, 2003.

[50] J. Heinzmann and A. Zelinsky, “A safe-control paradigm for human-robot interaction,” *Journal of Intelligent Robotic Systems*, vol. 25, pp. 295–310, 1999.

[51] K. Ikuta and M. Nokata, “General evaluation method of safety for human-care robots,” in *Robotics and Automation, 1999. Proceedings. 1999 IEEE International Conference on*, vol. 3, pp. 2065–2072, 1999.

[52] K. Ikuta, M. Nokota, and H. Ishii, “Safety evaluation method of human-care robot control,” in *Micromechatronics and Human Science, 2000. MHS 2000. Proceedings of 2000 International Symposium on*, pp. 119–127, 2000.

[53] M. Nokata, K. Ikuta, and H. Ishii, "Safety-optimizing method of human-care robot design and control," in *Robotics and Automation, 2002. Proceedings. ICRA '02. IEEE International Conference on*, vol. 2, pp. 1991–1996, 2002.

[54] M. T. Mason, "Compliance and force control for computer controlled manipulators," *Systems, Man and Cybernetics, IEEE Transactions on*, vol. 11, no. 6, pp. 418–432, 1981.

[55] N. Hogan, "Impedance control - an approach to manipulation. i - theory. ii - implementation. iii - applications," *ASME Transactions Journal of Dynamic Systems and Measurement Control*, vol. 107, pp. 1–24, 1985.

[56] D. E. Whitney, "Historical perspective and state of the art in robot force control," *The International Journal of Robotics Research*, vol. 6, no. 1, pp. 3–14, 1987.

[57] T. Yoshikawa, "Force control of robot manipulators," in *Robotics and Automation, 2000. Proceedings. ICRA '00. IEEE International Conference on*, vol. 1, pp. 220–226, 2000.

[58] P. Rocco, G. Ferretti, and G. Magnani, "Implicit force control for industrial robots in contact with stiff surfaces," *Automatica*, vol. 33, no. 11, pp. 2041–2047, 1997.

[59] M. H. Raibert and J. J. Craig, "Hybrid position/force control of manipulators," *Journal of Dynamic Systems, Measurement, and Control*, vol. 103, no. 2, pp. 126–133, 1981.

[60] S. Eppinger and W. Seering, "On dynamic models of robot force control," in *Robotics and Automation. Proceedings. 1986 IEEE International Conference on*, vol. 3, pp. 29–34, Apr. 1986.

[61] S. Eppinger and W. Seering, "Introduction to dynamic models for robot force control," *Control Systems Magazine, IEEE*, vol. 7, pp. 48–52, Apr. 1987.

[62] R. Volpe and P. Khosla, "A theoretical and experimental investigation of explicit force control strategies for manipulators," *Automatic Control, IEEE Transactions on*, vol. 38, pp. 1634–1650, Nov. 1993.

[63] H. Bruyninckx and J. De Schutter, "Specificatio of force-controlled actions in the"task frame formalism" - a synthesis," *Robotics and Automation, IEEE Transactions on*, vol. 12, pp. 581–589, Aug. 1996.

[64] T. Yoshikawa, "Dynamic hybrid position/force control of robot manipulators–description of hand constraints and calculation of joint driving force," *Robotics and Automation, IEEE Journal of*, vol. 3, no. 5, pp. 386–392, 1987.

[65] T. Yoshikawa and A. Sudou, "Dynamic hybrid position/force control of robot manipulators: On-line estimation of unknown constraint," in *Experimental Robotics I* (V. Hayward and O. Khatib, eds.), vol. 139 of *Lecture Notes in Control and Information Sciences*, pp. 116–134, Springer Berlin / Heidelberg, 1990.

[66] J. K. Salisbury, "Active stiffness control of a manipulator in Cartesian coordinates," in *Decision and Control including the Symposium on Adaptive Processes, 1980 19th IEEE Conference on*, vol. 19, pp. 95–100, 1980.

[67] H. Gomi and M. Kawato, "Human arm stiffness and equilibrium-point trajectory during multi-joint movement," *Biological Cybernetics*, vol. 76, no. 3, pp. 163–171, 1997.

[68] R. Shadmehr and F. Mussa-Ivaldi, "Adaptive representation of dynamics during learning of a motor task," *The Journal of Neuroscience*, vol. 14, no. 5, pp. 3208–3224, 1994.

[69] D. B. Debicki and P. L. Gribble, "Inter-joint coupling strategy during adaptation to novel viscous loads in human arm movement," *Journal of Neurophysiology*, vol. 92, no. 135, pp. 754–765, 2004.

[70] P. L. Gribble and D. J. Ostry, "Compensation for loads during arm movements using equilibrium-point control," *Experimental Brain Research*, vol. 135, pp. 474–482, 2004.

[71] M. Kawato, "Internal models for motor control and trajectory planning," *Current Opinion in Neurobiology*, vol. 9, no. 6, pp. 718–727, 1999.

[72] J. L. Pons, *Wearable Robots: Biomechatronic Exoskeletons; electronic version*. Chichester: Wiley, 2008.

[73] T. Tsuji, P. Morasso, K. Goto, and K. Ito, "Human hand impedance characteristics during maintained posture," *Biological Cybernetics*, vol. 72, pp. 475–485, 1995. 10.1007/BF00199890.

[74] J. Dolan, M. Friedman, and M. Nagurka, "Dynamic and loaded impedance components in the maintenance of human arm posture," *Systems, Man and Cybernetics, IEEE Transactions on*, vol. 23, no. 3, pp. 698–709, 1993.

[75] N. Hogan, "Adaptive control of mechanical impedance by coactivation of antagonist muscles," *Automatic Control, IEEE Transactions on*, vol. 29, pp. 681–690, Aug. 1984.

[76] F. A. G., "Once more on the equilibrium-point hypothesis (lambda model) for motor control.," *J Mot Behav.*, vol. 18, pp. 17–54, March 1986.

[77] T. Flash, "The control of hand equilibrium trajectories in multi-joint arm movements," *Biol. Cybern.*, vol. 57, pp. 257–274, November 1987.

[78] F. Mussa-Ivaldi, N. Hogan, and E. Bizzi, "Neural, mechanical, and geometric factors subserving arm posture in humans," *The Journal of Neuroscience*, vol. 5, no. 10, pp. 2732–2743, 1985.

[79] S. H. M. V. Mark W. Spong, *Robot Modeling and Control*. Emerald Group Publishing Limited, 2006.

[80] T. Tsumugiwa, R. Yokogawa, and K. Hara, "Variable impedance control based on estimation of human arm stiffness for human-robot cooperative calligraphic task," in *Robotics and Automation, 2002. Proceedings. ICRA '02. IEEE International Conference on*, vol. 1, pp. 644–650 vol.1, 2002.

[81] P. E. Mckenney, "Real time vs. real fast: How to choose?," 2009.

[82] J. Maass, T. Reisinger, J. Hesselbach, and W. Schumacher, "A versatile modular control architecture for sensor integrated assembly," in *37th Int. Symp. on Rob., ISR*, 2006.

[83] G. Pritschow, Y. Altintas, F. Jovane, Y. Koren, M. Mitsuishi, S. Takata, H. van Brussel, M. Weck, and K. Yamazaki, "Open controller architecture - past, present and future," *CIRP Annals - Manufacturing Technology*, vol. 50, no. 2, pp. 463–470, 2001.

[84] K. Kushal and J. Sneckenberger, "Implementation of a pc-based adaptive control scheme for an industrial robot," in *Intelligent Robots and Systems, 1992., Proceedings of the 1992 lEEE/RSJ International Conference on*, vol. 1, pp. 114–121, July 1992.

[85] K.-S. Hong, K.-H. Choi, J.-G. Kim, and S. Lee, "A pc-based open robot control system: Pc-orc," *Robotics and Computer-Integrated Manufacturing*, vol. 17, no. 4, pp. 355–365, 2001.

[86] P. Fiorini, H. Seraji, and M. Long, "A pc based configuratio controller for dexterous 7-dof arms," *Robotics Automation Magazine, IEEE*, vol. 4, pp. 30–38, Sept. 1997.

[87] A. Albu-Schaffer and G. Hirzinger, "State feedback controller for fl xible joint robots: a globally stable approach implemented on dlr's light-weight robots," in *Intelligent Robots and Systems, 2000. (IROS 2000). Proceedings. 2000 IEEE/RSJ International Conference on*, vol. 2, pp. 1087–1093, 2000.

[88] H. Kazerooni, "Open controller architecture - past, present and future," *IEEE Control Systems Magazine*, vol. 8, no. 1, pp. 21–25, 1988.

[89] O. Khatib, "Real-Time Obstacle Avoidance for Manipulators and Mobile Robots," *The International Journal of Robotics Research*, vol. 5, no. 1, pp. 90–98, 1986.

[90] M. Khatib and R. Chatila, "An extended potential fiel approach for mobile robot sensor-based motions," in *Proceedings of Int. Conf. on Intelligent Autonomous Systems*, pp. 490–496, 1995.

[91] M. Casadio, V. Sanguineti, P. G. Morasso, and V. Arrichiello, "Braccio di ferro: A new haptic workstation for neuromotor rehabilitation," *Technology and Health Care*, vol. 14, no. 3, pp. 123–142, 2006.

In: Robotics: State of the Art and Future Trends
Editors: G. Legnani and I. Fassi

ISBN: 978-1-62100-403-5

Chapter 11

ROBOTICS IN REHABILITATION - PART II: DESIGN OF DEVICES AND MECHANISMS

***Matteo Malosio*[1,*], *Nicola Pedrocchi*[1], *Federico Vicentini*[1],**
Lorenzo Molinari Tosatti[1], Marco Caimmi[1,2] and Franco Molteni[2]
[1]Institute of Industrial Technologies and Automation,
National Research Council, Italy
[2]Villa Beretta Rehabilitation Center, Ospedale Valduce, Italy

Abstract

Robotics in rehabilitation presents many aspects of physical human-robot interaction. In the companion paper (Part I, *this issue*) some general requirements derived from rehabilitation scenarios have been introduced. In that context, issues in the domain of robot control have been detailed, leaving for this part the discussion about the design process of manipulators or machines for rehabilitation. In particular, the design of upper limb solutions is particularly challenging from structural and mechanical points of view because of the complex kinematic topology and anatomy of the human arm. End-effector devices - sometimes anthropomorphic manipulators or, more often, machines with a limited number of degrees of freedom - do not constrain all the segments of a limb and therefore do not allow a complete control of the joints trajectories. Exoskeletons, on the contrary, completely wrap the arm and are therefore suitable for controlling all the limb joints along the kinematic chain, exchanging forces all along the limb. For this reason, the topology and the kinetostatic validation of exoskeletal robotic devices assume a very significant importance in order to prevent unwanted articular motions or damages due to exceeding torques. In particular, the analysis of singularities in robot kinematics is directly used for evaluating the behavior of a machine closely coupled with the human arm and is also preliminary for any robust implementation of control strategies. A case study of an exoskeleton design is presented along with singularity and kinetostatical analysis in order to display the effects of topological choices onto the actual dynamical behavior of the robot in interaction with living, yet impaired, limb.

*E-mail address: matteo.malosio@itia.cnr.it

PACS 05.45-a, 52.35.Mw, 96.50.Fm.
Keywords: Rehabilitation Robotics, Soft Robotics, Human Robot Interaction, Exoskeletons.

1. Introduction

Robot-assisted rehabilitation of the lower limb has long been promoted for the recovery of gait ability (*e.g.* the Lokomat) and has already gained a significant degree of reliability and maturity. Lower limb rehabilitation with robots is, in fact, common clinical practice because regaining abilities in standing and/or walking is considered of primary clinical importance. In addition, the lower limb kinematics, especially for walking exoskeletons that in general provide also a gravity discharge mechanism, is simpler than the upper limb kinematics. The upper limb is, in fact, an incredibly adaptive organ that allows to perform high-level interactive movements, capable of managing numerous functional tasks in an infinitive number of kinematic solutions (due to redundancy). Regarding the basic kinematics of the upper limb, it is worthy to firstly analyze which are the main functions of the different joints of the arm. Making a simplifying assumption, the shoulder may be considered as the set of joints that controls the orientation of the entire upper limb in its workspace, the elbow as the elongation-shortening element that allows reaching objects (and taking them towards the body) and finally, the hand as the *end effector*, the organ that enables direct interaction with the environment. Within this schematic model, shoulder and elbow movements are propaedeutic for the positioning of the hand in space [1].

Most of clinical methods for upper limb kinematic analysis separately evaluate the hand motion apart from the shoulder and the elbow joints. The kinematics of the hand is of such a complexity that it is almost impossible to instrumentally evaluate it during complex gestures. State-of-the-art rehabilitation routines do actually follow the same methodology, also in robot-assisted rehabilitation. Separate machines, in fact, should be designed for the treatment of the hand motor impairments.

The functional problem has therefore been reduced to the design of robotic devices for the recovery, maintaining and improvement of the shoulder and elbow 3D movement abilities. Although the problem has been simplified this is yet a very important rehabilitation goal as the shoulder and the elbow joint movements are fundamental prerequisites for the use of the hand in activities of daily living [1]. A robotic device with all the requirements above mentioned and capable to assist the patient in performing these fundamental movements would surely be a powerful rehabilitation machine.

A limited number of upper-limb devices has been introduced into clinical practice and very few of them display complex kinematics for training activities of daily living (ADL). Two major approaches in designing upper-limb robots have consolidated over time, *i.e.* *end-effectors* manipulators and *exoskeletons*. State-of-the-art upper-limb robots together with a detailed description of the design approaches are reported in Section 2. In short, the main difference is that the kinematical structures of end-effectors and exoskeletal devices do not or do replicate, respectively, the human arm kinematics while supporting its movements. Both exoskeletons and end-effector based devices show advantages and drawbacks and it is not likely possible to give a unique and ultimate indication for the use of any of them from a clinical viewpoint. End-effector based devices are surely more flexible and may be easily adapted for different anthropometric measures. The way they work is in line with the ultimate goal of the upper limb treatment that is to pre-position the hand in the operational space in order to prepare the interaction with objects. Careful attention has to be paid to the different configurations in which the hand may result, like for instance in the different approach direction in reaching for a bottle placed on a table or in reaching for picking up a pen from the same table. Different hand orientations imply different shoulder and elbow angle configurations. However, a drawback of end-effector based devices is that these angles are not directly controlled and for highly impaired patients this may represent a problem because of the very limited motion capability. In this cases a partial solution may be given by using properly designed hand splints that indirectly may influence the proximal joints position. On the contrary, the degree of freedom given to proximal joints configuration is avoided by exoskeletons, being the shoulder and elbow angles constrained. A great advantage, in fact, is represented by the availability of information on joints kinematics and dynamics parameters with enough accuracy to evaluate the motor performances of patients. At the same time, however, constraining the joints angles may represent a harmful situation in case the arm and the exoskeleton kinematics are not perfectly aligned. In Section 2., some design requirements are also derived with particular relevance to the problem of singularities and torques exchange. A case study for the design of an exoskeleton is therefore presented in Section 3. with reference to the kinetostatic analysis and its impact on the mechanical design.

2. Robots for Upper Limb Rehabilitation

From a mechanical point of view the close physical Human-Robot Interaction (pHRI) that characterizes rehabilitation devices leads to a series of requirements to be satisfied during the design phase. In fact, the development of a device mechanically constrained to human limbs requires that its kinematic structure and its movements are compatible with movements of human segments. Some detailed considerations about the two main approaches follow.

2.1. End-effector Devices

End-effector devices (*e.g.* standard SCARA or anthropomorphic robots) present a suitable mechanical interface where the end-effector of the human limb (*i.e.* wrist/foot for upper/lower human limbs) is mechanically constrained. Due to the fact that end-effector devices do not constrain the whole human kinematic chain, the arm is free to self adapt to external constraints and motions imposed by the end-effector position. Constraining only the wrist position or the forearm pose (wrist position and forearm orientation), without constraining the upper arm orientation, the humeral head results in a self-adapted position, exploiting the shoulder girdle degrees of freedom. Being the shoulder center (*i.e.* humeral head) not constrained, drawbacks due to shoulder over-constraining do not arise. On the other side, impaired patients tend to employ shoulder mobility to compensate and achieve otherwise not affordable movements and unreachable hand poses. On this aspect the debate is open in the medical field about the necessity of constraining the human shoulder or leaving it free to adapt. Moreover, from a mechanical point of view, end-effector devices

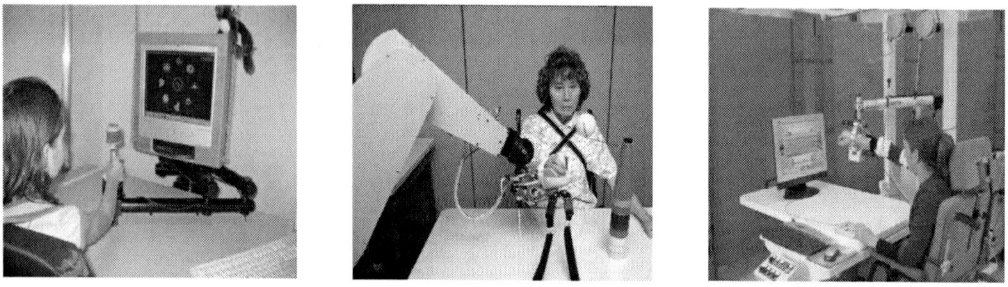

(a) Mit-Manus (b) MIME (c) Gentle/s

Figure 1. End-effector devices.

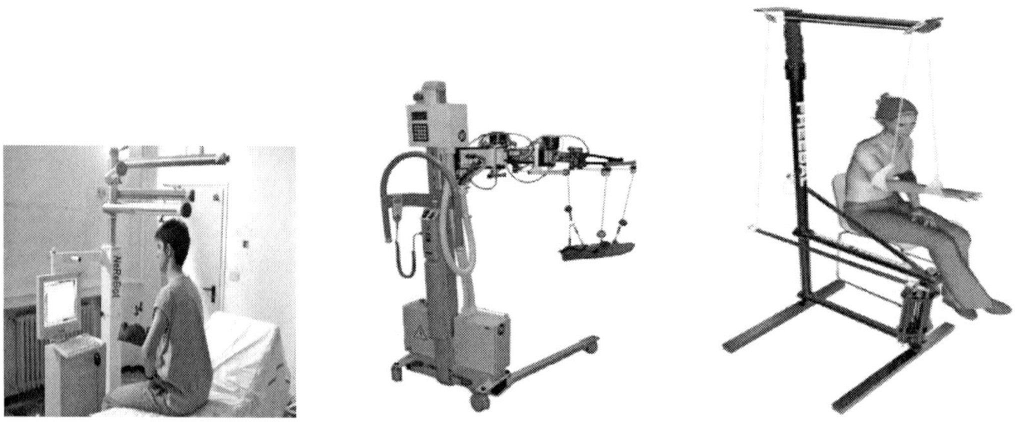

(a) NeReBot (b) MariBot (c) Freebal

Figure 2. End-effector cable-driven robotic solutions.

require large sizes and footprint dimensions to cover the full upper-limb range of motion (ROM) [2].

A number of solutions for robotized rehabilitation has been developed in less than two decades, based on simple kinematics or standard industrial robots suitably customized. Remarkable examples of developed prototypes are the MIT-MANUS [3], a planar assistive device largely used in rehabilitation facilities, the MIME [4], a PUMA-560 antropomorphic robot customized for spatial stroke patients treatments, and the Gentle/s project [5], based on a customized Haptic Master device (see Fig. 1). Some cable-driven robots have been developed, whose advantageous lighter architecture do not allow, on the other side, a good coverage of the complete upper-limb ROM [6, 7, 8] (see Fig. 2).

2.2. Exoskeletal Prostheses

Exoskeletons are typically designed to match and align their mechanical joints to human limb articulations and to constraint and/or support all the links of the human kinematic chain. Their kinematical structures are typically advantageous in terms of workspace coverage. Moreover, since they bear both the upper arm and the forearm, a better control and support of the human elbow and shoulder is allowed, if compared to end-effector devices.

Devices designed to operate kinematically in parallel with human limbs need to face the problem of a correct alignment between human articulations and mechanical joints. The parallel nature of the limb-device coupling can lead to unwanted mechanical stresses in limb articulations if a correct axes alignment is not achieved. Human-robot axes alignment can be achieved by a correct joint alignment in the setup phase, if the position/orientation of human joints is constant during the therapy, or introducing proper mechanical joint decouplings in order to guarantee a correct self-alignment of the mechanism, in case the considered limb joint position is not fixed during rehabilitation movements.

The most evident case of impact of alignment is in the human shoulder articulation, whose complex movement is due to a combination of the shoulder girdle movement and of the gleno-humeral joint movement [9, 10, 11, 12]. During physiological upper-limb movements the center of rotation of the humeral head translates with respect to exoskeleton reference frames, as function of the upper arm elevation angle (Fig. 3).

The simplest solution in setting the topology of the exoskeleton consists on designing the robotized shoulder as a spherical joint [13, 14, 15]. This solution correctly replicates the gleno-humeral articulation (at the humeral head), but it does not correctly compensate the complex movement of the shoulder girdle. This approach is acceptable for relatively small displacements of the human shoulder. Considering high amplitudes of arm elevation (greater than 90°) the displacement of the humeral head notably increases. As a consequence, arm movements are prevented the more the larger the amplitude of the movement, increasing in this way the level of discomfort of the patient (for a detailed analysis of human discomfort and pain levels refer to [16]).

Considering the effects and impact of shoulder displacement is very relevant for the development of rehabilitation exoskeletons because the patient's weakness and impairment emphasize the importance of wearability and human-robot kinematic compatibility.

To take into account this important aspect, advanced solutions have been in fact designed in order to properly compensate translational movements of the center of rotation of the humeral head [17, 18]. Moreover, methodologies for designing exoskeletons with improved ergonomic performances in terms of adaptability and human-robot joint self-alignment are presented in [19, 20, 21]. Furthermore, singularities can play a significant role in defining the actual exploitable range of motion of mechanical chains. Exoskeleton singularities typically occur if rotational joint axes are aligned during their movement. Some developed solutions have been optimized for handling and limit singularities drawbacks at the shoulder and for exploiting as much as possible the human shoulder ROM [15, 22]. This result is achieved positioning the singularities of the exoskeleton shoulder, which is typically obtained by three consecutive rotational axes, outside the ROM of the human shoulder. Another singular configuration may however happen at the elbow joint but very little emphasis has been so far given to it in literature. Experimental studies [23, 24] highlighted that some functional ADL movements (*e.g.* reaching movement) are characterized by a complete extension of the elbow articulation that has the very undesirable feature of being affected by an intrinsic singularity. If a complete human-robot joint alignment is achieved at the elbow (typical condition of a number of state-of-the-art exoskeletons) a double sin-

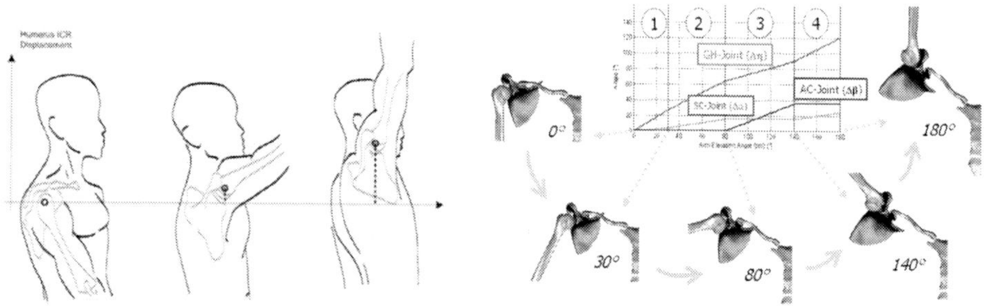

(a) Humeral head vertical displacement. Reproduced from ESA.

(b) Shoulder rhythm phases (with the permission of Tobias Nef and Robert Riener, ETH and University of Zürich, Switzerland).

Figure 3. Shoulder Instant Center of Rotation (ICR) motion.

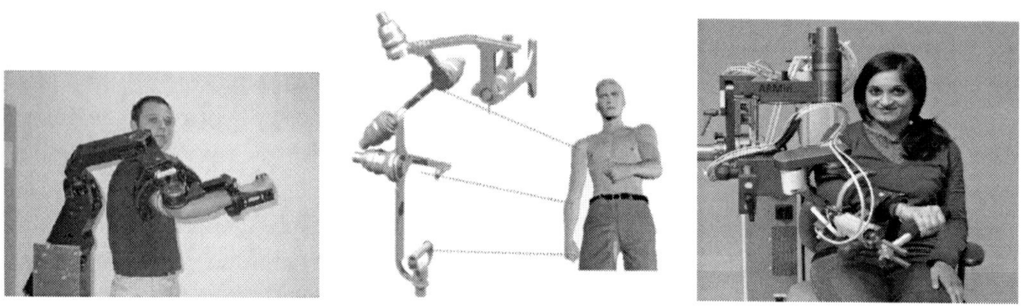

(a) L-Exos

(b) MGA

(c) ArminIII

Figure 4. Exoskeletons for upper-limb rehabilitation.

gularity affects both the exoskeleton and the human-arm kinematic chain in the extended elbow configuration (Section 3.4.).

Modern medical rehabilitation robotic devices are equipped with force feedback in order to interact with the patient on the basis of the actual exchanged force and proper control techniques. Force is typically measured by torque sensors in correspondence with robot joints, or measuring motors extra-currents. In both cases the mechanical kinematics and placement of joints need to guarantee that the force/torque generated in the human arm joint space are correctly transmitted to the robot joint space: torques generated by muscular activity need to be transmitted to robot joints. A correct human-robot joint alignment guarantees, in general terms, that torques generated by muscles act selectively on each single mechanical joint, but kinematic singularities can influence actual transmitted torques throughout the mechanical structure. Despite advantages related to axes alignment, problems can arise nearby singular configurations. This can cause problems and limitations in collaborative rehabilitation therapies where torques generated by patients (typically weak for impaired people) have to be transferred from human articulations to exoskeletal joints.

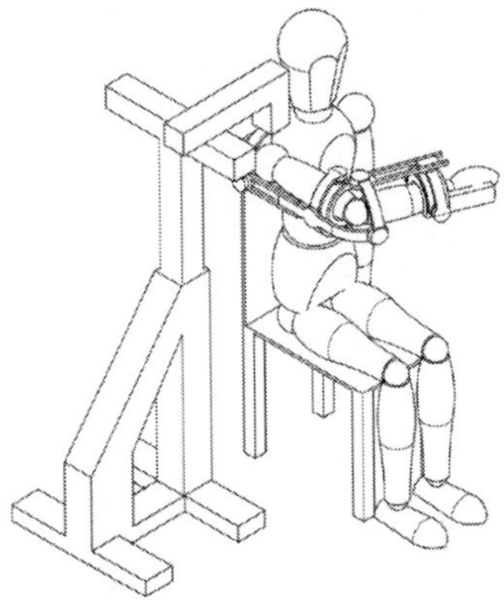

Figure 5. Exoskeleton setup: the gross position of the exoskeleton shoulder joint is defined by a Cartesian holder; the human shoulder center can float w.r.t. the holding structure thanks to its peculiar exoskeleton kinematics.

3. Case Study: Self-adaptable and Elbow-singularity-free Exoskeleton

A hybrid structure exoskeleton under development (Fig. 5) is presented hereafter. The mechanism design has been optimized to cope with two distinct and relevant aspects of upper limb rehabilitation:

- adaptability and compensation of shoulder displacements to prevent undesired shoulder internal stresses due to joint axes misalignments (*shoulder adaptability*);
- optimization of the elbow usable workspace and dexterity, limiting singularities drawbacks (*topology optimization*).

Shoulder adaptability is achieved by a hybrid kinematic mechanism and a proper set of compliant links (Fig. 7).

Topology optimization is achieved by a peculiar joint axes placement at the elbow level. Exoskeletons usually reproduce shoulder and elbow articulations by a spherical joint centered on the humeral head plus a revolute joint aligned to the elbow rotational axis. The solution here presented, on the contrary, features two Cardan joints centered respectively at the humeral head and in the elbow center. Details on advantages of the presented solution are given in Section 3.4.

3.1. Human Arm Kinematic Model

An introduction on the upper limb kinematic model is reported in order to correctly analyze kinematic interrelationships between human and robot kinematics and their compatibility. With reference to Fig. 6 the model includes: i) the gleno-humeral joint assigned as a spherical one, *i.e.* obtained by three serial revolute joints and reflecting acknowledged

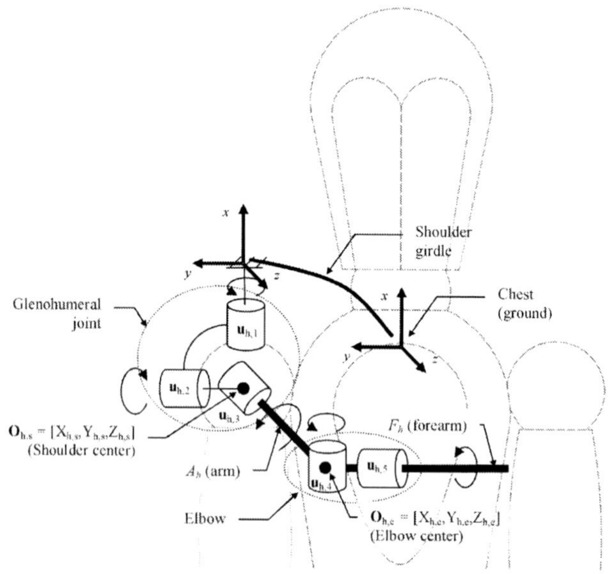

(a) Kinematic scheme

degree of freedom		Min	Max
Plane of elevation	$\{\phi_1, \mathbf{u}_{h,1}\}$	0	135
Angle of elevation	$\{\phi_2, \mathbf{u}_{h,2}\}$	-90	90
Internal rotation	$\{\phi_3, \mathbf{u}_{h,3}\}$	-55	70
Elbow flexion	$\{\phi_4, \mathbf{u}_{h,4}\}$	0	140
Angle of pronation	$\{\phi_5, \mathbf{u}_{h,5}\}$	5	160

(b) Joints range of motion

Figure 6. Human-arm model.

medical nomenclature; ii) the elbow joint assigned as a revolute one; iii) the forearm rotation (prono-supination) as a revolute one. This simplified shoulder-fixed model is enriched by a three degrees of freedom translational joint representing the displacement of the shoulder Instant Center of Rotation (ICR), *i.e.* the humeral head, due to the shoulder girdle motion [10]. Let denote by:

1. *xy*, *yz* and *xz* the coronal, transverse and sagittal planes, respectively.

2. $\mathbf{O}_{h,s} = (X_{h,s}, Y_{h,s}, Z_{h,s})$ and $\mathbf{O}_{h,e} = (X_{h,e}, Y_{h,e}, Z_{h,e})$ the shoulder and elbow centers. According to the introduced model, $\mathbf{O}_{h,s}$ can freely translate with respect to the chest, due to the shoulder girdle movement.

3. $\mathbf{u}_{h,i}$ the rotational axis of the i-th joint of the human limb chain and ϕ_i denotes the value of its angle. Refer to Fig. 6-b for joint names in medical nomenclature and their typical range of motion.

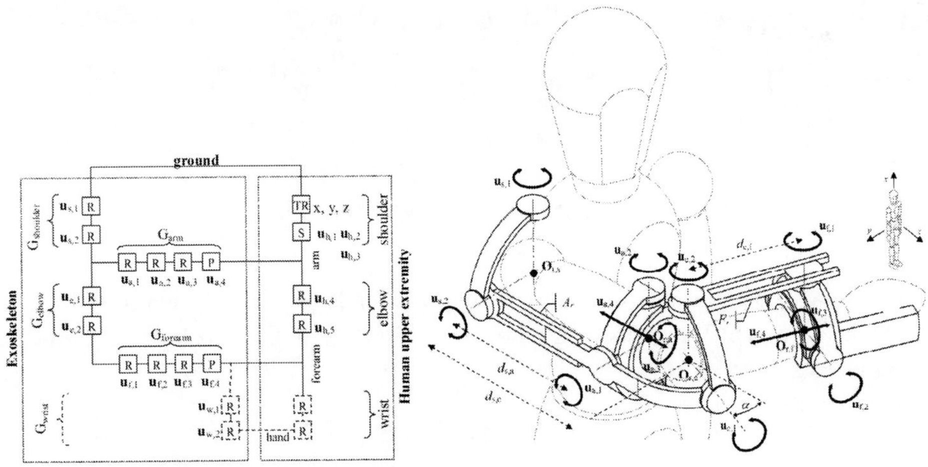

(a) (b)

Figure 7. The proposed hybrid kinematic architecture: (a) topological and (b) tridimensional exoskeleton representation.

Two singularity conditions may occur in the model:

- *shoulder singularity* ($\mathbf{u}_{h,1} // \mathbf{u}_{h,3}$) not representing an actual singularity in the human shoulder, but exclusively due to the representation of shoulder joint by three serial rotation axes;
- *elbow singularity* ($\mathbf{u}_{h,3} // \mathbf{u}_{h,5}$) actually occurring when the upper arm and the forearm are aligned (elbow completely extended).

The design choices that are more relevant for preventing or managing these singularities are detailed in Section 3.4.

3.2. Mechanism Description

The exoskeleton resembles an hybrid kinematics mechanism that is composed by four distinct serial chains denoted by $G_{shoulder}$, G_{arm}, G_{elbow} and $G_{forearm}$. Referring to Fig. 7, let $\mathbf{u}_{i,j}$ be the j-th rotational joint of the i-th serial chain, and $q_{i,j}$ be the corresponding joint position. The details of the four serial chains are then listed:

- the *exoskeleton shoulder* $G_{shoulder}$ is a Cardan joint made of the revolute joints $\{q_{s,1}, \mathbf{u}_{s,1}\}$ and $\{q_{s,2}, \mathbf{u}_{s,2}\}$ (2 degrees of freedom), intersecting in the exoskeleton shoulder center $\mathbf{O}_{r,s}$;
- the *arm connection* G_{arm} is a spherical joint made of the revolute joints $\{q_{a,1}, \mathbf{u}_{a,1}\}$, $\{q_{a,2}, \mathbf{u}_{a,2}\}$, $\{q_{a,3}, \mathbf{u}_{a,3}\}$ and the prismatic joint $\{q_{a,4}, \mathbf{u}_{a,4}\}$ (4 degrees of freedom), intersecting in $\mathbf{O}_{r,a}$;

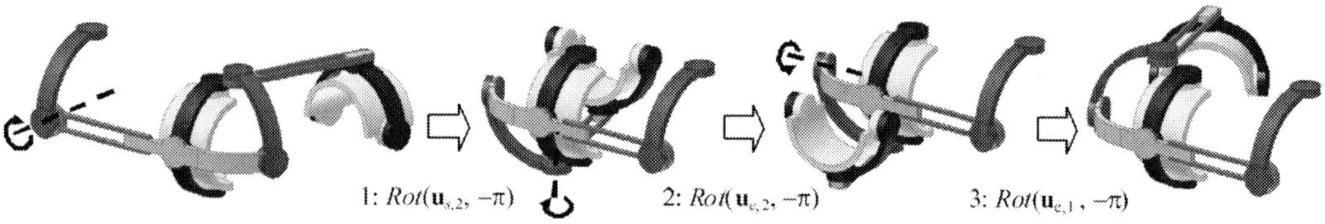

Figure 8. Exoskeleton bilaterality reconfiguration.

- the *exoskeleton elbow* G_{elbow} is a Cardan joint made of the revolute joints $\{q_{e,1}, \mathbf{u}_{e,1}\}$ and $\{q_{e,2}, \mathbf{u}_{e,2}\}$ (2 degrees of freedom), intersecting in the exoskeleton elbow center $\mathbf{O}_{r,e}$;
- the *forearm connection* $G_{forearm}$ is a spherical joint made of the revolute joints $\{q_{f,1}, \mathbf{u}_{f,1}\}$, $\{q_{f,2}, \mathbf{u}_{f,2}\}$, $\{q_{f,3}, \mathbf{u}_{f,3}\}$ and the prismatic joint $\{q_{f,4}, \mathbf{u}_{f,4}\}$ (4 degrees of freedom), intersecting in $\mathbf{O}_{r,f}$.

An additional serial chain G_{wrist} can be employed to control the two degrees of freedom of the wrist (flexion/extension, ulnar/radial deviation).

The exoskeleton kinematics and joint placement allow the human shoulder and elbow to be not coincident with the exoskeleton shoulder and elbow ($\mathbf{O}_{h,s} \neq \mathbf{O}_{r,s}$ and/or $\mathbf{O}_{h,e} \neq \mathbf{O}_{r,e}$).

Finally, let α be the angle defined as:

$$\alpha \triangleq \angle\left((\mathbf{O}_{r,s} - \mathbf{O}_{r,e}), \mathbf{u}_{e,1}\right). \tag{1}$$

This angle influences the transmission of torques at the elbow and the corresponding Jacobian indexes as detailed in Section 3.4.

A proper joints disposition allows the exoskeleton to be indifferently worn by both the right and the left upper extremity. Joints and elements symmetry have been arranged to allow structure reconfigurability (Fig. 8).

3.3. Kinematics

Functional and control issues require the correlation of desired human limb poses with the device joint kinematic profiles. The kinematic analysis is devoted to the identification and the analytical description of kinematical relationships among the exoskeleton links.

Under the assumption of $\mathbf{O}_{h,s} \equiv \mathbf{O}_{r,s}$, $\mathbf{O}_{h,e} \equiv \mathbf{O}_{r,e}$ and $\alpha = 0$ the exoskeleton joint movements are decoupled and each single joint is aligned to a single degree of freedom of the human arm. In particular:

- $q_{s,1}$ controls the plane of elevation $\forall\alpha$ $(\mathbf{u}_{s,1}//\mathbf{u}_{h,1})$;
- $q_{s,2}$ controls the angle of elevation $\forall\alpha$ $(\mathbf{u}_{s,2}//\mathbf{u}_{h,2})$;
- $q_{e,1}$ controls the arm rotation if $\alpha = 0$ $(\mathbf{u}_{e,1}//\mathbf{u}_{h,3})$;
- $q_{e,2}$ controls the elbow flexion if $\alpha = 0$ $(\mathbf{u}_{e,2}//\mathbf{u}_{h,4})$;

- $q_{f,3}$ controls the forearm pronosupination $\forall\alpha$ $(\mathbf{u}_{f,3}//\mathbf{u}_{h,5})$.

In general conditions, i.e. with $\mathbf{O}_{h,s} \neq \mathbf{O}_{r,s}$ and/or $\mathbf{O}_{h,e} \neq \mathbf{O}_{r,e}$ and/or $\alpha \neq 0$, the exoskeleton joints are coupled in defining the human arm configuration.

Referring to Fig. 7-b it is worth to note that:

- a suited actuation of $\{q_{a,1}, \mathbf{u}_{a,1}\}$, $\{q_{a,2}, \mathbf{u}_{a,2}\}$ and $\{q_{a,4}, \mathbf{u}_{a,4}\}$, allows the control of the center of rotation of the humeral head (see eq. (4));
- $\{q_{a,3}, \mathbf{u}_{a,3}\}$, $\{q_{f,1}, \mathbf{u}_{f,1}\}$, $\{q_{f,2}, \mathbf{u}_{f,2}\}$, $\{q_{f,4}, \mathbf{u}_{f,4}\}$, always unactuated, allow the exoskeleton to be compliant with respect to human limb relative displacements.

The direct kinematics of the mechanism comes straightforwardly out of the sequence of Cardan and prismatic joints. Referring to Fig. 7, let be:

- $d_{i,j}$ the distance $\|\mathbf{O}_{r,i} - \mathbf{O}_{r,j}\|$.
- $Rot(\mathbf{u}, q)$ and $Tr(\mathbf{u}, q)$ the homogeneous matrices corresponding to a rotation and a translation of q around the $\mathbf{u}$ axis respectively.
- $\overline{Rot}(\cdot)$ and $\overline{Tr}(\cdot)$ constant transformations.
- $\{0\}$ the reference frame, centered in the exoskeleton shoulder center $\mathbf{O}_{r,s}$, with the axes normal to the coronal, transverse and sagittal planes of the patient.
- $\{i\}$ the frame associated to the exoskeleton serial chain G_i, centered in $\mathbf{O}_{r,i}$.
- $\mathbf{M}_{ij}$ the transformation matrix from the frame $\{i\}$ to $\{j\}$.

Therefore, the exoskeleton direct kinematics equations are:

$$\begin{aligned}
\mathbf{M}_{0,s} &= Rot(\mathbf{u}_{s,1}, q_{s,1})\, Rot(\mathbf{u}_{s,2}, q_{s,2}),\\
\mathbf{M}_{s,a} &= \overline{Tr}(\mathbf{u}_{s,3}, d_{s,a})\, \overline{Rot}(\mathbf{u}_{s,3}, \pi/2)\, Rot(\mathbf{u}_{a,1}, q_{a,1})\cdot\\
&\quad Rot(\mathbf{u}_{a,2}, q_{a,2})\, Rot(\mathbf{u}_{a,3}, q_{a,3})\, Tr(\mathbf{u}_{a,3}, q_{a,4}),\\
\mathbf{M}_{s,e} &= \overline{Tr}(\mathbf{u}_{s,3}, d_{s,e})\, \overline{Rot}(\mathbf{u}_{s,3}, \pi/2)\, Rot(\mathbf{u}_{s,2}, -\alpha)\cdot\\
&\quad Rot(X, q_{e,1})\, Rot(Y, q_{e,2}),\\
\mathbf{M}_{e,f} &= \overline{Tr}(\mathbf{u}_{e,3}, d_{e,f})\, \overline{Rot}(\mathbf{u}_{e,3}, -\pi/2)\, Rot(\mathbf{u}_{f,3}, q_{f,1})\\
&\quad Rot(\mathbf{u}_{f,2}, q_{f,2})\, Rot(\mathbf{u}_{f,3}, q_{f,3})\, Tr(\mathbf{u}_{f,3}, q_{f,4}).
\end{aligned}$$

Finally, the absolute poses of links constrained to the human arm and forearm are:

$$\mathbf{M}_{0,a} = \mathbf{M}_{0,s}\mathbf{M}_{s,a}, \qquad M_{0,f} = \mathbf{M}_{0,s}\mathbf{M}_{s,e}\mathbf{M}_{e,f} \tag{2}$$

Two distinct hypotheses of actuation are proposed: *compliant-shoulder* and *actuated-shoulder*. In the former, the mechanism is left with some un-actuated joints and the human shoulder is free to adapt its position through the mechanism compliances. In the latter, the gleno-humeral joint position can be actively controlled by the device.

The *compliant-shoulder* configuration is characterized by the actuation of joints $q_{s,1}$, $q_{s,2}$, $q_{e,1}$, $q_{e,2}$ and $q_{f,3}$:

$$\left\{\begin{array}{l}\theta_1 = q_{s,1}\\ \theta_2 = q_{s,2}\\ \theta_3 = q_{e,1}\\ \theta_4 = q_{e,2}\\ \theta_5 = q_{f,3}\end{array}\right. \Rightarrow \begin{array}{rcl} M_{0,a} & = & M_{0,a}(\theta_1, \theta_2)\\ M_{0,f} & = & M_{0,f}(\theta_1, \theta_2, \theta_3, \theta_4, \theta_5)\end{array} \tag{3}$$

The *actuated-shoulder* configuration is characterized by the additional actuation of $q_{a,1}$, $q_{a,2}$ and $q_{a,4}$:

$$\left\{\begin{array}{l}\vdots\\ \theta_6 = q_{a,1}\\ \theta_7 = q_{a,2}\\ \theta_8 = q_{a,4}\end{array}\right. \Rightarrow \begin{array}{rcl} \mathbf{M}_{0,a} & = & \mathbf{M}_{0,a}(\theta_1, \theta_2, \theta_6, \theta_7, \theta_8)\\ \mathbf{M}_{0,f} & = & \mathbf{M}_{0,f}(\theta_1, \theta_2, \theta_3, \theta_4, \theta_5)\end{array} \tag{4}$$

denoting by θ_i the i-th actuated joint.

3.4. Kinetostatics

The kinetostatic analysis of the equivalent mechanism composed by the human upper limb and the exoskeleton through the study of geometric Jacobian $\mathbf{J}$ is presented; the geometric Jacobian matrices will be used also for the analysis of the mechanism singularities and stiffness.

Without loss of generality and for clarity of notation, the analysis will be performed considering the human shoulder and elbow centers coincident respectively with the exoskeleton shoulder and elbow centers ($\mathbf{O}_{h,s} \equiv \mathbf{O}_{r,s}$, $\mathbf{O}_{h,e} \equiv \mathbf{O}_{r,s}$). The objective in fact is to evaluate the impact of design choices on the Jacobian indexes and performances, focusing on the elbow dexterity.

Alternatively, the relation between the space of human coordinates $\Phi = \{\phi_i, i = 1...5\}$ and the space of machine coordinates $\Theta = \{\theta_i, i = 1...5\}$ is expressed by the implicit kinematic closure equation:

$$F(\Phi, \Theta) = 0, \quad \text{and} \quad \frac{\partial F}{\partial \Phi}\dot{\Phi} + \frac{\partial F}{\partial \Theta}\dot{\Theta} = 0, \tag{5}$$

for which the analytical Jacobians are defined as

$$\mathbf{J}^A_\Phi = \partial F / \partial \Phi \quad \mathbf{J}^A_\Theta = -\partial F / \partial \Theta \quad \rightarrow \quad \mathbf{J}^A = \mathbf{J}^{A^{-1}}_\Theta \mathbf{J}^A_\Phi. \tag{6}$$

The analysis of shoulder and elbow kinetostatics also requires the definition of the instantaneous velocity vector Ω of a spherical joint that is:

$$\begin{bmatrix}\omega_x\\ \omega_y\\ \omega_z\end{bmatrix} = \begin{bmatrix}\dot{\alpha} + \dot{\gamma}\sin(\beta)\\ \dot{\beta}\cos(\alpha) - \dot{\gamma}\cos(\beta)\sin(\alpha)\\ \dot{\beta}\sin(\alpha) + \dot{\gamma}\cos(\alpha)\cos(\beta)\end{bmatrix} \tag{7}$$

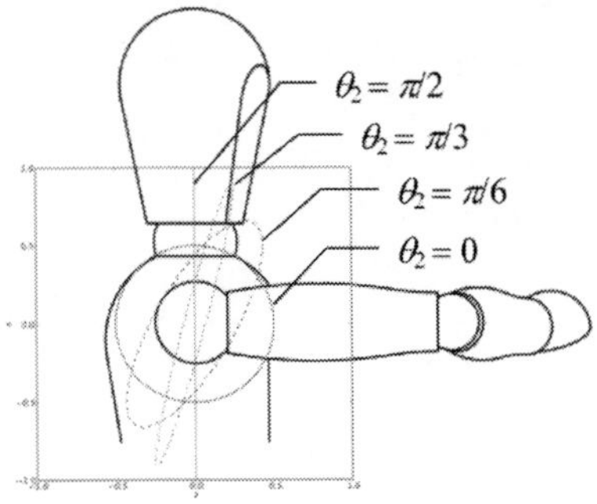

Figure 9. Stiffness ellipsoids of the shoulder mechanism. The stiffness is proportional to the ellipsoid radius along the considered direction. If the radius tends to zero, the mechanism mobility is critical.

where α, β and γ are the revolute angles of the joints around axes x, y and z. The corresponding Jacobian matrix, considering the joint space velocities $\begin{bmatrix}\dot{\alpha} & \dot{\beta} & \dot{\gamma}\end{bmatrix}^T$, is therefore:

$$\mathbf{J}_{sph}(\alpha, \beta) = \begin{bmatrix} 1 & 0 & \sin(\beta) \\ 0 & \cos(\alpha) & -\cos(\beta)\sin(\alpha) \\ 0 & \sin(\alpha) & \cos(\alpha)\cos(\beta) \end{bmatrix}. \tag{8}$$

When considering the **shoulder articulation**, the angular velocity of the upper arm bone (ω_a) w.r.t the reference frame $\{0\}$ results:

$$\omega_a = \mathbf{J}_{q,shoulder}(\theta_1, \theta_2)\begin{bmatrix}\dot{\theta}_1 \\ \dot{\theta}_2 \\ \dot{\phi}_3\end{bmatrix}$$

denoting with $\mathbf{J}_{q,shoulder}$ the Jacobian matrix of the shoulder spherical joint composed by $\mathbf{u}_{s,1}$, $\mathbf{u}_{s,2}$ and $\mathbf{u}_{h,3}$.

The singularity analysis highlights that the exoskeleton is in a singular configuration if $\mathbf{u}_{s,1}$ is aligned to $\mathbf{u}_{h,3}$. The corresponding stiffness ellipsoids are represented in Fig. 9.

When considering the **elbow articulation**, the mechanism is characterized by a misalignment α between $\mathbf{u}_{e,1}$ and $\mathbf{u}_{h,3}$, as mentioned in Section 3.2. This design feature imposes that the intra-extra rotation (ϕ_3) of the human arm and the elbow flexion (ϕ_4) are concurrently defined by both G_{elbow} joints (θ_3, θ_4). In details:

$$\begin{aligned} \dot{\phi}_3 &= \frac{\partial\phi_3(\theta_3, \theta_4)_{|\alpha}}{\partial\theta_3}\dot{\theta}_3 + \frac{\partial\phi_3(\theta_3, \theta_4)_{|\alpha}}{\partial\theta_4}\dot{\theta}_4, \\ \dot{\phi}_4 &= \frac{\partial\phi_4(\theta_3, \theta_4)_{|\alpha}}{\partial\theta_3}\dot{\theta}_3 + \frac{\partial\phi_4(\theta_3, \theta_4)_{|\alpha}}{\partial\theta_4}\dot{\theta}_4; \end{aligned} \tag{9}$$

Joints decoupling occurs if $\alpha = 0$ ($\mathbf{u}_{e,1}//\mathbf{u}_{h,3}$, $\dot{\phi}_3/\dot{\theta}_3 = 1$, $\dot{\phi}_4/\dot{\theta}_3 = 0$, $\forall\theta_3$).

The angular velocity of the forearm bone (ω_f) w.r.t. the exoskeleton upper arm is:

$$\omega_f = \mathbf{J}_s(\phi_3, \phi_4)\begin{bmatrix}\dot{\phi}_3 \\ \dot{\phi}_4 \\ \dot{\phi}_5\end{bmatrix} = \mathbf{J}_q(\theta_3, \theta_4 - \alpha)\begin{bmatrix}\dot{\theta}_3 \\ \dot{\theta}_4 \\ \dot{\theta}_5\end{bmatrix}, \tag{10}$$

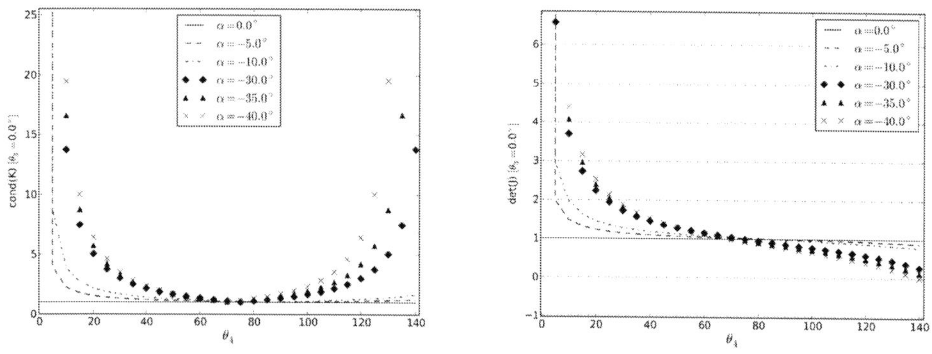

(a) (b)

Figure 10. Elbow condition number (a) and determinant (b) for different α values, as function of θ_4. The condition number tends to infinity if the human arm tends to its singular configuration (completely extended) or if the forearm tends to be aligned to $\mathbf{u}_{e,1}$.

where $\mathbf{J}_s$ and $\mathbf{J}_q$ are respectively the "human" Jacobian and "exoskeleton" Jacobian.

Denoting by $\mathbf{J}_{el}$ the elbow geometric Jacobian matrix:

$$\mathbf{J}_{el} = \mathbf{J}_s\,(\phi_3,\,\phi_4)^{-1}\,\mathbf{J}_q(\theta_3,\,\theta_4 - \alpha)$$

its related performance indexes (condition number and determinant) are represented in Fig. 10. As expected, the condition number tends to infinity: i) approaching the condition $\mathbf{u}_{h,3}//\mathbf{u}_{h,5}$ ($\theta_4 = 0°$) or ii) if the forearm ($\mathbf{u}_{h,5}$) tends to be aligned to $\mathbf{u}_{e,1}$ (*e.g.* $\theta_4 = 140°$, if $\alpha = -40°$).

To completely and correctly interpret $\mathbf{J}_{el}$ it is convenient to separately analyze $\mathbf{J}_q$ (mechanism space) and $\mathbf{J}_s$ (human arm space).

In the upper part of Fig. 12, $\det(\mathbf{J}_q)$ is plotted for values of elbow joint close to $\theta_4 = 0$ (complete extension), for different α values. If $\alpha = 0$, the plotted curve represents also $\det(\mathbf{J}_s)$. In the lower part of the figure, corresponding values of $\det(\mathbf{J}_{el})$ are reported (refer also to Fig. 10-(b)).

Three types of singularities can be identified:

I) $\det(\mathbf{J}_s) \neq 0$, $\det(\mathbf{J}_q) = 0$ ($\theta_4 = \alpha$): exoskeleton *gimbal lock* singularity where the mechanism loses one degree of freedom. This singularity must be outside the elbow range of motion.

II) $\det(\mathbf{J}_s) = 0$, $\det(\mathbf{J}_q) \neq 0$ ($\phi_4 = 0$): intrinsic upper limb singularity where the limb loses one degree of freedom (complete forearm extension $\phi_4 = 0$). An external force cannot induce the rotation of the forearm around the axis normal to $\mathbf{u}_{h,4}$ and $\mathbf{u}_{h,5}$. Upper limb articulations would not allow limb compliance and possible injuries can occur, in case the forearm is not correctly loaded. On the other hand, since the exoskeleton is not in a singular configuration, an elbow torque applied by the human from the human arm space

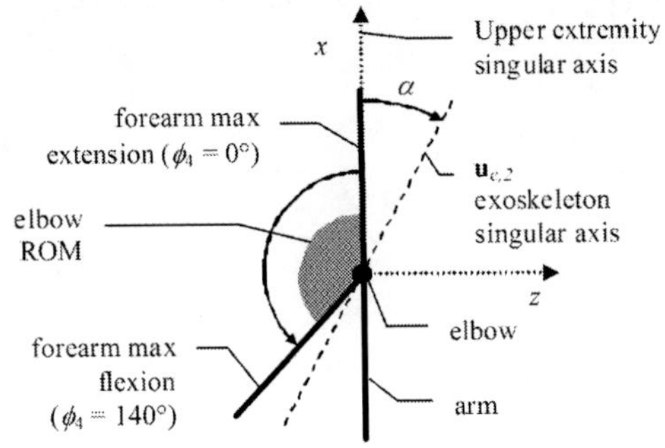

Figure 11. Elbow range of motion represented in elbow joint space O_e. Bold lines represent the limb, dashed line represents the exoskeleton singular axis.

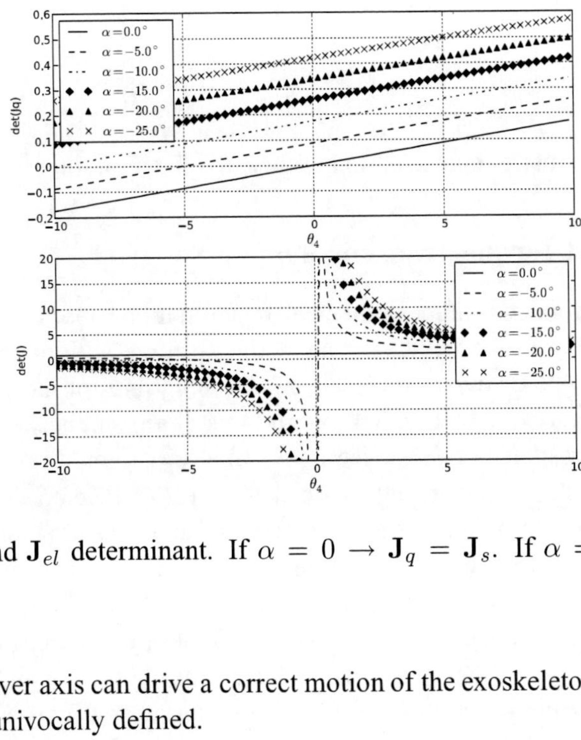

Figure 12. $\mathbf{J}_q$ and $\mathbf{J}_{el}$ determinant. If $\alpha = 0 \rightarrow \mathbf{J}_q = \mathbf{J}_s$. If $\alpha = 0, \theta_4 = 0 \rightarrow \mathbf{J}_{el}$ is undetermined.

around whichever axis can drive a correct motion of the exoskeleton elbow joint, because its position is univocally defined.

III) $\det(\mathbf{J}_s) = 0$, $\det(\mathbf{J}_q) = 0$ ($\alpha = 0$): both the exoskeleton and the elbow are in a singular configuration. The user cannot lead any motion in any direction, but only rotate around the prono-supination axis. The configuration of the exoskeleton is not univocally defined by the human arm configuration.

In conclusion, a correct choice of the α angle in the design phase of the exoskeleton can lead to an improved dexterity and control of the exoskeleton elbow by the user, also in the completely extended configuration. Moreover, typical elbow range of motion will not turn into singularities and related drawbacks. In details, referring to Fig. 6-(b) and Fig. 11, allowable α values for preventing exoskeleton singularities are $0° < \alpha < 40°$.

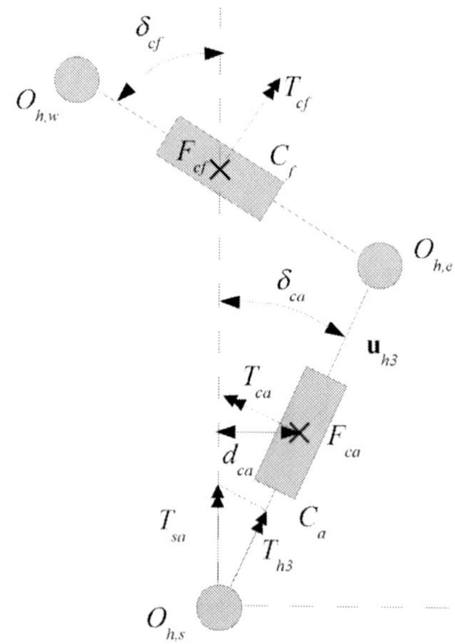

Figure 13. Schematic representation of the cuffs reaction forces and torques.

3.5. Human-robot Torque Transmission

As previously introduced, modern medical rehabilitation robotic devices are characterized by the capability of the device of interacting in force with the patient. Typically, force measurement is obtained by torque sensors at robot joints, or measuring motor over-currents. As demonstrated in Section 3.4., a joint mis-alignment between the human arm bone $\mathbf{u}_{h,3}$ and the corresponding exoskeleton joint axis $\mathbf{u}_{e,1}$ (namely α) is beneficial to reduce exoskeleton singularities and to univocally define the exoskeleton elbow configuration. In this Section a deeper analysis in terms of transmitted and measured torque is presented.

Both in case of sensorized joints and in case of torque estimation by motor over-currents measurement, a minimum torque transmitted to joints has to be guaranteed in order to make the resulting joints torque, due to torques generated by muscles, appreciable. This minimal level depends, in general, on the sensibility of the measuring system and mechanical frictions. Let denote this minimum torque for the i-th exoskeleton joint by $Tmin_{r,i}$. If the torque transmitted to the i-th mechanical joint is lower than $Tmin_{r,i}$ the capability of the exoskeleton of reacting to the torque generated at the limb is prevented.

In general terms, the importance of avoiding singularities is amplified by the effect of softness of the human tissues and the interfacing materials (*i.e. cuffs*). This is also reflected on joint torques transmission.

If the torque transmitted through cuffs from the human arm to exoskeletal mechanical joints is less than $Tmin_{r,i}$, a rotation between the human and the exoskeleton arm can occur and the kinematical model previously introduced (Fig. 7) is incongruent with the actual kinematic configuration.

Cuffs can be, simply but exhaustively, modeled as linear and torsional springs, to take into account the effect of soft tissues reactions and mechanical compliances. Referring to Fig. 13, C_a and C_f denote the cuffs constrained respectively to the arm and the forearm.

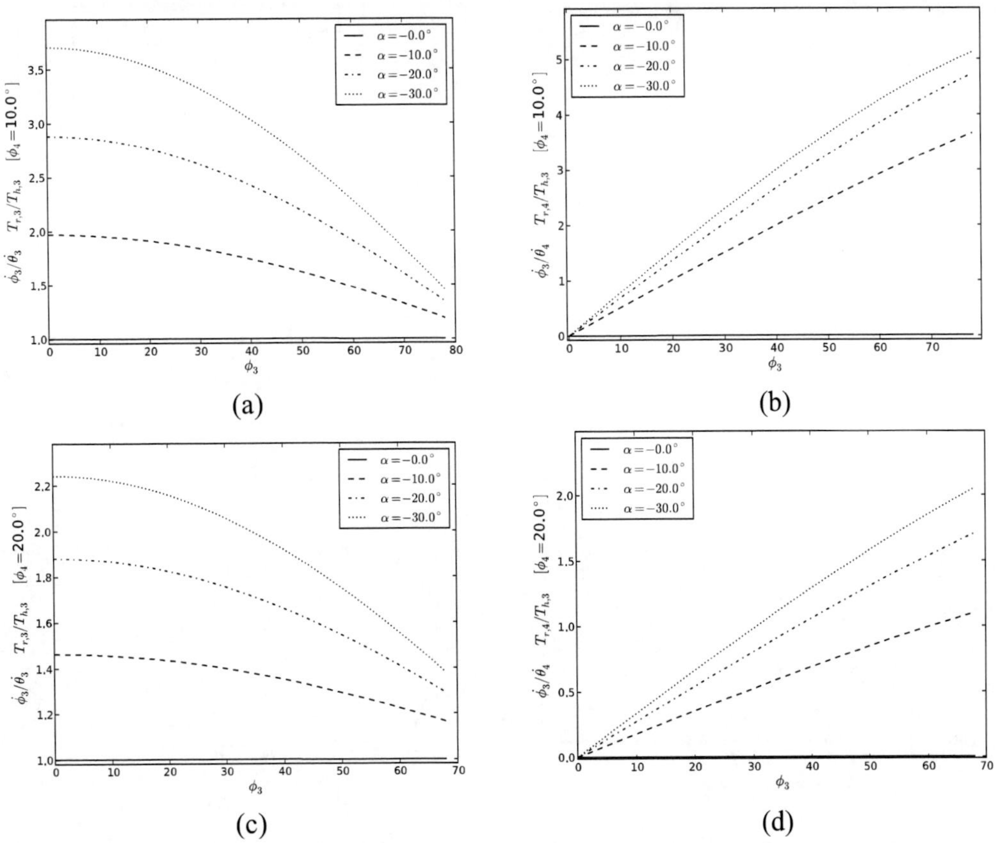

Figure 14. Torque transmission ratios of $T_{h,3}$ at different flexion angles: (a-b) $\phi_4 = 10°$; (c-d) $\phi_4 = 20°$.

Forces and torques exerted by cuffs are denoted respectively by $(F_{c,a}, T_{c,a})$ and $(F_{c,f}, T_{c,f})$ for the arm and the forearm:

$$F_{c,i} = kl_{c,i} dl_{c,i}$$
$$T_{c,i} = kt_{c,i} dt_{c,i}$$

where $kl_{c,i}$ and $kt_{c,i}$ denote the linear and torsional spring characteristics, $dl_{c,i}$ and $dt_{c,i}$ denote the linear and torsional displacements.

The total torque the cuffs can exert in constraining the human arm with respect to the exoskeleton structure is:

$$T_{s,a} = F_{c,a} d_{ca} + T_{c,a} \sin \delta_{ca} + T_{c,f} \sin \delta_{cf}$$

and, denoting by $T_{h,i}$ the torque of the i-th human joint, the resulting torque along the arm axis $\mathbf{u}_{h,3}$ is:

$$T_{h,3} = T_{sa} \cos \delta_{ca}$$

Torque components along the axes of the cuffs ($\mathbf{u}_{h,3}$ and $\mathbf{u}_{h,5}$) are negligible, due to the large compliance of the human soft tissues around the bones axes. Approaching to the elbow singular configuration, d_{ca}, δ_{ca} and δ_{cf} decrease and the total torque that the cuffs

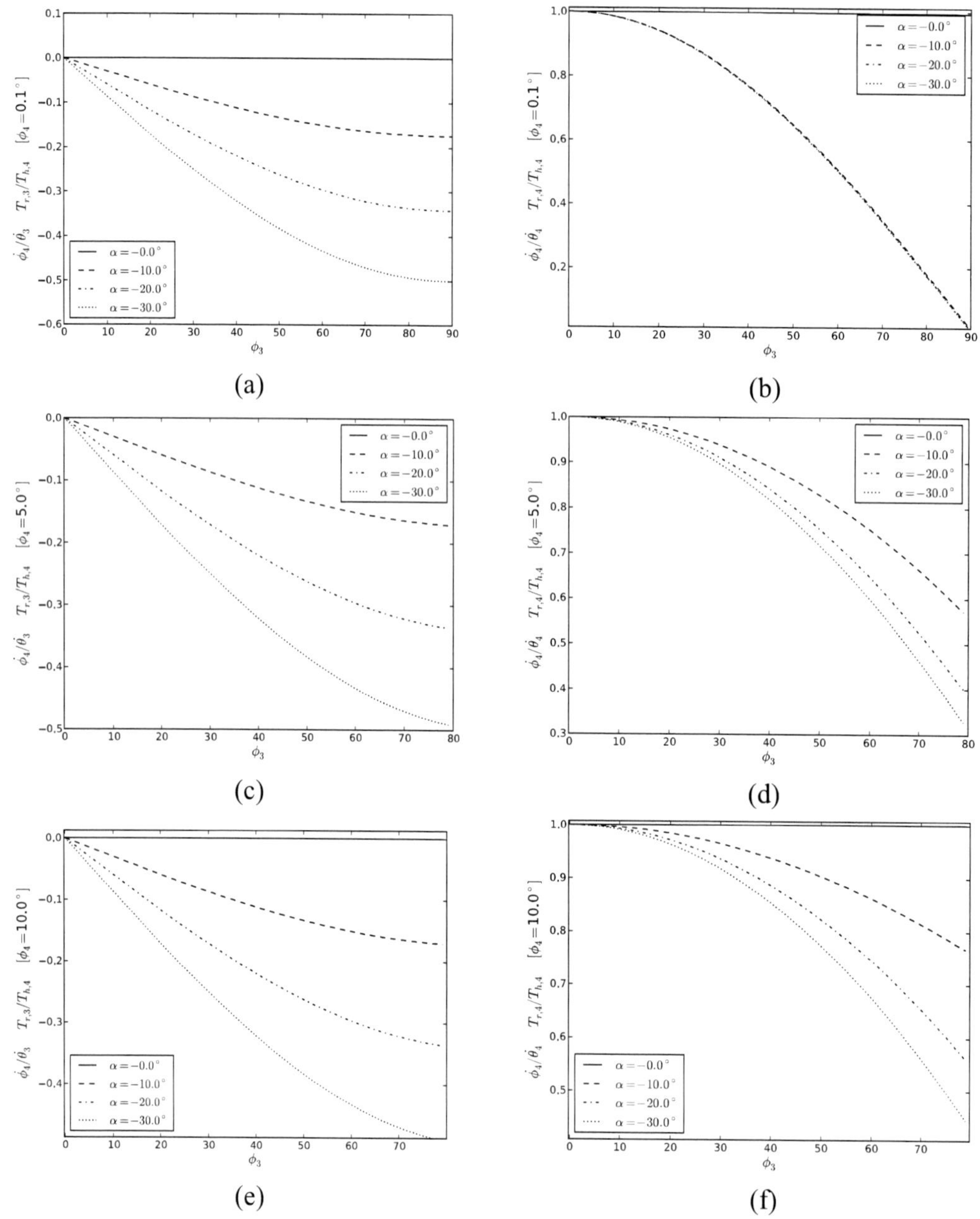

Figure 15. Torque transmission ratios of $T_{h,4}$ at different flexion angles: (a-b) $\phi_4 = 0.1°$; (c-d) $\phi_4 = 10°$; (e-f) $\phi_4 = 20°$.

can exert on the human arm consequently decreases. If a torque, exerted by the human, generates torques at exoskeleton elbow joints below their minimal sensitivity levels ($Tmin_{r,3}$, $Tmin_{r,4}$) a relative rotation between the human and the device can occur, due to mechanical compliances, without a correct rearrangement of the exoskeleton configuration.

In conclusion, singularity drawbacks occur not strictly if $\mathbf{u}_{h,3} // \mathbf{u}_{u,1}$ but its workspace influence is function of constraints and tissues uncompliances. They can lead to a relative intra-extra rotation of the human arm with respect to the exoskeleton arm, causing a misalignment between $u_{e,2}$ and $u_{h,4}$ and incoherency between human and robot joints.

This aspect emphasizes the importance, previously mentioned, of introducing the α angle misalignment to univocally define the elbow exoskeleton configuration during extended movements.

Fig. 14 and Fig. 15 display some analytical results about the torques transmitted to exoskeletal joints intersecting in the elbow ($\mathbf{u}_{e,1}$, $\mathbf{u}_{e,2}$). In particular, the charts display the influence of intra-extra rotations of the shoulder and flexo-extension movements of the elbow on transmitted torques, as function of the misalignment α. Torque transmission ratios, and not absolute values, are reported in order to generalize the impact of torque transmission over the design process, independently from the magnitude of the actual exerted torques and the sensibility levels.

Recall that $\mathbf{u}_{r,3}$ and $\mathbf{u}_{r,4}$ denote $\mathbf{u}_{e,1}$ and $\mathbf{u}_{e,2}$, respectively, in order to clarify the correspondence between human and robot joints.

In Fig.14 the torque transmission ratio to $\mathbf{u}_{r,3}$ and $\mathbf{u}_{r,4}$ due to a torque applied around the axis $\mathbf{u}_{h,3}$ is plotted. It is worth underlining that the elbow misalignment α allows the transmitted torque $T_{r,3}$ to be amplified, facilitating the exoskeleton elbow rearrangement and reducing the risk of incurring in joint mechanical blocks due to friction effects during shoulder rotational movements in configuration close to singularity. In Fig.15 the torque transmission ratio to $\mathbf{u}_{r,3}$ and $\mathbf{u}_{r,4}$ due to a torque applied around the axis $\mathbf{u}_{h,4}$ is plotted. These charts are useful to analyze what happens if a correct rearrangement of elbow joints does not occur; it can be due to an undesired rotational movement of $\mathbf{u}_{h,3}$ (with respect to $\mathbf{u}_{r,3}$) due to soft tissues and cuffs mechanical compliances nearby the elbow extended configuration. Increasing the α value will allow to avoid mechanical blocks with a proper torque trasmission from human elbow to exoskeleton elbow joints.

4. Conclusions

Most of currently available exoskeletons for upper limb neuro-rehabilitation are affected by shoulder noncompliance and lack of adaptability in terms of movements critical conditions (*i.e.* in singular configurations). The design of an exoskeleton based mechanism is advisably bound to the analysis of kinematical and kinetostatic aspects that generate such undesired behavior. In this work a design solution is proposed on the basis of the analysis of the effects of the arrangements of axes and the transmitted torques in singular or quasi-singular configurations. In particular, the designed solution prevents the human elbow singular configurations (*e.g.* complete extension) to exactly match the mechanism singularity, increasing in this way the exploitable functionalities with respect to preexisting exoskeletons. The kinematic and kinetostatic analyses have, in fact, given account of the impact of such design parameter over the whole motion performances. Regarding the relevant torques transmission aspects from the human articulations to the exoskeletal axes, the design choices are suited for leading to improvements in robot controllability by human exerted torques close to singular configurations. Modern exoskeletons, in fact, require to collaborate with the patient in achieving the final goal of the therapy and, consequently, a significant transfer of torques from the human articulations to the mechanical joints of the exoskeleton is required. Low level of torques transmitted to mechanical joints can lead to mechanical problems in following the active intention of the patient. This is due to mechanical frictions and intrinsic minimum level of sensibility of the joints. This particular aspect of pHRI interaction is determined by mechanical design choices in terms of axes placement, axes alignment and coupling interfaces between the limb and the machine. The influence of joints misalignment at the elbow, for instance, highlights the advantages in terms of torque transmission ratio to the exoskeletal joints close to the singularity (elbow completely extended). Near the singular configuration, the human arm can rotate with respect to the exoskeleton, due to mechanical compliances typical of the human tissues and the cuffs that are employed to constrain the arm to the exoskeleton. Compliances allow to have the human elbow axis misaligned with respect to the elbow axis of the robot. This undesired and unavoidable rotation can be correctly faced by a proper singularity-free mechanical elbow in the range of motion of the human arm. Such design choice, allowing a quite significant improvement in usability, controllability and safety of the exoskeletal structure, would have not been possible without the kinematical and kinetostatical analysis that have highlighted the source of misbehaviors and undesired features in the pool of design parameters.

References

[1] M. Caimmi, S. Carda, C. Giovanzana, E. S. Maini, A. M. Sabatini, N. Smania, and F. Molteni, "Using kinematic analysis to evaluate constraint-induced movement therapy in chronic stroke patients.," *Neurorehabilitation and neural repair*, vol. 22, no. 1, pp. 31–39, 2008.

[2] G. Arz and A. Toth, "Reharob: A project and a system for motion diagnosis and robotized physiotherapy delivery," in *Integration of Assistive Technology in the Information Age, Assistive Technology Research Series* (M. Mokhtari, ed.), vol. 9, pp. 93–100, IOS Press, Amsterdam, 2000.

[3] H. Krebs, N. Hogan, B. Volpe, M. Aisen, L. Edelstein, and C. Diels, "Overview of clinical trials with mit-manus: a robot-aided neuro-rehabilitation facility," *Technology and Health Care*, vol. 7, no. 6, pp. 419–423, 1999.

[4] C. Burgar, P. Lum, P. Shor, and H. Machiel Van der Loos, "Development of robots for rehabilitation therapy: the palo alto va/stanford experience," *J. Rehabil. Res. Dev.*, vol. 37, no. 6, pp. 663–673, 2000.

[5] R. Loureiro, F. Amirabdollahian, M. Topping, B. Driessen, and W. Harwin, "Upper limb robot mediated stroke therapy - gentle/s approach," *Autonomous Robots*, vol. 15, pp. 35–51, 2003. 10.1023/A:1024436732030.

[6] G. Rosati, P. Gallina, and S. Masiero, "Design, implementation and clinical tests of a wire-based robot for neurorehabilitation.," *IEEE Trans Neural Syst Rehabil Eng*, vol. 15, pp. 560–9, Dec. 2007.

[7] G. Rosati, P. Gallina, S. Masiero, and A. Rossi, "Design of a new 5 d.o.f. wire-based robot for rehabilitation," in *Rehabilitation Robotics, 2005. ICORR 2005. 9th International Conference on*, pp. 430–433, 2005.

[8] A. H. A. Stienen, E. E. G. Hekman, G. B. Prange, M. J. A. Jannink, F. C. T. van der Helm, and H. van der Kooij, "Freebal: Design of a dedicated weight-support system for upper-extremity rehabilitation," *Journal of Medical Devices*, vol. 3, no. 4, p. 041009, 2009.

[9] N. Klopčar and J. Lenarčič, "Kinematic model for determination of human arm reachable workspace," *Meccanica*, vol. 40, pp. 203–219, 2005.

[10] N. Klopcar and J. Lenarcic, "Bilateral and unilateral shoulder girdle kinematics during humeral elevation," *Clinical Biomechanics*, vol. 21, no. Supplement 1, pp. S20–S26, 2006.

[11] N. Klopcar, M. Tomsic, and J. Lenarcic, "A kinematic model of the shoulder complex to evaluate the arm-reachable workspace," *Journal of Biomechanics*, vol. 40, no. 1, pp. 86–91, 2007.

[12] B. Hingtgen, J. R. McGuire, M. Wang, and G. F. Harris, "An upper extremity kinematic model for evaluation of hemiparetic stroke," *Journal of Biomechanics*, vol. 39, no. 4, pp. 681–688, 2006.

[13] V. Zemlyakov and P. Mcdonough, "Upper extremity exoskeleton structure and method." U.S. Patent US 2003/0115954 A1, June 2003.

[14] M. Bergamasco, F. Salsedo, A. Dettori, M. Franceschini, A. Frisoli, and F. Rocchi, "Ekoskeleton interface apparatus." U.S. Patent US 2006/0150753 A1, July 2006.

[15] J. Perry, J. Rosen, and S. Burns, "Upper-limb powered exoskeleton design," *Mechatronics, IEEE/ASME Transactions on*, vol. 12, no. 4, pp. 408–417, 2007.

[16] J. L. Pons, *Wearable Robots: Biomechatronic Exoskeletons; electronic version*. Chichester: Wiley, 2008.

[17] C. Carignan, M. Liszka, and S. Roderick, "Design of an arm exoskeleton with scapula motion for shoulder rehabilitation," in *Advanced Robotics, 2005. ICAR '05. Proceedings., 12th International Conference on*, pp. 524–531, 2005.

[18] T. Nef, M. Guidali, and R. Riener, "ARMin III - arm therapy exoskeleton with an ergonomic shoulder actuation," *Applied Bionics and Biomechanics*, vol. 6, no. 2, pp. 127–142, 2009.

[19] A. Schiele and F. van der Helm, "Kinematic design to improve ergonomics in human machine interaction," *Neural Systems and Rehabilitation Engineering, IEEE Transactions on*, vol. 14, no. 4, pp. 456–469, 2006.

[20] N. Jarrasse and G. Morel, "A methodology to design kinematics of fixations between an orthosis and a human member," in *Advanced Intelligent Mechatronics, 2009. AIM 2009. IEEE/ASME International Conference on*, pp. 1958–1963, 2009.

[21] N. Jarrasseis and and G. Morel, "A formal method for avoiding hyperstaticity when connecting an exoskeleton to a human member," in *Robotics and Automation (ICRA), 2010 IEEE International Conference on*, pp. 1188–1195, May 2010.

[22] S. Ball, I. Brown, and S. Scott, "Medarm: a rehabilitation robot with 5dof at the shoulder complex," in *Advanced intelligent mechatronics, 2007 IEEE/ASME international conference on*, pp. 1–6, 2007.

[23] D. Magermans, E. Chadwick, H. Veeger, and F. van der Helm, "Requirements for upper extremity motions during activities of daily living," *Clinical Biomechanics*, vol. 20, no. 6, pp. 591–599, 2005.

[24] K. Petuskey, A. Bagley, E. Abdala, M. A. James, and G. Rab, "Upper extremity kinematics during functional activities: Three-dimensional studies in a normal pediatric population," *Gait & Posture*, vol. 25, no. 4, pp. 573–579, 2007.

In: Robotics: State of the Art and Future Trends
Editors: G. Legnani and I. Fassi
ISBN: 978-1-62100-403-5

Chapter 12

AN OVERVIEW ON ROBOTIC MICROMANIPULATION

***D. Sinan Haliyo*[*] *and Stéphane Regnier*[†]**
Université Pierre et Marie Curie, Paris 6, CNRS,
ISIR, Institut des Systèmes Intelligents et de Robotique, France

Abstract

In this chapter we will present a brief overview on specific issues in manipulation of parts and components in the sub-millimeter range, typically from 10^{-3} to 10^{-6} m. The scale reduction to these particular dimensions biases greatly the balance of forces present in the manipulation scene and requires a particular design philosophy in order to develop reliable robotic manipulation stations. This chapter will discuss the specific predominant forces at the microscale and propose practical models for basic usage scenarios. A complete robotic manipulation station will also be described in order to give to the reader the essential consideration to take into account when designing a similar system.

Introduction

There are two issues that set the micromanipulation fundamentally different from classic macro-scale manipulation. The first one is physical and is directly related to the scale reduction. As characteristic dimensions of objects tend towards zero, volumetric forces become negligible in comparison to surface or even intermolecular forces. Their predominancy over mass and inertia related forces leads to fundamental differences in the way the objects should be handled. On the contrary of the macroscale where underlying mechanics can be modeled and studied in a straightforward manner, a proper understanding of behavior in microscale requires the analysis of different physical phenomena in the study, in particular electrostatics, thermodynamics, fluids mechanics, material science, roughness, etc. Moreover, all this phenomena are generally nonlinear and extremely dependent on experimental and environmental conditions. The second issue comes from a technical and

[*]E-mail address: sinan.haliyo@upmc.fr
[†]E-mail address: stephane.regnier@upmc.fr

engineering point of view. At macroscale, real-time position and force feedback from the end-effector is considered granted in most cases. At the microscale, although MEMS and similar devices are advancing with a fast pace, there still aren't any sensors capable of 6 axis (or even 3) force measurement. This lack of a reliable and easy to integrate sensor systems makes the overall design and control of micromanipulators a tricky enterprise.

Those two issues combines into a third one, which is the wide range of forces and displacements to be considered for a micromanipulation task. Even in the frame of unique task, one should be able to treat force ranging from micronewtons to millinewtons and from submicrometer precision over centimeter range displacements. All these scale changes within a single task are assorted with changes in the balances of above-mentioned physical effects and technological bottleneck to design sensor feedback. Therefore, micromanipulation is a whole challenge beyond conventional robotics where, instead of bare miniaturization of classical techniques, new manipulation strategies are clearly called for.

In this chapter, an introduction to specificities of microscale manipulation will be presented to give some guidelines to the practitioner. Major physical phenomena occurring on the microscale with direct influence on a manipulation will be discussed and practical analytic models proposed, along with pointers for curious readers looking for a more in-depth understanding.

Lastly, a complete micromanipulation system will be described. This system introduces a new approach for pick-and-place of microscale components by taking advantage of adhesion phenomena. It will be used to illustrate the micromanipulation oriented design philosophy proposed here through its redundant kinematic architecture and control schemes adapted to particular vision force sensing capabilities. Additionally, enhanced user interaction is proposed through the use of haptic force feedback human interface, along with adapted coupling schemes. This macro/micro bilateral coupling involves scaling ratios in the $10^3 - 10^6$ range and their stability requires also special care. Experimental validation of the setup and several usage scenarios are presented to give the reader an overall idea of possibilities and limitations of the microscale robotic manipulation.

1. Physics at Microscale

Miniaturization of an object or process can prove complex, because the range of physical phenomena involved may not all change in the same manner as the scale is reduced. If, for example, we were to scale down a guitar, we would obtain a new guitar whose range of notes had become much higher. The resonant frequencies of the strings increase as their dimensions are reduced. In order to obtain a miniaturized guitar with the same range of notes as a conventional guitar, we would need to completely redesign the instrument. The same is true for most behaviors of a system: they will change as the scale is reduced. The impact of the scale change on physical phenomena is commonly known as the "scale effect".

The physical phenomena which dominate on a human scale, such as weight or inertia, are mostly volumic. In other words, they are directly proportional to the volume of the object under consideration. Thus, if we change between a cube of steel with sides whose lengths l are one centimeter and a cube with sides whose lengths are $l' = \frac{l}{10}$ (ten times smaller), the characteristic dimension l has been reduced by a factor of 10, whereas its mass changes from 7.9 grams to 7.9 milligrams and so has been reduced by a factor or $\left(\frac{l}{l'}\right)^3 = 10^3$.

Certain physical phenomena, generally ones that are less familiar in everyday life, are not volumetric. An example of this is the surface tension force. This is a length-based effect, and so its evolution is proportional to the scale under consideration. Hence the surface tension of a cube with sides of length l is directly proportional to this length. For a cube ten times smaller, with length l', the surface tension force is also simply divided by $\frac{l}{l'} = 10$. Hence, this effect decreases in strength much less rapidly during the miniaturization process then the weight, as stated above.

Consequently,three large classes of adhesive forces dominate on the microscopic scale. These forces are [1]:

- van der Waals forces, interaction forces between the molecules of two nearby bodies;
- electrostatic forces, classical Coulomb forces that depend on charges present on surfaces;
- capillary forces, whose existence is determined by thc environmental humidity conditions.

A classification of forces as a function of separation distance is given in [2] and is presented in Table 1. In general terms, adhesion between solids encompasses all the chemical binding effects which contribute to the cohesion of solids, such as hydrogen bonds and metallic, covalent and ionic bonds. Although the energy of these bonds is not negligible, their effects are not considered [3]. This is because, aside from extremely specific environments (ultra-vacuum), chemical bonds on object surfaces tend to be saturated by contaminants (oxidation, etc.). They cannot therefore form bonds when they come into contact with another body. The forces to be considered for microscale interactions are then those listed above and used in a number of reference works [4, 5, 6].

Table 1. Forces on the microscale

Interaction distance	Force
Infinite	gravity
few nm to 1 mm	capillary force
> 0.3 nm	electrostatic force
> 0.3 nm	van der Waals force
< 0.3 nm	molecular interactions
$0.1 - 0.2$ nm	chemical interactions

1.1. Contact at Microscale

The biased balance forces at the microscale has a direct influence on the contact because of preponderant adhesion phenomena. Prior to give more detailed description of above mentioned forces causing this adhesive behavior, a presentation of surface interactions from a thermodynamic point of view is required.

First, let's introduce the work of adhesion and cohesion. They respectively represent the energy required to separate to infinity two unit surfaces of media in contact, in a vacuum. For two different media, this energy is known as work of adhesion, noted W_{12}, whereas for identical media, it is known as work of cohesion, noted W_{11}.

The surface energy γ_1 represents the energy required to increase the free surface of the medium by one unit of area. Since this involves the separation of two surfaces in contact, this is equal to half the work of cohesion:

$$\gamma_1 = \frac{1}{2}W_{11}$$

This surface energy is often a function of the boiling point. Consequently, surfaces such as metals that have a high boiling point ($T_E > 2\,000^oC$) have high surface energies ($\gamma > 1\,000\ mJm^{-2}$).

When two media are in contact, their interface energy per unit area is known as the interfacial energy or interfacial tension γ_{12}. It can be calculated as follows:

$$\gamma_{12} = \frac{1}{2}W_{11} + \frac{1}{2}W_{22} - W_{12} = \gamma_1 + \gamma_2 - W_{12} \tag{1}$$

This equation is known as the Dupré equation.

An approximative combination equation exists [1] to calculate γ_{12} from γ_1 and γ_2:

$$\gamma_{12} = \gamma_1 + \gamma_2 - 2\sqrt{\gamma_1\,\gamma_2}$$

An extension to this approximation can be used to calculate the energy required to separate two media 1 and 2 immersed in a third medium 3:

$$W_{132} = W_{12} + W_{33} - W_{13} - W_{23} = \gamma_{13} + \gamma_{23} - \gamma_{12}$$

The work of adhesion (or cohesion) comes from the molecular interaction potential, whose origins are described through van der Waals forces.

1.2. Van der Waals Forces

Van der Waals forces were studied in the 1930s by Hamaker [7] and then developed in the 1950s by Lifshitz [8]. These forces depend on the materials in contact, through the Hamaker constant, and on the interaction distance. Evaluation of this constant, which determines the strength of the force, requires a wide range of physical data on the materials.

In a general approach, the van der Waals force derives from an interaction potential between two molecules or particles. This interaction potential, for two polar molecules interacting in vacuum can be written as:

$$\begin{aligned} w_{vdW}(r) &= -\frac{[C_{ind} + C_{orient} + C_{disp}]}{r^6} \\ w_{vdW}(r) &= -\frac{1}{(4\pi\varepsilon_0)^2 r^6}\left[(u_1^2\alpha_{01} + u_2^2\alpha_{02}) + \frac{u_1^2 u_2^2}{3kT} + \frac{3\alpha_{01}\alpha_{02}h\nu_1\nu_2}{2(\nu_1+\nu_2)}\right] \end{aligned} \tag{2}$$

where C_{ind}, C_{orient} and C_{disp} represent, respectively, the contributions of the induction,orientation and dispersion effects on the interaction potential. ν is the orbital frequency of the electron, $h\nu$ is the ionization potential (also written I) and α_{01}, α_{02} , u_1 and u_2 are, respectively, the electronic polarizabilities and the dipole moments for molecules 1 and 2. This interaction potential between atoms is more generally written as

$$w(r) = -\frac{C_{vdW}}{r^6} = \frac{C}{r^6} \tag{3}$$

This potential can also be written as a sum of attractive and repulsive potentials. Its most well-known form is Lennard-Jones [9], where the first term is the repulsive part :

$$w(r) = 4\varepsilon \left\{ \left(\frac{\xi_0}{r}\right)^{12} - \left(\frac{\xi_0}{r}\right)^{6} \right\} \tag{4}$$

with ε being the depth of the potential at its minimum and ξ_0 the equilibrium inter-atomic spacing. The repulsive part is hence on the scale of a few Angströms and can be safley ignored in the case of microscale operation. The remaining attractive part, as described also in 3 is sufficient to describe the van der Waals adhesion.

This potential is between two molecules. In order to calculate the interaction potential W between surface or volumes, one should integrate the above obtained w over volume elements. For example, in the case of an interaction between a sphere of radius R and an infinite half-plane with molecular densities ρ:

$$W_{sp}(z) = -\frac{\pi^2 C\rho^2}{6} \int_{\xi=0}^{\xi=2R} \frac{(2R-\xi)\xi}{(z+\xi)^3} d\xi = -\frac{\pi^2 C\rho^2}{6} \left[\ln \frac{z}{2R+z} + \frac{2R(R+z)}{z(2R+z)} \right] \tag{5}$$

If the radius R is much greater than the interaction distance z, this can be simplified as:

$$W_{sp}(z) = -\frac{\pi C\rho^2 R}{6z} \tag{6}$$

Similarly, the interaction potential between two planar surfaces can be expressed as:

$$W_{pp}(z) = -\frac{\pi C\rho^2}{12z^2} \quad \text{per unit surface.}$$

The van der Waals force can be then obtained by derivation of this expression:

$$F(z) = -\frac{dW(z)}{dz}$$

The following table gives some analytic expressions of the interaction potential for basic geometries as found in the literature.

The preferred way to describe this volumetric interaction potential is through the use of Hamaker constant A [7] as used in Table 1.2.. The advantage of the use of A is to separate the terms depending on the physical properties from terms coming from the geometric integration. A appears hence as independent of the interaction geometry and can be either analytically calculated or empirically estimated. Hamaker proposes the following expression for two interacting materials i and j :

Table 2. Van der Waals potentials over different geometries in the literature (z is the interaction distance and R the radius of the object)

Object 1	Object 2	Expression	Reference
Plane	Plane//	$W = -\frac{A}{12\pi z^2}; F = -\frac{A}{6\pi z^3}$ (per unit surface)	[10, 1, 11]
Cylinder	Cylinder //	$F = -\frac{3AL}{24\sqrt{2}z^{\frac{5}{2}}(\frac{R_1 R_2}{R_1+R_2})^{\frac{1}{2}}}$ (L, length of cylinder)	[1], [12]
Cylinder	Cylinder ⊥	$W = -\frac{A\sqrt{R_1 R_2}}{6z}; F = -\frac{A\sqrt{R_1 R_2}}{6z^2}$	[10, 1, 12]
Sphere	Plane	$W = -\frac{AR}{6z}; F = -\frac{AR}{6z^2}$	[10, 11]
Sphere	Sphere	$W = -\frac{AR}{12z}; F = \frac{AR}{12z^2}$	[10, 1, 11]

$$A_{ij} = \pi^2 C\rho_i\rho_j \tag{7}$$

where ρ is the atomic density and C is the atom-atom potential.

A then takes values in the range $[0.4 \sim 4].10^{-19}J$. This method of calculation provides good approximations for the constant in weakly polar materials, since it takes into account only the dispersion effect and is obtained by assuming additivity of the dispersion forces. In the converse case, it underestimates the value.

In certain cases it is possible to obtain approximate values for the Hamaker constants using combination equations. These equations are derived from the expression for A given by McLachlan in 1963 [13]. In this way, for two materials 1 and 2 interacting in a vacuum, it is possible to obtain A_{12} as a function of the constants A_{ii} for each material:

$$A_{12} \approx \sqrt{A_{11}A_{22}}$$

Similarly, A_{132}, for two materials 1 and 2 interacting through a third medium 3, is approximately

$$A_{132} \approx (\sqrt{A_{11}} - \sqrt{A_{33}})(\sqrt{A_{22}} - \sqrt{A_{33}})$$

These combination formula give very good approximations for A, except in the case of strongly polar media such as water.

Lifshitz [8] developed a more realistic theory, including strongly polar media, which included the influence of neighboring atoms on the pair under consideration for the calculation of A, where two media 1 and 2 are interacting in a medium 3:

$$\begin{aligned} A_{132} \approx\ & \frac{3}{4}kT(\frac{\varepsilon_1 - \varepsilon_3}{\varepsilon_1 + \varepsilon_3})(\frac{\varepsilon_2 - \varepsilon_3}{\varepsilon_2 + \varepsilon_3}) \\ + & \frac{3h\nu_e}{8\sqrt{2}} \frac{(n_1^2 - n_3^2)(n_2^2 - n_3^2)}{\sqrt{(n_1^2 + n_3^2)}\sqrt{(n_2^2 + n_3^2)}[\sqrt{(n_1^2 + n_3^2)} + \sqrt{(n_2^2 + n_3^2)}]} \end{aligned} \tag{8}$$

where k is the Boltzmann constant ($1.381 * 10^{-23} J/K$); T the temperature (^{o}K); ν_e the principal electronic absorption frequency (typically of the order of $3.10^{15} s^{-1}$); ε_i the dielectric permittivity and n_i i the refractive index.

Given the strong dependence of the modulus of the van der Waals forces on the Hamaker constant, many research groups have focused on the exact determination of this constant for various experimental environments [14, 15, 16, 17, 18, 19]. In all these articles, as well as in [1], physical values for materials and values of the Hamaker constant are given.

Please note that the van der Waals potential denoted here $W(z)$ is obtained from a volumetric integration hence is geometry dependent. In order to keep a generic approach for the work of adhesion (or cohesion) the approach proposed by Derjaguin approximation [20]is to use the van der Waals potential between planar surfaces W_{pp} to describe other geometries, notably sphere/plane and sphere/sphere interactions. The interaction between two spheres of radii R_1 and R_2 can hence be written as a function of W_{pp}

$$\begin{aligned} W_{pp} &= -\frac{A}{12\pi z^2} \\ F &= 2\pi \frac{R_1 R_2}{R_1 + R_2} W_{pp} \\ &= -\frac{R_1 R_2}{R_1 + R_2} \frac{A}{6\pi z^2} \end{aligned} \tag{9}$$

Pull-off force The ever-presence of adhesive effect implies a residual deformation on contact between two media. This deformation increases the contact area and in consequence the a certain amount of force, called "pull-off" is required to separate objects in contact. As classical Hertz contact model completely overlooks this phenomenon, alternate micro-scale dedicated modelings are proposed. Johnson, Kendall and Roberts' JKR theory [21] expresses the contact area radius a and deformation δ for the plane/sphere case as:

$$a^3 = \frac{R}{K}\left(F + 3\pi RW + \sqrt{6\pi RWF + (3\pi WR)^2}\right) \tag{10}$$

$$\delta = \frac{a^2}{R} - \sqrt{\frac{8\pi Wa}{3K}} \tag{11}$$

Consequently, the pull-off force is:

$$P = \frac{3}{2}\pi RW_{pp} \tag{12}$$

JKR theory particularly applies for strong adhesion energies, soft materials and large radii of curvature. Nevertheless, this method underestimates the surface charge and consequently exhibits an anomaly, from a theoretical point of view, in that it predicts an infinite constraint at the edge of the contact region. In 1975, Derjaguin, Muller and Toporov proposed the DMT theory [20], which avoids this issue by considering a Hertzian deformation of the contact region and an adhesion due to surface forces acting outside this region. The contact radius is then given by:

$$a^3 = \frac{R}{K}(F + 2\pi WR) \tag{13}$$

The pull-off force is given by:

$$P = 2\pi RW \tag{14}$$

This model is particularly suited to low adhesion energies and small radii of curvature. However, this theory underestimates the value of the contact radius due to its consideration of a Hertzian geometry.

It is general practice to consider the pull-off force in an interval whose bounds are defined by JKR and DMT theories. It has been shown in [22] then developed analytically in [23] that both theories give extreme cases which can be unified in a single model, called DUGDALE. However, this model is difficult to solve and its numerical approximation proposed in [24] is often preferred.

1.3. Capillary Forces

The presence of humidity can modify the interaction consequently. Note that in that case, the Hamaker constant should be considered in a three-media scenario, which result in a considerable decrease of its value. Table 3 compares Hamaker constants between different materials in vacuum and water.

Table 3. Variation of Hamaker constant between interactions in vacuum and in water

Material	A_{vacuum} $(10^{-20}J)$	A_{water} $(10^{-20}J)$
Polystyrene	≈ 7	$1,3$
Mica	10	2
Alumine	14	$5,3$
Metals (Au, Ag, Cu)	≈ 50	≈ 40

Although the resulting van der Waals interaction is diminished, the condensation of water creates a liquid bridge between two contacting surfaces. The pression difference between inside and outside of this liquid medium causes an additional adhesive effect to appear, called the capillary force. The resulting force depends on the shape of this bridge, hence on contact angles of the liquid on the surface through the surface tension γ, as depicted in fig. 1.

The radius of curvature can be obtained through the relative humidity $\frac{p}{p_{sat}}$ using Kelvin equation:

$$r_{menisc} = \frac{\gamma V}{RT \log(p/p_{sat})} \tag{15}$$

where γ is the surface energy of the liquid and V the molar volume ($\frac{\gamma V}{RT} = 0.54$ for water). The immersion distance d can be calculated using ϕ, the fill angle::

$$d = 2r_{menisc}\cos(\phi + \theta)$$

Considering the volume of the liquid constant, the resulting force can be written as:

$$F = \frac{4\pi R\gamma_{eau}\cos\theta}{1 + D/d} \tag{16}$$

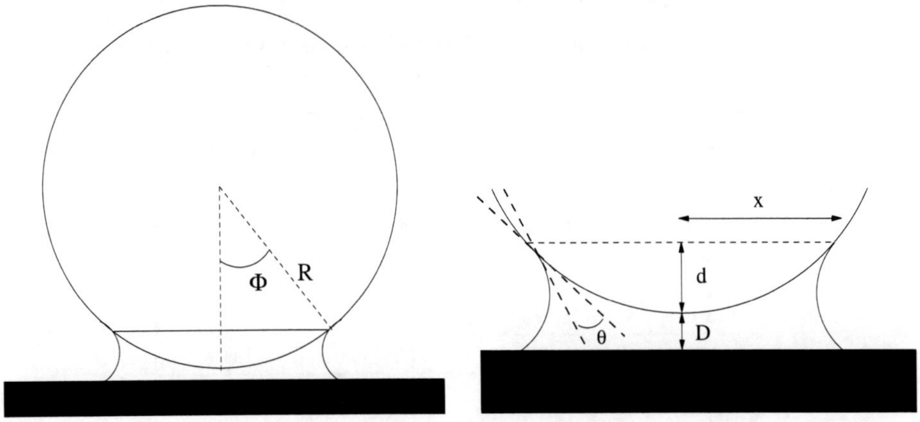

Figure 1. The geometric analysis capillary bridge.

In contact, where $D \simeq D_0$, this expression is simplified to :

$$F = F_{max} = 4\pi R\gamma_{eau}\cos\theta$$

Several values of surface energies and contact angles are given in tables 4 and 5 in the following.

Table 4. Surface energies γ of different materials ($mJm^{-}2$)

Material	γ on air	γ in vacuum
Teflon [25]	13.4	98.6
Polystyrene[26]	35.5	
Mica [25]	90	65
Carbon fibers [25]	42.5	28
Epoxy [25]	80	60
Silicium [25]	35	22.5
Gold[26]	1450	
Silver [26]	1450-1500	

1.4. Electrostatic Forces

Electrostatic forces occur in two specific situations [27]:

1. through the Coulomb interaction, in the presence of charged particles;
2. through Coulomb interactions caused by the appearance of charges generated by triboelectrification.

In the case of a charged particle interacting with a charged surface, the interaction force is described by Coulomb's law. The surface creates a field $E = \sigma/(2\varepsilon\varepsilon_0)$,, where σ is the surface charge density, ε is the relative permittivity of the medium and ε_0 is the vacuum permittivity. The electrostatic force, independent of the separation distance, is

Table 5. Contact angles for selected liquid-solid couples

Liquid	Solid	$\theta_A(^\circ)$	Reference
Water	Paraffine	110	[10]
Water	PTFE	98-112	[10]
Water	PET	79	[10]
Water	Polyethylene	88-103	[10]
Water	Platinum	40	[10]
Water	Silver iodide	17	[10]
Water	Glass	$\approx 0^\circ$	[10]
Water	Glass	0	[10]
Water	Steel	82-92	[12]
Water	Silicium	55	[12]
Silicone oil	Steel	23-34	[12]
Silicone oil	Silicium	29	[12]

$$F = qE \tag{17}$$

where q is the electric charge. For interactions between charged particles, the interaction force is equal to

$$F = \frac{q_1 q_2}{4\pi\varepsilon_0\varepsilon z^2} \tag{18}$$

where z is the distance between the two particles. [27] gives the maximum admissible charge for a particle as a function of its radius R as

$$\sigma = 30 \times (100R)^{-0,3} \quad \mu C.m^{-2} \tag{19}$$

Various different interactions can occur depending on the conducting or insulating nature of the materials involved. The contact effect that is best understood in terms of its formulation is the conductor-conductor interaction. For this class of contact, transfer of charge between the particle and the surface reduces the adhesive force [4]. Thus, when two different conducting materials are brought into contact, an electron transfer effect occurs if the separation distance is small. This transfer brings the materials into thermodynamic equilibrium, which tends to equalize their electrochemical potentials. The potential difference between the surfaces of materials 1 and 2, known as the contact potential, takes the following form [28]:

$$V_c = \frac{(\phi_1 - \phi_2)}{e} \tag{20}$$

with ϕ_1 and ϕ_1 being the work functions of each surface.

Under the hypothesis that:

1. smooth surfaces hence the topography is ignored;

2. the potential is uniformly distributed over the surface hence the electric field is normal to the surface;

3. the charges are only at the surface and there is no no charge present in the space between two interacting materials

it is possible to calculate the electrostatic interaction between simple geometries. Only plane/plane and plane/sphere cases are mentioned here.

These models are the most commonly encountered ones in the literature. The expressions for these forces are the derivatives of the electrostatic energy $W(d)$.

$$F_{elec}(z) = -\frac{\partial W(z)}{\partial z} = -\frac{1}{2}\frac{\partial C}{\partial z}U^2 \tag{21}$$

The simplest case [5] is that of plane-plane contact, where two smooth surfaces come into contact. The contact surface is denoted A, and the relative capacitance for this problem is determined from the simple case of a plane capacitor:

$$C(z) = \frac{\varepsilon_0 A}{z}$$

$$F_{plan} = \frac{\varepsilon_0 U^2 A}{2z^2} \tag{22}$$

This model gives the electrostatic pressure, and, if we know the contact area, we can determine the electrostatic force from this. Experiments have nevertheless shown that it is difficult to determine this contact region in the case of real configurations. The plane model is thus heavily limited in terms of its applications. The studied objects are rarely flat. In specific applications, this model can nevertheless be used at extremely small object separation distances in the cases where the contact can be treated as contact between planar surfaces. Sphere models were developed for more complex shapes and larger separation distances. Various authors such as [3] have used them in micromanipulation studies concerned with undesirable adherence phenomena. These models evaluate the electrostatic forces for contact between a conducting sphere and a conducting plane. As with the previous model, they use the derivative of the electrostatic interaction energy. The capacitance between a sphere and a plane is given by the following expression:

$$C_{sphere} = 4\pi\varepsilon_0 R \sinh(\alpha) \sum_{n=1}^{\infty} (\sinh n\alpha)^{-1}$$

where $\alpha = \cosh^{-1}((R+z)/R)$. This model is well established in the literature for the analysis of contact between the apex of an AFM probe and the surface in atomic force microscopy [4]. The expressions obtained depend on the separation distance and the ratio between the sphere radius R and the separation distance z. Three models have been developed based on the general expression given in [29], for different ranges of separation. For small separation distances, the electrostatic force is inversely proportional to the separation distance [30, 4, 31, 5]:

$$F_{sphere1} = \frac{\pi\varepsilon_0 R U^2}{z} \quad R >> z \tag{23}$$

For large separation distances, the electrostatic force is proportional to the inverse square of the separation distance [31, 32]:

$$F_{sphere2} = \frac{\pi\varepsilon_0 R^2 U^2}{d^2} \quad R << z \tag{24}$$

A general expression for arbitrary separation distances has been developed [33, 32] based on equations (23) and (24) :

$$F_{sphere3} = \pi\varepsilon_0 \frac{R^2 U^2}{z(z+R)} \tag{25}$$

These models are limited in the separation distances they can handle. They are often used to obtain a quantitative value for the electrostatic forces between the probe tool and the substrate in local probe microscopy. Similarly, in the case of microhandling using a local probe, they have proven themselves useful.

2. Pick-and-Place Using Adhesion Forces

As detailed above in this chapter, microscale manipulation has very specific constraints, because of both the physical interactions and technological limitations. Hence, the design of a micromanipulation system is radically different from its macroscale counterpart. Detailed description of such a system is presented here, with several specifications. First, the kinematic architecture is redundant and combines large range - low resolution actuators with low range - high resolution ones. Second, the pick and place operation relies solely on the use of adhesion forces and avoids a classical multifinger prehension mechanisms. Its control is also dedicated to microscale-specific vision and force sensing devices with associated control schemes. Additionally, an enhanced user interaction approach, using an haptic interface with force-feedback coupling is used to keep the operator in the control loop. This approach lets the user drive the manipulation task while supplying him with scaled-up force information through the haptic controller. Several experiments of the presented manipulation system are also discussed and analyzed.

2.1. Microscale Gripping Strategies

As detailed earlier in this chapter, adhesion is a predominant phenomena in micromanipulation. As contacting objects tends to stick together even under gravity, releasing a gripped object is the main issue. Under these circumstances, it is clear that classical multifinger gripper architecture is not well adapted for micro-manipulation, unless the adhesion could be considerably reduced, as the gripped objects stick to the gripper and the release is often hazardous. The reduction of adhesion can be achieved by choosing materials with weak Van der Waals potentials or by the use of rough surfaces but in most cases, it is not possible to guarantee a macroscale-like, i.e *adhesionless*, behavior.

However, it is possible to take advantage of the adhesion for gripping using a tool with high surface energy and low surface roughness. In this case, one can pick-up a microobject by simple contact. This approach is very interesting as it does not need a complex gripper architecture for grasping. The obvious problem in this case is the release, as it is necessary to overcome the adhesion between the gripping tool and the object. The use of the inertial force, in order to overcome the adhesion, is a possible solution. To study its feasibility, the dynamical analysis of the release task, including contact and adhesion forces, depicted in in Fig. 2, is performed.

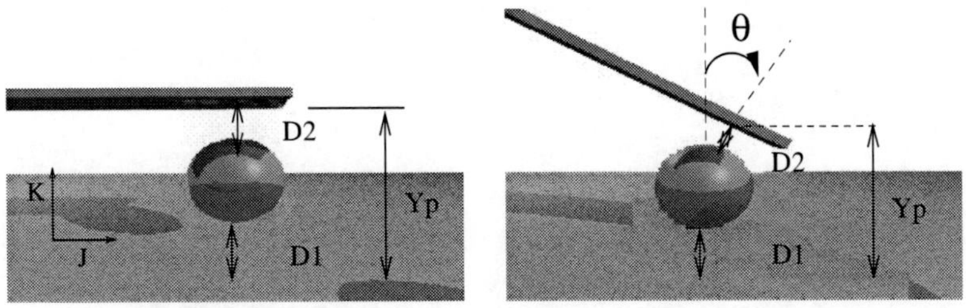

Figure 2. Modeling of a micro-scale pick-and-place task.

The established model is as follows:

$$m_t \ddot{Y}_p = F^t_{ext} - F^{ot}_{adh}(D_2) - m_t g \quad (26)$$

$$m_o \ddot{D}_1 = F^{ot}_{adh}(D_2) \cdot \cos\theta - F^{os}_{adh}(D_1) - m_o t \quad (27)$$

$$Y_p = D_1 + 2R_o + D_2 \qquad \ddot{D}_2 = \ddot{Y}_p - \ddot{D}_1 \quad (28)$$

m_t, m_o are respectively the mass of the gripping tool and the object, F^t_{ext} is the external force applied to the gripper and F^{ot}_{adh} and F^{os}_{adh} are respectively the adhesion forces between the object and the gripper and the object and the substrate, including Van der Waals, electrostatic and capillary forces. These forces are non-linear functions of distances $D_{1,2}$, as described earlier in this chapter. R_o is the radius of the object, which is supposed to be a perfect sphere.

Simulations of the dynamic model show the existence of a value of initial acceleration $\ddot{Y}$ of the gripper over which inertial effects overcome the adhesion between the tool and the object, causing the release. This acceleration depends on both the mass of the object and the angle of slope θ of the gripper, as illustrated in Figure 3 for a glass object of radius r initially adhering to a planar silicon gripper.

A close examination of these simulation results lead to the conclusion that in order to achieve the release accurately, accelerations ranging from 10^4 to 10^6 are needed.

Another way to release a spherical object is by rolling. If one is able to control the contact force, it is possible to move the object to the extremity of the gripper by switching between the sliding and rolling modes. At the boundary of the gripper, the contact area between the gripper and the object is very small but the substrate contact area is unchanged. One can then expect a higher adhesion between the substrate and the object to guarantee the release. The key component for this release strategy is hence the control of the contact force.

2.2. The Manipulator

The micro-manipulation system is built around an active gripper, which design is based on the adhesion phenomena. According to the study presented above, it includes two important capabilities: high acceleration generation for dynamical release and precise contact

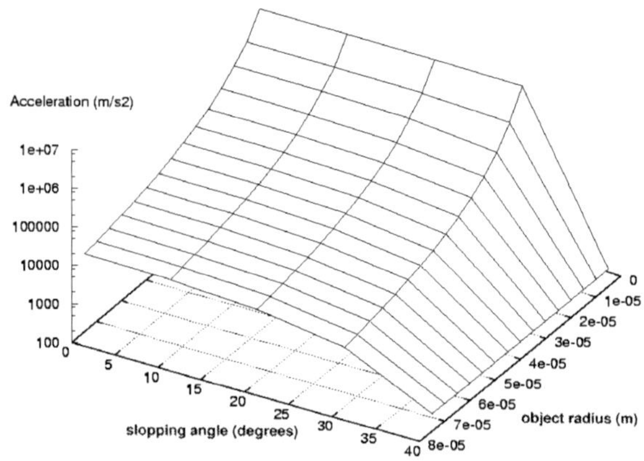

Figure 3. Minimum acceleration for release.

force control for rolling. This gripper is an AFM (atomic force microscopy) tipless probe, mounted on a piezoelectric ceramic which can produce impulses or high frequency sinusoidal waves, generating instantaneous accelerations as high as 10^6 ms^{-2} at the extremity of the AFM cantilever. These outstanding dynamical capabilities are used for release and characterization tasks. The gripper also provides micronewton resolution measurement of the contact force, due to the AFM probe. Its vertical displacement is provided by two serial actuators: a nanostage with 12 μm amplitude and a microstage with sub micrometer resolution over $2.5\,cm$. The contact force can thus be controlled by the motion of these actuators. The horizontal motion is produced by two identical microstages on $X-Y$ plane. Fig. 4(a) and 4(b) show the whole micromanipulator placed under an optical microscope and its kinematic configuration. The active gripper is shown in Fig. 5.

2.2.1. Force Sensing

The AFM tipless probe used as gripper is a fixed silicon rectangular cantilever of dimensions $600 \times 140 \times 10 \mu m^3$. Micro strain gauges are fixed at the fixed end, forming a Wheatstone bridge and measuring the local strain. According to its dimensions, the gripper can be modeled as a bending beam.

In the case of a force F_l applied to the beam at a distance l from the built-in end, the output voltage of the Wheatstone bridge is an instantaneous measure proportional to the rotation $\omega_l(\delta)$ of the first section located at distance δ from the built-in end. It is then amplified to produce a measurable voltage. Assuming that $\delta \ll L$, the ultimate measured voltage U is

$$U = A_U \left(\frac{F_l l \delta}{EI} \right) + U_0, \tag{29}$$

where A_U is the overall amplification gain and U_0 the measure offset and E and I are respectively the Young's modulus and the moment of inertia of the cantilever.

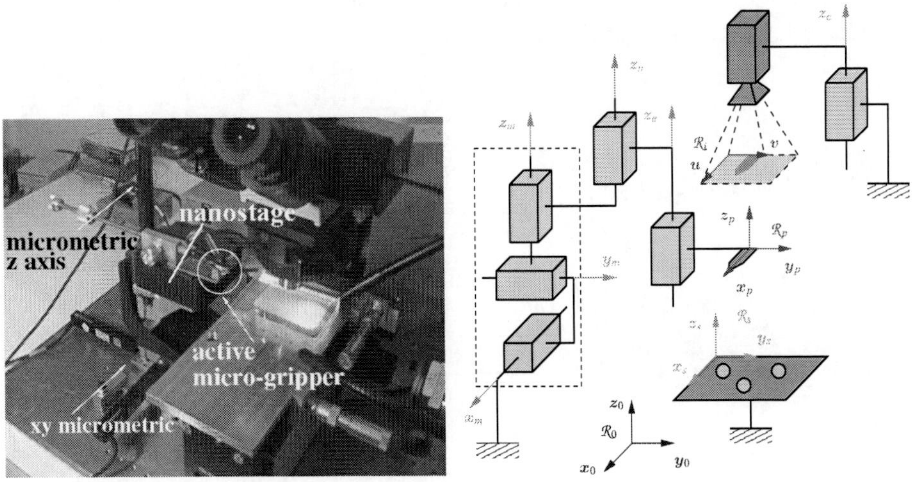

(a) The micromanipulator

(b) Kinematic configuration of the manipulator

Figure 4. The microscale manipulation setup.

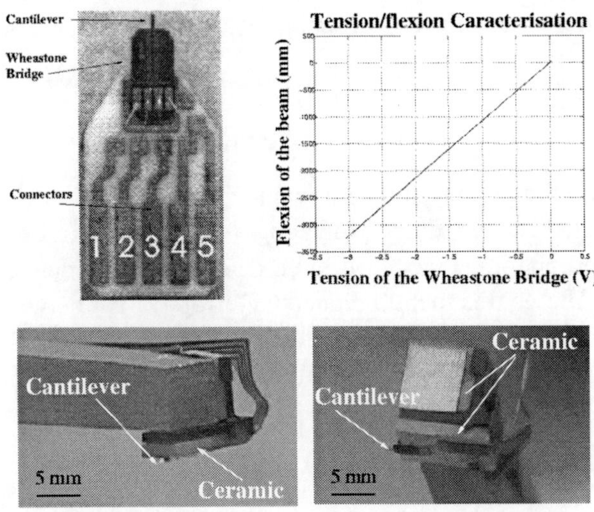

(a)

(b)

Figure 5. The active gripper for manipulation by adhesion: (a) AFM Cantilever, (b) end-effector with piezoceramic actuator.

2.2.2. Haptic User Interface

In order to control the overall motion of the manipulator, an interface with at least 3 DoF is needed. The chosen one is Virtuose 6D from Haption (*www.haption.com*). It is 6 DoF 6R arm with its own computing resources, in order to reduce the CPU load of the workstation. It is internally controlled based on force or position, with inertial and mass compensation. This control also includes inverse geometric and dynamic models, so that it can easily be controlled from an external application by sending set-point positions or forces. Its characteristics are given in Table 6.

Table 6. Specification of the Virtuose 6D master device

Geometric characteristics	
Maximal stroke	45 cm
Dynamic characteristics	
Maximum effort F_{macro}^{max}	35 N
Overall	
Apparent mass M_{Virt}	1 kg
Apparent inertia I_{Virt}	0.03 kgm^2

2.3. Control Implementation

For autonomous manipulation and user assistance, vision and force servoing are implemented in the manipulator. For horizontal plane motion, visual servoing based on the microscope top-view is used. For the vertical motion, due to the very small depth of field of microcope image, a focus-based control is implemented. Force control uses the grippers measurement capabilities with both vertical actuators, the nanostage and the microstage and takes advantages of this intrinsic redundancy.

2.3.1. Motion of the Gripper on Horizontal Plane

Position servoing of the gripper in the horizontal plane is done by image-based visual servoing using an external camera, described by Fig. 6. The gripper is detected by template matching in the microscope image and a desired contact point is defined by the user by clicking in the image.The goal is then to reduce the error $(\varepsilon_x\ \varepsilon_y)$ between the desired position of the gripper contact point $(u^*\ v^*)$ and its actual position $(u\ v)$ in the current image (expressed in *pixel coordinates*) by appropriately moving it in the workspace.

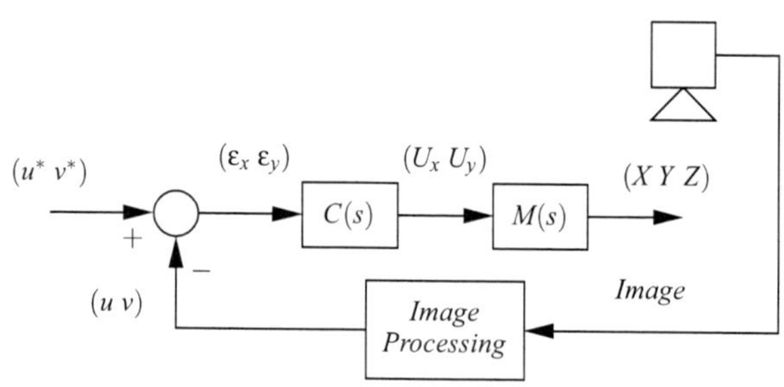

Figure 6. Image based visual servoing scheme.

Even if the static error of this servoing is null, there is a residual uncertainty about the absolute position of the gripper, which depends on the real area size covered by a pixel (i.e. on the zoom). In our case, this uncertainty is about $2\mu m$, which is not very accurate, but it is now stabilized with respect to the vertical motion. Indeed, if the optical and motion axes are not perfectly aligned, the induced deviation is permanently corrected. (For an alignment error of 1 degree and a vertical motion of $1\,mm$, the gripper would deviate by more than $17\,\mu m$.)

2.3.2. Focusing for Vertical Motion

The goal of an autofocus algorithm is to find the appropriate camera position so that a part of the image is focused. Moreover, as the vertical camera motion is precisely controlled, it can be used to measure the vertical gap between different features in the image and hence control the vertical gripper motion. Focus perception is not an absolute criterion. However, some image properties are affected by good or bad focus and can be used to build suitable criteria. Among these, it was observed that focused images have more high frequency components, more localized histograms, higher contrast, higher peaks and deeper valleys than blurred images. With a criterion based on one of these properties, the focusing problem can be processed as an optimization problem.

Some simple criteria have been tested, the chosen one being to quantify high frequency components of the image. For an image I_z, taken by a camera at position z, of width N and height M, the criterion is written as:

$$f(z) = \frac{1}{NM} \sum_{i=1}^{N-2} \sum_{j=1}^{M-2} \sqrt{|G_x(i,j)|^2 + |G_y(i,j)|^2}, \tag{30}$$

where G_x and G_y are the convolution of the initial image I_z with respective horizontal and vertical Sobel masks

$$S_x = \begin{bmatrix} -1 & 0 & 1 \\ -2 & 0 & 2 \\ -1 & 0 & 1 \end{bmatrix} \quad \text{and} \quad S_y = \begin{bmatrix} -1 & -2 & -1 \\ 0 & 0 & 0 \\ 1 & 2 & 1 \end{bmatrix}.$$

Fig. 7 illustrates a measure of focus versus camera position, using the described criterion.

Note that f is null for blurred images. In this case, the camera is driven in constant speed. When the criterion becomes non null, the desired camera motion speed is computed as:

$$v_c^* = K\Delta_n f \tag{31}$$

where K is the gain and $\Delta_n f$ is the average slope of f computed on the last n measures.

2.3.3. Force Servoing

As the AFM probe supplies only the measurement on the vertical axis, the force control is implemented for the vertical motion of the gripper through two redundant actuators, as described in Table 7. The given induced force resolution is theoretical because it is only computed from the travel range resolution.

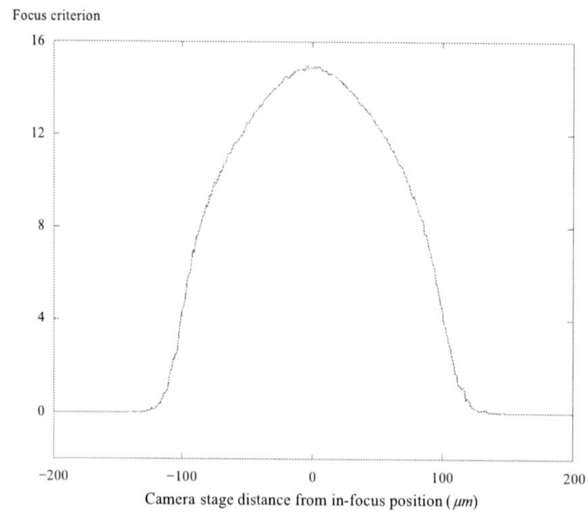

Figure 7. Focus criterion evaluation.

Table 7. Vertical axis specification

Axis type		Micro	Nano
Travel range	(μm)	$25 10^3$	12
Travel resolution	(nm)	50	2
Induced force range	(μN)	$526.5 10^3$	252.72
Induced force resolution	(nN)	$1.053\ 10^3$	42.12
Loop period limitation	(ms)	40	-

Basic loop A basic force servoing loop has been implemented to control the nanostage, thus the gripper is able to apply a precise contact force. Fig. 8 shows the force control scheme. F^* and F are respectively the desired and measured force, ε_n is the servoing error, U_n the input voltage and, z_n the nanostage position, $N(s)$ the transfer function of the low level controlled nanometric stage, $G(s)$ the transfer function of the beam and $C_n(s)$ the corrector term.

Since a proportional corrector is used, $C_n(s) = K_n$, and considering the gripper and object in contact, the transfer function of the system can be written as

$$H_f(s) = \frac{K_n K_g N(s)}{1 + K_n K_g N(s)}. \tag{32}$$

Owing to its travel range of $12\,\mu m$, the nanostage easily reaches its bounds. Hence, the gripper should be very close to the object (a few micrometers) before starting the servoing in order to keep as much travel range as possible. The position information given by focusing is not accurate enough to accomplish this operation. It is thus necessary to include the microstage in the force servoing loop.

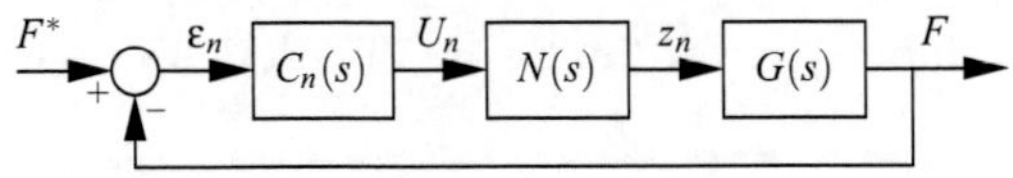

Figure 8. Basic force servoing loop scheme.

Enhanced loop An auxiliary loop controlling the microstage is associated with the basic one, whose goal is to maintain the nanostage in the middle of its travel range. Due to their different resolutions, it is important to define a dead zone around the desired nanostage position where the auxiliary loop has no effect in order to avoid undesirable oscillations. Fig. 9 illustrates the overall servoing scheme.

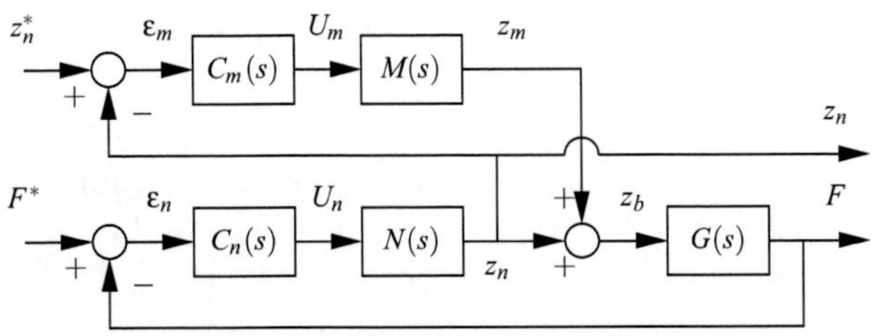

Figure 9. Enhanced force servoing loop scheme.

When a contact force F^* is desired, and the measured force F is null, the nanostage reaches its bottom bound due to the fast loop, and moves out of the dead zone. Consequently, the microstage moves down due to the auxiliary slow loop. When the gripper reaches the object, a non null contact force is measured and the former loop works in linear mode. The secondary loop will stop when the nanostage is in the dead zone and activate again if a new force command makes the nanostage go out of the dead zone.

2.4. Force Feedback Haptic Remote Handling

In the literature, several coupling methods have been developed for micro-teleoperation systems. When they include force feedback they are called *bilateral coupling*. The most natural and commonly used *bilateral coupling* method for micro-manipulation applications is direct homothetic **force-position** coupling. This method lacks passivity and robustness which makes that the overall teleoperated system can be unstable.

For general teleoperation applications, another commonly used coupling scheme is the bilateral **position-position** coupling which guaranties the robustness of the overall system by being passive. Moreover, for micro-teleoperation applications, it offers a modular architecture where both the slave or the master can be easily changed. The presented control scheme is based on that coupling. Its performances are then compared to the homothetic direct coupling.

The control scheme implemented for force feedback remote handling is called position-position bilateral control. It will be hereafter referred as '*PPB*'. Its design is based on passivity considerations for teleoperated systems [34]. The overall control scheme is split into local control loops. The unconditional stability of the control loop guarantees that the overall stability will not be affected when the system is coupled with passive components

[35]. The unconditional stability is verified using the Llewellyn's criteria, described 2.4.4..

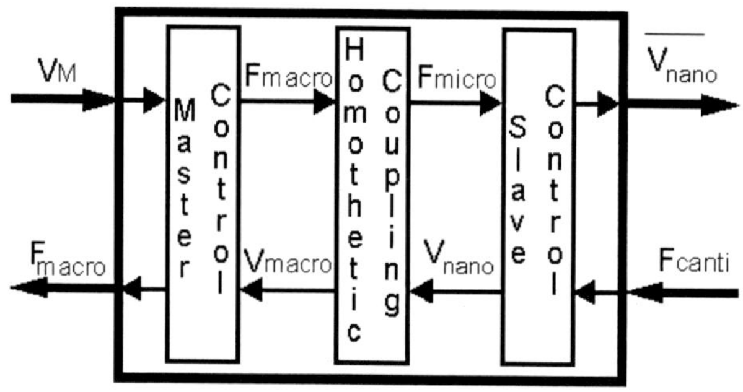

Figure 10. The *PPB* control scheme decomposition.

In this regard, the chosen architecture for the *PPB* control is modular. The global control is composed of three modules: the master control block, the homothetic coupling block and the slave control block (Fig. 10). Parameters used in this control are given Table 8.

Table 8. System parameters

P_M,V_M	Measured position/velocity of master
F_{macro}	Force on master
P_{nano},V_{nano}	Measured position/velocity of nanostage
$\overline{P_{nano}}$, $\overline{V_{nano}}$	Set-point position/velocity of nanostage
P_{macro}, V_{macro}	Position/velocity of nanostage translated in macroworld
F_{canti}	AFM gripper measured contact force

2.4.1. The Homothetic Coupling Block

The homothetic coupling block is used for macro-to-micro and micro-to-macro conversions, between force F, position P and velocity V data, using force and motion scaling ratios α_f and α_d, respectively. Scaling ratios are defined according to the master used for the teleoperation, manipulators available actuators, and the desired performances. The motion of micromanipulator is controlled by the haptic interface, thus the motion scaling is 'macro to micro'. On the other hand, the force is measured in the microworld and transmitted to the master haptic interface; thus the force scaling is 'micro to macro'.

The master is designed for a maximal force F^{max}_{macro}. Considering that the maximum flexion of the AFM cantilever is $12\mu m$, which is the maximum stroke C^{max}_{nano} of the nanostage, the maximal measurable force is given by:

$$F^{max}_{canti} = K_{canti}C^{max}_{nano} = 2.52 \times 10^{-4}N \tag{33}$$

Thus, the force scaling ratio is given by:

$$\alpha_f = \frac{F^{max}_{canti}}{F^{max}_{macro}} \tag{34}$$

For the motion scaling, the maximal vertical stroke C_z^{max} of the master is used. When the master's motion is unlimited, or when the master's bounds are too spaced out, virtual bounds have to be defined in order to limit the master's stroke C_{zvirt}^{max}. The bounds will correspond to $C_{nano}^{max} = 12\mu m$ of the nanostage motion. The motion scaling ratio α_d between the nanostage and the master is thus given by:

$$\alpha_d = \frac{C_{zvirt}^{max}}{C_{nano}^{max}} \tag{35}$$

The micro and macro homothetic ratios are then defined as follows:

$$P_{macro} = \alpha_d P_{nano} \tag{36}$$

$$V_{macro} = \alpha_d \overline{V_{nano}} \tag{37}$$

$$F_{micro} = \alpha_f F_{macro} \tag{38}$$

2.4.2. The Master Control Block

The master control block allows to compute the set-point force on the master F_{macro}, which is also used in the slave control block after using the force scaling ratio. It is a proportional-derivative control of the error on the master position with respect to the slave position converted into the macroworld. In a two port model it has the velocity of the nanostage converted into the macroworld V_{macro} and the velocity of the master V_M as inputs.

$$F_{macro} = K_p\left(P_M - P_{macro}\right) + K_d\left(V_M - V_{macro}\right) \tag{39}$$

with control parameters K_p and K_d chosen in accordance to the sampling period T_e. Considering the differential equation in P_M obtained with zero as reference position:

$$\frac{I_M}{K_p}\ddot{P}_M + \frac{K_d}{K_p}\dot{P}_M + P_M = 0 \tag{40}$$

The cut-off frequency ω_0 is then given by:

$$\omega_0 = \sqrt{\frac{K_p}{I_M}} \tag{41}$$

with $T_e = 2ms$, the sampling frequency is $f_e = 500Hz$. The bandwidth ω_{BP} is given by:

$$\omega_{BP} = \frac{2\pi f_e}{10} = 314 rad.s^{-1} \tag{42}$$

Since it is necessary for $\omega_0 < \omega_{BP}$, it has been thus chosen $\omega_0 = 100\, rad.s^{-1}$. Accordingly, K_p is computed as follows:

$$K_p = I_M \omega_0^2 \tag{43}$$

Choosing the damping ratio $\zeta = 1$ in order to limit the over-shooting, K_d is then given by:

$$\frac{2\zeta}{\omega_0} = \frac{K_d}{K_p} \Longrightarrow K_d = \frac{2\zeta K_p}{\omega_0} \tag{44}$$

2.4.3. The Slave Control Block

The slave control block allows to compute the set-point velocity $\overline{V_{nano}}$ of the nanostage, which is also sent back to the homothetic coupling block to be used in the master control. The calculation of $\overline{V_{nano}}$ is based on the comparison of the master force translated in the micro world F_{micro}, and the cantilever contact force F_{canti} (45). As the nanostage is controlled on position, $\overline{P_{nano}}$ is computed by integrating $\overline{V_{nano}}$ with saturation levels at 0 and $12 \times 10^{-6} m$, travel limits of the nanostage.

$$\overline{V_{nano}} = K(F_{micro} + F_{canti}) \tag{45}$$

where K is the enslaving gain of the force loop. Actually, the gain K must be small to ensure the unconditional stability of the overall system and a good time response less than the human reflex time, which is about 10 ms. Simulations using the parametric model obtained with the given transfer functions show that these performances are achieved with $K = 3$.

2.4.4. Llewellyn's Unconditional Stability Criteria

It is possible to represent a dynamic system by a two-port model considering velocities V_1, V_2 and forces F_1, F_2 as inputs and outputs.

According to the choice of inputs and outputs, three representations can be defined: impedance **Z** (46), admittance **Y** (47) or hybrid **H** (48).

$$\begin{bmatrix} F_1 \\ F_2 \end{bmatrix} = \begin{bmatrix} Z_{11} & Z_{12} \\ Z_{21} & Z_{22} \end{bmatrix} \begin{bmatrix} V_1 \\ V_2 \end{bmatrix} \tag{46}$$

$$\begin{bmatrix} V_1 \\ V_2 \end{bmatrix} = \begin{bmatrix} Y_{11} & Y_{12} \\ Y_{21} & Y_{22} \end{bmatrix} \begin{bmatrix} F_1 \\ F_2 \end{bmatrix} \tag{47}$$

$$\begin{bmatrix} F_1 \\ V_2 \end{bmatrix} = \begin{bmatrix} H_{11} & H_{12} \\ H_{21} & H_{22} \end{bmatrix} \begin{bmatrix} V_1 \\ F_2 \end{bmatrix} \tag{48}$$

The unconditional stability of the system can thus be verified using the Llewellyn's criteria. The two-port system is passive, and unconditional stability is ensured if all three following criteria are true:

$$Re(Q_{11}) \geq 0 \tag{49}$$

$$Re(Q_{22}) \geq 0 \tag{50}$$

$$2Re(Q_{11})Re(Q_{22}) - |Q_{12}Q_{12}| + Re(Q_{12}Q_{12}) \geq 0 \tag{51}$$

where **Q** = **Z**, **Y** or **H** in accordance with the considered two-port system.

The Llewellyn's criteria for *PPB* coupling is computed according to Eqn. 46 with:

$$F_1 = F_{op},\ F_2 = F_{canti},\ V_1 = V_B \text{ and } V_2 = V_{canti}.$$

By inserting the dynamics equations of each of the blocks in (46), the two-port equation of the overall system can be calculated. The chosen frequency range is $[0\ 50] Hz$ ($[0\ 300] rad.s^{-1}$), according to the bandwidth ω_{BP} given in (42). Each criterion is plotted in Fig. 11.

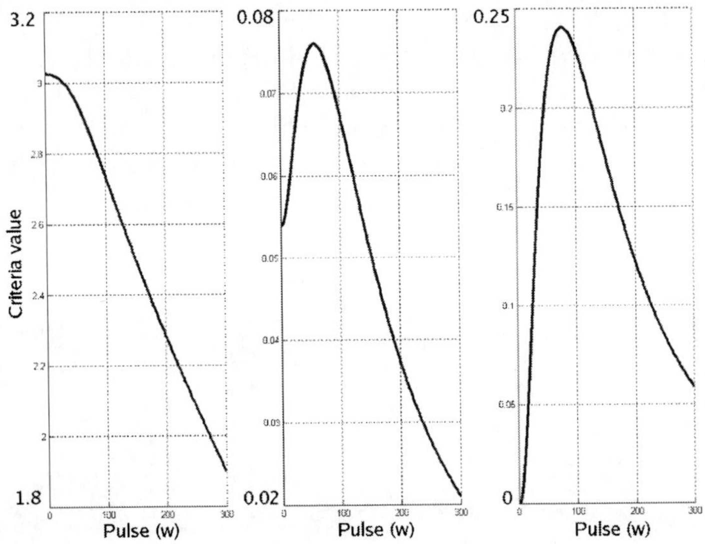

Figure 11. Llewellyn's criteria for the *PPB* control: first row $Re(Y_{11})$, second row $Re(Y_{22})$ third row $2Re(Y_{Y11})Re(Y_{22}) - |Y_{12}Y_{21}| - Re(Y_{12}Y_{21})$.

Fig.11 shows that for the considered frequency domain the coupling with the proposed *PPB* control always verifies the three Llewellyn's conditions: they are all positive, and the coupling is then passive. Thus it is unconditionally stable. Moreover, analytical expressions of 3 criteria show that the stability is not affected even if the scaling ratios α_f and α_d are changed.

2.5. Experiments and Usage Scenarios

The above described functionalities of the system are experimentally validated, starting from the principle of using adhesion for pick-up and release, selective release for sorting, and then advanced modes requiring more complex control: rolling and remote force-feedback handling. All these experiments are achieved using the AFM based active gripper. Its experimental calibration is shown in Fig. 12.

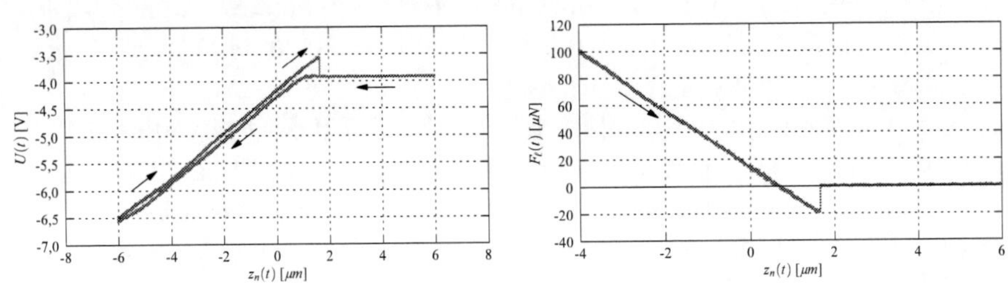

Figure 12. Calibration of the AFM gripper (a) Output voltage vs. distance and (b) Force vs. distance.

2.5.1. Adhesion Forces

Static pick-up and release A glass sphere of 50 μm in diameter lying on a polystyrene substrate has been captured by simple contact with the gripper (Fig.13). The estimated adhesion forces on substrate/object and gripper/object interfaces are respectively $F_{adh}^{so} = 4.70\ \mu N$ and $F_{adh}^{go} = 6.80\ \mu N$. The release is then accomplished by sloping the end effector by 45°, decreasing the adhesion between the end-effector and the object. Alternatively, it is possible to release the object on a different target substrate, whose adhesion energy is higher than the gripper.

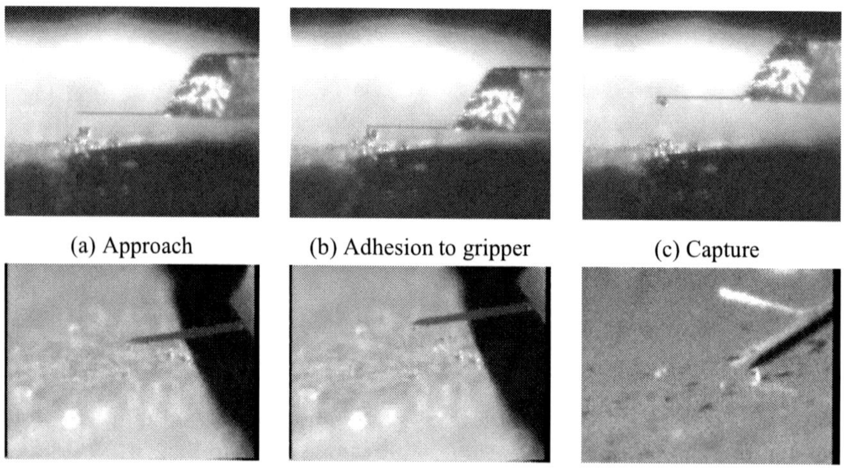

(a) Approach (b) Adhesion to gripper (c) Capture

(a) Approach (b) Adhesion to substrate (c) Release

Figure 13. Capture and static release by adhesion of 50 μm diameter object.

This mode does not take advantage of the dynamic properties of the gripper. Only the high adhesion of the AFM cantilever is employed. It is obvious that in this case, the success of the release is limited to spherical objects, by sloping the gripper, or to applications where the target substrate has an higher adhesion energy.

Dynamic Effects In this second example, the capture operation is achieved by adhesion, as in the previous case. The release task on the contrary is accomplished by inertial effect. The necessary acceleration to produce this inertial force is estimated around $10^6\ m/s^2$. This acceleration can be reached by exciting the piezoceramic actuator with a 120 V 1 μs impulse. The fig 14 shows the output signal of the AFM device for this case.

This output signal is proportional to the inflection of the AFM cantilever. The high magnitude of the impulse causes important vibrations of the AFM cantilever, as observed on the output signal. In the case of the 80 V excitation, the resulting acceleration does not produce sufficient inertia force to overcome the end-effector/object adhesion. A Fourier analysis of the output signal showed the existence of two superimposed modes, at 20 kHz and 200 kHz. The first mode at 20 kHz is the first inflection mode of the beam with the mass of the object at its extremity, as verified by a finite element simulation. The second mode at 200 kHz is either the second inflection mode of "beam+object" system or possibly the first torsion mode about the central axis of the beam generated by the eccentricity of the object.

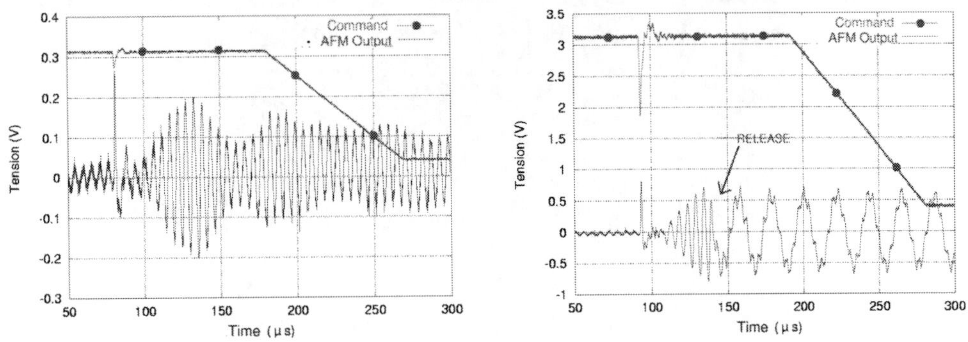

(a) $80V$, no release (b) $120V$, release

Figure 14. Output signal of AFM during the dynamic release of a 40 μm radius glass sphere.

The presence of this two modes proves that the object is still adhered to the gripper and the release is ineffective. In the case of 120 V excitation, a change of the oscillation frequency of the cantilever is visible at $t = 150\,\mu s$. Prior to this instance, the vibrational mode is similar to the one observed in the first excitation. After this instance, the cantilever oscillates at 40 kHz, which is its first free inflection mode of the cantilever alone. We can deduce that the object adheres to the gripper until this instant, and is then released, changing the natural frequency of the system. It is then possible to determine the release by mere observation of the output signal of the AFM, without needing any other external sensor.

The dynamic mode, as described in this example, allows to release of the captured object independently of external factors, such as the geometry of the object or the target substrate.

2.5.2. Sorting by Selective Release

The selective release is an extension of the dynamic release mode by use of vibrational behavior of the AFM cantilever. It was first observed when experimenting a gripper whose AFM cantilever was gold coated as shown in Fig. 15-a. Although the gold has a higher adhesion energy, it is observed that the resulting adhesion forces between the gripper and the objects are considerably reduced. This decrease, also observed in [36], is due to improper coating and surface conditions. In fact, the oxidation of the thin gold film and its imperfections increase the roughness, reducing the adhesion drastically [37]. It has been observed that the 50 μm diameter glass beads, easily captured by the non-coated gripper as described in section 2.5.1., would not adhere to the gold coated gripper. Smaller, thus lighter objetcs are then used in this experiment.

Ambrosia pollens, approximately 20 μm in diameter are introduced. These specimens come in powder and they form a very dense and homogeneous distribution on the substrate. It is quite difficult, if not impossible, to pick an isolated pollen. Thus, in the approach-and-capture cycle, several pollens are captured. We report an experiment with 2 pollens caught. In order to release those objects, the piezoceramic actuator is excited with a sinusoidal signal with frequency variation. The chosen start frequency for this signal is 15 kHz, which is well below the natural frequency of the system. Next, this frequency is slowly

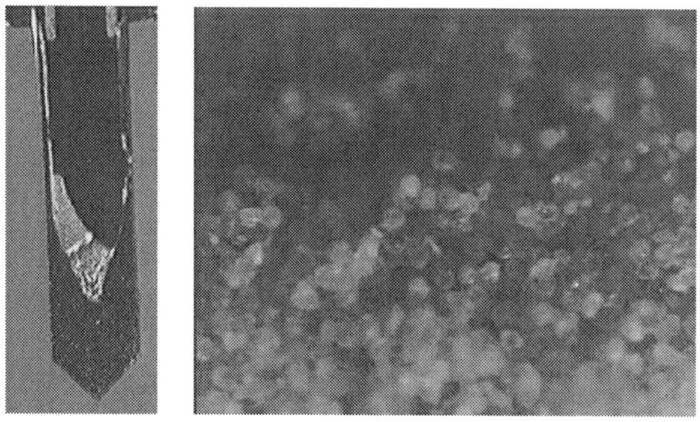

Figure 15. (a) Gold coated end-effector (b) Ambrosia pollens on the substrate.

increased. When it reaches the natural frequency of the "beam + 2 objects" system, the vertical displacements, thus the instantaneous acceleration is maximum. Accordingly, the object closer to the tip, where the acceleration is higher, is released. Following the ejection of the first object, the natural frequency of the system changes to a higher value, preventing the release of the second object. In order to release this one, the excitement frequency must be increased to reach the natural frequency of the "beam + object" system. As the pictures of Fig. 16 illustrate, it is then possible to release the objects captured in random one by one.

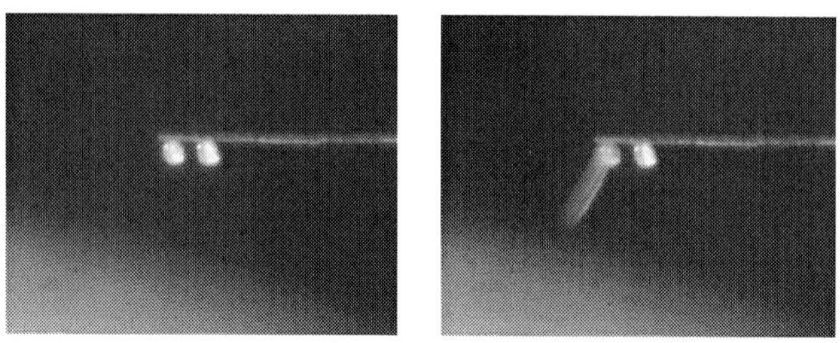

(a) Release of the first object

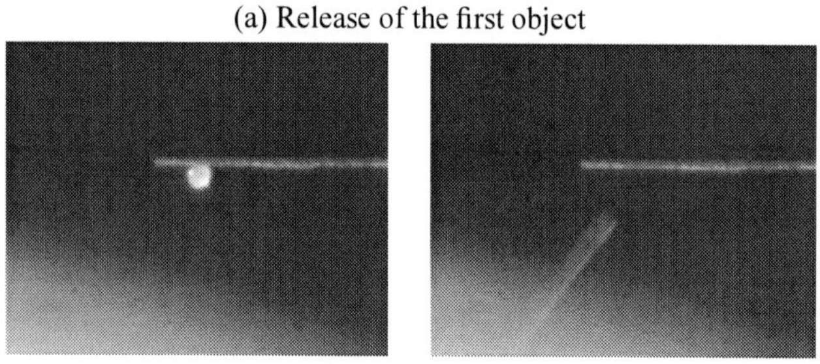

(b)Release of the second object

Figure 16. Selective release operation.

This operational mode basically allows to isolate a micro-object from an initial large group.

2.5.3. Rolling

Comparing the variation of l with the gripper vertical motion, we are able to detect the current mode. In practice, it is easy to roll the object, as expected. The observed rolling is not perfect, but it is efficient to release objects. Moreover, knowing by vision the initial contact point on the beam and the size of the object, it is possible to estimate the distance required to release, and thus to release the object accurately. Figure 17(a) gives experimental results. The gripper and the object are in contact and the basic force servoing loop is running. The gripper starts to move along its X axis at time t_1. Due to the servoing loop, the measured voltage is steady while the nano-stage moves downwards in order to keep applied force constant. This behavior proves that the contact point l moves to the extremity of the beam, i.e. the object rolls. Between t_2 and t_3, the end-effector follows the curve of the object and after t_3, there is no more contact. An estimation of the ratio l/L is given in the bottom plot. Figure 17(b) shows an autonomous manipulation task of a glass micro-ball with a $50\,\mu m$-diameter from the microscope view.

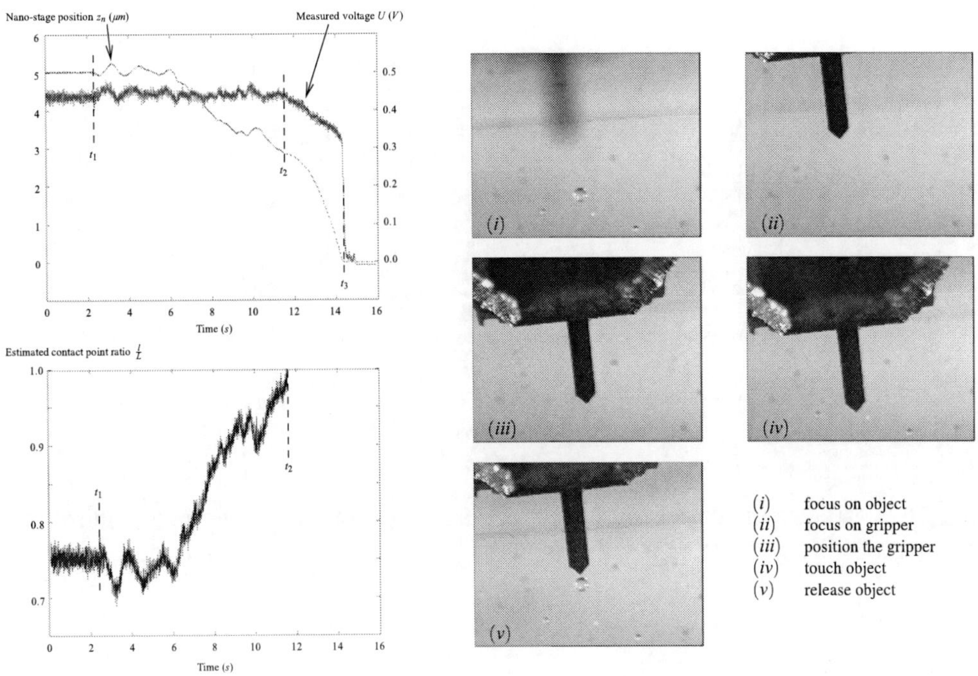

(a) Force sensing and contact point estimation

(b) Top view of release

Figure 17. Release by rolling.

2.5.4. Teleoperation

The experimental behavior and performances of the overall system are then studied, first with the above chosen force scaling ratio (34), and second when scaling ratios are changed. The force scaling ratio is divided by 10: $\alpha'_f = 8.8 \times 10^{-6}$, thus the contact force feeling is amplified on the master (38). The proposed control PPB is compared to a classical direct homothetic force-position coupling.

In the first case, both controls have a stable behavior and results are almost identical. However, the force feeling on the master is weak; for example the pull-off phenomena is nearly unnoticed (Fig. 18). Typically in such cases, the operator may require an amplification of the force feeling. The results for the modified scaling ratio α'_f are shown on Fig.19(b) for *PPB* control and Fig. 19(a) for force-position coupling. For the *PPB* control, as seen clearly in the motion plots, the feeling of force is firm and well transmitted even in case of the weak pull-off force, and the system is perfectly stable. On the other hand, Fig. 19(a) shows that for the force-position control, instability occurs as soon as the measured forces are non-zero and the change of the scaling ratio is consequently not possible.

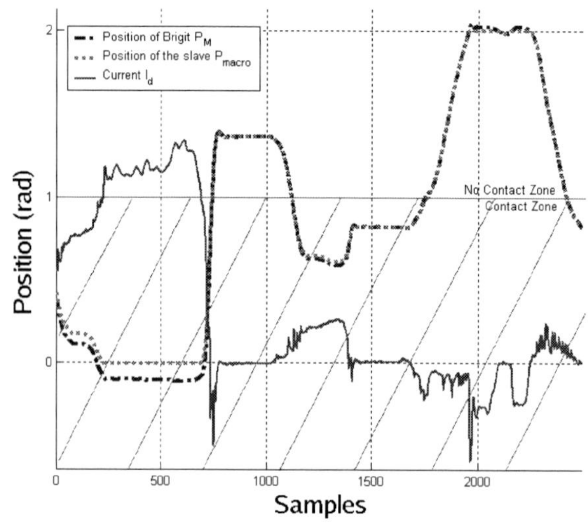

Figure 18. Position and current (proportional to force feedback) for the *PPB* control for an arbitrary trajectory with ($P_M < 1rad$) and without ($P_M > 1rad$) contact of the slave.

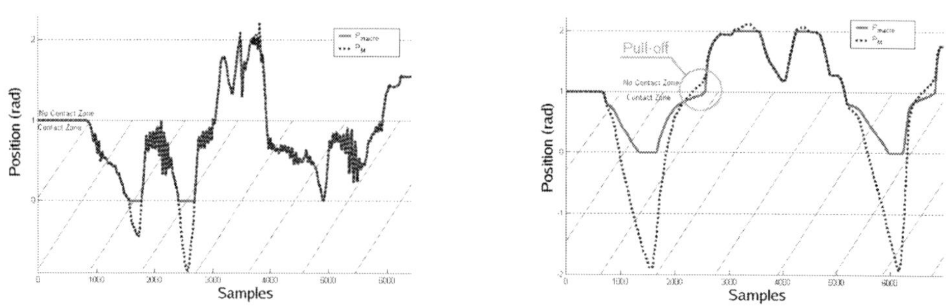

(a) Force-position control: instability occurs when the slave is in contact ($P_M < 1$)

(b) *PPB* control amplifies the feeling of the pull-off forces when contact is broken

Figure 19. Influence of the force scaling ratio α_f on force-position and *PPB* couplings.

An other issue to discuss is the transparency. In the case of the *PPB* control, the system is not transparent by design, as the force felt by the operator is proportional to the position error between the master and the slave, on contrary of the direct homothetic coupling which reproduces solely the force measured by the AFM probe. Although it seems to be a drawback, this particularity brings some considerable advantages. In direct homothetic coupling, all external factors influencing the motion of the slave without producing any effect on the force sensor are unnoticed by the user. In *PPB* coupling the user, feels the following error and can adapt his motion for a better grip in the manipulation task. A good example of this case is when the nanostage reaches its bounds, which is completely unfelt in direct homothetic case, but clearly apparent in *PPB* control (fig. 18). Moreover, as the microscale physical phenomena are quite different from classical macroscale manipulations due to the predominance of the surface forces, this loss of transparency is rather welcome as it isolates the user from unexpected phenomena and smooths its operation.

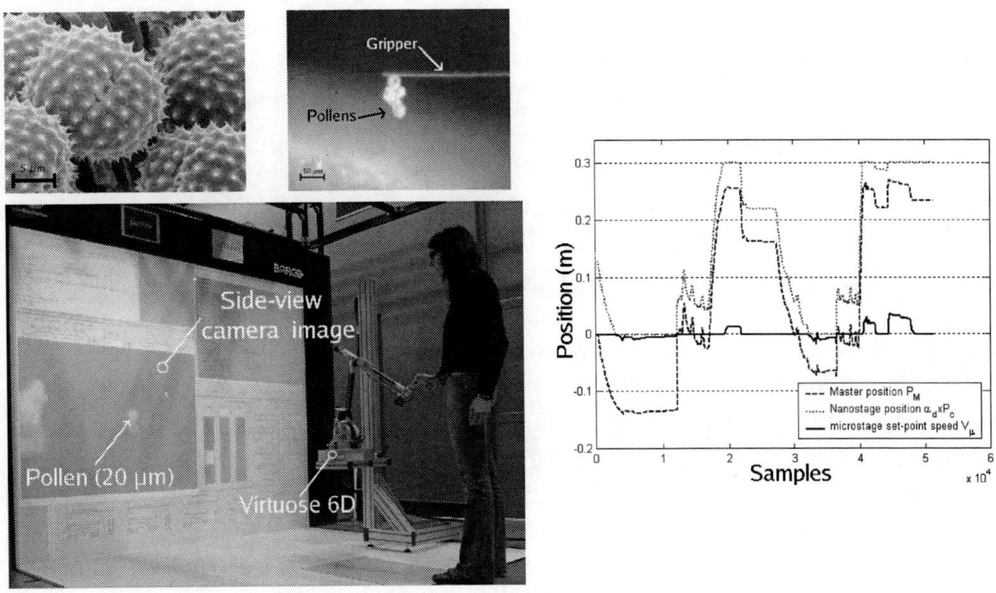

Figure 20. (a)Manipulation of ragweed pollens, (b) Redundant coupling with Virtuose 6D.

The system is also experimented in teleoperation using the proposed *position-position* coupling, which was compared to a basic force-position coupling scheme. Pick-up, rolling and release operations of ragweed pollens (diameter=20 μm) are experimented (Fig. 20(a)). During this experiments, users is perfectly able to feel stiffness of the pollens when compressed and rolled, as well as the pull-off and adhesion forces during release. *Position-position* coupling shows better performances such as the feeling of the bounds and the damping coefficient between the master and the slave. Fig. 20(b) illustrates the master's position vs. nano and microstage motions. Moreover, it is even possible to change the scaling ratios on the fly (under some restriction on energy dissipation), as the stability of the system is unaffected. It is then possible to adapt the coupling to a given phase of the manipulation task which would need more or less precision or travel range using the same haptic interface.

3. Final Words and Further Reading

Manipulation of microscale components is a new challenge in the light of recent developments in production technologies, where in the search for maximum efficiency component dimensions are constantly sought to reduce. A condensed overview on the robotic issues to handle those small parts were presented here. The main issue comes from the biased balance of forces and the resulting adhesion phenomena. Origins of these mechanical behavior lay on the scale reduction and the predominance of surface forces such as van der Waals interaction, capillary effects and electrostaticity. Practical models for simple cases were presented above. This mechanical analysis lead to the requirement to specific designs for microscale dedicated manipulation systems. Such a setup is described here in order to propose different solutions in terms of mechanical design, control implementation, pick-and-place strategies and human interaction paradigms.

The main principle of the proposed manipulator is inspired of atomic force microscopy. As of today, the AFM is the most widely used tool in micromanipulation as it combines high precision sensing to mechanical action. Several AFM based system, for basic 2D operation [38], carbon nanotube characterization [39, 40] or biologic manipulations [41] are proposed in the literature. In order to overcome the limitation of using a single cantilever, systems combining the use of several probes is also proposed [42]. In this case, it is also possible to use one of the probes for AFM imaging while the other one is dedicated to the manipulation [43].

An alternate approach to avoid adhesion problems is to manipulate in completely immersed environment. In this case, van der Waals forces are greatly reduced, capillary forces do not apply because there isn't any formation of a liquid bridge and water distributes the electrostatic charges. This is also a quite adapted environment for manipulation of biologic materials. Optical tweezers [44] and magnetic traps [45] are particularly adapted tools for in liquid manipulation. In these cases, a spherical object is manipulated using a focused laser beam, or magnetic forces. This trapped object is then used as the manipulation tool to act on other objects in the environment. In the case of optical tweezers, several tools can be trapped simultaneously [46]. Some recent works includes haptic coupling to such indirect manipulation systems using vision and force model based sensing approaches [47].

References

[1] J. Israelachvili. *Intermolecular and Surface Forces*. ACADEMIC PRESS, 1991.

[2] L.-H. Lee. The chemistry and physics of solid adhesion. In L.-H. Lee, editor, *Fundamentals of Adhesion*. Plenum Press, 1991.

[3] H. Krupp. Particle adhesion: theory and experiments. *Journal of Advances in Colloid*, **1**:111–140, 1967.

[4] R. Allen Bowling. A theoretical review of particle adhesion. In *Proc. of Symposium on particles on surfaces 1: Detection, Adhesion and Removal*, pages 129–142, San Francisco, 1986.

[5] R. S. Fearing. Survey of sticking effects for micro parts handling. In *Proc. of IEEE/RSJ Conf. on Intelligent Robots and Systems*, pages 212–217, 5-9 August 1995.

[6] L. Hecht. An introductory review of particle adhesion to solid surfaces. *Journal of the IES*, **33**(2):33–37, 1990.

[7] H.C. Hamaker. The london van der waals attraction between spherical particles. *Physica*, **10**:1058–1072, 1937.

[8] Evgenni Mikhailovich Lifshitz. unknown title. *Sov. Phys. - JETP*, **73**(2), 1956.

[9] L.M. Fok, Y.H. Liu, and J.L. Wen. Modeling of haptic sensing of nanolithography with an atomic force microscope. In *Proc. of the IEEE International Conference on Robotics and Automation*, pages 2457–2462, April 2005.

[10] Arthur W. Adamson and Alice P. Gast. *Physical Chemistry of Surfaces*. John Wiley and Sons, 6th edition, 1997.

[11] M. van den Tempel. *Adv. Colloid Interface Sci.*, **3**(137), 1972.

[12] Pierre Lambert. *A Contribution to Microassembly: a Study of Capillary Forces as a gripping Principle*. PhD thesis, Universit libre de Bruxelles, Belgium, 2004.

[13] A. McLachlan. Three-body dispersion forces. *Mol. Phys.*, **7**:423–427, 1964.

[14] L. Bergstrm, A. Meurk, H. Arwin, and D.J. Rowcliffe. Estimation of hamaker constant of ceramic materials from optical data using lifshitz theory. *Journal of the Am. Ceram. Soc.*, **79**(2):339–348, 1996.

[15] I. Dzyaloshinskii, E. Lifshitz, and L. Pitaevski. Van der waals forces in liquid films. *Soviet Physics*, **37**(10):161–170, 1960.

[16] D. B. Hough and L. R. White. cited by [adamson97], p.239. *Adv. Colloid Interface Sci.*, **14**(3), 1980.

[17] V. A. Parsegian and G. H. Weiss. cited by [adamson97], p329. *J. Colloid Interface Sci.*, **81**(285), 1981.

[18] P.H.G.M. Van Blokland and J.G.T. Overbeek. Van der waals forces between objects covered with a chromium layer. *Journal of Chemical Science*, **74**(11):2637–2651, 1978.

[19] E. Zaremba and W. Kohn. Van der waals interaction between an atom and a solid surface. *Physical Review*, **13**(6):2270–2285, 1976.

[20] Boris Vladimirovich Derjaguin, V. M. Muller, and Yu. P. Toporov. Effect of contact deformations on the adhesion of particles. *J. Colloid Interface Sci.*, **53**(2):314–26, 1975.

[21] K. L. Johnson, K. Kendall, and A. D. Roberts. Surface energy and the contact of elastic solids. *Proc. R. Soc. London Ser. A*, **324**:301, 1971.

[22] D. Tabor. Friction - the present state of our understanding. *Journal of Lubrification Technology*, **103**:169–179, 1981.

[23] D. Maugis. Adhesion of spheres: the jkr-dmt transition using a dugdale model. *J. Colloid Interface Sci.*, **150**(1):243–269, 1992.

[24] R. W. Carpick, D. F. Ogletree, and Salmeron M. A general equation for fitting contact area and friction vs. *J. Colloid Interface Sci.*, **211**:395–400, 1999.

[25] Olivier Pietrement. *Imagerie et caractrisation nanomcanique des surfaces par microscopie force atomique*. PhD thesis, Universit de Reims Champagne-Ardennes, Reims, France, 2000.

[26] Y. Rollot. *Micro-manipulations par adhsion: Modlisations dynamiques et exprimentation*. PhD thesis, Universit Pierre et Marie Curie, Paris, France, 2000.

[27] D.A Hays. Electrostatic adhesion of non-uniformly charged dielectric sphere. *Int. Phys. Conf. ser. No. 118 : section 4*, pages 223–228, 1991.

[28] L. Lee. Dual mechanism for metal-polymer contact electrification. *Journal of Electrostatics*, **32**:1–29, 1994.

[29] E. Durand. *Electrostatique: Tome II Problmes gnraux, Conducteurs*. Masson, 1966.

[30] S. Belaidi, P. Girard, and G. Leveque. Electrostatic forces acting on the tip in atomic force microscopy: modelization and comparison with analytic expressions. *J. Appl. Phys.*, **81**(3):1023–1029, 1997.

[31] Brunero Cappella and G. Dietler. Force-distance curves by atomic force microscopy. *Surf. Sci. Rep.*, **34**:1–104, 1999.

[32] S. Hudlet, M. Saint Jean, and J. Berger. Evaluation of the capacitive force between an atomic force microscopy tip and a metallic surface. *Eur. Phys. J. B*, **2**:5–10, 1998.

[33] Hans-Jrgen Butt, Brunero Cappella, and Michael Kappl. Force measurements with atomic force microscope: Technique, interpretation and applications. *Surf. Sci. Rep.*, **59**:1–152, 2005.

[34] R. J. Adams and B. Hannaford. Stable haptic interaction with virtual environments. *IEEE Trans. on Robotics and Automation*, **15**(3):465–474, 1999.

[35] R.J Adams and B. Hannaford. A two-port framework for the design of unconditionally stable haptic interfaces. In *Proc. of the IEEE/RSJ International Conference on Intelligent Robots and Systems, Victoria, B.C.*, pages 1254–9, 1998.

[36] Y. Zhou and B.J. Nelson. Adhesion force modeling and measurement for micromanipulation. In *Proc. of SPIE Microrobotics and Micromanipulation, November 1998*, pages 169–180, 1998.

[37] F. Arai and T. Fukuda. Microrobotics based on micro physics-design and control strategy based on attractive force reduction. In *Proc. of SPIE Conf. on Microrobotics: Components and Applications*, volume 296, pages 184–195, December 1996.

[38] Metin Sitti and Hideki Hashimoto. Controlled pushing of nanoparticles: Modeling and experiments. *IEEE-ASME Trans. Mechatron.*, **5**(2):199–211, 2000.

[39] P.M. P. M. Albrecht and J. W. Lyding. Lateral manipulation of single-walled carbon nanotubes on h-passivated si(100) surfaces with an ultra-high- vacuum scanning tunneling microscope. *Small*, **3**(1):146–152, 2007.

[40] E. Tranvouez, A. Orieux, E. Boer-Duchemin, C. H. Devillers, G. Comtet V. Huc, and G. Dujardin. Manipulation of cadmium selenide nanorods with an atomic force microscope. *Nanotechnology*, **20**(16):165304, 2009.

[41] M. Sitti. Atomic force microcope probe based controlled pushing for nanotribological characterization. *IEEE/ASME Transactions on Mechatronics*, **9**(2):343–349, 2004.

[42] H. Xie, S. Haliyo, and S. Rgnier. A versatile atomic force microscope for 3d nanomanipulation and nanoassembly. *Nanotechnology*, **20**:215301 (9pp), 2009.

[43] H. Xie and S. Rgnier. Development of a flexible robotic system for multiscale applications of micro/nanoscale manipulation and assembly. *IEEE/ASME Transaction on Mechatronics*, 2010.

[44] A. Ashkin, J. M. Dziedzic, J. E. Bjorkholm, and Steven Chu. Observation of a single-beam gradient force optical trap for dielectric particules. *Optics Letters*, **11**(5):288–290, 1986.

[45] E. van West, A. Yamamoto, and T. Higuchi. The concept of haptic tweezer, a non-contact object handling system using levitation techniques and haptics. *Mechatronics*, **(17)**:345-356, 2007.

[46] F. Arai, D. Andou, and Y. Nonoda. Integrated microendeffector for micromanipulation. *IEEE-ASME Trans. Mechatron.*, **3**, March 1998.

[47] C. Pacoret, Bowman R., Gibson G., S. Haliyo, D. Carberry, A. Bergander, S. Rgnier, and M. Padgett. Touching the microworld with force-feedback optical tweezers. *OPTICS EXPRESS*, **17**(12):10260, 2009.

INDEX

D

E

F

G

H

I

J

K

L

M

N

O

P

Q

R

S

T

V

W

X

Y